Retrying Galileo, 1633–1992

Maurice A. Finocchiaro

UNIVERSITY OF CALIFORNIA PRESS

Berkeley   Los Angeles   London

University of California Press
Berkeley and Los Angeles, California

University of California Press, Ltd.
London, England

First paperback printing 2007
© 2005 by the Regents of the University of California

Library of Congress Cataloging-in-Publication Data

Finocchiaro, Maurice A., 1942–
    Retrying Galileo, 1633–1992 / Maurice A. Finocchiaro.
        p.   cm.
    Includes bibliographical references and index.
    ISBN 978-0-520-25387-2 (pbk. : alk.)
        1. Galilei, Galileo, 1564–1642—Trials, litigation, etc.   2. Religion
and science—Italy—History—17th century.   3. Science—
Philosophy.   I. Title.
QB36.G2F56   2005
520'.92—dc22   2004001861

Manufactured in the United States of America

16   15   14   13   12   11   10   09   08   07

10   9   8   7   6   5   4   3   2   1

Printed on Ecobook 50 containing a minimum of 50% postcon-
sumer waste, processed chlorine free. The balance contains virgin
pulp, including 25% Forest Stewardship Council Certified for no
old-growth tree cutting, processed either TCF or ECF. The sheet is
acid-free and meets the minimum requirements of ANSI/NISO
Z39.48–1992 (R 1997) (Permanence of Paper).

CONTENTS

PREFACE AND ACKNOWLEDGMENTS

I first became seriously interested in the Galileo affair in October 1980, by way of the fortuitous coincidence of two events: the Vatican announcement that a papal commission was being appointed to reexamine the affair and the publication of my *Galileo and the Art of Reasoning*. That book was an analysis of Galileo's *Dialogue on the Two Chief World Systems* (1632), the work that triggered his condemnation in 1633, and so I felt able to make a contribution. However, I knew little about the facts of the trial, and to fill that lacuna I decided to master the relevant documents; thus I produced *The Galileo Affair: A Documentary History* (1989). Moreover, my earlier work analyzed Galileo's *Dialogue* from the point of view of logic and scientific methodology, but to better relate it to the trial I needed a more down-to-earth analysis; so I created *Galileo on the World Systems: A New Abridged Translation and Guide* (1997). Next, I wanted a firmer grasp of the history of the various interpretations and evaluations of the trial, in order to avoid their limitations and utilize their insights; thus the present book came into being. With these three foundations, the last remaining step will be to work out my own historical and critical account.

This book is a survey of the Galileo affair from the time of his condemnation by the Inquisition in 1633 to his alleged rehabilitation by Pope John Paul II in 1992. A key recurring question has been whether, how, and why the condemnation was right or wrong, and that is what the title *Retrying Galileo* is meant to convey. The survey is introductory in the sense that it contrasts to both a narrative history and a critical assessment: it has elements of both but aims to perform the more fundamental task of presenting the primary sources, historical facts, and controversial issues. Moreover, the survey is document-based in the sense that it stresses the texts that make up the primary sources, and the facts and issues are made to emerge from them. My threefold concern—with sources, facts, and issues—is reflected in the chapter and section titles, most of which contain dates (at the end), references to sources (after the colon), and interpretive or evaluative issues or themes (before the colon).

Although the *original* Galileo affair (1613–1633) is one of the most studied events in Western culture, until now the *subsequent* Galileo affair (1633–1992) has never been surveyed in a systematic manner. So this book is directly and explicitly about the latter but, because of their connection, it is also indirectly and implicitly about the original trial.

The primary sources for the subsequent affair have offered opportunities, presented challenges, and occasioned decisions that need explanation. On the one hand, because of the length of the period I examine, the interdisciplinary subject matter, the international perspective, and the involvement of several languages, the primary sources are so numerous that stringent criteria are needed to judge their relative importance. Often a source deserves no more than a bibliographic mention. However, a comprehensive bibliography would list about 2,500 entries. So, in the interest of producing a book of manageable proportions, I decided to include here a select bibliography, thus reducing the list to about 1,100 entries.

On the other hand, some primary sources are so valuable—because they are especially informative, insightful, eloquent, or otherwise revealing—that I decided to let them tell part of the story by quoting their full text or long excerpts. Although they are all from printed sources, they are all relatively inaccessible. They have never been translated into English from their original Italian, French, or Latin. About twenty of them are private letters; about ten are Church documents; and another ten are interpretive or critical accounts published, or intended for publication, by their authors. They range in length from less than a page to about twenty pages.

Throughout the book, I have also used several conventions. In presenting the long quotations, I have inserted numerals in brackets to refer to the pagination of the quoted source. In my commentary after such quotations, numbers in parentheses refer to such page numbers from the quoted source. In the notes, I use an author-date style of reference, but occasionally I use it also in the body of my exposition, especially for simple citations that involve just authors' names and book titles or dates of publication. In the notes and in cross-references within the bibliography, references to the National Edition of Galileo's works edited by Antonio Favaro (1890–1909) are given using only Favaro's name followed by volume number, colon, and page numbers. Finally, all translations are my own unless otherwise noted.

In researching and writing this work, I have received valuable assistance from many persons, institutions, and works, and here I should like to express my thanks.

For constant and multifaceted support since the conception of this project, I am indebted to Mario Biagioli, John Heilbron, Albert van Helden, and William Wallace. For reading the final manuscript, I am grateful to

John Brooke, Paolo Galluzzi, Heilbron, Helden, Jürgen Renn, Michael Segre, and Robert Westman. For various specific forms of assistance, I thank Ugo Baldini, Antonio Beltrán Marí, Francesco Bertola, David Cahan, I. Bernard Cohen, James Colbert, George Coyne, Albert Di Canzio, Owen Gingerich, Margaret Jacob, Nicholas Jardine, Michel-Pierre Lerner, Carlos López Beltrán, James MacLachlan, Franco Motta, Mauro Pesce, Theodore Porter, Rose-Mary Sargent, Philip Shoemaker, Leen Spruit, and Jon Topham. My student Aaron Abbey provided invaluable assistance by reading the manuscript and playing the role of intelligent consumer and perceptive nonexpert reader. My colleagues James Frey, Todd Jones, Mary Phelps, and Paul Schollmeier provided important institutional support.

Acknowledgments are also due to the Department of History of Science at Harvard University for appointing me visiting scholar for thirteen months in 1998–1999, during which I did the bulk of my research; the Guggenheim Foundations for a one-year fellowship held during the same period; the Program in Science and Technology Studies of the National Science Foundation for a Scholar's Award with a four-year grant (SBR 9729117) in 1998–2002; the University of Nevada, Las Vegas, for a research leave and a Barrick Distinguished Scholar Award in 1998–1999 and a sabbatical leave in 2001–2002; the Vatican Library for consultation privileges in April 1999; and the British Library for consultation privileges in June 2000.

I also wish to express my indebtedness to several published works and their authors that have been indispensable, even though my notes also contain the standard citations. Without the documents provided in these works and the archival research of these authors, my book would not be what it is; certainly my documentary survey would have been impossible. I am referring to Baldini 2000b, Bertolla 1979, Brandmüller and Greipl 1992, Maccarrone 1979, Maffei 1987, Motta 2000, Mayaud 1997, Monchamp 1893, and Simoncelli 1992. For completeness's sake, I should add Favaro 1887b, besides of course his edition of Galileo's works; but this addition would be axiomatic for Galilean scholars, who know the giants on whose shoulders they stand.

Finally, acknowledgments are due to several publishers of works containing texts which I have translated into English: Edizioni dell'Arquata di M. R. Trabalza (Foligno) for Olivieri's (1820) summary as found in Maffei's (1987) edition; Società Editrice Vita e Pensiero (Milan) for two passages from Gemelli's (1942b) article; Edizioni La Civiltà Cattolica (Rome) for two passages from Soccorsi's (1947) book; Cooperativa Libraria Editrice Università di Padova for Lazzari's (1757) report as found in Baldini's (2000b) *Saggi*; Editrice Pontificia Università Gregoriana (Rome) for four passages reproduced in Mayaud's (1997) book; Deputazione di Storia Patria per il Friuli (Udine) for one of Paschini's letters reproduced in Bertolla's (1979) contribution to the *Atti del convegno di studio su Pio Paschini nel centenario della nascita;* Edizioni Studium (Rome) for a passage from Paschini's (1943) article; the journal *Lateranum* (Rome) for one of Paschini's letters reproduced in Maccarrone's (1979b) article; FrancoAngeli (Milan) for two of Paschini's letters reproduced in Simoncelli's (1992) book; and the Pontifical Academy of Sciences (Vatican City) for three passages from Paschini's (1965) book, as reprinted by Herder Editrice e Libreria (Rome).

Introduction

The Galileo Affair from Descartes to John Paul II

A Survey of Sources, Facts, and Issues

In 1633 the Inquisition condemned Galileo for holding that the earth moves and the Bible is not a scientific authority. This condemnation ended a controversy that had started in 1613, when his astronomical ideas were attacked on scriptural grounds and he wrote a letter of refutation to his disciple Benedetto Castelli. This was a controversy involving issues of methodology, epistemology, and theology as well as astronomy, physics, and cosmology: whether the earth is located at the center of the universe; whether the earth moves, both around its own axis daily and around the sun annually; whether and how the earth's motion can be proved, experimentally or theoretically; whether the earth's motion contradicts Scripture; whether a contradiction between terrestrial motion and a literal interpretation of Scripture would constitute a valid reason against the earth's motion; whether Scripture must always be interpreted literally; and, if not, when Scripture should be interpreted literally and when figuratively.

Although the 1633 condemnation ended the original Galileo affair, it also started a new controversy that has continued to our own day—about the facts, causes, issues, and implications of the original trial. This subsequent controversy reflects in part the original issues, but it has also acquired a life of its own, with debates over whether Galileo's condemnation was right; why he was condemned; whether science and religion are incompatible; how the two do or should interact; whether individual freedom and institutional authority must always clash; whether cultural myths can ever be dispelled with documented facts; whether political expediency must prevail over scientific truth; and whether scientific research must bow to social responsibility.

Besides such *controversial issues,* the subsequent Galileo affair has two other strands. First, the *historical aftermath* of the original episode consists of facts and events stemming from it and involving actions mostly taken by the

Catholic Church, such as the partial unbanning first of Galileo's *Dialogue on the Two Chief World Systems* (1632) and later of Copernican books in general during the papacy of Benedict XIV (1740–1758); the total repeal of the condemnation of the Copernican doctrine in the period 1820 to 1835; the implicit theological vindication of Galileo's hermeneutics in Pope Leo XIII's encyclical *Providentissimus Deus* (1893); the beginning of the rehabilitation of Galileo himself, occasioned by the commemoration in 1942 of the tricentennial of his death; and most recently the further rehabilitation of Galileo by Pope John Paul II (between 1979 and 1992). But the historical aftermath also includes such actions as Descartes's decision (in 1633) to abort the publication of his own cosmological treatise *The World;* the Tuscan government's reburial of Galileo's body in a sumptuous mausoleum in the church of Santa Croce in Florence (1737); Napoleon's seizure of the Vatican file of the Galilean trial proceedings and his plan to publish its contents (between 1810 and 1814); and the publication of those proceedings by lay scholars in France, Italy, and Germany between 1867 and 1878.

Second, the past four centuries have produced vast amounts of *reflective commentary* on the original trial, consisting of countless interpretations and evaluations advanced by astronomers, physicists, theologians, churchmen, historians, philosophers, cultural critics, playwrights, novelists, and journalists. These comments have appeared sometimes in specialized scholarly publications, sometimes in private correspondence or confidential ecclesiastical documents, and sometimes in classic texts; among these are Milton's *Areopagitica,* Pascal's *Provincial Letters,* Leibniz's *New Essays on Human Understanding,* Voltaire's *Age of Louis XIV,* Diderot and D'Alembert's French *Encyclopedia,* Comte's *Positive Philosophy,* John Henry Newman's writings, Leo XIII's *Providentissimus Deus,* Brecht's *Galileo,* and Koestler's *Sleepwalkers.*[1]

Now, whereas the original trial is one of the most studied events of Western cultural and intellectual history, the subsequent affair has not received the attention it deserves. Almost all books on the original episode do contain some account of the trial's aftermath, of the reflective commentary, and of subsequent issues; indeed, some of these accounts are very useful.[2] There are also detailed and excellent studies of particular episodes of the subsequent affair, on which I draw.[3] However, the whole story of the aftermath and of the repercussions has never been told; the rich variety of reflections on the trial have never been collected or catalogued, let alone systematically and critically examined; and the main questions, issues, and problems of the subsequent affair have hardly been formulated, let alone answered or solved. The subsequent affair deserves more study partly because it has acquired an autonomous existence whose fascination rivals that of the original trial; partly because it constitutes a uniquely instructive example of the interaction between science and religion; partly because it provides an excellent instance of the rise, diffusion, and development of cultural myths; partly because the enormously voluminous literature on the trial is a key aspect of the subsequent affair and embodies a potentially very fruitful case study in historiography or metahistory; and partly because the systematic critical examination of that literature can provide the basis for a new and better historical and critical account of the original episode, avoiding the weaknesses and incorporating the insights of previous accounts.

Of course, the events of the subsequent controversy are often by-products of the original episode, and the subsequent issues tend to reflect the original ones. But this is not always the case. For example, a novel aspect of the subsequent affair is the fascinating story of the compilation in 1755 of the special Vatican file of Inquisition proceedings of Galileo's trial; its removal from Rome to Paris by Napoleon between 1810 and 1814; its disappearance in 1814; the Church's indefatigable but unsuccessful efforts to retrieve it from 1814 to 1817; the publication in 1821 of parts of it from the incomplete French translation ordered by Napoleon; its retrieval and return to Rome in 1843; and its gradual publication between 1867 and 1878. And with regard to the issues, I have already mentioned several novel ones. In particular, some of the subsequent issues are necessarily novel insofar as they concern the original trial; this is clearly the case when one asks whether the condemnation of Galileo was right or wrong, and if so in what sense—theological, scientific, philosophical, legal, moral, pastoral, practical, political. This issue, which of course has many ramifications and is very widely encompassing, has perhaps been the key issue. But there are other, similar questions, such as whether that condemnation refutes the Catholic doctrine of papal infallibility;[4] whether Galileo was tortured, and whether such torture would have been legally improper; whether the extant trial documents in the archives were tampered with by clerical officials; and whether a legal impropriety occurred at the 1633 trial in regard to the special injunction which the Inquisition claimed to have issued to Galileo in 1616.

Moreover, even when the questions span or apply to both affairs, they acquire different meanings in the two contexts. For example, consider the implications of Galileo's trial for the relationship between science and religion. As traditionally interpreted, Galileo's trial epitomizes the conflict between science and religion.[5] At the opposite extreme is the revisionist thesis that the trial illustrates the *harmony* between them.[6] However, I would argue that the trial had both conflictual and harmonious aspects when viewed in terms of science and religion, but that these are elements of its surface structure and that its most profound deep-structure lies rather in the clash between cultural conservation and innovation. A key conflictual element stems from the contradiction between Copernican astronomy and Scripture alleged by the institutions and persons that opposed Galileo. A key harmonious element is the fact that Galileo and his supporters did not see a contradiction between Copernicanism and the Bible. Thus, an irreducible conflict in the trial was the clash between those who affirmed and those who denied that Copernicanism was contrary to Scripture. However, both camps included clergymen, scientists, and clerical and secular institutions; thus the conflict was not between an ecclesiastic monolith on one side and a scientific monolith on the other, but rather between two attitudes that crisscrossed both science and religion.[7] I believe the most fruitful way of describing the two camps is to label them conservatives or traditionalists on one side and progressives or innovators on the other. In this sense, Galileo's trial illustrates the clash between conservation and innovation and constitutes one battle that the conservatives happened to win. This conflict is also evident in other domains of human culture, such as politics, art, economy, and technology. It cannot be eliminated, on pain of stopping cultural development; it is a moving force of human history.[8]

By contrast, even those who advocate the harmonious view of the original trial acknowledge that the key feature of the subsequent affair was indeed a conflict between science and religion;[9] such a conflict is precisely what they bemoan and want to put an end to. As regards this subsequent controversy, it thus seems clear that the science-versus-religion conflict is indeed an essential feature of it, and that this conflict is more of an integral part of it than of the original trial. However, again, underlying such surface structure there is a cultural deep-structure, which here lies (I believe) in the phenomenon of the birth and evolution of cultural myths and their interaction with documented facts.[10]

Consider now the question, "When and how did the Galileo affair start, and who started it?" With regard to the original episode, this question is one of the most frequently discussed issues. It may be taken to concern the timing, manner, and identity of the factor that *precipitated* the controversy, while recognizing that its original roots go back to Copernicus's theoretical system elaborated in his *Revolutions* of 1543; to Galileo's telescopic discoveries in his *Sidereal Messenger* of 1610; and to the theologians' responses to these works. A common answer (and the one that comes closest to the truth) is that the original Galileo affair was precipitated by the conservative clerics Niccolò Lorini and Tommaso Caccini in 1612–1614, when they charged Galileo with heresy for his Copernican inclinations; then, Galileo's discussion of the biblical objection, of the scientific authority of the Bible, and of the compatibility between the Bible and Copernicanism can be seen as a legitimate attempt to defend himself and an astronomical theory from irrelevant and illegitimate attacks and criticism.[11] Others blame Galileo himself for having unnecessarily become involved in questions of biblical interpretation.[12] And other scholars blame instead Aristotelian professors of philosophy (such as Ludovico delle Colombe and Cosimo Boscaglia) for having been the first to criticize Galileo on biblical grounds.[13] Clearly in such discussions one must avoid the genetic fallacy of equating the nature of a controversy with its origin; nevertheless, the question of the origin is extremely important.

With regard to the subsequent affair, however, this question has not even been asked, let alone answered. My investigation will provide evidence relevant to this question. I suggest that, on the one hand, the roots of the subsequent controversy lie in Pope Urban VIII's decision to give unique publicity to the condemnation of Galileo; in the decision of an international group of Protestants to publish (in Strasbourg in 1635–1636) a Latin edition of Galileo's *Dialogue*, together with several theological essays (including his Letter to Christina), arguing that Scripture is not an astronomical authority and does not contradict Copernicanism; and in the Jesuit Giovanni Battista Riccioli's publication of his *Almagestum Novum* (1651), which contains an explicit justification of the Inquisition's condemnation of Galileo on the grounds that he was wrong in astronomy (because the Tychonic system was the correct one) as well as in theology and methodology (because biblical statements on all subjects must be held to be literally true). On the other hand, the debate was relatively subdued until 1784, when Jacques Mallet du Pan published a new justification of the Inquisition and a new indictment of Galileo, claiming that he had been condemned not for being a good (Copernican) astronomer but for being a bad theologian (supporting astronomical propositions with biblical statements). This account was repeated more or less uncritically by many others for almost a century; it also encouraged the explicit formulation of other proclerical accounts, for example, those of Girolamo Tiraboschi in 1793 and Peter Cooper in 1838; and these developments in turn led many to come to Galileo's defense and attack the Church, for example, Giambattista Venturi in 1818–1821 and Giulio Libri around 1841.

While such theses are interesting and important (and will have to be elaborated in some future work), my primary concern in this book is not to fully articulate and defend them but rather to suggest them and to undertake a more fundamental task. It is the following.

Although the literature on the original affair is enormous, the story of the aftermath has never been told; the reflective commentary has never been systematically examined; and the subsequent controversial issues have never been contextualized in the story or anchored in the textual sources. Thus, the more fundamental task is to provide an introduction to, and survey of, the textual sources, the chronological facts, and the controversial issues of the Galileo affair from 1633 to 1992. To this end, the book incorporates many long excerpts from the most important and manageable documents and texts; all these texts are from *printed sources*, but they have never before been collected and assembled together, let alone translated into English.[14] An important subset of these sources consists of the essential documents of Galileo's trial; these are of course well known and now widely available and so are not reproduced here, but as the story unfolds they are mentioned or summarized in the context in which they were first discovered, publicized, published, or discussed. The book also aims to establish

inquisitors usually met two or three times a week. One of these meetings was usually presided over by the pope, as was fitting for a department regarded as the single most important administrative unit of the Church. Moreover, the decisions reached when the pope was not present normally had to be ratified by him; and in this particular case, the pope was directly involved in the proceedings. Nevertheless, the Inquisition and the pope were distinct, and the decisions of the former could not be formally equated with those of the latter. In short, Galileo was being condemned neither by a formal decree of the pope nor by the whim of a lone inquisitor, but by a collective body that was the supreme tribunal of the Catholic Church.

The substantive part of the sentence began by going back to the year 1615 and describing several charges advanced against Galileo then.[3] Five distinct accusations were mentioned: (1) holding the truth of the earth's motion; (2) corresponding about this doctrine with some German mathematicians; (3) publishing a book titled *Sunspot Letters* that explained the truth of the doctrine; (4) answering scriptural objections against the doctrine by elaborating personal interpretations of Scripture; and (5) writing a letter to a disciple containing various propositions against the authority and the correct meaning of Scripture. In accordance with standard inquisitorial practice and with the purpose of a sentence, no names or details were mentioned.

Today we know that these charges had been made by two Dominican friars: Niccolò Lorini, who had written a letter of complaint to a cardinal-inquisitor, and Tommaso Caccini, who had made a personal appearance before the Roman Inquisition.[4] We also know who Galileo's German correspondents were: Johannes Kepler, the mathematician to the Holy Roman Emperor, who in 1610 had written a public letter endorsing the telescopic discoveries announced in Galileo's *Sidereal Messenger;* and Christopher Scheiner, the Jesuit mathematician with whom Galileo had exchanged the sunspot letters in 1612–1613. We know, too, that the disciple to whom Galileo had (on 21 December 1613) written a refutation of the scriptural objection to Copernicanism was Benedetto Castelli, a Benedictine friar and a professor of mathematics at the University of Pisa and later at Rome.[5] And indeed, as suggested in the last two charges, there were two ways in which Galileo had replied to the objection that Copernicanism must be wrong because it contradicts many scriptural passages: first, even if Copernicanism contradicts Scripture, that is irrelevant because Scripture is not a philosophical (or scientific) authority, but an authority only on questions of faith and morals; and second, it is questionable whether Scripture contradicts Copernicanism any more than it contradicts the Ptolemaic system, for an analysis of the relevant scriptural passages shows that they are more in accordance with the earth's motion than with its rest.

The historical account continued by relating that after those charges were made against Galileo, the Inquisition consulted its experts requesting an opinion on the Copernican doctrine.[6] For this purpose, two parts were distinguished in this doctrine: the thesis of heliocentrism or heliostaticism, and the geokinetic thesis. Both theses were judged false and absurd from a philosophical point of view, namely the point of view of natural philosophy, physics, astronomy, or science. Theologically speaking, heliocentrism was declared formally heretical, on the grounds that it was explicitly contrary to Scripture, whereas geokineticism was determined to be at least erroneous in faith.

The sentence did not elaborate these judgments. But the philosophical falsity of these propositions was grounded on the many astronomical, phys-ical, metaphysical, and epistemological objections that for centuries had accumulated against the earth's motion and seemed very cogent.[7] The difference in the theological evaluation of the two propositions was prob-ably due to the degree of conflict with Scripture: heliostaticism seemed directly contrary to such statements as that "the sun also riseth, and the sun goeth down, and hasteth to the place where he ariseth" (Ecclesiastes 1:5); whereas geokineticism seemed to be contrary only to what is implied by such statements.

The account continued with the story of cardinal-inquisitor Robert Bel-larmine's warning and the Inquisition's special injunction to Galileo.[8] It recalled that in the earlier phase of the proceedings, the Inquisition had decided to treat Galileo with benign consideration. Consequently, at the Inquisition meeting of 25 February 1616 chaired by the pope, it was decided that Bellarmine would privately and informally warn Galileo to abandon Copernicanism; that if he refused, the Inquisition's commissary would give Galileo a formal injunction not to hold, teach, defend, or dis-cuss it in any way whatever; and that if he did not acquiesce, he would be imprisoned. The next day, we are told, Bellarmine gave Galileo the friendly warning; the commissary gave him the injunction; and Galileo agreed.

These developments raise many questions. For now, suffice it to stress what the sentence said: that Bellarmine's warning was supposed to be dis-tinct from the special injunction; and that the injunction was supposed to be contingent on Galileo's rejection of Bellarmine's warning, just as impris-onment (or prosecution) was to be contingent on Galileo's rejection of the injunction. The sentence did not explicitly say that Galileo rejected Bel-larmine's warning, and so it is unclear whether the procedure on 26 Febru-ary followed the Inquisition's orders of the day before.

Next,[9] the sentence mentioned the anti-Copernican decree issued by the Congregation of the Index, the department of the Church in charge of book censorship. This consisted of another committee of cardinals that operated in a manner analogous to the Inquisition. We are told that this decree declared the earth's motion false and contrary to Scripture and prohibited books discussing it.

Because this decree was a public edict, unlike both Bellarmine's warning and the special injunction, we will soon consult it and get a firsthand look at what it said and whether the sentence was summarizing it correctly.

The next development mentioned[10] occurred sixteen years after the anti-Copernican decree and one year before the sentence itself: Galileo's publication in 1632 of a book titled *Dialogue on the Two Chief World Systems, Ptolemaic and Copernican*. The Inquisition was soon informed that this book was causing the dissemination and establishment of the geokinetic doctrine. An examination of the book was ordered, revealing that it defended the earth's motion and so was an explicit violation of the special injunction.

This finding led the Inquisition to summon Galileo, examine him under oath, and obtain the following confession.[11] Galileo admitted that he wrote the book after being issued the special injunction; that he requested the imprimatur without disclosing the special injunction; and that the book defended the earth's motion, in the sense that it gave the reader the impression that the arguments in favor of this false doctrine were stronger than those favoring the geostatic view; and that this transgression was due to literary vanity (the natural tendency to show off one's cleverness) rather than intentional.

Then the document summarized Galileo's defense,[12] which hinged on Bellarmine's certificate and Galileo's denial of a malicious intention. Galileo presented a brief memorandum written by Cardinal Bellarmine, which Galileo had obtained in 1616, soon after those earlier proceedings had been concluded, to clarify the situation; it stated that Galileo had not been formally condemned or put on trial but only notified of the Church's decision that the earth's motion was contrary to Scripture and so could be neither held nor defended. Galileo pointed out that Bellarmine's certificate did not contain the prohibition "to teach in any way whatever," and for this reason he eventually forgot about this part of the special injunction and felt no need to mention the injunction at all when requesting the imprimatur. All this was meant not to deny or excuse the error (of defending the earth's motion) but to show that it was due to literary vanity rather than to a malicious intention.

Bellarmine had died in 1621 and so was unavailable to resolve the matter at the 1633 trial. The certificate he wrote[13] for Galileo was not an official document but an informal one written at the request of Galileo, who wanted not only to dispel rumors that he had been tried and condemned but also to clarify the import of Bellarmine's warning, the special injunction, and the Index's Decree. The certificate could be regarded as Bellarmine's understanding of the warning and/or injunction issued to Galileo. Given Bellarmine's authority and his direct involvement in the earlier proceedings, the certificate could not be dismissed.

The Inquisition was, however, quick to rebut this defense.[14] It held that Bellarmine's certificate in fact aggravated Galileo's situation because it stated that the earth's motion was contrary to Scripture, and yet Galileo had dared to defend it. Furthermore, the imprimatur obtained for the book was invalid because Galileo did not disclose the special injunction.

Regarding the question of the intention, the Inquisition was unconvinced by Galileo's denial of malice, and to resolve its doubts it conducted a "rigorous examination" of the accused.[15] Galileo apparently passed this test.

The term *rigorous examination* was the standard inquisitorial jargon for torture. Hence, this passage of the sentence generated some of the most hotly debated questions of the Galileo affair: whether Galileo was indeed tortured; if so, why; and whether torture was legally or morally justified in his case (see chapters 11 and 12). For now, the important point is that one does not have to look at other documents or rely on innuendos to be faced with the issue; it arises immediately out of the official text of the sentence. Moreover, there is no denying that this text states in no uncertain terms that it was judged "necessary to proceed against you by a rigorous examination."[16] It is also undeniable that "rigorous examination" referred to torture. However, the crucial distinction that must be made at this point is that there were many categories of "torture." It could be actual or merely threatened: that is, a rigorous examination could be an interrogation under the verbal threat of physical torture. The actual torture involved several steps, at any one of which it could be stopped: taking the accused to the torture chamber and showing him the instruments; undressing the accused, as if to proceed with the administration of the physical pain; tying the undressed accused to the instrument; and operating the instrument so as to inflict physical pain. The last step, which could be taken to be "actual torture" in the primary sense, was in turn susceptible of various degrees of severity, naturally subdivided into mild, ordinary, and severe. Finally, there was the possibility that actual torture might be applied on more than one day or session, thus generating the category of "repeated torture." We need not delve further into these nauseating details, but, given these distinctions, when the sentence asserted nonchalantly that Galileo was subjected to a rigorous examination, we can take it to mean that he underwent at least an interrogation under the verbal threat of torture, although it does not necessarily imply that he was subjected to any step or degree of actual torture.

After this summary of the proceedings since 1615, the sentence assured that the inquisitors had carefully considered the merits of the case, the confessions and excuses of the accused, and all other relevant evidence; that they had asked for the guidance of Jesus Christ and the Virgin Mary; and that they had consulted with the experts in the fields of theology, canon law, and civil law.[17] On this basis, they reached the following verdict.

The verdict[18] was that Galileo had been found guilty of "vehement suspicion of heresy." Two main errors, and not just one, were mentioned. The first involved holding a doctrine that was false and contrary to Scripture, namely the heliocentric and geokinetic theses. The second imputed error was the principle that it is permissible to defend a doctrine contrary to Scripture.

The notion of "vehement suspicion of heresy" embodies the complexity of the theological concept of heresy and of the Inquisition's antiheretical practices.[19] Galileo was being convicted of a religious transgression that, although falling short of the most serious possible crime, was nonetheless regarded as a religious crime. The most serious such crime would have been "formal heresy"; below formal heresy there was the crime of "suspected heresy," which was in turn subdivided into three kinds: strong suspicion of heresy, vehement suspicion of heresy, and mild suspicion of heresy. There were thus four subtypes of heresy. And besides heresies, there were lesser religious crimes: beliefs or behavior could be erroneous, scandalous, temerarious, and dangerous. So Galileo was being found guilty of a relatively serious type of religious crime, clearly not the most serious one (formal heresy), nor even the second most serious offense (strong suspicion of heresy), but equally obviously above the nonheretical transgressions and the slight suspicion of heresy.

It is also crucial to note that two suspected heresies were being attributed to Galileo. The first was a doctrine about the location and behavior of natural bodies; four particular theses were distinguished, namely heliocentrism, heliostaticism, geokineticism, and the denial of geocentrism; but the respective status of these four elements was not being censured in any more nuanced fashion. The second suspected heresy was a methodological or epistemological principle, a norm about what is right or wrong in the search for truth and the quest for knowledge; it was a rule that declared Scripture irrelevant to physical investigation; the principle denied the authority of Scripture in natural philosophy (physics, astronomy, science). These two suspected heresies were the main charges of which Galileo had been accused in 1615 by Caccini and Lorini; thus the sentence was presenting the condemnation of 1633 as a resumption of the proceedings temporarily suspended in 1616 and as a demonstration of the essential correctness of those charges, now demonstrated by the publication of the *Dialogue*.

The verdict was followed by a list of penalties.[20] The first was that Galileo was to immediately recite an "abjuration" of the "above-mentioned errors and heresies." Second, the *Dialogue* was to be banned by a public edict. Third, Galileo would be imprisoned indefinitely. Fourth, he would have to recite the seven penitential psalms once a week for three years. Finally, the Inquisition declared that it reserved the right to reduce or abrogate any or all of these penalties.

Although the abjuration was a penalty in the sense that it was a great humiliation to have to recant one's views, it was also a procedural step that allowed the culprit to gain absolution for the sin of heresy. The book's prohibition took effect immediately, although a year passed before it would be included in a formal decree of the Index that included other books as well. The imprisonment stipulated here never did involve detention in a real prison, although it did last for the rest of Galileo's life. In fact, the formal imprisonment was abrogated as follows: for two more days he was confined to the prosecutor's apartment at the Inquisition palace, which had been his place of detention for certain periods of the trial (April 12–30 and June 21–22); for about two weeks (June 24 to July 6) he was under house arrest at Villa Medici, the sumptuous palace in Rome belonging to the Tuscan grand dukes; then his place of detention was commuted to the residence of the archbishop of Siena, where he stayed from early July to early December; from 17 December 1633 onward, he was under house arrest at his own villa in Arcetri near Florence. However, because these details were not generally known at the time, on the basis of the sentence one could justifiably conclude that Galileo had been put in prison as a result of his condemnation. The last penalty in the list was a "spiritual" penance for the good of his soul.

The sentence ended with the signatures of seven cardinal-inquisitors.[21] That is, of the ten cardinals mentioned at the beginning, Francesco Barberini, Gasparo Borgia, and Laudivio Zacchia did not sign. The explanation of this fact has become one of the many controversial questions in the Galileo affair. Two obvious possibilities are that the lack of these signatures reflected a significant disagreement among the ten judges, and that the three cardinals were not present at the meeting when the document was signed.

In conclusion, on the one hand, the sentence contains more information than one might expect, for we find not only an announcement of the verdict but also a statement of the penalty and a historical account of the proceedings; these proceedings go back to the years 1615–1616, when they had reached a temporary resolution without a formal trial or condemnation of Galileo himself. Once we distinguish these parts of the sentence, it should be clear that the verdict possesses an intrinsic validity that neither the account of the proceedings nor the list of penalties does; that the account of the proceedings and the list of penalties are susceptible of an additional evaluation, a retrospective-historical kind for the former and a prospective one for the latter; and that the missing justification provides a degree of speculative freedom for its reconstruction which we do not have for the parts that are present.

To elaborate, this document's *saying* that the verdict and penalty were as indicated *makes* them so, whereas its account of the proceedings may or may not reflect what really happened and what other documents reveal. Of course, these documents were not publicly available then; it took centuries for them to be unearthed, and it is the purpose of this book to tell the latter story. But even then it was self-evident that those documents might or might not correspond to the sentence. By contrast, there were no analogous doc-

being heliocentrism, heliostaticism, geokineticism, and nongeocentrism.[27] This was followed by the admission that this verdict was right, when he spoke of "this vehement suspicion, rightly conceived against me."[28] Here we have a second new confession, which of course could not have been made during the (presentencing phase of the) trial; he was also confessing that he was guilty as judged.

After these confessions, we come to the abjuration proper, where the culprit "with a sincere heart and unfeigned faith" abjured and cursed the above-mentioned errors and heresies.[29] This part of the document has led to the question of whether such Galilean abjuring amounted to perjuring— another classic controversy of the Galileo affair.

The culprit then made a series of solemn promises about future thought and behavior: that he would never again hold any such beliefs; that he would denounce to church officials any heretics or suspected heretics; that he would comply with the penalties imposed on him by the judges; and that he would submit to further penalties if he failed to comply with the current ones.[30] An important aspect of these affirmations is the explicit distinction between heresy and suspected heresy; in the complex taxonomy of theology and canon law, suspected heresy might have been a lesser crime than formal heresy, but it was a crime nonetheless. The sentence had declared Galileo to be a suspected heretic; the abjuration here repeated this characterization.

The two final paragraphs suggested something that was entirely normal for Inquisitorial practice: at the final session of the trial Galileo read this document word for word and affixed his own handwritten signature to the end, but the text had been compiled by the Inquisition's clerks.

Whether this abjuration was extorted from Galileo during the "rigorous examination"—perhaps in exchange for relief from additional "rigors," as some scholars have claimed[31]—is a controversial question that we can simply add to our accumulating stock. However, it is certainly striking that in this abjuration Galileo was not only being made to retract earlier beliefs and behavior but also to plead guilty to the verdict already announced by the judges and to confess to a transgression not confessed earlier.

1.3 "Suspended until Corrected": The Index's Anti-Copernican Decree (1616)

One of the few available and public documents mentioned in the sentence was the anti-Copernican decree of the Index of 5 March 1616. It represented one of the outcomes of the earlier Inquisition proceedings that set the stage for the later ones. Thus the learned world of the time had this other item on the basis of which it could understand better and perhaps assess critically the 1633 condemnation of Galileo.[32]

This document was a decree emanating from the Congregation of the Index, the department of the Church in charge of book censorship. Such censorship had been practiced for a long time, and the first main edition of the *Index Librorum Prohibitorum* had been published in 1564 as a result of the Council of Trent. However, this congregation, whose name was usually shortened to "Index," had been formally established in 1571–1572. Like the Inquisition, the Index consisted of about ten cardinals and was similarly organized. For example, there was a cardinal-chairman (the "prefect") and a secretary (who was usually a lesser clergyman), and its decrees were normally signed only by these officials (not by all cardinal-members of the congregation); in the present case, both signed it, Cardinal Paolo Sfondrati and the Dominican friar Francesco Maddaleni Capiferreo. Although the Index was regarded as an important department, it did not enjoy as much power, prestige, and authority as the Inquisition; it did not have the massive bureaucratic apparatus of local chapters that the Inquisition possessed; nor did it inspire as much fear.

In any case, there was some overlap between the two congregations, with respect to both membership and jurisdiction. In fact, at the time of this decree, three cardinals (including Bellarmine) were members of both the Index and the Inquisition.[33] And, as reported in the 1633 sentence, the Index had become involved in the censure of Copernican books as a result of the Inquisition investigation of the charges advanced against Galileo in 1615. Thus, even from the point of view of that sentence, let alone from the vantage point of someone (like Galileo himself) directly involved in the proceedings of 1615–1616, it was quite extraordinary that neither Galileo's name nor any of his works were mentioned in this Index decree. In the context of our discussion, this absence is perhaps the most striking feature of the decree.

Even a casual reader would also notice that this decree had two distinct parts. The first banned five books dealing with topics involving theology, morals, law, and ecclesiastical history. Catholics were reminded of what such banning implied; that is, they could not only not read these books but were not even permitted to own them. No explanation or justification was given of exactly what was wrong with them.

The expository style of this first part was entirely typical for decrees of the Index. It was thus very striking that this particular decree had a second part where three other publications were banned and where not only was the kind of applicable prohibition more nuanced, but an explanation and a justification were given. These features of the second part of this document clearly suggested that one was dealing with a very special situation.

The decree announced that at issue was an ancient doctrine that went back to the Pythagoreans and that had recently been revived in Copernicus's *Revolutions*, Diego de Zúñiga's *Commentaries on Job*, and Paolo A. Foscarini's *Letter on the Pythagorean Opinion*. This doctrine was succinctly summarized as the proposition that the sun is motionless at the center of the universe and the earth revolves around it. The congregation regarded the doctrine as both false and contrary to Scripture and so was very concerned that it was gaining wider acceptance.

We have seen that this was the assessment reported in the 1633 sentence as having been made in 1616 by the Inquisition consultants during the first phase of the proceedings against Galileo. In fact, here the Index's decree was appropriating that assessment. It is even more important to note that this decree appropriated only part of that assessment, for there was no explicit talk of heresy here. The doctrine was being censured both philosophically and theologically, but the theological assessment was not "heretical" but rather "contrary to Scripture." In the subsequent history of the Galileo affair, this has created the problem and the controversy of why one should talk of heresy in the 1633 trial, when the 1616 Index decree stayed away from such stronger censure. On the other hand, it must be pointed out that this part of this decree represented a doctrinal pronouncement. So what we have here is not merely the condemnation of some books, to which we will come presently, but the condemnation of a doctrine. It was this doctrinal condemnation, in fact, that provided the basis for the censure of the three Copernican books.

Two of these books (by Copernicus and Zúñiga) were "suspended until corrected," whereas Foscarini's work was "completely prohibited and condemned." The stronger censure for Foscarini's book was due to the fact that, besides trying to show that heliocentrism was physically true, it primarily argued that it was compatible with Scripture; thus it violated both of the negative assessments of the doctrine, especially the more directly relevant theological assessment. On the other hand, Copernicus and Zúñiga were given the milder censure apparently because they merely taught the (physical, philosophical, or scientific) truth of the doctrine.

This impression is certainly correct for the case of Copernicus's *Revolutions,* which was a book of technical mathematical astronomy. Published in 1543, it was in a sense a modern revival of a Pythagorean doctrine, as the decree suggests; but the book was original in that it provided a new argument in favor of this ancient idea, namely a detailed demonstration that the known observational details about the motion of the heavenly bodies could be explained on the basis of the heliocentric geokinetic hypothesis, and that such an explanation was in fact simpler and more coherent than the Ptolemaic one. Although it was dedicated to Pope Paul III, Copernicus's book hardly mentioned scriptural, theological, or religious matters.[34]

On the other hand, the treatment of Zúñiga as similar to Copernicus must have been puzzling. In fact, Zúñiga's *Commentaries on Job* was a massive work of some 850 pages, first published in Toledo in 1584 with the approval of the local inquisitor, the local archbishop, and the provincial head of the order of Augustinian friars to which Zúñiga belonged; the book was then reprinted in Rome in 1591 without difficulties.[35] Only a brief passage of two pages dealt with the earth's motion, in connection with the verse in Job 9:6, "He shakes the earth out of its place, and the pillars beneath it tremble." Zúñiga suggested that this verse implied that the earth is in motion. Such a Copernican interpretation of this scriptural passage was in fact far-fetched, because it is more likely that this passage is talking about earthquakes (a far cry from the Copernican terrestrial motion!). Still, Zúñiga (insofar as his work was relevant) was trying to reconcile Scripture and Copernicanism, and so was doing something more analogous to Foscarini than to Copernicus. To be sure, these facts would be known only to persons acquainted with the details of all three books, and so the average reader of the Index decree need not have been puzzled. On the other hand, all such knowledge was available at the time.

Perhaps the puzzle is solved by saying that both Copernicus's and Zúñiga's books were extremely valuable works in their respective fields, and both could be easily "corrected," if need be by deleting the two problematic pages from Zúñiga's work and the few problematic references in Copernicus's. By contrast, Foscarini's book dealt very centrally with the reconciliation of Copernicanism and Scripture, and so, given the authorities' judgment that the two were incompatible, there was nothing to do but to condemn and completely prohibit it. At any rate, the Index did publish the "corrections" of Copernicus relatively promptly (in 1620), but never did the same for Zúñiga's work. Before we examine those Copernican corrections, we must discuss a last item in the decree.

The last substantive clause in the document formulated a censure that involved "all other books which teach the same."[36] It was clearly a reiteration that the decree was a general doctrinal condemnation and not just the prohibition of three particular books. And it seemed to repeat the nuance of the previous distinction between "suspension until corrected" and "complete prohibition and condemnation"; the suggestion was apparently that any work similar to Copernicus's *Revolutions* was similarly suspended until corrected, and that any work like Foscarini's *Letter* was likewise completely prohibited and condemned.

However, that last substantive clause also contained considerable ambiguity and obscurity. When it spoke of what the present decree "prohibits, condemns, and suspends . . . respectively,"[37] it was confusingly reversing the order of the announced censures, which was (first) suspension for Copernicus and Zúñiga, and (second) prohibition and condemnation for Foscarini. When it mentioned all other books that teach the same, "the same" could refer to the same thing as Foscarini taught, which was the immediately preceding referent; or it could refer to the same Pythagorean-Copernican doctrine, which was the general referent in the paragraph. The difference is that Foscarini's main point was that it is not contrary to Scripture to say that the earth moves and the sun stands still, whereas Copernicus's main point was that the earth moves and the sun stands still; in other words, Foscarini argued chiefly that heliocentrism is reconcilable with Scripture (a theological-sounding proposition), whereas Copernicus argued mainly that heliocentrism is physically true (a proposition in natural philosophy). Although the decree made it clear that Foscarini's theological claim was being given a harsher condemnation than Copernicus's physical assertion, it was relatively unclear whether the general condemnation applied only to the more problematic theological claim, or whether it applied also to the more modest physical-philosophical proposition.

This question was partly resolved three years later. On 28 February 1619, the Congregation of the Index decided to prohibit and condemn Kepler's *Epitome of Copernican Astronomy,* which had been published the previous year.[38] Moreover, a new supplement appeared in Rome to update the main current edition the *Index of Prohibited Books* that had been published in 1596 under Pope Clement VIII; titled *Edictum Librorum Qui post Indicem Felicis Recordationis Clementis VIII Prohibiti Sunt* (1619), it included not only the books by Copernicus, Zúñiga, and Foscarini banned in 1616 and Kepler's *Epitome* but also listed the entry "all books that teach the motion of the earth and the immobility of the sun."[39]

1.4 "Hypothesis versus Assertion": The Index's Correction of Copernicus's *Revolutions* (1620)

If the events of 1619 seemed to represent a hardening of Catholic censure, let us now examine the corrections to Copernicus's *Revolutions* published on 15 May 1620, and see whether they may have softened or otherwise clarified the situation by spelling out the conditions under which that book could be read and, by implication, what aspect of the doctrine was not condemned or prohibited.

The document spelling out the changes consisted of an introduction giving a general interpretation and clarification of the situation and of eleven specific emendations.[40] It will be useful to go through the emendations before examining the introduction. I present the substitutions by both quoting Copernicus's original wording and also giving the Index's revised wording; to improve readability, I have struck through the parts of the original passages that were revised and underlined the parts that were added by the censors.

The first emendation amounted to deleting the first several sentences of the last paragraph of Copernicus's preface, except for its final clause:

~~Perhaps there will be babblers who claim to be judges of astronomy although completely ignorant of the subject and, badly distorting some passages of Scripture to their purpose, will dare to find fault with my undertaking and~~

~~censure it. I disregard them even to the extent of despising their criticism as unfounded. For it is not unknown that Lactantius, otherwise an illustrious writer but hardly an astronomer, speaks quite childishly about the earth's shape, when he mocks those who declared that the earth has the form of a globe. Hence scholars need not be surprised if any such persons will likewise ridicule me. Astronomy is written for astronomers. To them~~ For the rest, my work too will seem, unless I am mistaken, to make some contribution also to the Church, at the head of which Your Holiness now stands.[41]

The second emendation changed the following passage in book 1, chapter 5:[42] "To be sure, there is general agreement among the authorities that the earth is at rest in the middle of the universe. They hold the contrary view to be inconceivable and downright silly. Nevertheless, if we examine the matter more carefully, ~~we shall see that this problem has not yet been solved, and is therefore by no means to be disregarded.~~ we think it is immaterial whether the earth is placed at the center of the world or away from the center, so long as one saves the appearances of celestial motions."[43]

Third, in book 1, chapter 8, the following passage needed revision: "Why then do we still hesitate to grant it the motion appropriate by nature to its form rather than attribute a movement to the entire universe, whose limit is unknown and unknowable? ~~Why should we not admit, with regard to the daily rotation, that the appearance is in the heavens and the reality in the earth? This situation closely resembles what Virgil's Aeneas says:~~ Why not grant that the things which appear in heaven happen in the same manner as expressed by Virgil's Aeneas: 'Forth from the harbor we sail, and the land and the cities slip backward.' "[44]

The fourth correction involved the following sentence in book 1, chapter 8: "Besides, ~~it would seem quite absurd to attribute motion to the framework of space or that which encloses the whole of space, and not, more appropriately, to that which is enclosed and occupies some space, namely, the earth.~~ it is no more difficult to attribute motion to that which is in a place in a container, namely to the earth, than to the container."[45]

Fifth, a passage on the same page (in book 1, chapter 8) had to be deleted: "You see, then, that all these arguments make it more likely that the earth moves than that it is at rest. This is especially true of the daily rotation, as particularly appropriate to the earth."[46]

The sixth change concerned the first sentence in book 1, chapter 9: "~~Accordingly, since nothing prevents the earth from moving, I suggest that we should now consider also whether several motions suit it, so that it can be regarded as one of the planets.~~ If, then, I assume that the earth moves, I think that we now have to see also whether several motions can belong to it."[47]

The next two changes occurred on the same page in book 1, chapter 10, and each consisted of replacing a single word in a sentence. The seventh emendation involved the passage: "Hence I feel no shame in ~~asserting~~

<u>assuming</u> that this whole region engirdled by the moon, and the center of the earth, traverse this grand circle amid the rest of the planets in an annual revolution around the sun. Near the sun is the center of the universe."[48] The eighth revision applied to the next sentence: "Moreover, since the sun remains stationary, whatever appears as a motion of the sun is ~~really~~ <u>consequently</u> due to the motion of the earth."[49]

Ninth, the last sentence of book 1, chapter 10 had to be deleted: "So vast, without any question, is the divine handiwork of the most excellent Almighty."[50]

The tenth emendation was to change the title of chapter 11 of book 1, from "Proof of the Earth's Triple Motion"[51] to "On the Hypothesis of the Earth's Triple Motion, and Its Demonstration."[52]

Finally, the title of chapter 20 in book 4 had to be changed: "The Size of ~~These Three Heavenly Bodies,~~ Sun, Moon, and Earth, and a Comparison of Their Sizes."[53] The reason for the deletion, as the document states, is that "the earth is not a heavenly body, as Copernicus would have it."[54]

Of the eleven emendations, thus, one came from the Preface; nine involved the relatively untechnical book 1; and one applied to book 4. Most seemed to involve the replacement of realist, descriptive language by instrumentalist, hypothetical language, of which more presently. Two of the changes (the first and the ninth) amounted to the deletion of religious references, one to Scripture and the other to God.

Copernicus's reference to Scripture in the Preface was particularly revealing. It represented a preemptive answer to the anticipated objection that his system contradicted Scripture. By dismissing such an objection, he was implicitly saying that Scripture is irrelevant to astronomy, physics, or natural philosophy; that is, he was denying the scientific (or philosophical) authority of Scripture. Copernicus's justification of such a denial was also merely indirect: he referred to Lactantius's scriptural objection to the roundness of the earth, thus suggesting that biblical arguments against the earth's motion are no more relevant. On the other hand, by deleting this passage the Index was signaling its prohibition and condemnation of the principle that Scripture is not an authority on natural philosophy (science).

But what was the upshot of the other nine revisions? Here the introduction to the document describing these changes is important:

> The Fathers of the Holy Congregation of the Index decreed that the writings of the distinguished astronomer Nicolaus Copernicus, *On the Revolutions of the World*, were to be absolutely prohibited, because he does not treat as hypotheses, but advances as completely true, principles about the location and the motion of the terrestrial globe that are repugnant to the true and Catholic interpretation of the Holy Scripture; this is hardly to be tolerated in a Christian. Nevertheless, since Copernicus's work contains many things that are very useful generally, in that decision they were pleased by unanimous con-

sent to allow it to be printed with certain corrections according to the emendation below, in places where he discusses the location and motion of the earth not as a hypothesis but as an assertion. In fact, copies to be subsequently printed are permitted only with the above-mentioned places emended as follows and with this correction added to Copernicus's preface.[55]

The censors were here distinguishing between treating the key elements of the Copernican system as "hypotheses" and treating them as "assertions." This distinction did indeed correspond to most of the emendations, which involved changing the language from assertive to hypothetical. So presumably, the assertion of the earth's motion was prohibited, but its hypothetical discussion was allowed. As a general clarification of the situation, this was helpful as far as it went.

However, it did not go far, and the clarification was insufficient. In fact, the formulation embodied the conflation of two other distinctions. For a "hypothesis" could mean either a proposition that describes physical reality but whose truth is not known with certainty and can only be supported with some probability; or it could mean a proposition which only "saves the appearances," which is not meant to be a description of physical reality but only an instrument for mathematical calculation and observational prediction, and which consequently is not susceptible of being either true or false but only more or less useful, convenient, or efficient. And an assertion could mean either a description of physical reality that can be demonstrated with certainty to be physically true; or it could mean a proposition that is not merely an instrument of calculation and prediction, but also a potentially true (though as yet undemonstrated) description of physical reality. The ambiguity of both terms yielded two distinctions: the contrast between a hypothesis and an assertion could mean the dichotomy between a description of physical reality (true or false) and an instrument of calculation or prediction (more or less useful and convenient); and the distinction could also refer to the difference between a demonstrably true description of reality and a potentially true description of reality (demonstrable but not yet demonstrated).

When the censors said that the hypothetical discussion of the earth's motion was allowed, it was clear that at least they were allowing treating the earth's motion as a useful and convenient instrument of calculation and prediction; this usefulness was in any case completely undeniable and universally accepted. But were they also allowing the earth's motion to be treated as a realistic description that could be true, though it could not yet be demonstrated; that is, were they also allowing a realistic interpretation (as long as it was conjoined to probabilism or fallibilism)? This was unclear. Similarly, when they said that it was "hardly to be tolerated in a Christian" to advance the earth's motion as "completely true" or as an assertion, at the very least they were prohibiting the claim that the proposition of the earth's motion had been conclusively proved or was a demonstrated truth. But were they also prohibiting treating the earth's motion as a realistic description but not yet fully demonstrated assumption; that is, were they prohibiting the realistic interpretation of heliocentrism and so dictating an instrumentalist interpretation? This was also unclear.

We can speak of a weaker and a stronger interpretation of the anti-Copernican decree. The stronger prohibition would prohibit any realistic interpretation of the earth's motion (whether combined with demonstrativism or probabilism), and allow only instrumentalism. The weaker prohibition would prohibit only demonstrativism and would allow not only probabilism but also realism (besides, of course, instrumentalism).

These distinctions may shed some light on the initial remark in this document, to the effect that in 1616 the Index decreed that Copernicus's *Revolutions* be "absolutely prohibited," a remark that was echoed in the Inquisition sentence's account to the effect that "the Index issued a decree which prohibited books treating of such a doctrine."[56] This claim was puzzling, since that censure said that this book was "suspended until corrected."[57] There was here an apparent contradiction. If we want to resolve it, we could say that Copernicus's book, insofar as it advanced a realistic interpretation of the earth's motion, was to be absolutely prohibited. In other words, what was being prohibited was the proposition that the earth's motion is a physical truth.

Such an "absolute prohibition" could also be regarded as a reiteration of a point made in the earlier decree, to the effect that Foscarini's book was to be completely prohibited and condemned not only because it argued that terrestrial motion is reconcilable with Scripture, but also because it claimed that the earth's motion "is consonant with the truth."[58] Generalizing Foscarini's condemnation, one got the impression that the prohibition involved at least two claims: that the earth's motion is physically true; and that it is compatible with Scripture.

Finally, the condemnation of the physical truth of the earth's motion connects with a puzzling aspect of the Inquisition's condemnation of Galileo in 1633. In several places, the sentence faulted Galileo for his probabilism. In its account of the proceedings, the sentence recognized that his discussion of the earth's motion in the *Dialogue* tried "to give the impression of leaving it undecided and labeled as probable; this is still a very serious error since there is no way an opinion declared and defined contrary to divine Scripture may be probable."[59] And criticizing Galileo's defense that Bellarmine's certificate contained no special injunction but only a declaration that the earth's motion was contrary to Scripture and so could be neither held nor defended, the sentence claimed that the certificate aggravated Galileo's case because, despite the latter prohibition, he "dared to treat of it, defend it, and show it as probable."[60] Finally, in stating the verdict, the sentence attributed to Galileo the suspected heresy "that one may hold and defend as probable an opinion after it had been declared and defined contrary to Holy Scripture."[61] Presumably, Galileo should have known that in condemning the physical truth of the earth's motion, the authorities had condemned its probable truth as well as its certain truth.

Chapter 2

Promulgation and Diffusion
of the News (1633–1651)

In the summer of 1633 all papal nuncios in Europe and all local inquisitors in Italy received from the Roman Inquisition copies of the sentence against Galileo and his abjuration, together with orders to publicize them. Such publicity was unprecedented in the annals of the Inquisition and never repeated.[1] As a result, many manuscript copies of Galileo's sentence and abjuration have survived in European archives.[2] By contrast, no copies of the full text of the Inquisition's sentence against Giordano Bruno survive, even though his crime (formal, unrepented, and obstinate heresy) and his penalty (to be burned alive at the stake) were much more serious; consequently, not only was his execution obviously public (in Campo de' Fiori in Rome, on 17 February 1600) but so was his sentencing, when the sentence was read to him at a public ceremony at the palace of one of the cardinal-inquisitors (on 8 February);[3] Galileo's sentencing instead took place at the convent of Santa Maria sopra Minerva, at a ceremony which was formal but private.[4]

2.1 Nuncios and Inquisitors:
Pope Urban VIII's Orders (July 1633)

The Inquisition's orders were contained in a letter signed by Antonio Barberini, who held the title of Cardinal of Sant'Onofrio. He was a brother of the reigning Pope Urban VIII, as well as a trusted and influential member of the Inquisition and one of the trial judges who signed the sentence. We ought not to confuse this cardinal with Francesco Barberini, who was a nephew of the pope and the Vatican secretary of state and who held the title of Cardinal of San Lorenzo in Damaso; he was also another member of the Inquisition, though his signature is one of those missing from the sen-

26

tence.[5] Here is the text of Antonio Barberini's letter to nuncios and inquisitors, dated 2 July 1633:[6]

> The Congregation of the Index had suspended Nicolaus Copernicus's treatise *On the Revolutions of the Heavenly Spheres* because that book maintains[7] that the earth moves, and not the sun, which is the center of the world, an opinion contrary to Sacred Scripture; and several years ago this Sacred Congregation of the Holy Office had prohibited Galileo Galilei of Florence from holding, defending, or teaching in any way whatever, orally or in writing, the said opinion. Nevertheless, the same Galileo has dared to write a book titled *[Dialogo di] Galileo Galilei Linceo*,[8] without revealing the said prohibition, he has extorted the permission to print it and has had it printed; claiming at the beginning, within the body, and at the end of that book to want to treat hypothetically of the said opinion of Copernicus (although he could not treat of it in any manner), he has however treated of it in such a way that he became vehemently suspected of having held such an opinion. Thus, he was tried and detained[9] in this Holy Office, and the sentence of these Most Eminent Lords condemned him to abjure the said opinion, to stay under formal arrest subject to the wishes of their Eminences, and to do other salutary penances. Your Reverence can see all that in the attached copy of the sentence and abjuration; this document is sent to you so that you can transmit it to your vicars and it can be known by them and by all professors of philosophy and of mathematics; for, knowing how the said Galileo has been treated, they can understand the seriousness of the error he committed and avoid it together with the punishment they would receive if they were to fall into it. By way of ending, may God the Lord preserve you.[10]

We know today that such a promulgation of Galileo's condemnation had been decided at the Inquisition meeting of 16 June 1633, presided over by Pope Urban VIII; this was the same meeting at which Galileo's trial was discussed and the pope reached a decision on its conclusion, the verdict, and the penalty.[11] Thus, the promulgation was not an afterthought but part of a well-considered plan. In fact, the plan was reaffirmed at the meeting of June 30, when the pope was again presiding over the Inquisition meeting and was a little more explicit about its details.[12] Cardinal Barberini's letter followed immediately thereafter.[13]

From the replies of the nuncios and inquisitors, there is concrete evidence that the sentence circulated in the manner intended. Letters of reply have survived from the nuncios to Naples, Florence, Venice, Vienna, Paris, Brussels, Cologne, Vilnius, Lucerne, and Madrid, and from the inquisitors of Florence, Padua, Bologna, Vicenza, Venice, Ceneda, Brescia, Ferrara, Aquileia, Perugia, Como, Pavia, Siena, Faenza, Milan, Crema, Cremona, Reggio Emilia, Mantua, Gubbio, Pisa, Novara, Piacenza, and Tortona.[14]

The most common reply was a brief acknowledgment of receipt and a promise that the orders would be carried out. However, in this case the standard response was not sufficient for the Inquisition. It expected to be notified that the orders had in fact been carried out. Those who did not send such a follow-up letter were soon reprimanded and had to write back to Cardinal Barberini to explain the oversight or the delay.[15] For example, on 28 September 1633, the inquisitor of Pavia wrote that he had not yet notified Rome because classes at the university were in recess until autumn and consequently he had not yet notified the professors; but he stated that he had notified his subordinates, and in fact he attached a printed flyer which he had sent them.[16]

2.2 Professors of Mathematics and Philosophy: Guiducci's Report (August 1633)

The quickest promulgation occurred in university circles.[17] In Florence it happened almost immediately, on July 12. In fact, on August 27 Mario Guiducci wrote the following letter from Florence to his friend Galileo, who was under house arrest at the residence of the archbishop of Siena:

[240] Before last week,[18] I had never written to Your Lordship that I was present at the public reading of the sentence because the occasion had not arisen [241] and because it did not seem a good idea to inform you of something that might disgust you. Now that you have expressed the desire to know what happened, I shall tell you what I remember.

In the month of July, late one day the Father Deputy Inquisitor came to my house and in the name of the Father Inquisitor invited me to be present at a meeting that would take place at the Holy Office on the 12th of the said month; he did not want to tell me what it was about. I went on that day, arriving as they were about to begin. In attendance were the consultants, some lawyers, and other clergymen. I also found there Mr. Filippo Pandolfini, Mr. Aggiunti, Mr. Francesco Rinuccini, and Mr. Dino Peri, who had received invitations similar to mine.[19] After we all sat down, the Father Inquisitor said he had orders from the Congregation to read the sentence and abjuration, etc., in the presence of those invited; and he had the secretary, who is a friar of the same order, do the reading. The latter read about Galileo Galilei at the age of seventy years: that a decree had been issued in the year 1615; that a particular and special injunction not to hold or teach that doctrine had been given him in Rome by the commissary in the presence of Cardinal Bellarmine; that despite these things, he wrote a book titled *Dialogues* etc., and he fraudulently extorted the permission to print it by not disclosing that he was under such a special injunction; that in this book he advanced arguments for the opinion that the sun does not move from east to west (which is heretical) and for the mobility of the earth (which is erroneous and against good philosophy), without destroying and refuting them; that therefore he had rendered himself vehemently suspected of such a heresy; that he had been condemned to prison at the pleasure of the same Congregation, with the possibility of a reduction of this penalty; and also that as a salutary penance, he was required to recite the seven penitential psalms once a week for three years. Then the secretary read the abjuration, which said that the author had supported such an opinion not because he held it to be true, but to show off his cleverness; that he now held it to be false and hated and cursed it; that he submitted to the penalty of perpetual imprisonment if he were to do otherwise; and that he obliged himself always to report on anyone who he might know holds such a despicable opinion. In short, this was the content. As regards obtaining a copy, a consultant who had not been present, due to his being away from Florence, was curious to hear it, and it was read to him; but, having asked for a copy, he was unable to get it. I was curious to know the reason why I had been invited, and the father deputy told me that they had orders from Rome to summon as many mathematicians and philosophers[20] as possible.

. . . [21][242] On other matters, if Your Lordship yearns to return to your studies, here your acquaintances and friends want you equally; but we fear that it will not be granted you as quickly as you are requesting. However, you should try to accommodate yourself to the wishes of the rulers, and in the meantime enjoy the tranquillity and love you are receiving from Msgr. the Archbishop.[22] To him I send my reverence, and for Your Lordship I kiss your hands and end by praying for your complete happiness.[23]

Although Guiducci's report was largely accurate, it did contain several misrepresentations of the sentence. However, we can hardly blame him, given that he did not have a copy and had heard it read to him just once. His account is a good example of how information spreads, with inaccuracies or variations introduced at each step, especially when the process is oral. The oral dissemination of this information was better than nothing; but it was no substitute for the written word. Anyone directly involved with this affair would certainly have wanted to lay hands on a copy of the actual text.

In fact, Galileo himself soon became curious about the text of the sentence which had been read to him and of the abjuration which he had recited on June 22, but of which he had been given no copies. In September 1633, Galileo obtained copies of both through the efforts of Giovanfrancesco Buonamici, a diplomat who was a good friend and a distant relative.[24]

Although the Florentine academics may have been the first to receive the details of Galileo's condemnation (and the implicit warning it contained), they were not the only ones. On 1 September 1633, Fabio di Lagonissa, nuncio to Brussels, wrote to Cornelius Jansenius (professor of Sacred Scripture at the University of Louvain) and Matthew Kellison (rector at the English College of Douai) requesting that they inform their respective institutions about Galileo's condemnation and the importance of avoiding Copernicanism; whereas the response from Louvain was cool, the response from Douai was enthusiastic.[25] On 17 September 1633, Padua's inquisitor reported that Fortunio Liceti, senior professor of philosophy at the University of Padua, had surrendered the copy of the *Dialogue* Galileo had given him.[26] And on 11 November, the nuncio to Madrid wrote the Roman Inquisition that he had notified of Galileo's condemnation not only all Spanish bishops, but also the Universities of Salamanca and Alcalà.[27]

2.3 Printed Posters and Flyers:
Carafa's Liège Notification (September 1633)

Besides ordering that professors be notified about Galileo's condemnation, Barberini's letter made it clear it was to be promulgated in other ways. Of these, the most important action was perhaps that taken by the nuncio to Cologne, Petrus Carafa, who compiled a summary of the trial in Latin and had a poster printed for public notification. His jurisdiction covered the Rhineland and Lower Germany and included the principality of Liège, which at that time was politically independent of Belgium and ecclesiastically a part of the archdiocese of Cologne; moreover, at the time the archbishop of Cologne happened to be the same person as the ruling prince of Liège and so resided in Liège; thus the nuncio too decided to reside there.[28] Consequently, Carafa's poster gave Liège as the place of publication, and this was the poster that was seen by Descartes while he was living in Holland. This is the text:

[14] Petrus Aloysius Carafa, by the grace of God and of the Holy Apostolic See bishop of Tricarico, and nuncio with the rank of special ambassador of His Holiness Pope Urban VIII for the Rhineland and Lower Germany.

The Sacred Congregation of the Most Eminent and Most Reverend Cardinals of the Holy Roman Church charged with the Index had suspended Nicolaus Copernicus's treatise *On the Revolutions of the Heavenly Spheres* because it asserts that the terrestrial globe is in motion but the sun is motionless and is the center of the world, which is an opinion contrary to Sacred Scripture. Furthermore, the other Congregation of the Most Eminent [15] and Most Reverend Cardinals of the Holy Roman Church, General Inquisitors against heretical depravity in all of Christendom, had forbidden the Florentine Galileo Galilei to follow or explain to others (as he had done previously) the above-mentioned opinion of Copernicus. Indeed it had also declared that Galileo's assertion that the sun is the center of the world and does not move is philosophically absurd and false and formally heretical because it is explicitly contrary to Holy Scripture; and it had judged that his other assertion that the earth is not the center of the world and does not stand still is philosophically equally absurd and false and theologically at least erroneous in faith. The Most Eminent Cardinal Bellarmine had added some salutary warnings. In the year 1616, the commissary of the Holy Inquisition had issued Galileo an injunction forbidding him henceforth to hold or teach such opinions, and the Sacred Congregation of the Index had prohibited his books on this topic, since they contained a doctrine that is false and completely contrary to Sacred Scripture. [16] Nevertheless, before long there appeared in Florence a book titled *Dialogue of Galileo Galilei on the Two Chief World Systems, Ptolemaic and*

Copernican, in which Galileo seemed to again propagate and confirm that false doctrine. Because of this, Galileo was summoned to the Sacred Tribunal of the Inquisition, interrogated, and detained in prison; after examination, he confessed and seemed almost to be still of the same opinion, although he pretended to propose it in the manner of a hypothesis. Thus, the same Most Eminent Cardinals General Inquisitors, sitting as a tribunal, having carefully considered the case, pronounced and declared that the same Galileo appeared to be vehemently suspected of heresy, insofar as he had followed a doctrine that is false and contrary to the sacred and divine Scriptures (namely that the sun is the center of the world and does not move from east to west, but that the earth moves and is not the center of the world), and insofar as he had thought that his doctrine could be defended as probable even though it had been declared contrary to Scripture. They also pronounced that the same Galileo had incurred the censures and penalties prescribed by the Sacred Canons and the other general and particular laws. [17] Consequently, Galileo, at the age of seventy, kneeling before the Most Eminent Cardinals General Inquisitors, in prescribed words, and with a sincere heart and unfeigned faith, abjured and cursed that opinion; and he promised under oath that henceforth he would never again assert such an opinion and that he accepted the penalties imposed on him, both the imprisonment at the pleasure of their Most Eminent Cardinals and the recitation of the seven penitential psalms once a week for three years.

By order of the same Most Eminent Cardinals, we have wanted to report and promulgate in the provinces under our jurisdiction a summary of Galileo's trial, so that it is known everywhere and especially in the universities, and most importantly so that the seriousness of Galileo's error will convey to all students and professors of philosophy and mathematics not to want to follow or teach others anything besides doctrines that are sound and in accordance with Holy Scripture. Dated in Liège, on 20 September of the year 1633, the eleventh of the papacy of His Holiness mentioned above.

Liège, Press of Léonard Streel, Licensed Printer, 1633.[29]

Carafa's announcement was an account of the trial intermediate in detail between the original sentence and Cardinal Barberini's July 2 memorandum. Clearly, the nuncio was not just uncritically repeating the cardinal's letter but seemed to have studied the attached documents and used his own judgment in compiling the summary. However, the nuncio introduced several discrepancies.[30] Two in particular were especially significant. First, the poster made it sound as if it was the Congregation of the Inquisition itself that had declared heliocentrism heretical and geokineticism theologically erroneous, whereas the sentence states that this judgment had been made by the consultants and that the Inquisition was endorsing only the judgment that those doctrines were contrary to Scripture. The second discrepancy hinges on the content of the special injunction. Carafa described it as a prohibition "to follow or explain to others"[31] the Copernican doctrine, or "henceforth to hold or teach"[32] it; he did not include the clause "in any way whatever, either orally or in writing,"[33] which the sentence used but which Galileo tried unsuccessfully to dispute. Both discrepancies were also contained in Guiducci's letter, but whereas Guiducci's oversight is relatively insignificant because he was not in possession of a copy of the sentence and spoke from memory, Carafa's reading was much more significant; for he had the text of the sentence as well as Barberini's letter, which on both points faithfully reflects the sentence.[34] Carafa's conflation of "heretical" and "contrary to Scripture" was the first sign of how easy it would be to come to think that Copernicanism had been declared heretical,[35] which was to become one of the most persistent myths in the subsequent controversy. And his conflation of the special injunction and Bellarmine's warning may have reflected his judgment that the judges had been unfairly harsh; it may perhaps be regarded the first salvo in the long controversy about the "validity" and "authenticity" of the special injunction.

Finally, it is puzzling that the Liège poster did not mention the banning of Galileo's *Dialogue*. However, this omission was probably inconsequential because the book was included in the next decree of the Congregation of the Index, issued on 23 August 1634.[36] This decree was a typical one, with a brief introduction reminding Catholics of the rules about prohibited books followed by a long list of works. A total of twenty-seven books were listed: most in Latin, several in Italian, and one in Spanish. Their subjects were mostly religious and theological, but many dealt with political, civil, and military history. Six books were marked as "suspended until corrected." One curiosity is a work by one Giovanni Battista de Vilela whose title means *Practical Rules to Die Well, Even for Those Who Can Only Read, and to Learn How to Live Well from the Things That Must Be Done at the Time of Death*. Besides the *Dialogue* there seem to be only two other books listed dealing with questions of natural philosophy: *Naturae Constantia* and *Thaumatographia Naturalis* by one Ioh. Ionstonus Polonus.

Carafa's poster was the most important of the written notifications by church officials that followed the Inquisition's orders to publicize Galileo's condemnation, not only because it had an international context but also because it affected Descartes, as we will see in the next chapter. However, it was not the only such notification. Even earlier, on 7 August 1633, the inquisitor of Pavia (in Lombardy) had sent to his deputies a printed notification in Italian of the condemnation of Galileo, which contained a one-page summary of the sentence; this was the first printed announcement of the condemnation.[37] On 13 September, the bishop of Cortona (in Tuscany) published an edict containing a brief summary in Italian of Galileo's sentence and abjuration;[38] this edict shows that the nuncios and inquisitors acting on the Inquisition's orders were transmitting the information not only to their deputies but also to other regular officials in the ecclesiastical hierarchy within their jurisdictions. Indeed, the nuncio to Madrid, in his letter replying to the Inquisition's orders, stated explicitly that he had notified Spanish bishops as well as professors.[39] And on 27 September 1633,

Tommaso P. Belli, the inquisitor of Rimini (in Romagna) circulated a printed broadside in Italian announcing generally that books defending or discussing the earth's motion were prohibited; that in particular this prohibition applied to Galileo's *Dialogue;* that professors of philosophy and of mathematics should avoid holding or teaching this opinion either in writing or orally; and that anyone found in violation would be punished as a heretic.[40]

2.4 Private Correspondence: Buonamici's Account (July 1633)

Besides the Inquisition's sentence and Galileo's abjuration, another important contemporary source of information about the trial was an account written by Giovanfrancesco Buonamici (1592–1669).[41] A native of the Tuscan town of Prato, in the early part of his life Buonamici followed a diplomatic career that led him to travel widely in Germany, Austria, and Spain, and later he was a government official in Prato. When Galileo's son Vincenzio married Sestilia Bocchineri in 1629, he became the brother-in-law of Buonamici, who was married to Sestilia's sister Alessandra; Buonamici and Galileo thus became relatives of sorts. In 1633, during Galileo's trial, Buonamici happened to be living in Rome, and so it is likely that he learned about the trial firsthand, from Galileo himself as well as from the many connections Buonamici had in Rome, including Cardinal Francesco Barberini. After writing this account, Buonamici sent a copy to Galileo as well as to several persons in Germany, Spain, and Flanders.[42] The account thus complements the Inquisition's sentence with regard to both content and circulation. It is worth reproducing:[43]

[407] Galileo Galilei, a Florentine, professor of philosophy and mathematics, nicknamed professor of the spyglass or telescope,[44] is too well known to the world for me to have to give an account of his personal life on the occasion of relating the long molestation he suffered because of the system of Nicolaus Copernicus. Many decades ago, he wrote a book[45] on the constitution of the universe in which he contradicted Aristotle and Ptolemy; he asserted it is not true that, as they say, the earth is motionless, nor that it is the center of the world, nor that the immense structures of the planets and the heavens turn around this miniscule terrestrial globe in the period of twenty-four hours by being carried by the imagined sphere of the *primum mobile.* [408] Instead he said that the sun is the center of the world and motionless in regard to displacement but moving by rotation on itself; that the planets we see moving in the heavens revolve around it according to their periods; that the earth does the same by an annual motion in the plane of the ecliptic, being located between Venus and Mars; and that it also rotates on itself by a diurnal motion from west to east, which allows it to see in twenty-four hours all the heavens, stars, and planets. Few have believed or noticed this opinion of Copernicus on account of both its extravagance and the implausibilities involving human

sense experience which it seems to imply. However, the sensible demonstration of the new spyglass or telescope seemed to flatten many difficulties and implausibilities which natural vision cannot grasp; thus, many thinkers, and especially the said Galileo, argued that the system of Copernicus deserved better consideration than had been the case in the past; at the same time, they greatly admired his mind because, without having the convenience of the telescope, he still was able to understand some properties and features of the planets that strengthened his opinion and cannot be discerned by the natural eye without using the telescope; for example, when Venus and Mars are nearer the earth they are seen, respectively, 40 and 60 times larger than when they are farther and appear smaller; and when Venus is near conjunction with the sun, it appears sickle-shaped, as the moon does when it is new.

These and other sensible demonstrations, which Galileo with the benefit of the telescope discovered in the heavens before anyone else, excited envy in many persons; being envious for his glory and unable to contradict the manifest truth of the discoveries made in the heavens, they started to persecute him; this was especially true for some Dominican friars who took the road of the Inquisition and Holy Office in Rome, complaining that he attributed stability to the sun and mobility to the earth against the words of Sacred Scripture. Thus Paul V, instigated by the same friars, would have declared this Copernican system erroneous and heretical, insofar as it is contrary to the teaching of Scripture in various places and especially in Joshua, had it not been for the opposition and defense of Lord Cardinal Maffeo Barberini (today Pope Urban VIII) and of Lord Cardinal Bonifazio Caetani. [409] However, these cardinals argued, first, that Nicolaus Copernicus could not be declared heretical for a purely natural doctrine, without eliciting the laughter of the heretics, who do not accept the reform of the calendar of which he was the principal master; second, that it did not seem prudent to assert on the authority of Sacred Scripture in purely natural subjects that something is true which (with the passage of time and by means of sensible demonstrations) could be discovered to be false, for even in subjects concerning Faith (which is the principal if not only purpose of Sacred Scripture) it is frequently necessary to understand that it speaks in accordance with our abilities, otherwise if one wanted to abide by the pure sound of the words one would end up in errors and impieties (such as that God has hands, feet, emotions, etc.). Thus, these cardinals dissuaded Paul V from the ruling that the said friars had come close to extracting from him. These friars have shown themselves to be persecutors more of the person than of the opinion; for after it was asserted by Copernicus, no one persecuted it in the period of so many years, whereas when Galileo merely discussed it they made him appear before the dreadful tribunal of the Holy Office. At that time the pontifical decree was reduced to ordering that the system of the sun's stability and the earth's mobility could be neither held nor defended because it appeared to be contrary to the account in Sacred Scripture.

Thereafter Galileo, obeying this order, no longer gave any thought to this subject, until the year 1624 when Lord Cardinal Hohenzollern provided the following encouragement: he said that he had spoken with the current pontiff about this opinion; that His Holiness had recalled defending Copernicus at the time of Paul V; and that His Holiness assured him that he would never allow this opinion to be declared heretical, if for no other reason than for the veneration rightly due to the memory of Nicolaus Copernicus. Spurred by this, Galileo started writing a book in the form of a dialogue; in it he examined the foundations and reasons of the two different systems, Aristotelian and Copernican, and without leaning more on one side than on the other, he left the matter undecided. He himself brought this book to Rome in the year 1630 and placed it in the hands of His Holiness, who with his own hand corrected something in the title; then it was examined by the Master of the Sacred Palace and returned with his endorsement [410] and with a preface produced and compiled by order of His Holiness, which the printed book indeed carries. Approved in the manner described above, this book was printed in Florence. It awakened once again his old persecutors, who were joined by those who quarreled with them about de auxiliis,[46] out of personal spite between someone from their order and Galileo about who was the first to discover sunspots;[47] and so they brought new complaints before the same tribunal, which is always open to accusations and ready to fulminate censures and excommunications against free thought. There was in addition a hatred and persecution (typical of friars) by Father Firenzuola, the Commissary of the Holy Office much loved by His Holiness more for knowing civil engineering and accounting than for preaching or theology, against Father Mostro,[48] Master of the Sacred Palace and approver of the book. The pope was unwilling to prevent Firenzuola from allowing the filing of complaints against Galileo in order to harm Father Mostro and Ciampoli,[49] another friend and supporter of Galileo; and the pope allowed that Galileo be summoned and forced to come to Rome, despite the plague in Florence, the harshness of the winter, and the age of 60 years.[50]

Galileo obeyed, against the opinion and advice of his truest friends, who argued that he should move abroad, write an apology, and not expose himself to the impertinent and ambitious passion of a friar. He came to Rome, and they held him for two months at the house of the Tuscan ambassador without ever telling him anything other than that he should not go out or engage in conversations. Finally, they made him go to the Holy Office; they kept him in free custody[51] for eleven days; and they examined him only about the imprimatur and approval of the book. He said he received it from the Master of the Sacred Palace. Then they sent him back to the house of the same ambassador, with the same order not to go out or socialize. Thus they turned the persecution against Father Mostro: he first excused himself by saying that His Holiness himself had ordered him to approve the book; but when the pope denied it and became angry, Father Mostro said that it was secretary Ciampoli who had conveyed to him the order of His Holiness; the pope replied that there was no such order; finally, Father Mostro produced a note by Ciampoli stating that His Holiness was ordering the book's approval (and was present while Ciampoli was writing). Seeing therefore that Father Mostro could not be successfully attacked, the commissary did not want to seem to have conducted the proceedings in vain and became more influenced by the strong requests of the old enemies of Galileo and of the new pretenders to the discovery of sunspots; and so Galileo was made to appear before the Congregation of the Holy Office and to formally abjure the opinion of Copernicus, even though this was superfluous for him since he had not [411] held or defended it but had only discussed it. Galileo thus saw himself forced to do something he would have never believed, especially because the commissary Father Firenzuola in their conversations had never mentioned such an abjuration. However, he pleaded with the Lord Cardinals that since they were proceeding with him in that manner, they could make him say whatever their Eminences wanted, except only two things: one, that he would not have to say that he was not a Catholic, for such he was and wanted to die, in spite of and out of spite for his detractors; the other, that he also could not say that he had ever deceived anyone, especially in the publication of his book, which he had submitted for ecclesiastical censorship and had had printed with a legitimate approval. After this protestation, he read what Father Firenzuola had written. Later, with the permission of His Holiness, he left toward Tuscany, having learned from experience that perhaps it would have been better to follow the advice of his friends than to obey the angry persecutions of his enemies.[52]

Clearly, Buonamici's account was not merely factual but interpretive and evaluative as well. For the first phase of the proceedings (1615–1616), Buonamici reported the very interesting detail that Cardinal Maffeo Barberini (the future Pope Urban VIII) argued successfully against a formal declaration that Copernicanism was heretical, and attributed to him arguments that had an unmistakable Galilean flavor and content.[53] For the outcome of those earlier proceedings, Buonamici mentioned only the prohibition to hold or defend geokineticism and said nothing about the special personal injunction not to discuss the subject at all; since he was in possession of a copy of the sentence,[54] his choice was the result not of ignorance but rather of a judgment to question its existence or validity, analogous to Nuncio Carafa's interpretation.

Buonamici's account of the 1632–1633 proceedings contained a wealth of details which were (and remain) impossible or difficult to verify independently. Moreover, he ignored the verdict of "vehement suspicion of heresy" and described the abjuration as the formal rejection of a doctrine Galileo had never really held or defended but only discussed. Thus, he was suggesting that Galileo was not even guilty of transgressing Bellarmine's warning (a prescription which Galileo admitted), let alone guilty of violating the special injunction (whose existence Galileo had questioned). This question became a key part of the subsequent cause célèbre.

Throughout his account, Buonamici discussed the psychological motivation and personal factors underlying actions, thus suggesting that the Galileo affair had a human element involving personal frailties and political intrigue. This suggestion was important and offered a valuable supplement to the legalistic account found in the sentence. It was a prototype of what I call the circumstantialist approach.

2.5 Newspapers and Books:
From Renaudot's Abridgment (1633)
to Riccioli's Documents (1651)

The first newspaper account of Galileo's condemnation was published in French in Théophraste Renaudot's *Gazette*, no. 122, Paris, December 1633.[55] It reads more like an abridgment of the Inquisition's sentence than a real summary:

[413] We, Gasparo Borgia, with the title of the Holy Cross in Jerusalem; Fra Felice Centino, with the title of Santa Anastasia, called d'Ascoli; Guido Bentivoglio, with the title of Santa Maria del Popolo; Fra Desiderio Scaglia, with the title of San Carlo, called di Cremona; Fra Antonio Barberini, called di Sant'Onofrio; Laudivio Zacchia, with the title of San Pietro in Vincoli, called di San Sisto; Berlinghiero Gessi, with the title of Sant'Agostino; Fabrizio [414] Verospi, with the title of San Lorenzo in Panisperna, of the order of priests; Francesco Barberini, with the title of San Lorenzo in Damaso; and Marzio Ginetti, with the title of Santa Maria Nuova, of the order of deacons; by the grace of God Cardinals of the Holy Roman Church, and especially commissioned to be General Inquisitors of the Holy Apostolic Faith:

Whereas you, Galileo, son of Vincenzio Galilei, Florentine, aged seventy years, were denounced to the Holy Office in the year 1613 because you held as true the false doctrine taught by some that the sun is the center of the world and motionless and the earth is not there but moves with diurnal motion; whereas you taught this doctrine to your disciples and wrote about it to the mathematicians of Germany, your correspondents; and whereas you had a book on sunspots printed and other writings published containing the same doctrine, which is also that of Copernicus;

Whereas the theologians and doctors have found this opinion to be not only false and absurd in philosophy, but also at least erroneous in faith;

Therefore, on 29 February 1616, at the Sacred Congregation held before His Holiness, it was decided that the Most Eminent Cardinal Bellarmine would order you to abandon this false opinion completely, in default of which the Commissary of the said Office would give you an injunction forbidding you to ever teach it to anyone or to support it, on pain of prison; to execute this decision the following day, after some benign and friendly remonstrances made to you by Cardinal Bellarmine at his house, the said Commissary (assisted by a notary and witnesses) issued you the above-mentioned injunction and prohibitions; having promised to obey them, you were dismissed; and this Congregation issued a decree censuring books that treat of such a doctrine contrary to Holy Scripture.

Nevertheless, a book has appeared lately printed in Florence under your name, entitled *Dialogue on the Two World Systems of Ptolemy and Copernicus,* in which you still defend the same opinions. That is why we have summoned you again.

Based on your confessions, admissions, and exhibits; with the advice and counsel of the Reverend Fathers Masters[56] of Sacred Theology and of the

Doctors of both laws; after invoking the Holy name of Jesus and of His glorious Mother ever Virgin; with a final sentence rendered by our tribunal in the case between the Magnificent Carlo Sinceri, Doctor of both laws and Prosecuting Attorney of this Holy Office, on one side, and you, Galileo, accused and here present, on the other; we say, pronounce, and sentence that you, Galileo, have rendered yourself vehemently suspected of heresy, having held this false doctrine of the earth's motion and sun's rest, and that one may defend as probable an opinion after it has been declared contrary to Scripture.

Consequently, you have incurred all the censures and penalties of the sacred canons. But we absolve you of them provided that from this moment, with a sincere heart and unfeigned faith, in front of us you abjure, curse, and detest the above-mentioned errors and heresies, and every other error and heresy contrary to the Church.

However, so that your great error does not [415] remain completely unpunished, and so that you will be more cautious in the future and serve as an example to others, we order that the above-mentioned *Dialogue* be prohibited by public edict and that you be imprisoned in the prison of said Holy Office at our pleasure; and as a salutary penance we enjoin you to recite the seven penitential Psalms once a week for three years. We reserve the authority to moderate, change, or condone wholly or in part the above-mentioned penalties and penances.

To this, the said Galileo acquiesced the same day, abjuring, cursing, and detesting the above-mentioned errors, orally and in writing, at the convent of the Minerva; and he promised on his knees and with his hand on the Holy Gospels, to never go against the above-mentioned sentence.[57]

It is obvious that whoever compiled this text must have been working from a copy of the Inquisition's sentence. This fact makes it all the more puzzling why it should give some inaccurate dates; perhaps these were mere typographical errors. The abridgment omitted references to the Index, thus conflating this congregation with the Inquisition; this misrepresentation had already been advanced in Nuncio Carafa's notification. Also like the Liège poster (and Guiducci's letter earlier), this article weakened the content of the special injunction by describing it as an "injunction forbidding you to ever teach it to anyone or to support it."[58] If Renaudot's article was compiled just from the text of the Inquisition's sentence, independent of Carafa's poster, that would be significant, since it would suggest a widespread perception that the Inquisition, even by its own account, might have been unjustifiably harsh. On the other hand, of course, Renaudot may have very well read the Liège poster before making his abridgment. By contrast, the Parisian newspaper did (properly) mention the banning of the *Dialogue*, which was overlooked by Carafa.

Renaudot's article may have been the first publication of Galileo's condemnation in a print medium; but other summaries, abridgments, translations, and editions soon followed. In 1634, Libert Froidmont published in

Antwerp a two-page Latin summary in a book whose very title mentioned the condemnations of 1616 and of 1633: *Vesta, sive Ant-Aristarchi Vindex, Adversus Iac. Lansbergium Philippi F[ilium] Medicum Middelburgersem; in Quo Decretum S. Congregationis S. R. E. Cardinalium Anno MDCXVI & Alterum Anno MDCXXXIII Adversus Copernicanos Terrae Motores Editum, Iterum Defenditur.* In this work, Froidmont was defending the two ecclesiastical condemnations in the context of a defense of his own earlier book, *Ant-Aristarchus, sive Orbis Terrae Immobilis* (1631), from criticism by Jacob van Lansbergen in a 1633 book that was in turn a defense of his father Philip's 1630 book on the motion of the earth.[59] Embedded in such an apologetic and controversial context, this printing of the sentence is perhaps more indicative of reception and reaction to it than of its diffusion; nevertheless, Froidmont's summary appeared in the book's preface as a freestanding item, and so such separate display, together with the advertisement in the book's title, gave it also the character of dissemination stressed here.

If the context of this publication sounds complicated enough, the next publication was no less so, even though it involved a single book. In fact, the first publication of the full text of the sentence and abjuration appeared in French in 1634 in Paris in a book by Marin Mersenne titled *Les Questions Theologiques, Physiques, Morales, et Mathematiques.*[60] This work had forty-six chapters (called "questions") dealing with the miscellaneous topics suggested in the title; the sentence and abjuration were inserted in chapter 45. The wording and the translation are basically accurate, with two minor and inexplicable omissions. Where the sentence reported that in the earlier phase of the proceedings against Galileo two Copernican propositions were evaluated by its consultants,[61] in Mersenne's book we find neither the statement nor the assessment of the first proposition but only those of the second;[62] the first proposition was the heliocentric and heliostatic thesis, the second proposition was the a-geocentric and geokinetic thesis. And in the abjuration, where Galileo recited that he had been judicially instructed to abandon both propositions,[63] Mersenne's text mentions the first but not the second.[64] We may presume these were merely typographical errors, especially in light of the following machinations and manipulations, which complicate considerably the picture I have just given and which here can only be alluded to, without going into all the details.[65]

There seems to have been more than one version of Mersenne's book circulating in 1634. All versions contained the French translation of the sentence and abjuration at the end of chapter 45. The work originally included a discussion of the earth's motion and of Galileo's *Dialogue* in four of the chapters (34, 37, 44, and 45). Although the book had received the proper imprimaturs, after publication it elicited some criticism at the Sorbonne. Thus, to avoid difficulties, Mersenne produced an expurgated version of the book for circulation in more sensitive circles. He completely replaced the topic of discussion in the first three of these four chapters, and he reworked the discussion in the last. Thus, chapter 34 originally discussed the question, "What Reasons Does One Have to Prove and to Make Persuasive the Motion of the Earth on Its Axis in the Period of Twenty-four Hours?";[66] but in the expurgated version this chapter dealt with "Whether One Can Establish a New Science of Sound, Which Might Be Called Psophology, or Some Such Name."[67] The original topic of chapter 37 was the question, "What Reason Can One Have to Believe that the Earth Moves around the Sun, Which One Places at the Center of the World?";[68] the expurgated version examined "How Much Must One Rise above the Surface of the Earth or above Any Other Body One Wishes (Whether Larger or Smaller) to See a Given Space."[69] Chapter 44 originally treated of "What Is Most Noteworthy in the *Dialogue* That Galileo Has Written on the Motion of the Earth (This Chapter Dealing with Its First Day)";[70] the new topic was "What Should Be the Strength of the Voice in Order for It to Be Transported to and Heard (Whether Naturally or Artificially) at the Moon, Sun, and Firmament."[71] Finally, the original focus of chapter 45 was, "What Is Remarkable in the Second Day of Galileo's *Dialogue*";[72] in the expurgated book the focus of this chapter changed to, "Is it Permitted to Maintain that the Earth Is in Motion?"[73] although (as mentioned) both discussions were followed by the French translation of the sentence and abjuration.

Besides the original and the fully expurgated versions of Mersenne's book, there appears to have been a partially expurgated version, in which the titles in the table of contents corresponded to the expurgated version but the actual content of the chapters corresponded to the original version.[74] However, the full story of this book need not concern us any further; suffice it to show that it holds an important place in the immediate diffusion of Galileo's condemnation.

In 1636, another French newspaper was instrumental in publicizing the condemnation of Galileo. The *Mercure François* reprinted Renaudot's abridgment of the sentence from the December 1633 issue of the *Gazette*.[75]

In the year of Galileo's death (1642), by fortuitous coincidence I suppose, the Latin text of his abjuration was published in a book by Jean Baptiste Morin published in Paris: *Tycho Brahaeus in Philolaum pro Telluris Quiete.* This was a work against Copernicanism in general, and in particular against a book published anonymously in Amsterdam in 1639: *Philolai, sive Dissertationis de vero Systemate Mundi, Libri IV;* its author was Ismaël Boulliau, who had been afraid to show his Copernican leanings. Thus, obviously, here Galileo's abjuration was being exploited in the context of a controversy.[76]

In accordance with inquisitorial practice, the language used in sentences and abjurations was the native language of the accused person, and so in Galileo's case this was Italian. In 1644, Giorgio Polacco put his hands on a copy of the original Italian text of these documents and included them in a book he published in Venice. Entitled *Anticopernicus Catholicus, seu de Terrae Statione et Solis Motu, contra Systema Copernicanum, Catholicae Assertiones,* this work continued the controversy over the motion of the earth, and so the context of his publication of these documents was analogous to that of Morin's.[77]

Seven years after Polacco, and eighteen after Galileo's condemnation, this genre of publication reached its climax in Giovanni Battista Riccioli's *Almagestum Novum*. In this monumental work of two ponderous volumes in folio, this Jesuit author found a place[78] to include an important collection of Latin documents on Galileo's condemnation: the anti-Copernican decree of the Index of 1616; the 1620 warning of the Index detailing the corrections to Copernicus's *Revolutions;* the Inquisition's sentence against Galileo of June 1633; Galileo's abjuration; and Cardinal Barberini's letter (of July 1633) to the inquisitor of Venice announcing Galileo's condemnation and abjuration. This small collection of documents was for more than two centuries the most complete and authoritative source of information about the trial. Its canonical status continued even after the publication in the 1870s of the proceedings of the trial from the manuscripts of the Vatican Secret Archives because, as I discuss later, the sentence and abjuration are missing from those manuscripts and thus posed a different type of challenge. To be sure, Riccioli developed his own account, interpretation, and evaluation of the Galileo affair; and it was important enough that I discuss it later (in chapter 4.3). However, even those who did not accept his account could and did use these texts.

In summary, after Galileo's condemnation, information about the trial was first promulgated by officials of the Catholic Church through readings at meetings of professors of philosophy and mathematics, and through summaries circulated on printed posters and flyers. Originally, the information was also spread privately through the efforts of Buonamici. And soon the diffusion occurred in printed media, such as books and at least two newspapers. By 1651 the educated public of Europe had available the full text of the sentence and abjuration in three languages: French, Italian, and Latin. The material basis had thus been laid for what was to become a key cause célèbre in Western culture.

Substantively speaking, two important and surprising findings emerge. First, although the terms *special injunction* and *Bellarmine's warning* were not employed in these documents with technical rigor, if we bypass the terminology we can say that almost everyone who read the sentence failed to make a distinction between the prescription not to discuss the earth's motion in any way whatever and the prescription not to hold or defend the geokinetic opinion; they ignored the former and spoke as if Galileo had been issued only the latter; and so presumably Galileo was condemned only for defending Copernicanism. Such an interpretation was advanced not only by Galileo's friends Guiducci and Buonamici, but also by an authoritative churchman, Carafa, and by someone who was perhaps relatively neutral, Renaudot. In other words, there seemed to be an unwillingness to admit the existence of the special injunction and its role in Galileo's condemnation, even though the text of the sentence, as well as that of the abjuration, states clearly that he was issued the special injunction and that he was condemned in part for violating it.

Second, there was a common tendency to interpret the sentence incorrectly as containing an official declaration that the Copernican doctrine was heretical, not merely contrary to Scripture. This tendency was less strong and less pervasive than the tendency to overlook the special injunction: Buonamici, for example, explicitly distinguished between what is heretical and what is contrary to Scripture. But the heresy interpretation was exemplified by Carafa (among others), and if this nuncio could not make such a fine but important distinction, it is easy to understand how the myth of Copernicanism as heresy developed in the history of the Galileo affair.

Chapter 3

Emblematic Reactions

Descartes, Peiresc, Galileo's Daughter (1633–1642)

Reception of and reactions to Galileo's condemnation continue to this day. Although this process is obviously subject to chronological periodization and analytical subdivision, it is only at the end of our historical survey that we can be sure of those chronological stages and analytical principles. In the meantime, the survey must proceed in a somewhat ad hoc manner, although not a random one. Thus, I begin with the period from 1633 to approximately 1642—the period of Galileo's life after the trial. I focus on the reactions of four individuals that for various reasons have emblematic significance: Galileo himself; Nicholas Claude Fabri de Peiresc (1580–1637), an enlightened French Catholic who was one of the leading intellectual and cultural politicians of the time; Galileo's elder daughter, Sister Maria Celeste, who has no claim to fame beyond her sheer humanity, her great soul, and her loving heart; and René Descartes, who needs no introduction but whose reaction was perhaps the most sensational and important of them all, for which reason I discuss it first.

3.1 The End of the World: Descartes (1633–1644)

In November 1633, while living in Deventer, Holland, René Descartes sent the following letter to Marin Mersenne in Paris:

[340]I had planned to send you my *World* as a New Year's present, and only two weeks ago I was still very determined to send you at least a part if the whole could not be transcribed in time. But I will tell you that recently I made inquiries in Leiden and Amsterdam about whether Galileo's *System of the World* was available, for I seemed to remember that it had been printed in Italy last year; I was told that indeed it had been printed, but that all copies had been simultaneously burned in Rome and he had been condemned to some

penalty. This [341] has shocked me so much that I have almost decided to burn all my papers, or at least not to let anyone see them. For I surmise that he, who is Italian and (as I understand) well liked by the pope, was convicted for no other reason than that he undoubtedly wanted to establish the earth's motion, which I know well was formerly censured by some cardinals, although I think I have heard it said that afterwards one did not stop teaching it publicly, even in Rome; and I confess that if it is false, so are also all the foundations of my philosophy; it is easily demonstrated from them, and it is so connected with all parts of my treatise that I would not know how to detach it without rendering the rest flawed. However, just as I would not want for anything in the world to produce an essay containing the least word that was disapproved by the Church, so I would rather suppress it than publish it maimed. I have never had the inclination to produce books; and I would have never come to the end if I had not made a promise to you and some other friends, thinking that the desire to keep my word would oblige me to study that much more. Now after all, I am sure you will not send me a policeman to coerce me to keep my promise, and perhaps you will be very glad to be exempt from the pain of reading evil things. There are already so many opinions in philosophy which have plausibility and can be supported in disputes that, if mine do not have greater certainty and cannot be endorsed without controversy, I do not want to ever publish them. However, I would be ungracious if after my promise so long ago I thought of repaying your kindness in this manner, and so I shall not fail to let you see what I have done as soon as I can; I ask you, if you don't mind, another year of delay to revise and polish it. You reminded me of the words of Horace, "Let it be suppressed till the ninth year,"[1] and it is only three years since I started the treatise which I am thinking of sending you. I also beg you to send me what you know of the affair of Galileo.[2]

Using for the first time a label that eventually became generally adopted, Descartes was eager to know more about what he called the "affair of Galileo." But more than two months elapsed without any news from Mersenne, and so in February 1634 Descartes wrote to him from Amsterdam:

Although I have nothing in particular to send you, nevertheless because it has been more than two months since I received news from you, I thought I should not wait any longer to write you; for if I had not had too many long-standing proofs of the goodwill with which you favor me, for me to have any reason to doubt it, I would almost fear that is has cooled a little since I failed to keep the promise I made to you to send you some sample of my philosophy. On the other hand, the knowledge I have of your virtue makes me hope that you will have an even higher opinion of me when you see that I have decided to entirely suppress the treatise I had written and lose almost all my work of four years in order to render full obedience to the Church, insofar as it has prohibited the opinion of the earth's motion. However, because I have not yet seen that either the pope or a Council has ratified this prohibition that was issued by the Congregation of Cardinals in charge of book censorship, I would be very pleased to learn what one thinks about it in France nowadays, and if their authority is sufficient to make it an article of faith. I have been told that the Jesuits assisted in the condemnation of Galileo; and the entire book of Father Scheiner shows sufficiently that they are not his friends. But on the other hand, the observations contained in this book provide so many proofs for taking away from the sun the motions commonly attributed to it that I cannot believe Father Scheiner himself in his heart does not accept Copernicus's opinion; this bothers me so much that I do not dare write out how I feel. As far as I am concerned, I only strive for rest and tranquillity of mind, which are goods that cannot be had by those who have animosity and ambition; but I do not live doing nothing, and instead for now I think only of educating myself and consider myself hardly capable of helping to teach others, least of all those who, having already acquired some credit for false opinions, perhaps fear losing if the truth is discovered.[3]

Here Descartes was raising an issue that went to the heart of the affair: the geokinetic idea had been condemned merely by a congregation of cardinals, but this condemnation had not been approved by the pope speaking ex cathedra or by an ecumenical Church council; so he was (correctly) questioning whether geostaticism had been made an article of faith and was asking Mersenne what they thought about the issue in France. Thus he also made it clear that his decision to put an end to his *World* had a temperamental motivation as well as a religious one.

In reporting the rumor that the Jesuits had been instrumental in the condemnation of Galileo, Descartes was echoing a point mentioned the previous year in Buonamici's account. This Jesuit conspiracy theory was destined to become one of the great controversial points in the Galileo affair. The evidence Descartes mentions, the Jesuit Christopher Scheiner's *Rosa Ursina* (1630), made it obvious that Scheiner and Galileo were involved in controversy over the priority of discovery and the correct interpretation of sunspots. At the time of this letter (February 1634) Descartes had not yet read or seen Galileo's *Dialogue* (as he would be able to do by August),[4] but a reader of the latter could have added another piece of evidence for this conspiracy theory: when Galileo discussed sunspots in the Third Day,[5] without mentioning Scheiner's name or book explicitly, he made clear that he was rebuffing his rival; moreover, in many other passages[6] discussing other topics, Galileo criticized explicitly Ioannes G. Locher's *Disquisitiones Mathematicae de Controversiis ac Novitatibus Astronomicis* (Ingolstadt, 1614), and Locher was a student of Scheiner and had written the book under his supervision. To be sure, these facts are just the tip of the iceberg of the controversy over the Jesuits' role in the Galileo affair, but the tip was already visible when Descartes was writing.

Despite his efforts, Descartes was not to achieve peace of mind about Galileo's trial any time soon. In April 1634, still from Amsterdam, he wrote again to Mersenne in Paris:

[88] Undoubtedly you know that a short time ago Galileo was reproved by the Inquisitors of the Faith and that his opinion on the earth's motion was condemned as heretical. Now, I will tell you that all things I explain in my treatise, including also this opinion of the earth's motion, depend so much on one another that it is sufficient to know that one of them is false to realize that all the reasons I employ have no force at all; and although I think they are based on demonstrations that are very certain and very evident, nevertheless I would not want for anything in the world to maintain them against the authority of the Church. I know well that one could say that nothing decided by the Inquisitors of Rome is thereby automatically rendered an article of faith, and that it is necessary that it first be approved by a Council. But I am not so enamored of my own thoughts as to want to use such exceptions to have the means of maintaining them; and desiring to live quietly and to continue the life I have started by taking as my motto "He lived well who lived a well-sheltered life,"[7] I am more relieved to be free of the fear of coming to know too many people through my book than I am angered for having wasted the time and effort to write it. . . .

As for the experiments by Galileo which you wrote me about, I deny them all, but do not thereby judge that the earth's motion is any less probable. I do not deny that the motion of a chariot, boat, or horse still remains [89] in some fashion in the stone thrown by someone riding in them; but there are other reasons that prevent it from remaining too strongly. Regarding the cannon ball shot from a high tower, it must take much longer to reach the ground than if it were allowed to fall from top to bottom, for in its path it meets more air that hinders it not only from moving parallel to the horizon but also from descending.

As for the motion of the earth, I am surprised that a churchman should dare to write about it in such a way as to make excuses for himself;[8] for I have seen a poster on the condemnation of Galileo printed in Liège on 20 September 1633 that contains the words, "although he pretended to propose it in the manner of a hypothesis,"[9] and so it seems that one is forbidden even to use this hypothesis in astronomy. This holds me back to the point that I do not dare send him any of my thoughts on this subject. However, seeing that this censure has not yet been authorized by the pope or by a Council, but only by a particular Congregation of Cardinal Inquisitors, I am not entirely giving up hope that it will not happen as it did with the antipodes, which were once condemned almost in the same way; thus, my *World* might see the light of day in due course, in which case I would myself need to use my arguments.[10]

The most striking remark in this letter is perhaps the initial sentence, in which Descartes claimed that the Inquisition had declared the geokinetic opinion a heresy. In his earlier letters, he had not described the anti-Copernican censure this strongly, saying rather that Galileo's book had been burned (in his November 1633 letter) and that the geokinetic opinion had been prohibited (in his February 1634 letter). In the meantime Descartes had read Nuncio Carafa's summary printed in Liège. As chapter 2 shows, one of Carafa's misrepresentations was to attribute the heresy declaration to the Inquisition, rather than to the advisory committee of consultants. Thus, Descartes was probably echoing the Liège poster's interpretation on this point.

Moreover, Descartes's judgment was also natural and understandable by way of the following reasoning. His thinking at this time reflected the effect of Galileo's condemnation on the anti-Copernican censure. The point is that, although the Index's decree of 1616 said only that Copernicanism was "contrary to Scripture" (without using the assessment of "heretical"), and although the Inquisition's sentence of 1633 reiterated this milder judgment, the fact remained that Galileo had been found "vehemently suspected of heresy" on account of his Copernicanism (among other reasons); and although suspected heresy was not technically the same as formal heresy, or even just plain heresy, one could hardly resist the inference that since Galileo was a suspected heretic for his Copernicanism, Copernicanism must be a suspected heresy and thus (loosely or unpedantically speaking) a heresy.

Another example of Cartesian exaggeration involved the clause Descartes quoted from the Liège poster. Carafa had reported that at the 1633 trial the Inquisitors established to their own satisfaction that in the *Dialogue* Galileo "seemed almost to be still of the same opinion, although he pretended to propose it in the manner of a hypothesis";[11] and that after arriving at this finding, they felt justified in reaching the verdict. To Descartes this apparently suggested that even the hypothetical discussion of the earth's motion had been generally condemned and prohibited.

Once again, Descartes's impression was justifiable. The Liège summary ignored the special injunction allegedly issued to Galileo in 1616; only on the basis of the special injunction could it have been a transgression for Galileo to engage even in a hypothetical discussion. But the special injunction applied only to Galileo personally, and not to Catholics generally, as did the Index's decree and its implied prohibition "to hold or defend." With no access to the text of the Inquisition's sentence, and no knowledge of the special injunction, Descartes had no way of knowing that the prohibition of hypothetical discussion was not general. In the last paragraph of this letter, Descartes was simply drawing this plausible conclusion from Carafa's summary.

In this letter, Descartes reiterated his earlier point that the claims advanced in his *World* were so interconnected that he would not know how to take out the geokinetic thesis alone. Thus his decision to leave the work unpublished remained firm. However, he said something more specific about the logical structure of his aborted treatise, namely that its demonstrations were "very certain and very evident."[12] Presumably, such demonstrations included those proving the earth's motion. So Descartes seemed to be claiming that he had conclusively proved the geokinetic thesis. What is surprising is not so much this boast but his intention nevertheless to submit to ecclesiastical authority. There would be nothing strange in deferring to this authority if one felt that one's arguments, however plausible and probable, were less than certain. But Descartes was apparently saying that he intended to do something much more difficult: bow to Church authority, which dictated that the earth stands still, while holding demonstrative proofs that the earth moves. This was not only a difficult feat but perhaps an impossible one. One could respond that either his proofs were not really conclusive, or he would not continue his submission to Church authority on this point.

At any rate, as Descartes learned more about the situation, his worst fears must have subsided. Whether Galileo's condemnation implied a prohibition of the hypothetical discussion of the earth's motion was relatively easy to ascertain. Mersenne must have questioned the Cartesian reference to the Liège poster, for, in a letter dated 14 August 1634, Descartes included a quotation of ten lines from that poster[13] to provide a context for the clause he had quoted in his April letter. Since we know that Mersenne had access to the text of the Inquisition's sentence (and published it in his own *Questions Theologiques, Physiques, Morales, et Mathematiques* of 1634), it is likely that he explained to his correspondent that, despite the implication of the Liège poster, the actual sentence did not suggest any prohibition of hypothetical discussion. It is also possible that Descartes himself was soon able to read that text.

Whereas there is no direct evidence, we do have documentary evidence (in the same letter to Mersenne dated 14 August 1634) that Descartes finally had been able to read Galileo's *Dialogue*. By that time this book was already a collector's item. Although it was not literally true, as Descartes stated in his November 1633 letter, that copies of this book had been ceremoniously burned in Rome, the book had been confiscated in the summer of 1632, but even then there were hardly any copies left to confiscate. Later, the trial and condemnation spurred interest in the book, and so Descartes was hopelessly late in the autumn of 1633 to try to buy a copy. However, in August 1634 he was able to borrow one from his friend Isaac Beeckman for thirty hours. Although Descartes had many objections of detail, on the key issue his opinion was very revealing: "His reasons proving the earth's motion are very good; but it seems to me that he does not present them as one must in order to be persuasive, for the digressions which he intermingles in their midst have the result that one no longer remembers the first by the time one reads the last."[14] Thus Descartes apparently felt he could do a better job of proving the geokinetic thesis than Galileo had done. Not that he would want to re-present Galileo's arguments. Instead he would want to present in the proper "persuasive" manner his own "very certain and very evident" arguments.

Descartes also soon became convinced that he could, after all, detach the geokinetic thesis from other parts of his *World*. After some indications in his correspondence that he was working on such a revision,[15] in 1637 he published in French a book containing treatises on optics, geometry, and meteorology, preceded by a *Discourse on the Method of Conducting One's Reason Well and of Seeking the Truth in the Sciences*. This *Discourse* described a method which he claimed to have followed in his three particular inquiries, and he discussed their relationship to his unpublished *World*. The last section of the *Discourse* (part 6) explained why he had not published the original work and why he was publishing the present one instead.

Descartes's original motivation was that he had "discovered a path which appears to me to be of such a nature that we must by its means infallibly reach our end,"[16] the end being "to render ourselves the masters and possessors of nature,"[17] especially practical knowledge about physical and mental health. This path was such that Descartes could usually explain everything that was observed, could do so in several ways, and thus needed experimentation to decide which explanation was correct. But the experiments to be performed were so numerous that other persons had to be involved in the effort.[18]

Later he reasoned that publication might not be such a good idea. This reflection started "when I learned that certain persons, to whose opinion I defer, and whose authority cannot have less weight with my actions than my own reason has over my thoughts, had disapproved of a physical theory published a little while before by another person. I will not say that I agreed with this opinion, but only that before their censure I observed in it nothing which I could possibly imagine to be prejudicial either to Religion or the State, or consequently which could have prevented me from giving expression to it in writing, if my reason had persuaded me to do so."[19] This condemnation convinced Descartes that his treatise on *The World* would have been highly controversial and that the resulting controversies would make him "lose the time which I meant to set apart for my own instruction";[20] and in his experience, controversy and critical discussion produced little if any enlightenment. Moreover, regarding potential benefit to others, "no one can so well understand a thing and make it his own when learnt from another as when it is discovered for himself";[21] for serious learners, "it is not necessary that I should say any more than what I have already said in this Discourse."[22] Finally, as regards the experiments that others might perform, such assistance would hardly be worth the trouble, either because other people would not follow instructions exactly, or because they would constantly ask for additional explanations, or because experiments already performed would need to be interpreted and evaluated; so the only real help might be financial. These considerations had led him three years earlier to decide not to publish his treatise.

However, since then two other reasons had made him decide to publish certain parts of it, such as those in the present *Discourse*. One reason was to prevent misunderstanding of the failure to publish; the other was to

Even before Dava Sobel's popularization of the life of Sister Maria Celeste Galilei and her relationship to her father,[37] no one who read her letters to her father could help being impressed by her warmth, love, intelligence, sensitivity, and unassuming eloquence. Although none of Galileo's letters to her have survived, one hundred and twenty-four of her letters to him have done so. In the first, dated 10 May 1623, she expressed condolences for the death of Galileo's sister Virginia, after whom she had been named.[38] The last is dated 10 December 1633; in it she expressed her joy that Galileo had finally been given permission to return to his villa in Arcetri and would soon move there from Siena.[39] Once he had returned to Arcetri (on 17 December), father and daughter lived within walking distance of each other, and there was no longer a need for letters. Unfortunately, Maria Celeste died on 2 April 1634, after a brief illness, and the old man was devastated.[40]

The most touching of her letters is perhaps the one in which she reported (on 3 October 1633)[41] that she had finally been able to read the text of the Inquisition's sentence and had decided to assume onto herself the burden of her father's salutary penance to recite the seven penitential psalms once a week for three years. Of course, it was not legally or theologically possible to substitute one person's penance with that of another, but that was irrelevant. The point was to share her father's burden by taking on an equal burden herself, so that the old man would have a companion in his misery.

3.3 "The Mirrour of True Nobility & Gentility": Peiresc's Plea (1634–1635)

In 1657 there appeared in London the English translation of a book by Pierre Gassendi first published in Paris in Latin in 1641. It was a biography of Nicholas Claude Fabri de Peiresc, and its English title conveys a good idea of the type of person Peiresc was: *The Mirrour of True Nobility & Gentility*. The nobleman had died in 1637, and so the rapidity with which he became the subject of biographies also indicates the respect and renown he had enjoyed.[42]

Peiresc was a lawyer, politician, diplomat, and amateur natural philosopher. Like Marin Mersenne, he was also a kind of clearinghouse for correspondence from all parts of Europe, as well as North Africa, the Middle East, and Asia; his correspondence takes up ten volumes of about one thousand pages each.[43] He became acquainted with Galileo in Padua between 1599 and 1602 while he studied law at the university there. Peiresc took an active interest in Galileo's telescopic discoveries, so much so that immediately after the publication of the *Sidereal Messenger* in 1610, he had an observatory built in his house in Aix-en-Provence; and when he heard that

Galileo was writing the *Dialogue,* Peiresc was eager to obtain a copy and even wrote to him by way of intermediaries. However, it was not until thirty-five years after their first meeting, after an equally long period of distant admiration, that Peiresc wrote to Galileo directly. Peiresc had, of course, heard about the Inquisition's condemnation and had had no difficulty obtaining a copy of the text of the sentence.[44]

Peiresc was in a good position to try to help Galileo because he was on very good terms with the Barberini family and had hosted at his own house Cardinal Francesco Barberini, the pope's nephew and the Vatican secretary of state, when the latter was returning from Paris on a special diplomatic mission in 1625. Thus, on 26 January 1634, he finally wrote a long letter to Galileo reminding him of their earlier acquaintance, mentioning specific names and details, offering words of encouragement, and expressing an eagerness to help.[45] What Peiresc had in mind was to obtain a pardon for Galileo. And so in due course he made a plea to Cardinal Francesco Barberini; Peiresc hoped to be more effective by including the plea in the second half of a very long letter that discussed many other, more pleasant topics about which he and Francesco Barberini frequently corresponded.[46] This is the text of the plea, dated 5 December 1634:

[169] Finally, I have to make a plea to Your Eminence, and I beg you (as much I know how and am capable of) to excuse the daring of this most faithful servant of yours and to blame on the usual confidence you show me the hope I nurture from the sublime goodness of Your Eminence: [170] my hope is that you will deign yourself to do something for the consolation of a good old septuagenarian, who is in ill health and whose memory will be difficult to erase in the future. He may have erred in regard to some proposition (as human nature allows), but without showing obstinacy of opinion and instead underwriting the opposite opinion in accordance with the orders received; so please do not keep him in the kind of confinement which I understand is being applied in his case, if it is possible to obtain some relaxation, as the natural kindness of Your Eminence makes me hope. I met him thirty-four or more years ago at the University of Padua and at the most beautiful conversations we enjoyed at the house of Giovanni Vincenzio Pinelli (God bless his soul), together with Messrs. Aleandro and Pignoria[47] (God bless their souls). It will be difficult for posterity not to show him eternal gratitude for the admirable novelties he discovered in the heavens by means of his telescope and his extremely penetrating mind. Recall that in regard to Tertullian, Origen, and many other Church Fathers who fell into some errors due to simple-mindedness or other reason, the Holy Church (like a good mother) has not failed to show them great veneration on account of their other ideas and other indications of their piety and zeal in the service of God; on the contrary, she would regard as perverse and would reproach the zeal of whoever would want to punish them with the same severity with which obstinate heretics are punished, or administer to them those penalties that can be applied to persons guilty of some great error or crime; such is human weakness, which may have been the source of some of their sins, and their fragility is sometimes worthy

of excuse and forgiveness, as has happened in more serious cases involving persons who hold the first places among the saints. Similarly, future centuries may find it strange that, after he retracted an opinion which had not yet been absolutely prohibited in public and which he had proposed merely as a problem, so much rigor should be used against a pitiable old septuagenarian as to keep him in a (private if not public) prison: he is not allowed to return to his city and his house or to receive visits and comfort from friends; he has infirmities that are almost inseparable from old age and require almost continuous assistance; and often this help cannot be subject to the delay due to the distance from the villa to the city, when there is an emergency or an immediate remedy is needed. I say all this out of compassion for the pitiable good old man Mr. Galileo Galilei, to whom I had wanted to write lately; thus, having asked the advice of a Florentine friend to know where he could be found, I was told that he was confined to his villa near a monastery, where one of his daughters (his only consolation) was a nun and had recently died, and that he was forbidden to receive visits and correspondence from friends and to go to his city and his house; this shook my heart and drove me to tears, as I considered the vicissitudes of human affairs and the fact that he had had so many uncommon honors and accomplishments, whose memory will last for centuries. I know that painters who excelled in their art have had pardoned extremely serious sins whose enormity was most horrifying, in order not to let their other merit go to waste; so, is it not the case that so many inventions (the most noble that have been discovered in so many centuries) deserve a display of indulgence toward a philosophical play[48] in which he never categorically asserted that his own personal opinion was the one that had been disapproved?

[171] This will really be deemed most cruel everywhere, and more by posterity, for it seems that in the present century everyone neglects the interests of the public (especially of the disadvantaged) in order to look after his own. And indeed it will be a stain on the splendor and fame of this Pontificate, if Your Eminence does not decide to take some precaution and some particular care, as I beg and implore you to do most humbly and with the greatest zeal and urgency which I can legitimately display toward you; I also beg you to pardon the liberty I take, which may be too great, but it is important that sometimes your faithful servants be allowed to give you these indications of their faithfulness, for I do not think that the other servants around you would dare to reveal to you in this manner the thoughts which they have in their heart and which touch the honor of Your Eminence much more than it may seem to many.[49]

Francesco Barberini acknowledged receiving this plea, and did promise that he would talk to the pope about the matter; but he was otherwise cryptic, hinting that Inquisition matters were covered by secrecy.[50] To make sure the cardinal would not forget, Peiresc repeated and expanded his plea in this letter dated 31 January 1635:

Finally, I cannot conceal from you the fact that if Your Eminence will deign yourself to obtain from His Holiness some consolation for the venerable old man Mr. Galilei, I would not take this favor from your immense goodness as any less than if it were for my own father (may he rest in peace). I stoop before you with the greatest submission of which I am capable, in order to present to you my most humble plea for him; I am much more jealous of the honor and reputation of this Pontificate and of the most prudent direction and administration of Your Eminence than of the preservation of my life; and I am sure that the indulgence which you will procure for his sin of human frailty will be in accordance with the desire of the noblest minds of the century, who feel so sorry for the severity and prolongation of his punishment. However, a contrary outcome would run the risk of being interpreted and perhaps compared some day to the persecution of the person and wisdom of Socrates in his own country, so much reproached by other nations and by the very descendants of those who caused him so much trouble. Your Eminence, please excuse my daring and order me to be absolutely silent if it displeases you, for I am equipped to obey in every way I can; but I hope rather for the desired concession of the pardon from the mercy and most powerful intercession of Your Eminence.[51]

The cardinal never did answer this second plea, and in fact nothing came of Peiresc's valiant effort. However, these letters are important both because of their content and because they give us a glimpse of the way enlightened Catholics[52] felt about Galileo's condemnation.

Peiresc was obviously aware that Galileo was not being held in a real jail but rather was under house arrest and in seclusion at his villa in Arcetri. In effect he was asking for Galileo's complete freedom. One of Peiresc's arguments was that Galileo's crime was intrinsically minor because it stemmed from having written and published the Dialogue, and for Peiresc this book was a "philosophical play" (scherzo problematico);[53] this was a memorable interpretation that deserves to be stressed. The comparison of Galileo with Socrates was an obvious one that was to become an important theme of the subsequent Galileo affair, and Peiresc was perhaps the first to make it. That he had the courage to say it explicitly to the cardinal is an indication of his seriousness.

Although Peiresc's plea is the one for which we have the best documentary evidence, other influential figures also attempted to have Galileo pardoned; their efforts were equally unsuccessful with the pope and equally unsolicited by Galileo. At about the same time that Peiresc wrote his first letter to Cardinal Barberini, François Count de Noailles, the French Ambassador to the Holy See from 1634 to 1636, during a papal audience on 8 December 1634, pleaded with the pope and with the cardinal for Galileo's liberation; the plea was later denied, even though Urban VIII probably owed to French support first his appointment to cardinal in 1605 and then his election to pope in 1623.[54]

King Ladislaus IV of Poland also tried to obtain a pardon for Galileo. Ladislaus was a Catholic but had decreed freedom of conscience for his subjects; he was so well disposed toward Galileo that Giovanni Pieroni, who was trying to get the *Two New Sciences* published in central Europe (see chapter 4.1), suggested to Galileo that he dedicate the book to him. But Ladislaus's efforts (between 1636 and 1638) were unsuccessful, partly because the Tuscan grand duke refused to cooperate as a result of political and diplomatic differences, and partly because Ladislaus had political disagreements with the pope that led to a break of diplomatic relations in 1642.[55]

3.4 "No Pardon to Innocents": Galileo (1634–1642)

How did Galileo himself react to his condemnation? Some of his subsequent letters are revealing in this regard. One was addressed to Elia Diodati (1576–1661), a French Protestant born in Geneva, who was a lawyer for the Paris Parliament and an old friend and strong supporter of Galileo.[56] On 7 March 1634, Galileo wrote to Diodati in Paris:

> [58] I now come to your letter. Since you repeatedly ask me for some account of my past troubles, I must (summarily) tell you this. From the time I was summoned to Rome until the [59] present, my health has (thank God) been better than it had been in the last several years. I was detained in Rome in prison for five months, and my prison was the house of the Lord Ambassador of Tuscany;[57] he and the Ladyship his wife treated me in such a way that they could not have treated their own fathers with greater affection. After my trial was concluded, I was condemned to prison at the discretion of His Holiness. For a few days, the prison was the palace and garden of the grand duke at Trinità dei Monti.[58] Then it was commuted to the house of Monsignor the Archbishop[59] in Siena, where I stayed five months and was treated like a father by His Most Illustrious Lordship, with constant visits by the nobility of that city; there I wrote a treatise on a new topic, pertaining to mechanics and full of many intriguing and useful speculations. From Siena I was allowed to return to my villa, where I still am, being forbidden to go into the city; this prohibition is made in order to keep me away from court and princes. However, since I returned to the villa while the court was in Pisa, two days after the grand duke returned to Florence he sent me a member of his staff to notify me that he was on his way to visit me; half an hour later he arrived in a very small carriage accompanied by a single gentleman, and he came into my house and stayed almost two hours talking to me with extreme kindness. Thus I have not suffered at all in the two things that should be esteemed above all others, that is, health and reputation; for in regard to the latter, I am reassured by the redoubled affection of the rulers and of all my friends. The wrongs and injustices which envy and ill will have perpetrated against me have not bothered and do not bother me. Instead, given that my health and honor remain unharmed, the greatness of the abuse is rather comforting to me and represents a kind of revenge; the infamy reverts back to the traitors and those who are in the highest state of ignorance, which is the mother of ill will, envy, anger, and all other wicked and ugly vices and sins. The friends who are not

here must be satisfied with these generalities because the particulars, which are very many, greatly exceed my ability to include them in a letter. Your Lordship should be satisfied with this much, and rest assured that I am still in such a state as to be able to bring my other works to fruition and publish them.

[60] The news you have from Strasbourg pleased me a lot, and I am honored by your intervention and unwavering vigilance. I wish my *Dialogue* had reached Louvain and passed through the hands of Froidmont,[60] who among nonmathematical philosophers seems to me more flexible. In Venice one Antonio Rocco[61] has published in defense of Aristotle and against those considerations which I oppose to him in the *Dialogue;* he is a very pure Peripatetic and very far from understanding anything in mathematics or astronomy, as well as full of stings and insults. I also understand a Jesuit has printed in Rome a work to prove that the proposition of the earth's mobility is absolutely heretical;[62] but I have not yet seen it.[63]

Here Galileo gave an informative account of how his prison sentence was being implemented. He noted that, during his five-month exile in Siena, he had been treated extremely well by Siena's archbishop. In fact, he had been treated so well by everybody in Siena that in February 1634 (unbeknownst to him) an anonymous complaint had been filed with the Roman Inquisition against Galileo and the archbishop of Siena, about such treatment.[64] Indeed his treatment had become common knowledge even abroad; for example, in autumn 1633, on his way from Rome to France, Gerard Marc Anthony Saint-Amant visited Galileo in Siena and reported that he was lodged in rooms elegantly decorated with damask and silk tapestries.[65] The Inquisition took no action, perhaps because by then Galileo had returned to Arcetri and so the issue was moot. This complaint may, however, have hardened the Inquisition with regard to its next decision, which is one of the things Galileo related in the next significant letter to Diodati, dated 25 July 1634:

[115] I hope that, after you hear about my past and present troubles together with other suspected future ones, Your Lordship and the other friends and patrons near you will excuse me for my delay in answering your letters and my total silence regarding theirs, as Your Lordship can make them aware of the sinister direction in which my affairs are going at this time.

In my sentence in Rome, I was condemned by the Holy Office to prison at the discretion of His Holiness, who assigned me as prison the grand duke's palace and garden at Trinità dei Monti. This happened last year in the month of June, and I was [116] given to understand that after that month and the following one, if I had petitioned for a pardon and total liberation, I would have obtained it; but in order not to have to stay there the whole summer and also part of the autumn (and suffer the restrictions of such a season), I obtained a commutation to Siena, where I was assigned the archbishop's house. I lived there for five months, after which the prison was commuted to the limits of this small villa, a mile away from Florence, with a very strict prohibition against going into the city, holding conversations or meetings of many friends together, or inviting them to banquets. I was living here very quietly, frequently visiting a nearby monastery where my two daughters were nuns; I loved them very much, especially the elder, a woman of exquisite mind and singular goodness and extremely attached to me. Due to an accumulation of melancholy humors during my absence (which she considered to be troublesome) and to a sudden dysentery acquired later, my elder daughter died within six days at the age of thirty-three and left me in extreme grief. This was doubled by another unfortunate event: I was returning home from the convent in the company of the physician who had examined my sick daughter just before she passed away and who was telling me that her condition was quite hopeless and that she would not have lasted more than the following day, as indeed it turned out; when I arrived home I found the inquisitor's deputy, who had come to convey to me an order from the Holy Office in Rome received by the inquisitor in a letter from Lord Cardinal Barberini; I was ordered to stop submitting any more petitions for a pardon and for permission to return to Florence, otherwise they would recall me to Rome and hold me in the real prison of the Holy Office. This was the reply given to the memorandum which, after nine months of exile on my part, the Lord Ambassador of Tuscany had presented to that Tribunal. From this reply, I think one can conjecture that very likely my present prison will not end, except by turning into the one that is universal, most strict, and eternal.

From this and other incidents which it would take too long to write about, one sees that the anger of my very powerful enemies is constantly becoming exacerbated. They finally have chosen to reveal themselves; for about two months ago a good friend of mine was [117] in Rome talking to Father Christopher Grienberger, a Jesuit and a mathematician at that College; when they came to discussing what happened to me, the Jesuit said these exact words: "If Galileo had been able to retain the affection of the Fathers of this College, he would be living in worldly glory; none of his misfortunes would have happened; and he could have written at will on any subject, I say even of the earth's motions, etc." So Your Lordship sees that it is not this or that opinion which has provoked and provokes the war against me, but to have fallen out of favor with the Jesuits.

I have other indications of the vigilance of my persecutors. One example involves a letter written to me from a transalpine country and sent to me in Rome, where the writer must have thought that I was still residing; it was intercepted and brought to Lord Cardinal Barberini; according to what was later written to me from Rome, fortunately the letter was not a reply but a first contact, and it was full of praises for my *Dialogue;* it was seen by many persons, and I understand that copies are circulating in Rome; and I have been told that I will be able to see it. Add to this other mental troubles and many bodily ailments; at my age over seventy, they keep me depressed in such a way that any small effort is tiring and hard. Thus for all these reasons my friends must sympathize with me and forgive my failures, which look like negligence but are really impotence; and since Your Lordship more than anyone else is partial toward me, you must help me to maintain the goodwill of those there who are well disposed toward me.[66]

In contrast to the cheerful and optimistic mood of the earlier letter, this one shows how depressed and pessimistic Galileo had become on account of the double blow of the death of his beloved daughter and of the Inquisition's ultimatum refusing a pardon. This letter also added to the evidence for a Jesuit conspiracy theory, which, as we know from Descartes' February 1634 letter to Mersenne, was spreading throughout Europe. Coming from Galileo's own letter, this evidence may be given more weight than that coming from Descartes; on the other hand, Galileo was reporting only indirect or hearsay evidence.

At any rate, the condemned man's spirits were lifted when he learned of Peiresc's efforts on his behalf. A distant relative named Roberto Galilei, living in Lyons, had told Galileo about them; in fact, he had sent Galileo copies of Peiresc's first letter and of a cryptic sentence from Barberini's reply.[67] Galileo immediately wrote to thank Peiresc in this letter, dated 21 February 1635:

[215] I could never by means of a pen express to Your Most Illustrious Lordship the joy given to me by the reading of the most formal and most prudent letter which you wrote on my behalf, and a copy of which was sent me by my relative and patron Mr. Roberto; I received it only yesterday. My pleasure was and is infinite, not because I hope for any improvement, but because I saw a Lord and Patron of such excellent qualities empathize with my condition with such tender affection, and be moved to attempt with such ardent spirit and with such generous as well as moderate daring an undertaking that has rendered silent so many others who are favorably inclined toward my innocence. Oh, if my misfortunes produce for me such sweetness, may my enemies engage in new machinations, for I will always be thankful to them!

I said, my Most Illustrious Lord, that I do not hope for any improvement, and this is so because I did not commit any crime. I could hope to obtain clemency and pardon if I had erred, because mistakes are the subject matter over which princes can exercise the power of reprieve and pardon; but in regard to someone who has been unjustly condemned, it suits them to maintain rigor, as a cover for the legality of the proceedings. And believe me, Your Most Illustrious Lordship, also for your own consolation, that this rigor afflicts me less than what others may think because there are two comforts that constantly assist me. One is that in reading all my works, no one can find the least shadow of anything that deviates from piety and reverence for the Holy Church. The other is my conscience, which is fully known only by me on earth and by God in heaven; it understands perfectly that in the controversy for which I suffer, many people might have been able to proceed and to speak more knowledgeably, but no one (even among the Holy Fathers) could have proceeded and spoken more piously, or with greater zeal toward the Holy Church, or in short with a holier intention than I did. This most religious and most holy [216] intention of mine would appear all the more pure if one were to expose into the open all the calumnies, scams, stratagems, and deceptions that were used in Rome eighteen years ago to cloud the vision of the authorities! But there is at present no need for me to give you any other greater justifications of my sincerity; for you have been gracious enough to read my writings and may very well have understood what was the true, real, and first motive that under a feigned religious mask has been waging a war against me, constantly laying siege to me and blocking all roads in such a way that neither can help come to me from the outside, nor can I any longer get out to defend myself. In fact, a direct order has gone out to all inquisitors not to allow the reprinting of any of my works that have already been published years ago, nor to license any new one which I might want to publish; thus not only am I supposed to accept and ignore the great many criticisms dealing with purely natural subjects made against me to suppress the doctrine and advertise my ignorance, but also I am supposed to swallow the sneers, bites, and insults recklessly thrown at me by people more ignorant than I. But I want to stop making complaints, although I have barely scratched the surface. Nor do I want to keep any longer Your Most Illustrious Lordship or bother you with distasteful things; instead I must beg you to excuse me if, drawn by the natural relief which those who suffer get occasionally when they unburden themselves to their most trusted confidants, I have taken too much liberty to annoy you. There remains for me to convey to you with the power of my heart those thanks which I could never convey with words for the humane and compassionate task you undertook on my behalf; you were able to present the case so effectively that, if I do not benefit, we can be pretty sure that it must have caused some twinge of remorse in the recipients, who, being men, cannot be devoid of humanity. I continue to be your most obliged and most devout servant. May God the Lord repay the merit of the charitable deed you performed, and I bow to you with reverent affection.[68]

The attitude first expressed by Galileo here was the one he would retain as long as he lived. He had given up hope for a pardon and formulated the following interpretation: he would not be pardoned precisely *because* he was innocent, *because* an injustice had been committed against him; and the pope and the Inquisition had to cover up a miscarriage of justice, displaying firmness and rigidity to try to prove that they had been right to condemn him.

Besides refusing a pardon, the judges seemed to increase Galileo's punishment. The additional penalty seemed to be an injunction not to publish any of his works. Galileo was referring to something he had just learned from Fra Fulgenzio Micanzio, the official theologian of the Republic of Venice, a very independent-minded Venetian and a strong and helpful supporter of Galileo. Micanzio had offered to help Galileo publish in Venice the new work on mechanics and physics which he had started writing in Siena. On 10 February 1635, the friar had written Galileo that the Venetian inquisitor had clearly and explicitly told him that there was a ban on the publication of any of Galileo's writings.[69] Although this did not deter the fearless Micanzio, he pointed out that publishing the *Two New Sciences* in Venice might cause Galileo unnecessary trouble.

In his letter to Peiresc, Galileo was taking this additional penalty as further evidence of the Church's overcompensating rigidity, and consequently as further confirmation of his innocence, by the explanatory interpretation he had come to formulate. Thus, his interpretation also helped Galileo to justify himself. His self-justification in this letter also included an account of his demeanor and motivation throughout the whole affair: he maintained an attitude of piety and reverence toward the Church; his motivation was a sincere zeal to help her.

With such an interpretation and justification Galileo seemed to have reached a state of imperturbable serenity. Of course, pleasant and unpleasant events continued to happen, and good and bad news continued to reach him, but none were such as to alter significantly his self-assured resignation and his inner peace.

On the positive side, Galileo continued his efforts to publish his new work on motion, exploring next the possibility of publishing it in Germany with the assistance of a former student named Giovanni Pieroni, a Tuscan-born military engineer in the service of the Holy Roman Emperor.[70] Then, due to delays and difficulties there, a better opportunity came along with the Elzevier publishing company, based in Holland but having offices and connections across Europe. Thus, eventually (in 1638) the book came out in Leiden with an Italian title that means *Discourses and Mathematical Demonstrations on Two New Sciences Pertaining to Mechanics and Local Motion.* Galileo dedicated the book to François Count de Noailles, the French ambassador to the Holy See, a patron who had tried unsuccessfully to convince the pope to grant a pardon; and in the preface, Galileo made it sound as if the book had been published without his knowledge, from a manuscript he had given to the count. This story was mostly a subterfuge, although it is true that on 16 October 1636 Galileo traveled to the Tuscan town of Poggibonsi to meet the ambassador on his way back to France, and that for such a meeting the count had obtained a special dispensation to release Galileo from his house arrest.[71] Perhaps because of this subterfuge, or perhaps because the Venetian inquisitor's words to Micanzio about a general injunction against any Galilean publications was not exactly true, Galileo did not seem to suffer any harm from the Inquisition for publishing this book; indeed, in January 1639 the book reached Rome's bookstores, and all available copies (about fifty) were quickly sold and everyone seemed to like it.[72]

Moreover, other works of Galileo were published during this period, but such publications were indeed due to the efforts of other people. He was pleased, of course, and they lifted his spirits. In 1634, Mersenne published in Paris *Les Mécaniques de Galilée,* a French translation of an unpublished manuscript Galileo had written in Italian in his university days. In 1635, a Latin translation of Galileo's *Dialogue* was published in Strasbourg, titled *Systema Cosmicum,* under the general supervision of Elia Diodati, the translation having been done by Matthias Bernegger (see chapter 4.2). In 1636, also in Strasbourg, Galileo's Letter to the Grand Duchess Christina, written in 1615, was published for the first time in an edition that contained both the original Italian text and a Latin translation by Diodati. In 1639, Mersenne edited and translated into French Galileo's just-published *Two New Sciences* and published it in Paris under the title *Les Nouvelles Pensées de Galilei, Mathematicien et Ingenieur du Duc de Florence.* In 1640, there was a reprint in Padua of Galileo's booklet on the calculating device known as the proportional compass, entitled *Operazioni del compasso geometrico e militare.*[73] In 1641, the Latin *Systema Cosmicum* was reprinted in Lyons.

On the other hand, Galileo also suffered some setbacks. In December 1635 a new calumny against him emerged: the Peripatetic character in the *Dialogue,* named Simplicio, was alleged to be a caricature of Pope Urban VIII, presumably on the grounds that at the end of the book Simplicio was the one who gave a statement of the pope's favorite argument against Copernicanism, the objection from divine omnipotence.[74] This was the argument that since God is all-powerful, He could have created any one of a number of different worlds, including one in which the earth is motionless; therefore, regardless of how much evidence there is in support of the earth's motion, one can never assert that this must be so, for such an assertion would be an attempt to limit God's power to do otherwise.[75] In the subsequent history of the Galileo affair, this allegation of caricature has been often mentioned as one of the original charges against the book in 1632 that antagonized Galileo's former patron and admirer and led to the trial and condemnation the following year. Although this claim is plausible, the fact remains, as Sante Pieralisi argued in 1875,[76] that this accusation is not mentioned in any documents prior to December 1635; it is not mentioned even in Magalotti's letter to Guiducci of 7 August 1632, which mentions the rumor that the three dolphins of the publisher's trademark were a caricature of Urban's nepotism.[77] Thus, it seems more accurate to regard the Simplicio-as-Urban allegation as a new slander against Galileo. It took six months before Castelli and the French ambassador Noailles succeeded in convincing the pope that nothing could have been further from Galileo's mind.[78]

Another setback was that nothing came of the project of publishing a Latin translation of Galileo's collected works, even though he himself spent considerable time, effort, and money on it. Although this project was initiated by others, it became dear to his heart, and at one point he was paying an assistant to translate his Italian works into Latin. At another point, when the Elzeviers seemed interested in the idea, he expressed a willingness to contribute to the expenses and promised to purchase at least one hundred copies.[79]

Furthermore, Galileo had several near misses and close encounters with the Inquisition in 1638. In December 1637, Galileo had become completely blind.[80] For this and other reasons, in January 1638 he petitioned the Inquisition for his complete freedom.[81] In April 1639 this request was at last forwarded to the pope, who denied it,[82] although without the punishment of real imprisonment threatened in 1634. However, in view of his blindness and of other ailments that needed medical attention, in March 1638 he was allowed to move to his house in Florence and even to leave the house to attend religious services at the nearest church; but he was supposed to stay in seclusion and to refrain totally from discussing the earth's motion.[83]

Galileo ended up spending the rest of the year in the city, and the interaction with the Inquisition continued. Since August 1636 he had been negotiating with the Dutch government over his proposal for determining longitude at sea by the observation of Jupiter's satellites.[84] Both sides were aware of the complexity of the problem and of the fact that much more work needed to be done. In 1637, the Dutch government decided to show its preliminary appreciation by sending Galileo a gold chain worth two hundred scudi;[85] to get an idea of the value of this sum, we should note that Galileo's annual salary amounted to an unprecedented 1000 scudi. Even before the Dutch government's representative reached Florence, the Inquisition learned something about it and ordered Galileo not to meet with him if he was a non-Catholic or came from a non-Catholic country.[86] Galileo did meet with the representative and accepted a letter from the Dutch government, but he declined the gift, giving as a reason the fact that his method for determining longitude had not yet been fully worked out.[87] In the autumn of that year, at the request of the grand duke, the pope consented to let Galileo's favorite disciple, Benedetto Castelli, go to Florence for a few months, partly to discuss the longitude problem and partly to allow Galileo to share his last thoughts with Castelli;[88] at the time, Castelli was a professor of mathematics at the University of Rome. Finally, as Galileo's health improved, in January 1639 he returned permanently to Arcetri, perhaps on orders from the Inquisition, or perhaps because the house arrest was more tolerable in the country villa than in the small house in the city.[89]

It is also likely that after the Inquisition's interference with the visit of the Dutch government's representative, Galileo concluded that, despite the identity of the formal conditions, he was actually freer to receive visitors in Arcetri because the ecclesiastical authorities could not watch over him there as easily as they could in the city. In fact, in the previous four years, he had had little difficulty receiving the visits of the following persons: Grand Duke Ferdinand II on 25 December 1633 (as we have seen); Pierre Carcavy in December 1634;[90] Jean Baptiste Haultin and Jean Bire on 8 October 1635;[91] Louis Henselin and one Dr. Maucourt on 29 October 1635;[92] Jean de Beaugrand three times in October and November of the same year;[93] Thomas Hobbes that same November;[94] Louis Elzevier in May and July 1636;[95] Anthony Kester in June of the same year;[96] Bonaventura Cavalieri for a whole week in July;[97] Arrigo Robinson in November;[98] Prince Giovanni Carlo de' Medici and the poet Giovanni Carlo Coppola in January 1637;[99] Dino Peri in May of that year;[100] Ferdinand II on 2 September;[101] and Pierre Michon on 20 September.[102] Indeed, the abbé Girolamo Ghilini was perhaps not exaggerating when he stated, in a biographical dictionary published in 1647, that after the 1633 condemnation no important personage traveling through Tuscany failed to try to pay a visit to Galileo, who was thus considered to be a major tourist attraction.[103]

Similarly, after returning to Arcetri in 1639, the blind man was able to enjoy the companionship and use the services of several persons. Father Clemente Settimi, a member of a charitable religious order, became his personal assistant and occasionally even stayed overnight. The young Vincenzio Viviani began studying with Galileo and assisted him with correspondence, for which he received a modest salary from the grand duke.[104] Benedetto Castelli was able to visit him several times.[105] And Evangelista Torricelli began living in Galileo's house as his research assistant in October 1641.[106] The Inquisition by and large acquiesced with such practical and ad hoc arrangements. In December 1641 Father Settimi was tried by the Florentine Inquisition partly on charges that he was too close to Galileo; but he was acquitted, and in any case the trial also involved an internal struggle within his order that led to a dispute between Tuscany and the Holy See and in 1646 to the degradation of the order to a lay congregation.[107]

On 8 January 1642, "from the prison of Arcetri" (as he sometimes labeled his residence at the end of his letters just before his signature),[108] Galileo passed into the one he had described to Diodati as "universal, most strict, and eternal."[109]

Chapter 4

Polarizations

Secularism, Liberalism, Fundamentalism
(1633–1661)

Descartes's reaction to Galileo's condemnation points up an important and clear distinction between condemnations issued by a Roman congregation of cardinals and pronouncements of either the pope speaking ex cathedra or an ecumenical Church council. The former did not have the binding status of either of the latter. Nevertheless, as Descartes also exemplified, most Catholics felt some kind of duty toward the former; for example, they felt obliged to refrain from publicly opposing congregational decisions.

Another important distinction was that between the condemnation of persons as heretics (whether formal, relapsed, obstinate, or suspected) and the condemnation of doctrines and books (as heretical, erroneous, temerarious, etc.). Normally, the condemnation of an individual applied only to the particular case and did not have the generality which condemnations of doctrines did. On the other hand, since normally an individual was condemned for holding or defending a condemned doctrine, the condemnation of the individual had an indirect kind of generality deriving from the doctrinal censure under which he was condemned. However, Galileo's case was exceptional because his condemnation helped to redefine, clarify, and reinforce the kind and degree of anti-Copernican censure, as well as the content of the prohibited doctrine. If before 1633 a Catholic was prohibited to hold or defend the scriptural compatibility and the physical truth of the earth's motion, after the trial the prohibition extended to such things as the probability of (the physical truth of) the earth's motion, the denial of the philosophical (or scientific) authority of Scripture, and the superiority of the geokinetic hypothesis over the geostatic hypothesis. The Church's unprecedented effort to promulgate Galileo's sentence and abjuration is evidence of the attempt to generalize Galileo's case, to derive general prescriptions from his condemnation.

4.1 States versus Church

A third distinction was that between political and ecclesiastical jurisdictions. The more autonomous and powerful a state or civil government was,
the more likely it was to claim for itself the right to ratify decisions emanating from Rome. This was clearly true in France, Spain, and the Venetian
Republic, where even doctrinal decisions from Rome had no validity unless
they were ratified through a local institutional process. Yet even for apparently nondoctrinal condemnations (like Galileo's), the Church was able to
communicate with the local bishops independently of state authority,
through its nuncios. Let us examine how some state authorities reacted to
Galileo's condemnation.

As regards France, there is a story first told in 1670 by Father Nicolas-
Joseph Poisson in a commentary to Descartes' method.[1] In 1635, Cardinal
Armand Jean de Richelieu, chief minister of France, tried to have the geokinetic thesis condemned by the Sorbonne along the lines of the Index and
Inquisition in Rome. He apparently used all his influence to accomplish his
goal, although Favaro plausibly conjectured that Richelieu did this not out
of any geostatic enthusiasm but rather "in order to win some favor with the
Roman Court and to be able to resist to it in matters of greater importance."[2] He would have certainly succeeded had it not been for a dissenting
professor, probably Jean de Launoy.[3]

This dissenter argued as follows. He began by reminding his colleagues
that they had no scruples about teaching Aristotelian doctrines, even
though these doctrines had at one time been forbidden by some Church
Councils.[4] He proceeded to remind his colleagues that Church Councils
have greater authority and demand greater deference than the Inquisition.
Then he argued that since the Copernican doctrine was condemned by the
Inquisition, and since one had the freedom of teaching Aristotelian doctrines once censured by an authority higher than the Inquisition, therefore
one should have the freedom of teaching and following Copernicanism;
thus a condemnation of Copernicanism by the Sorbonne would not have
the aim of promoting the truth but at best that of preventing factional
divisions.

From this story, it appears that the Sorbonne never did approve the con-
demnation of Copernicanism or the consequent condemnation of Galileo.
Since France had the policy that decisions of the Roman congregations
were not valid unless endorsed by the Sorbonne and the Parliament of
Paris, it seems that those condemnations had no formal legal standing in
France.

In Spain,[5] when the papal nuncio to Madrid received the Inquisition's
orders from Rome to promulgate the 1633 condemnation of Galileo, he
directly notified all Spanish bishops as well as the universities of Salamanca and Alcalà (as we saw in chapter 2.2).[6] This direct notification was

of questionable legal validity, and the difficulty emerged explicitly the following year in connection with the 1634 Index's decree banning the *Dialogue,* among other books. The nuncio, Cardinal Cesare Monti, again transmitted the decree to the Spanish bishops. But this time there was a special difficulty.

Besides Galileo's *Dialogue,* the Index's decree included twenty-six other books. One of these was a work published in Sicily that advanced antipapal and pro-Spanish ideas.[7] Perhaps for this reason, the Spanish Inquisition denounced the nuncio's action as a violation of Spanish law, which gave it jurisdiction over book censorship and required its ratification of the Roman Inquisition's decrees. Thus the 1634 Index's decree was not published in Spain, and the *Dialogue* was not included in the next Spanish edition of the *Index* (in 1640) or in subsequent editions, which were published separately from those of Rome.

It should be mentioned that Copernicus's *Revolutions* was also never included in the Spanish *Index,* although for different, yet still extrinsic reasons. Through an oversight it was omitted from the 1632 edition of the Spanish *Index.* On the other hand, Copernicus's name was mentioned in the table of contents and in connection with an entry on Rheticus; and many other Copernican works, especially those of Kepler, were included.

Thus the omission of Copernicus's and Galileo's books from the Spanish *Index* cannot be taken as a sign of religious tolerance or freedom of thought in Spain. The Spanish response was, rather, an indication of nationalism and the interaction between religion and political power.

The response of the Holy Roman Empire was less clear, but we can get a glimpse of it from the efforts, activities, and reports of Giovanni Pieroni. Born in Tuscany, he had studied with Galileo in Padua. After the trial, Pieroni was a military engineer in the service of the emperor, who had promised him a printing press for use in his work. Thus, in the summer of 1635,[8] after Galileo had learned from Fulgenzio Micanzio that the Venetian inquisitor had said that there was a total ban on any Galilean publications, the possibility emerged that Pieroni would publish Galileo's new work, either on his own press or somewhere else in the empire.

However, Pieroni did not get his printing press, and so he explored other avenues for publication. Two years later, he had not yet managed to have the book published, but the situation looked promising despite the difficulties he had encountered. On 9 July 1637 he wrote to his former teacher from Prague:

[130] I was most pleased that I happened to be in Prague when the Most Serene Prince Matthias arrived, so that I could pay my respects and be of service to him; so that I could have news from such a prince about Your Most Excellent Lordship; and in particular so that I could have the opportunity to tell His Most Serene Highness what I now intend to tell also to Your Most Excellent Lordship in regard to the printing of your book.

After Your Most Excellent Lordship wrote me to return it to you (thinking that I myself was going to return there), I was unable to come so soon because there were delays in obtaining the permission, and in fact I did not obtain it. Thus I decided to go ahead with the plan to print it. I saw that the avenue for doing this in Vienna was closed because Father Scheiner was there and the [Jesuit] Fathers have to approve whether or not any book is printed in Vienna, and so I thought that he might be asked to review it, or that if he found out about it he might try to prevent the printing there and later everywhere else. Thus, since the request for my own printing press has not yet received final approval, I resorted to Lord [131] Cardinal Dietrichstein,[9] who welcomed the project; he favored publishing it in Olmütz and being reviewed there by a Father of another order, so that we did not have to fear being discovered by the said Father Scheiner and his confreres, as I had warned the cardinal. And that is what he did. He got the book and sent it to a Dominican Father, who approved it,[10] as Your Most Excellent Lordship will see.[11] But before the process was completed, the Lord Cardinal died. Thus I went to Olmütz to get the book and have it printed; there the newly appointed bishop, a prelate of great wisdom, approved and signed the imprimatur, leaving some space blank to insert the book's title, which he regarded as essential.[12] Because of this small difficulty, and because I did not much like the quality of that press, and because I learned that Father Scheiner in the meantime had been sent to live in Nissa[13] in Silesia, I retook possession of the book and returned to Vienna to print it there, where I had my residence and so enjoyed greater convenience. But here the previous approval was not sufficient, and the new one could not be obtained without the [Jesuit] Fathers; so I exploited the friendship I have with a [Jesuit] Father who is a theologian and an important professor; he reviewed and approved the book himself[14] and also obtained the imprimatur of the rector of the university.[15] Thus I could have already begun the printing, except that Father Scheiner came again to Vienna to print a book of his that will soon appear; so, in order not to take any risk, I thought I should let him leave first, and I have heard that he will finish in a few weeks and will have to leave. In the meantime I received orders from His Majesty to come here to Prague, just as during the period of the events I have narrated I was sent to Styria and other provinces for months at a time in the service of His Majesty. Thinking that perhaps I would have to remain here for some time, I brought the book with me to print it here if need be; here the Lord Cardinal D'Harrach,[16] in accordance with my previous request, has offered me to use the press which he built for this university. However, the Lord Cardinal is not here at the moment; and I have been informed that in any case I would have to obtain a new review and approval here; and soon I must return to Vienna because of my work; so I am inclined to start the printing there immediately, if Your Most Excellent Lordship so desires and does not order me otherwise. I say this because the Most Serene Prince told me that I should not do it without a new order by Your Most Excellent Lordship, because you are having it printed elsewhere.[17]

Vienna, Olmütz, Vienna, Prague, Vienna, and on and on. Of course, we know that the book would not be printed by Pieroni in the Holy Roman

Empire but by the Elzeviers in Holland. Moreover, we can easily sympathize with Galileo's impatience at such delays and his preference to opt for the high reputation and professionalism of the Elzeviers. However, what I want to stress in this context is that an imperial official was apparently having little difficulty obtaining imprimaturs and borrowing printing presses to publish the book of a convicted (suspected) heretic; and that suggests that in the German-speaking lands of central Europe, state institutions did not feel bound to comply with the condemnation in Rome.

To be sure, Pieroni had to be careful not to cross Scheiner's path; but here Scheiner emerged as a self-interested party with an axe to grind. So, while his behavior was certainly indicative of human nature and of the larger picture, it did not reflect the reaction of the state apparatus. Interestingly enough, it did not even reflect the attitude of the Jesuit order, for one of the most significant details of this story is the fact that the Vienna imprimatur for the *Two New Sciences* was underwritten by the Jesuit Gualterus Paullus, professor of sacred theology and interim dean of the Imperial Jesuit College of Vienna.[18]

The response from Poland is interesting for a still different reason. Far from enforcing the condemnation of Galileo and the Copernican doctrine, the Polish king, Ladislaus IV, apparently tried to have Galileo pardoned (see chapter 3.3). Although there is no explicit evidence for this, we do have circumstantial evidence: Ladislaus decreed freedom of conscience for his subjects; he had many Italians at his court; and Galileo was defending the doctrine of a Pole.[19]

Moreover, there is some indirect evidence. Some of it is contained in Pieroni's correspondence. In August 1635, Pieroni tried to convince Galileo to dedicate the *Two New Sciences* to Ladislaus, saying, for example, that "the King of Poland has excellent taste, especially regarding such matters; he is not excessively scrupulous or fond of them [the Jesuits]; out of respect for him, Rome would not (I am sure) abhor or harm anything placed under his protection. He already holds Your Lordship in great esteem, and your reputation enables you to appeal to him, who does not know you personally, as much as you appeal to those who have known you for a long time."[20] And when in the same letter Pieroni commented on Galileo's condemnation, he stated he wanted to do everything he could to have him freed, adding: "I hope to be able to obtain any favor for you from the King of Poland; let me know if you want me to do it, and when and how, and I shall try; I expect him to be particularly effective and forceful in obtaining your liberation and other things you may desire."[21]

Another piece of indirect evidence is the letter the king wrote to Galileo in the spring of 1636.[22] The letter primarily asked for some good telescopic lenses, but it seems likely that the gesture was meant primarily to signal another message: that Ladislaus not only was undeterred by Galileo's condemnation and continued to hold him in high esteem, but that he was ready to help. Galileo lost no time in replying,[23] sending three (and not just one) pairs of telescopic lenses and relating a little about the trial and his current state; he did not explicitly ask the king to intercede, but the subtext was clear.

Ladislaus's efforts were unsuccessful for several reasons.[24] Galileo was increasingly arousing suspicion because of his contacts with Holland about the method for determining longitude. Protestants were beginning to exploit his condemnation in their anti-Catholic struggles. Ladislaus became implicated in the controversy over whether the Italian city of Urbino should belong to the Duchy of Parma or to the Papal States. The king had a disagreement with the pope about the appointment of a Polish cardinal that became so bitter as to lead to a break of diplomatic relations in 1642.

The situation adumbrated for the reactions by France, Spain, the Holy Roman Empire, and Poland becomes much more explicit and concrete for the Republic of Venice. There, the local inquisitor and the papal nuncio did their share to attempt to publicize Galileo's condemnation. But these efforts reflected primarily the attitude of officials whose allegiance was primarily to the Church and to Rome. A more reliable barometer of the state's reaction was the official theologian of the republic, Fra Fulgenzio Micanzio, a disciple of Paolo Sarpi and his successor in that position.

After the trial, Micanzio and Galileo exchanged frequent correspondence, with the friar doing the bulk of the writing. For the first few years after Galileo's return to Arcetri, Micanzio wrote him almost weekly. This pace slackened somewhat in later years, but in all Galileo received from Micanzio no fewer than 139 letters. The friar always offered words of encouragement and provided all kinds of assistance, from help in ensuring that one of Galileo's pensions (from a religious institution in Brescia) was paid, to assistance in exploring the publication of his new book in Venice, to acting as an intermediary in arranging for its publication by the Elzeviers in Holland.

This correspondence leaves no doubt that the Venetian Republic did not endorse and would not enforce the implications of Galileo's condemnation. For example, after learning that the Inquisition had ordered Galileo to stop petitioning for a pardon (in early spring 1634),[25] Micanzio observed: "I empathize with and intensely feel your fortunes and misfortunes. I do not think the house arrest is that bad, as least as compared to that barbaric injunction that makes imploring a crime. In any case, one must have courage and seize liberty oneself; after all, even a prisoner in irons tinkers with them to be more comfortable; one must enjoy whatever one can for the present and hope for the best. I am surprised that such an insignificant friar should execute the wishes of others against such an employee of his Prince. In other countries this certainly would not be done, or he would do it at his own risk."[26] The "insignificant friar" was the Florentine inquisitor, and Micanzio found it improper that he should have the power or authority to annoy with impunity the favorite subject of the Florentine ruler. Micanzio was also encouraging Galileo to disregard the terms of his sentence.

After Micanzio was told by the Venetian inquisitor (in early 1635) that there was a ban on the publication of any of Galileo's works, old or new,[27] Micanzio wrote to Galileo that the Republic planned to take action to bring such abuses under control: "I see that a remedy is being prepared so that the inquisitor here does not disturb the printing presses, in violation of the law and of the orders; this is in the interest of commerce, for he interferes too much and arbitrarily denies the imprimatur to works that do not concern religion in any way."[28]

And when in 1638 Galileo, in accordance with the Inquisition's orders, refused to accept the gift of the gold chain which the Dutch government had offered him,[29] Micanzio on three different occasions[30] tried to convince him to change his mind. On 17 September 1639, Micanzio wrote:

[104] In regard to that gift, I replied once before. I repeat now that absolutely I do not feel that you should refuse it at all; nor can I imagine any reason that could move you to do this. You are dealing with a state, and a great and powerful state, which would surely regard it as an affront; for they could not imagine it to be anything other than a reproach for their religion, which has nothing to do with it (and I would like the most scrupulous person in the world to tell me otherwise). Your Prince, the Most Serene Grand Duke (may God fill him with happiness, as I unceasingly pray) conducts commerce with them and receives their ships in his ports; [105] the Most Serene Republic, the Most Christian King,[31] and all states that are not at war with them exchange ambassadors with them; there is really no impediment. What is Your Lordship afraid of? It should be reassuring that they are a republic, which is not accustomed to secret deals because these are appropriate only when there is a single ruler. Thus religion is not a problem. In secular terms, what are your motives? The fact that, because of your ailments, you have not perfected the method? I do not accept that, because the point at which Your Lordship has arrived is such that that republic could not sufficiently reward you even by giving you a whole city. Nor should Your Lordship doubt that the minds of that nation will discover the instruments to enjoy the fruit of an invention that has made the greatest intellects sweat in vain and has been abandoned by them as a desperate or impossible undertaking.[32]

We know that Micanzio did not succeed in changing his revered friend's mind. However, his failure was certainly not due to lack of effort but probably to the fact that Galileo's piety, reverence, and zeal for the Church may have exceeded those of the friar; Galileo may not have shared Micanzio's view of the relationship between church and state. As the official theologian for the Venetian Republic, Micanzio, like his mentor Sarpi, believed in the separation of church and state.

In fact, Sarpi (who was dead by this time) had had occasion to comment explicitly on the anti-Copernican decree of 1616 when he was consulted on the matter by the Venetian government. In an opinion dated 7 May 1616,[33] Sarpi was willing to go along with the prohibition of the Copernican books; but on the legal and jurisdictional question, he wanted to uphold the supremacy of the state over the church stipulated by the treaty of 1596. Thus he recommended that to be valid, the anti-Copernican decree had to be reissued by the state council of heresy, as an attachment to its own decree of endorsement. To my knowledge, this procedure was not followed, and so one could question the legal validity in the Venetian Republic of the anti-Copernican decree of 1616. After the 1633 trial Micanzio, as the new official theologian, must have been thinking of Sarpi's opinion, besides other considerations, in responding to Galileo's condemnation as he did.

In conclusion, Catholic states in Europe by and large did not react favorably to Galileo's condemnation. They tended to see it as an abuse of power by Rome and did not cooperate in its enforcement. On the other hand, non-Catholic states did not need to make an official response, and so they did not.

4.2 "Philosophic Freedom": From Strasbourg (1635–1636) to London (1644–1661)

In January 1633, as he was about to leave Florence to go to Rome to stand trial, Galileo wrote to Elia Diodati, to whom he had long owed a letter. Among many other points, Galileo briefly related his troubles to Diodati, stressing two developments. The earlier one, about the Letter to Christina, was that "many years ago, at the beginning of the uproar against Copernicus, I wrote a very long essay showing, largely by means of the authority of the Fathers, how great an abuse it is to want to use Holy Scripture so much when dealing with questions about natural phenomena, and how it would be most advisable to prohibit the involvement of Scripture in such disputes."[34] The more recent development was that "I am about to go to Rome, summoned by the Holy Office, which has already suspended my *Dialogue*. From reliable sources I hear the Jesuit Fathers have managed to convince some very important persons that my book is execrable and more harmful to the Holy Church than the writings of Luther and Calvin."[35]

Acting on his own initiative, Diodati apparently lost no time in arranging for a Latin translation of the *Dialogue*. By the time the trial ended, he had almost completed the arrangements. In August 1633, Matthias Bernegger had agreed to do the translation, received a copy of the book, and begun working.[36] Bernegger was a Protestant born in Austria who had moved to Strasbourg in his youth.[37] Strasbourg was at that time a free city federated with the Holy Roman Empire.[38]There he studied languages and mathematics, became a professor of history and politics, and later was appointed rector of the university. In 1612, he had published a Latin translation, with extensive annotations, of Galileo's booklet on the proportional compass.

Diodati also enlisted the support of the one of the most reputable Protestant publishers, the Elzeviers of Holland. They found a good printer in Strasbourg, the Protestant David Hautt. The project had such a high priority that printing started before the translation had been finished. But that is not to say that the job was being carelessly done. On the contrary, as Bernegger proceeded, he often wrote Galileo and Diodati for advice on the translation of difficult words and phrases. Printing was finally completed in March 1635, and the book was offered for sale the following month.[39]

The book bore the title *The Cosmic System, a Discussion in Four Dialogues of the Two Chief World Systems, Ptolemaic and Copernican, and of Their Philosophical and Natural Arguments Proposed Indeterminately*.[40] Thus, although the initial term had changed (from *Dialogo* to *Systema Cosmicum*), the rest of the title conveyed the same information as the original: dialogue form, critical discussion, Ptolemy versus Copernicus, natural or philosophical and not theological or scriptural arguments, and a nonbinding proposal. In the Preface, Bernegger avoided any reference to Galileo's condemnation; he made it sound as if the book was being published without the author's permission or knowledge and attempted to justify this action. This story was told to avoid causing further problems to Galileo; there is, however, no doubt that Galileo participated indirectly and discreetly in the project.[41]

Although Diodati and Bernegger were careful not to implicate Galileo in any way, it was also clear that they knew what they were doing and what they wanted to accomplish. Their aim began to emerge on the title page, which exhibited two quotations. One was from Alcinous, in Greek, and means: "One must be mentally free if one wants to become a philosopher."[42] The other was a Latin sentence from Seneca that means: "It is especially among philosophers that one must have equal liberty."[43] The liberal intention was clear. But this was still window dressing, relatively speaking.

The title page also announced that the book had two appendixes discussing the role of Scripture. In fact, following the text of the *Systema Cosmicum*, there was a five-page selection from Kepler's introduction to his *Astronomia Nova* of 1609 and a Latin translation of Paolo A. Foscarini's *Letter* of 1615 (thirty pages long).[44] These appendixes were preceded by a one-page introduction in which Bernegger stated that Foscarini's *Letter* had been translated from the Italian by one David Lotaeus.[45] This was a pseudonym for Elia Diodati, whose name and role in the project are therefore nowhere revealed in the volume—another precaution to protect Galileo.

Diodati and Bernegger apparently gave considerable thought to the question of how to supplement the *Systema Cosmicum*. At one point, Bernegger thought of appending Tommaso Campanella's *Apologia pro Galilaeo*,[46] which had been written in 1616 in Rome and published in 1622 in Frankfurt. And they considered including Galileo's own Letter to Christina of 1615, which they eventually decided to publish separately.

The appendixes they finally included were intended to show that Copernicanism was compatible with Scripture. This claim was announced on the title page, two lines of which indicated that the book included "a double appendix in which the statements of Sacred Scripture are reconciled with the Earth's motion."[47] This purpose was also announced in the Preface, at the end of which Bernegger asserted, "The truth has won, and it will win more broadly provided one prevails in our cause over those Cleantheses who, deceived by thoughtless piety, think that this theory undermines Holy Scripture."[48]

In fact, Foscarini's *Letter* was the book condemned and prohibited by the anti-Copernican decree of 1616 for its main conclusion that the geokinetic thesis was compatible with Scripture. And in the Introduction to the *Astronomia Nova*, Kepler had argued in support of the same conclusion, although in a slightly different way.[49] His argument was that the scriptural passages that talk about the sun and the earth do not contradict the astronomical opinion of the earth's motion because the former speak in accordance with the order of appearances and common observation, whereas the latter speaks according to the order of reality; since speaking in accordance with common appearances is appropriate in many contexts, it is *not* correct to say that the earth's motion would "give the Lye to the Holy Ghost speaking in the Scriptures";[50] and the context of Scripture is the order of appearances because, for example, when the author of Genesis wrote about creation, "it was his intent to extol things known, and not to dive into hidden matters, but to invite men to contemplate the benefits that redound unto them from the works of each of these days."[51]

Since the main reason for condemning the Copernican doctrine in 1616, and a main reason for condemning Galileo in 1633, had been that Copernicanism was contrary to Scripture, the appendixes attempted to provide a vindication and justification of Galileo. Thus, the Protestants responsible for the project were publishing not only the work for which Galileo had been tried and condemned but also two additional writings arguing that the key intellectual basis for the verdict was untenable.

The story of the dissemination of the *Systema Cosmicum* provides some revealing details.[52] A relatively large number of copies were printed (six hundred by a conservative estimate), and they were quickly distributed, although this happened more by way of informal contact than commercial transaction. Many of these copies found their way into the hands of Catholic clergymen and institutions, thus showing the limitations of the 1633 condemnation of its author and the 1634 prohibition by the Index.

The Strasbourg apology for Galileo went further the following year, when the same group of Protestants published his Letter to Christina. In 1636, again in Strasbourg, and from the same printer (Hautt) and the same publisher (Elzevier), a small book appeared with this title: *Nov-antiqua Sanctissimorum Patrum, & Probatorum Theologorum Doctrina, de Sacrae Scripturae Testimoniis, in Conclusionibus Mere Naturalibus, Quae Sensata Experientia et Necessariis Demonstrationibus Evinci Possunt, Temere non Usurpandis.* The book contained the Italian text of the letter in the right column of each page, along with a Latin translation in the left column. The translation had been done by Diodati, who once again did not reveal himself; this time he used the name Robertus Robertinus Borussus, which was the name of a real person, Robert Roberthin, a poet from Königsberg.[53] There was a preface consisting of a five-page letter from Robertinus to Bernegger, and a shorter (one-page) letter from Bernegger to Robertinus.[54] And again there was an appendix, consisting of a four-page excerpt from Diego de Zúñiga's *Commentaries on Job*, suggesting a geokinetic interpretation of the biblical passage Job 9:6; this was the part of Zúñiga's book that had caused its suspension in the anti-Copernican decree of 1616.

The Latin title could be translated as *New and Old Doctrine of the Most Holy Fathers and Esteemed Theologians on Preventing the Reckless Use of the Testimony of the Sacred Scripture in Purely Natural Conclusions That Can Be Established by Sense Experience and Necessary Demonstrations.* The title conveyed several suggestions that are worth stressing. One was that the view Galileo was propounding was both old and new; new, but rooted in the tradition of the Church Fathers and the arguments of traditional theologians. Another was that Galileo was objecting to what he saw as a widespread abuse, namely using the words of Scripture to prove or disprove conclusions in astronomy. And the conclusions on which he focused were those that were *capable* of being conclusively proved by sense experience and necessary demonstrations, not merely those that had already been so conclusively proved; for in regard to the latter (e.g., the fact that the earth was spherical) there was no controversy, and nobody would dream of trying to overturn them on the basis of biblical texts.

This descriptive title did indeed correspond to the content of Galileo's essay. Its key thesis was that Scripture is not an authority on philosophical (astronomical) questions but only on questions of faith and morals. This principle would imply that those who advanced the scriptural argument against Copernicanism were committing a non sequitur or were reasoning irrelevantly; for they argued that heliocentrism must be rejected because it contradicts scriptural passages, but, given the principle, scriptural passages cannot be properly used to support astronomical conclusions. Galileo formulated his "new-old" principle in a memorable aphorism that he attributed to Cardinal Cesare Baronio: "The intention of the Holy Spirit is to teach us how one goes to heaven and not how heaven goes."[55]

Of course, Galileo could not simply state his principle and apply it against his opponents. The bulk of the *Nov-antiqua* consisted of arguments designed to justify it. They cannot be discussed here.[56] Instead here I want to stress that this booklet added a second dimension to the Strasbourg apology. The appendix to the *Systema Cosmicum* had stressed that the geokinetic hypothesis was not contrary to Scripture. The *Nov-antiqua* was stressing that Scriptural assertions were not relevant to astronomical investigation. The result was a double criticism of the biblical argument that Copernicanism is false because it contradicts Scripture: first, Copernicanism does not really contradict Scripture (see Kepler's and Foscarini's appendixes); second, even if it did, that finding could not be properly counted as evidence against Copernicanism (see the *Nov-antiqua*).[57]

Besides such a reinforcement of the criticism of the geostatic argument from Scripture, the "new-old" principle advocated in the *Nov-antiqua* also defended Galileo against one of the two suspected heresies attributed to him in the 1633 sentence. As we saw in chapter 1.1, while the verdict faulted him in part for holding (or tending to hold) the Copernican doctrine, it also faulted him for believing "that one may hold and defend as probable an opinion after it has been declared and defined contrary to the Holy Scripture."[58] At the trial Galileo could have hardly raised such issues unless he had wanted to commit suicide. In 1636, the Strasbourg editors enabled him to defend himself on the world stage.

To be sure, they again avoided implicating Galileo by continuing the fiction that the book was being published without his knowledge. In the Preface, Robertinus (Diodati) claimed that he was the one who had provided the manuscript, from a copy he had acquired during one of his trips to Italy. However, by contrast with the *Systema Cosmicum,* the Preface to *Nov-antiqua* is openly critical of Galileo's condemnation and explicitly praises his moral character. Robertinus called him the "new father of astronomy"; he blamed Galileo's troubles on jealous and hateful rivals and their maneuverings; and he commented on the importance of Galileo's Letter to Christina, claiming that it advanced sound advice on the topic and proved the depth of his Catholic piety and faithfulness. These remarks were taken essentially from the opening paragraphs of the Letter to Christina and from Galileo's letters to him (Diodati). Other references to the trial are cryptic and of questionable accuracy.[59] Isabelle Pantin does not exaggerate when she states that this Preface was meant to be "a conclusive document for Galileo's rehabilitation."[60]

A key aspect of the Strasbourg Galilean publications was to advance the "liberal"[61] cause that freedom is essential in philosophy, namely that freedom of thought, of speech, of inquiry, and of the press is extremely important in the search for truth. This liberal principle was echoed and also connected to Galileo by John Milton in 1644 in his *Areopagitica*. Milton was reacting against a censorship law enacted by the English Parliament in 1643, after a brief period of relative freedom that had followed the abolition of the royal Star Chamber in 1641.[62]

Milton gave several wide-ranging and nuanced arguments: that such censorship should be repealed because it stems from the Inquisition; because virtue and vice are interrelated, and so such censorship will not accomplish its aim of making people virtuous; because it will cause harm to learning; and because it is a violation of human liberty.[63] In the context of his argument about harm to learning, he gave some contemporary historical evidence by describing the situation in Italy and citing Galileo's condemnation as supporting evidence.[64] Milton's account was perhaps the first to suggest that the trial harmed philosophic and scientific progress, especially in Italy and more generally in Catholic countries. This was destined to become a much-discussed issue in the subsequent Galileo affair.

Milton spoke from direct experience, for he had visited continental Europe in 1638–1639, spending most of his time in Italy, including about two months in Tuscany in late summer 1638, and then a month in early spring 1639. In fact, in the same passage of the *Areopagitica,* he stated that "there it was that I found and visited the famous Galileo grown old, a prisoner to the Inquisition, for thinking in astronomy otherwise than the Franciscan and Dominican licensers thought."[65] Although there is no other direct evidence and the question is a much-discussed topic among Milton scholars,[66] this sentence immortalized Galileo's image as a symbol of the struggle between individual freedom and institutional authority.

An important event in the original liberal reception of Galileo's condemnation occurred in 1649.[67] A friend of Pierre Gassendi (Michel Neuré, the pseudonym of Laurent Mèsme) published in Lyons, without Gassendi's knowledge, a work by Gassendi entitled *Apologia in Io. Bap. Morini Librum, Cui Titulus, Alae Telluris Fractae, Epistola IV de Motu Impresso a Motore Traslato.* As this title indicates, this book was Gassendi's defense against the criticism advanced in 1643 in Jean Baptiste Morin's *Alae Telluris Fractae,* which in turn was a critique of the book published by Gassendi in 1642 entitled *De Motu Impresso a Motore Traslato;* and the latter work was for its part a defense of Copernicus and Galileo from the criticism advanced in Morin's *Responsio pro Telluris Quiete* (1634) and *Famosi et Antiqui Problamatis de Telluris Motu* (1631). However, in order not to further inflame their dispute, Gassendi in his 1642 work had not mentioned Morin's name and later decided not to publish his rebuttal to Morin's 1643 book. Neuré may have wanted to help his friend's reputation, although Morin was not silenced and replied in 1650 with a *Résponse à une Longue Lettre de Monsieur Gassend.*[68]

However, Neuré may have had ulterior motives, for in Gassendi's *Apologia* he included an appendix containing four items of Galilean "apology": a reprint of the Latin text of Galileo's Letter to Christina and of the excerpt of Zúñiga's *Commentaries on Job* from the 1636 Strasbourg edition of the *Novantiqua;* and Galileo's Letter to Benedetto Castelli of 21 December 1613 and Letter to Monsignor Dini of 23 March 1615, both published for the first time, also in Latin translation.[69] Galileo's Letter to Castelli was the writing that had started the earlier Inquisition proceedings against Galileo and had been explicitly mentioned in the 1633 sentence. It may be regarded as a first draft of the Letter to Christina; but it is really more significant than that because it is much shorter (seven pages as compared to thirty-nine),[70] and its greater succinctness makes it more incisive and enhances the persuasive power of the main argument.

The Letter to Dini was one Galileo wrote from Florence after he learned that in early 1615 he had been denounced to the Inquisition in Rome. Piero Dini (1570–1625) was a minor Vatican official then living in Rome with his uncle, Cardinal Ottavio Bandini. Dini was so well connected that he had been able to talk to several important persons, such as Cardinal Bellarmine and the Jesuit Father Christopher Grienberger, a professor of mathematics at the Roman College. When Dini reported to Galileo some of their concerns,[71] Galileo tried to address them in a letter of 23 March. In this letter Galileo did primarily two things: he criticized the instrumentalist interpretation of Copernicanism and explained and defended a realist construal; and he elaborated a Copernican interpretation of the passage in Psalms 19:1–6,[72] which according to Dini had been explicitly mentioned by Bellarmine as problematic.

Thus, Neuré's appendix to Gassendi's *Apologia* was not only a reaffirmation of Galileo's own earlier Strasbourg "apology" and "liberal" ideology but also an amplification of it. Moreover, since the context was that of Catholic France, Neuré's effort shows that the liberal reception of Galileo's condemnation was not just a Protestant phenomenon; it reflected not only the struggle between Protestants and Catholics but a wider and perhaps more fundamental development, although the liberal movement was probably stronger in the Protestant than in the Catholic world.

Finally, the original liberal reaction to Galileo's condemnation reached a kind of climax in 1661 with the English translation and publication in London of the first volume of Thomas Salusbury's *Mathematical Collections and Translations*. This volume included full translations of all the Strasbourg texts: the *Dialogue*, the Letter to Christina, Kepler's introduction to the *Astronomia Nova*, Zúñiga's commentary on Job 9:6, and Foscarini's *Letter*.[73] It also had a foreword that was openly critical of the Catholic Church and supportive of Galileo; it left little doubt that Salusbury could be "counted happy to be born in such a place of philosophic freedom, as they supposed England was,"[74] to quote Milton's *Areopagitica*.

In this foreword,[75] Salusbury described the *Dialogue* as a "singular and unimitable piece of reason and demonstration"; this judgment had constantly gained currency in the thirty years since its original publication. Less universal, but widely shared, was Salusbury's judgment that the work was objective, balanced, and unbiased: it had been "with all the veneration valued, read & applauded by the iudicious yet . . . with much detestation persecuted, suppressed & exploded by the superstitious"; and the author had been "indefinite in his discourse." Most important, Salusbury was the first one to formulate publicly the thesis that the root cause of the trial was Pope Urban's personal anger about being caricatured by the character of Simplicio: "His Holiness thereupon conceived an implacable displeasure against our author, and thinking no other revenge sufficient, he employed his apostolical authority . . . to condemn him and proscribe his book as heretical; prostituting the censure of the Church to his private revenge." As we saw in chapter 3.4, although this allegation is plausible, there is no documentary evidence that it played a role in the trial, as it first emerged in a document two years after the trial;[76] now, after three decades, it was making its first appearance in print. And Salusbury concluded ironically, "I shall not presume to censure the censure which the Church of Rome past upon this doctrine and its assertors"; but for the sake of pious souls he was including several appendixes that reconciled that doctrine with Scripture.

4.3 Illegitimate Births, Burials, and Books: Various Retrials to Riccioli's Apology (1651)

For about twenty years, until about 1651, the condemnation of Galileo led to an expansion and intensification of the Copernican controversy, as shown by the growth of books on the topic. Of course, much of this work generated more heat than light and was the result of the momentum which the controversy had already possessed before the trial, but it would be cynical and incorrect to deny that the condemnation was a factor and that some progress was made as a result.

In fact, at least sixty books were written between 1633 and 1651,[77] although a few remained unpublished then.[78] The debate focused on the question of the earth's motion, perhaps because this thesis was the key distinguishing characteristic between the Copernican and Tychonic systems (that is, from the point of view of the geokinetic issue, the Ptolemaic and Tychonic systems were indistinguishable, and the real innovation was Copernicus), but also because the question of the earth's motion was connected with the question of the laws of motion in general, which had great and independent importance. However, some works also focused on the methodological and theological question of whether Scripture ought to be regarded as an authority in astronomy and natural philosophy.[79] The great majority of these works were in Latin, but a few appeared in the vernacular (Italian, English, and French).[80] Although most such works came from Catholic countries or authors, many did not;[81] England seemed the most active among the Protestant nations. Although most participants are obscure, many are important philosophical, scientific, or cultural figures: they include Tommaso Campanella, Bonaventura Cavalieri, Pierre Gassendi, Thomas Hobbes, Marin Mersenne, Giovanni Battista Riccioli, Evangelista Torricelli, John Wilkins, and Descartes, who belongs in a class by himself.

Despite some scientific progress and value, a major strand of this literature (on the anti-Galilean side) was to criticize the ideas and arguments in the *Dialogue* and thus to discredit the book and the author that had been condemned by the Inquisition and consequently to vindicate the condemnation. Curiously enough, however, the reverse argument was also occasionally advanced; that is, some authors attempted to discredit the book by taking at face value the author's abjuration. A good example is provided by a passage in Alexander Ross's *The New Planet No Planet, or the Earth No Wandring Star Except in the Wandring Heads of Galileans* (1646), which was a rebuttal to John Wilkins's *A Discourse concerning a New Planet, Tending to Prove that 'Tis Probable Our Earth Is One of the Planets* (1640). Ross was taking issue with some remarks made by Wilkins about the number of thinkers who were followers of Copernicus: "And yet of these five you muster up for your defense, there was one, even the chiefest, and of longest experience, to wit, *Galileus,* who fell off from you; being both ashamed, and sorry that he had been so long bewitched with so ridiculous an opinion."[82]

One of the most relevant and important issues of this period was the correctness, interpretation, status, and implications of Galileo's science of motion. It is important because, after Galileo's *Dialogue* (1632) and *Two New Sciences* (1638), it was becoming increasingly obvious that Copernicanism and the earth's motion required a new physics, and the outlines of such a physics needed clarification. And this controversy is very pertinent to our theme because, as Paolo Galluzzi has stated, the dispute may be aptly called "*l'affaire Galilée* of the laws of motion"[83] and a "second 'trial' "[84] of Galileo.[85]

The discussion started when Gassendi's *De Motu Impresso* (1642) elaborated the connection between the new Galilean physics of motion and Copernican cosmology, strengthening each. The focal point was Mersenne, whose correspondence and contacts facilitated and encouraged discussion; he started out relatively pro-Galilean but ended up on the other side, concluding that no science of motion was possible without knowledge of causes and encouraging an instrumentalist attitude toward Galileo's laws of motion. The anti-Galilean critics were: Morin, who stressed the heretical implications of Galileo's ideas; the Jesuit Pierre Le Cazre, who questioned the causal explanation and metaphysical basis of Galileo's laws and replaced the Galilean law of odd numbers (that the distances traversed by a falling body in successive equal times increase as the odd numbers from unity) with one of constantly doubling geometrical progression (1, 2, 4, 8, 16, etc.); the Jesuit Honoré Fabri, who was also concerned with causal explanation and metaphysical foundations and replaced the law of odd numbers with one involving a progression of natural numbers (1, 2, 3, 4, 5, etc.); a disciple of Fabri named P. Mousnier; a Genoese Jesuit named Niccolò Cabeo; and Giovanni Baliani, who revised the first edition of his relatively pro-Galilean *De Motu* (1638) into a relatively anti-Galilean edition in 1646, one which incorporated Fabri's formula and thus involuntarily fell victim to the Jesuits' pressure. The main pro-Galilean exponents were Gassendi himself, who however withdrew from the debate after his *De Proportione Qua Gravia Decidentia Accelerantur* of 1646; L. A. Le Tenneur, a Toulouse senator, who provided cogent replies to Fabri's criticism; to a lesser extent Evangelista Torricelli, who, like the other Italian Galileans, was suspicious of such criticism and hesitated to get involved in what they saw as a continuation of the Church's persecution of Galileo; and Christiaan Huygens, whose analysis impressed but did not convince Mersenne.

From the number of Jesuits among Galileo's critics, one might form the impression that this was another instance of Jesuit conspiracy and persecution. It was, however, the pro-Galilean Gassendi who started the discussion in 1642, and so perhaps the Jesuits merely seized the opportunity. At any rate, the Galilean laws of motion (though not Galileo himself, who had just died) survived this attack, and indeed emerged triumphant. Thus, by the time this controversy petered out six years later, everyone knew that the Galileo affair was not going to go away with his death, any more than the issues had gone away with his condemnation in 1633.

If this was the first major posthumous retrial of Galileo, there were several other minor ones that immediately followed his death. He died at Arcetri on 8 January 1642.[86] The following day he was quietly buried at the church of Santa Croce in Florence, not in the family tomb within the church proper but in a grave without decoration or inscription, located in an out-of-the-way room behind the sacristy and under the bell tower, next to the chapel dedicated to Saints Cosma and Damian.[87] There was considerable interest in Florence to build a sumptuous mausoleum for him in the same church; in fact, Vincenzio Viviani was quickly able to raise three thousand scudi (in pledges) from the Florentine intelligentsia and elites for the project.[88] To be sure, Galileo's enemies raised the question of whether it was permissible in canon law to erect an honorary mausoleum for a man condemned of vehement suspicion of heresy; but a formal legal opinion was written, concluding that it was not illegal.[89] At any rate, within days of Galileo's death, the Florentine nuncio informed Rome that Galileo was dead and that the grand duke was planning to build a mausoleum for him in the church of Santa Croce.[90] Thus, on 25 January 1642 the pope and the Inquisition let the grand duke know that it would not be pious or proper to build a mausoleum for a condemned heretic like Galileo.[91] The grand duke acquiesced,[92] and as we shall see later (in chapter 6.2), it would be about a century before the mausoleum project was accomplished.

Another minor posthumous skirmish involved Galileo's last will and testament. His enemies raised the question of whether someone convicted of suspected heresy had the right to have his will executed. Again, a formal legal opinion concluded in his favor.[93]

Next, in 1643 the legitimacy of Galileo's birth was impugned. In a biographical dictionary titled *Pinacotheca Imaginum Illustrium Virorum,* Janus Nicius Erythraeus (also known as Giovanni Vittorio de' Rossi) stated that Galileo's parents were not formally married at the time of his birth or conception![94] Viviani refuted the claim in his biography of Galileo, written in 1654 but not published until 1717.[95] But the misinformation began to circulate widely, and was repeated even by a careful scholar such as Johann J. Brucker in the first edition of his monumental history of philosophy.[96] However, he corrected himself in the second edition, and noted the error.[97] Others were not so careful; and even someone as well disposed toward Galileo as Jean D'Alembert repeated it in 1751 in his entry on astronomy in the French *Encyclopedia.*[98]

It is amusing to see how angry a Florentine could become when he heard the myth of Galileo's illegitimate birth repeated. This is what Giovanni Battista Clemente Nelli had to say about D'Alembert's repetition of the myth: "I judge the celebrated author just mentioned on the one hand inexcusable, but on the other hand pathetic. For since he was a profound philosopher and mathematician and well versed in all types of literature and erudition, he should have read the works of the Florentine philosopher and consequently his biography found in the Preface and written by his illustrious disciple Vincenzio Viviani;[99] there he conclusively proves the legitimate birth of his master. Then I judge the case pathetic because it seems to me that, if Mr. D'Alembert knew of the legitimate birth of Galileo, he may have wanted to assert the contrary in order to have him similar to himself in birth as well as in doctrine."[100] In a footnote Nelli added: "See the book entitled *L'Observateur Anglois,* vol. 3, p. 119, where the illegitimate birth of Mr. D'Alembert is discussed."[101] Amusement, gossip, and salaciousness aside, the myth of Galileo's illegitimate birth, however minor in itself, is important insofar as it is a microcosm of the whole Galileo affair.

Moving on to matters of greater weight, we now come to the two enormous folio volumes of Jesuit Giovanni Battista Riccioli's *Almagestum Novum* (1651), which may be regarded as the climax of the original reception of Galileo's condemnation. There is no denying the book's ambitious and encyclopedic scope. The reader is impressed, indeed overwhelmed: 764 pages in volume 1, 676 in volume 2, and 66 pages of front matter. The material was subdivided into ten "books": spherical astronomy, terrestrial elements, sun, moon, eclipses, fixed stars, planets, comets and new stars, systems of the world, and general problems. Book 9, on the systems of the world, was by far the longest and covered 342 pages.[102]

The front matter included a chronological list of persons who had contributed to astronomy, beginning in 1990 B.C. (with Zoroaster) and ending with the year 1651 (for which about a dozen contributors are listed).[103]

For 1543, there was an entry saying that Copernicus died at the age of seventy-one. The only mention of Galileo in this chronology was for the year 1633, and it said: "Galileo is forced to abjure the opinion of the earth's motion."[104]

There followed an alphabetical glossary of names, with a paragraph each about the individual's life and work. The entry on Galileo read as follows: "Galileo Galilei, a Florentine of competent mind in geometry, and very well known in astronomy, and even more for attempting to promote the opinion of the earth's motion as a mere hypothesis in his *Dialogues on the System of the World,* while at the same time he had been warned by Cardinal Bellarmine to obey the holy censures; nevertheless, subsequently, he had to publicly abjure that opinion. To him we owe, if not the invention, certainly the improvement and use of the telescope, and the discovery by its means of many things that were invisible to the ancients. He wrote on sunspots and on other things detected with the telescope in *The Sidereal Messenger* and other works."[105] The entries on other major astronomers were much longer.

The crucial book 9 was subdivided into five sections of many chapters each. Section 1 dealt with the substance and accidents of heavenly bodies. Section 2 treated of the heavenly movers and motions. Section 5 discussed the "harmonic" system of the world, including aspects of Kepler's. Section 3 was on the geostatic systems of the world, such as those of Eudoxus, Ptolemy, and Tycho. Riccioli preferred the latter, according to which the earth stands still at the center of the universe, and the planets revolve around the sun, while the sun moves diurnally and annually around the motionless earth; but he modified the Tychonic systems somewhat, making only Mercury, Venus, and Mars revolve around the sun, but Jupiter and Saturn revolve around the earth.[106]

Section 4 (of book 9, in volume 2) treated of the system of the moving earth. It contained forty chapters and covered 210 pages. Riccioli examined scores of astronomical, physical, observational, and philosophical arguments both in favor of and against the earth's motion. By one count, no fewer than forty pro-Copernican arguments were stated and criticized, and no fewer than seventy-seven anti-Copernican arguments were presented and developed.[107] As a result of such a critical examination, Riccioli had no doubt that the earth was motionless.[108]

In chapters 36–39 of section 4, Riccioli took up the role of Scripture and elaborated explicitly and systematically a very conservative version of biblical literalism.[109] According to Riccioli, the literal meaning of biblical statements is physically true and scientifically (philosophically) correct, and it takes precedence in cases of conflict with doctrines in natural philosophy.[110] One interesting consideration he advanced is the following holistic argument: "If the liberty taken by the Copernicans to interpret scriptural texts and to elude ecclesiastic decrees is tolerated, then one would have to fear that it would not be limited to astronomy and natural philosophy and that it could extend to the most holy dogmas; thus it is important to maintain the rule of interpreting all sacred texts in their literal sense."[111] He also examined many particular biblical passages to show they accorded with the geostatic system.

The upshot of Riccioli's monumental effort, as regards Galileo's trial, was to provide the first explicit justification of his condemnation. The justification was twofold. On the astronomical issue, according to Riccioli, Galileo was wrong because the evidence favored the (modified) Tychonic rather than the Copernican system. Galileo was also (allegedly) wrong on the key methodological and theological issue: Scripture is and must be regarded as an authority in astronomy and natural philosophy, and not merely for questions of faith and morals. The Inquisition thus acted justly and prudently insofar as it was upholding this principle and attempting to prevent the disregard of Scripture and of ecclesiastic decrees from spreading beyond astronomical subjects.

Riccioli's apology for the Inquisition not only was implicit in what he said and did in section 4 of book 9 of the *Almagestum Novum* but corresponded to an explicit intention on his part. In fact, he concluded this section of his book with the confession that he had wanted to give "an *apologia* for the Sacred Congregation of the Cardinals who officially pronounced these condemnations, not so much because I thought such great height and eminence needed this at my hands but especially in behalf of Catholics; also out of love of truth to which every non-Catholic, even, should be persuaded; and from a certain notable zeal and eagerness for the preservation of the Sacred Scriptures intact and unimpaired; and lastly because of that reverence and devotion which I owe from my particular position toward the Holy, Catholic and Apostolic Church."[112]

Finally, let us recall (from chapter 2) that Riccioli performed an invaluable service by including a small collection of documents about Galileo's trial that for at least two centuries constituted the single best and most authoritative source. They were given on the last four pages of section 4, but the folio size of the book enabled him to squeeze into those four pages the (second part of the) anti-Copernican decree of the Index of 1616, the 1620 Index's warning on Copernicus's *Revolutions*, the Inquisition's sentence condemning Galileo in 1633, Galileo's abjuration, and Cardinal Barberini's letter to nuncios and inquisitors of 2 July 1633.

In conclusion, the twenty-eight years from Galileo's condemnation to the first volume of Salusbury's *Mathematical Collections and Translations* embody a second wave of reactions to Galileo's condemnation (besides the first wave of emblematic individual reactions that go from the condemnation to Galileo's death and that were discussed in the last chapter). In this second wave, the reactions may be regarded as more societal, because of the number of persons involved, and relatively polarizing, because of the way the issues were formulated. Catholic states had to decide whether to ratify and enforce this verdict and the earlier anti-Copernican one on which it was largely based; and, generally speaking, they did not. Protestants and progressive and liberal-minded Catholics came to Galileo's defense and started using his arguments and image in the struggle for individual freedom. Official Catholic institutions and traditional-minded Catholics, especially the Jesuits, attempted to delegitimize his scientific work and to justify the condemnation; the attempt at delegitimization quickly backfired, while the apologetic effort started a series of virtual retrials that were to continue for a long time. Practitioners of astronomy, mathematics, and natural philosophy became polarized into pro-Galilean and anti-Galilean camps, but, despite the polarization, progress was made that might have taken longer without the motivation of wanting to understand the intellectual issues underlying the condemnation.

By the end of this period, several key issues had been formulated. There was the philosophical or scientific question of the reality or physical truth of the earth's motion. There was the methodological or theological issue of whether Scripture is an authority in natural philosophy; this contained the seeds of the science-versus-religion conflict in the form of a conflict between those who affirmed and those who denied this principle, namely between those who affirmed and those who denied a conflict between Scripture and nature (natural philosophy). There was the methodological or epistemological question of instrumentalism versus realism, that is, whether physical theories in general and the geokinetic one in particular were descriptions of physical reality or merely instruments for saving the appearances; this question was connected with the fact that there seemed to be no prohibition on the "hypothetical" treatment of the earth's motion, and yet it was not clear whether a hypothesis meant an instrument of calculation and prediction or a description likely to be true but not yet demonstrated conclusively. And there were several questions of a historical nature that were pregnant with implications in other areas: the role of the Jesuits in Galileo's trial; whether Copernicanism had been declared heretical or merely contrary to Scripture; whether Galileo had been condemned for defending or for discussing the earth's motion, or for both; and whether the Simplicio character in Galileo's *Dialogue* was a caricature of Pope Urban VIII and whether this was the root cause of the condemnation.

Chapter 5

Compromises

Viviani, Auzout, Leibniz (1654–1704)

In this chapter I examine what might be called a third wave of reactions to Galileo's trial, covering the period between 1654 and 1704 and most significantly represented by the figures of Vincenzio Viviani (1622–1703), Adrien Auzout (1622–1691), and Gottfried W. Leibniz (1646–1716). In calling this the third wave, I am taking the first to have been the series of responses corresponding approximately to the rest of Galileo's life (1633–1642) and most importantly exemplified by Descartes, Galileo's daughter, Peiresc, and Galileo himself; this was treated in chapter 3. And I take the second wave to have been the reception during the period from 1633 to 1661, consisting of the secularist responses by nation-states and political powers; of the liberal responses centered in the free city of Strasbourg and in London; and of the conservative and traditionalist reactions culminating with Riccioli's *Almagestum Novum;* this was discussed in chapter 4. Generally speaking, we might say that the first wave produced reactions whose immediacy and authenticity give them emblematic and universal (besides, of course, historical) significance. The second wave generated a polarization of the two sides: a progressive, liberal, pro-Galilean, and anticlerical side holding that Galileo was right not only scientifically (philosophically) but also theologically; and a conservative, authoritarian, anti-Galilean, and proclerical side claiming that he was wrong not only theologically or religiously, but astronomically and physically as well. The third wave is characterized mostly by attempts at compromise. As long as we do not take such dates and themes too strictly or pedantically, and allow for overlapping, qualifications, and exceptions, they can serve as useful guideposts in our story.

5.1 Galileo "Human Not Divine":
Vincenzio Viviani (1654–1693)

In January 1639, Vincenzio Viviani began studying with Galileo and assisting him by reading to him and taking dictation, since Galileo had become completely blind; for this service Viviani received a modest salary from Grand Duke Ferdinand II.[1] The duke had been impressed by Viviani's mathematical talent when he was presented at court the year before. The teenager soon impressed Galileo himself when he mustered the courage to ask the master for a clearer confirmation of one of the basic unproved postulates of the *Two New Sciences:* that is, the principle that two bodies falling along inclined planes of different inclinations acquire the same terminal velocity as long as their elevations are the same. Galileo was sleepless for the next few nights trying to think of a demonstration, and when he found one, he communicated it to the young man and asked him to write it down. In the summer of that year, Viviani started living in Galileo's house and continued assisting him until his death.

Viviani went on to acquire the reputation of being one of the leading mathematicians of the age. He was appointed mathematician to the grand duke when Galileo's immediate successor in that position (Torricelli) died in 1647. He was invited, but declined, to serve in a similar capacity for the king of Poland and the king of France. The latter nevertheless gave him a sizable pension and made him a foreign associate of the Paris Academy of Sciences.

However, Viviani's favorite self-image was that of "the last disciple of Galileo." Indeed, he dedicated his life to preserving and disseminating the legacy of his master. He collected as much material as he could regarding Galileo's life and work; we owe to him the preservation of Galileo's papers and of some correspondence. He planned an annotated edition of Galileo's collected works, but this never came to fruition. He attempted to have the condemnation of Galileo and of his *Dialogue* repealed, but all he could accomplish in this regard was to display on the facade of his own house in Florence a lengthy inscription praising his master. He wrote the first biography of Galileo, although it was a pale shadow of the ambitious work he had planned on the topic and was not even published during his lifetime. Viviani's last will and testament stipulated that he should be buried next to Galileo and that his heirs should use the resources of the estate to build a mausoleum for Galileo as soon as conditions allowed.

Viviani's historical account of Galileo's life was written in 1654 in the form of a letter addressed to prince Leopold de' Medici, but it was not published until 1717.[2] It is a relatively short essay, about thirty pages, and it has all the historical value and the scholarly limitations one might expect from a biography written by one's last disciple.[3] Viviani's account focused on

Galileo's work in astronomy and physics and on his personality. The trial is discussed only in the following paragraph:

> For his other admirable speculations Galileo had been raised to heaven with immortal fame, and for his many discoveries he had been regarded by men as a god; thus, the Eternal Providence allowed him to prove himself human by letting him commit an error when, in discussing the two systems, he showed himself more inclined to believe the Copernican hypothesis, which had been condemned by the Church as incompatible with Divine Scripture. Because of this, after the publication of his *Dialogue* Galileo was called to Rome by the Congregation of the Holy Office. Having arrived there around 10 February 1633, through the great generosity of that Tribunal and of the Sovereign Pontiff Urban VIII (who already knew him as highly meritorious in the republic of letters), he was kept under arrest at the residence of the Tuscan ambassador in the delightful palace of Trinità dei Monti. Having been shown his error, he quickly retracted this opinion like a true Catholic. As a punishment his *Dialogue* was banned. After this five-month detention in Rome (while the city of Florence was infected with the plague), he was generously assigned for house arrest the residence of Monsignor Archbishop Piccolomini, who was the dearest and most esteemed friend he had in the city of Siena. He enjoyed the latter's highly cordial conversation with so much ease and emotional satisfaction that he resumed his studies and discovered and demonstrated most of his mechanical conclusions on the resistance of solids, among other speculations. After about five months, when the plague in his homeland had completely ceased, at the beginning of December 1633 His Holiness commuted his house arrest from the restriction of that residence to the freedom of the countryside, which he so much enjoyed. So he returned to his villa in Arcetri, where he had been living already most of the time on account of the healthy air and the great accessibility to the city of Florence, and where consequently he could easily receive visits by friends and relatives, which always brought him great comfort and consolation.[4]

It is difficult to take Viviani's account of Galileo's trial at face value; and the fact that he was brief and cryptic suggests that he really did not want to talk about it. The account seems to have a rhetorical and mythological quality. Viviani certainly did not want to give a negative portrayal of his master; but apparently he also did not want to, or was in no position to, criticize the Church. We might categorize his account as both pro-Galilean and proclerical and thus contrast it to both the Strasbourg account (which might be labeled pro-Galilean and anticlerical) and Riccioli's account (which might be regarded as anti-Galilean and proclerical). At the same time, Viviani tried to derive some useful, positive lesson from the tragedy, and he found it in the providential reminder of human fallibility. Thus we might classify his account as constructive and appreciative. Nevertheless, Viviani seemed to accept the incompatibility between Copernicanism and Scripture, although it is unclear whether this was lip service or reflected his real thinking. This issue is further clouded by evidence that in the same year that he wrote his biographical essay, he requested but was denied a waiver that would have allowed him to read the banned *Dialogue*.[5]

At any rate, the writing of the biographical essay at the request of Prince Leopold was only Viviani's first explicit encounter with the Galileo affair. Besides his own internal motivation, he was the natural person to receive requests for information and enlightenment on the matter. For example, in 1662 Robert Southwell wrote to Viviani expressing encouragement and support for his project of publishing Galileo's unpublished writings and his biography; Southwell stated that "the world is eager to know the truth about the affair."[6]

Viviani's next explicit encounter occurred in 1678 in correspondence with the Jesuit Antonio Baldigiani, professor of mathematics at the Roman College.[7] On May 26, Baldigiani wrote to Viviani saying that his colleague and confrere Athanasius Kircher was about to finish a book on Tuscany to be titled *Iter Hetruscum* in which he was planning to include biographical sketches of various great Tuscans, including Galileo, Torricelli, and Viviani himself; he asked for more information, especially a complete list of their works.[8] There followed an exchange of nine letters in which Viviani tried to be cooperative, although he was also somewhat embarrassed concerning his own inclusion. The final results are unknown because Kircher's book apparently was never published and the manuscript was lost.

With regard to Galileo's biographical entry, on 14 June 1678 Viviani commented on the proposed sketch.[9] Viviani stated that Galileo's title should be added after his name, namely "Florentine Patrician and Chief Philosopher and Mathematician" to Grand Dukes Cosimo II and Ferdinand II. He suggested that some of Galileo's particular discoveries should be mentioned, especially the astronomical ones. And he asked for the deletion of a passage beginning with the words "who, if he had been more cautious in some things . . ."[10] Viviani did not explain his reasons for this deletion, but said only "in part for reasons involving all of you gentlemen,"[11] which was a reference to the Jesuits.

We do not have the full text of that objectionable passage, but it did apparently contain a criticism of Galileo as reckless, imprudent, incautious, or careless. This evaluation was similar to the one advanced in 1616 in a report by Tuscan ambassador Piero Guicciardini to Grand Duke Cosimo II, but it was independently formulated, since Guicciardini's report was not published until 1773.[12] The criticism was, however, destined to become widely debated in the subsequent Galileo affair. We can get a better glimpse of Baldigiani's point from his comments in his letter to Viviani of 18 July 1678.[13]

Baldigiani claimed that he had been instrumental in convincing Kircher to revise his original passage by eliminating the more offensive remarks and leaving only the statement in question. Baldigiani argued that the assertion was really a favorable judgment because it suggested that Galileo would not have been condemned if he had adopted a more cautious manner of expression, and hence that his crime was not really that of heresy. "This means transferring the trial from the criminal to the civil domain,"[14] he added. Moreover, one could not very well say, especially in Rome, that Galileo "was completely innocent; that an entire Congregation erred; and that the holiest of tribunals was unjust."[15] Even if one believed this, one could not say it. It was better to say that Galileo gave some reason to be condemned insofar as he should have been more prudent. With this argument, Baldigiani seemed to be willing to attribute some fault to Galileo in order to exonerate him from the allegedly more serious charge of heresy.

Although the correspondence continued, there was no more discussion of Galileo's trial at that time. Instead the exchange focused on the content of the biographical entry for Viviani himself.

Viviani's pro-Galilean efforts also continued, but we do not have any explicit and concrete evidence of them until December 1689, when Leibniz visited Florence. He discussed with Viviani what could be done to lead the Church to repeal or ease its anti-Copernican and anti-Galilean censures; he encouraged Viviani to do something during his forthcoming visit to Rome, which concerned official business involving water management issues at the border between Tuscany and the Papal States;[16] and he may have asked Viviani to deliver to Baldigiani his latest version of a plea why the anti-Copernican censures should be repealed.[17] Moreover, at about the same time Baldigiani was appointed consultant to both the Inquisition and the Index.[18] Thus on 22 August 1690 Viviani finally decided to act: after first sending Baldigiani an exploratory letter to arrange for a confidential method of correspondence, Viviani sent him the following plea:

[153] I render infinite thanks to Your Most Reverend Paternity for being agreeable to letting me reveal to you the feelings of that Lord, and for the means which you tell me I should use in sending my letters without risks and which I shall use exactly as you prescribe them. Your Most Reverend Paternity should know then that it is desired most ardently that Galileo's *Dialogue on the Two Chief World Systems* be freed from any prohibition; it is believed that this could be obtained with the correction of those passages which the Sacred Congregation of the Index will deem to need it; one thinks that they will be very few and such that they can be revised with great ease without counterfeiting the work and without removing or decreasing whatever it contains that is beautiful or good; that is how Copernicus's book *On the Revolutions of the Heavenly Spheres* was corrected many years after it had been printed and dedicated to the Supreme Pontiff Paul III, even though the book from beginning to end has primarily the aim of systematically supporting the stability of the Sun and the triple motion of the Earth; on the contrary, Galileo claims explicitly in several places, in the preface and in the body of the work, to be proposing indeterminately the reasons on one side

and on the other and to show them incapable of proving one hypothesis more than the other.

In order to bring such an undertaking to a good ending, one thinks it is necessary to have someone who is authoritative, worthy, learned, and expert on the subject, and who would promote it and apply himself to it with effectiveness and persuasiveness. Your Most Reverend Paternity is considered today to be the only person in the world of scholarship to possess these qualities to the highest degree, to the exclusion of anyone else who should pretend to have them; thus I must beg you to reflect on this proposal, embrace such a commission, and seriously apply yourself [154] to its conclusion. If this succeeds, as one hopes, besides the joy that would undoubtedly come to the person who is so concerned about it, you should consider this: when the world learns of the role you played, it will know how false is the common belief that that prohibition was stimulated by some persons of your order, and it will also know that there are still some who retain a favorable attitude toward that author and his works; and when it becomes generally known that any one of you had more than enough right reasons for an action that after all is so Christian, heroic, and worthy of Your Paternity, consider how much praise would accrue to the same order, which in every other regard is rightly venerated above the others.

Here we are dealing with a book that was not published in a clandestine manner, but first licensed in Rome by the Master of the Sacred Palace appointed by Pope Urban, and then with the permission of other officials printed in Florence; a book written by a Catholic and pious author; dedicated to a most religious grand duke of Tuscany; treating not of doctrines that are clearly against the faith, but of subjects that are merely philosophical, mathematical, astronomical, and physical and that have been written about, circulated, and discussed by extremely distinguished persons, such as Gassendi, Riccioli, Tacquet, and a hundred others. Finally, if in that work the author perhaps appeared to be more inclined toward the Copernican than the Ptolemaic hypothesis, against the injunctions he had received, later he recanted, confessed his error, was absolved, and lived and died in a most holy manner with all the assistance of the Church; I know this because I was present together with two priests, and we assisted him in the illness that led to his death and that was endured with total trust in God, alertness until the last moment, and edification for those who were present; among these were also the son, his wife, their whole family, and Torricelli.

I will not relate to Your Most Reverend Paternity, who knows it better than I do through other channels, what I have heard more than one of your Fathers say on various occasions: that the prohibition decreed by the sacred Congregation of Cardinals shook all scientists[19] on the other side of the Alps; that nowadays, through so many new discoveries made with the telescope, they subscribe more to Copernicus's opinion than to any other; and that the same prohibition about a merely natural subject motivated the ultramontanes to question the other determinations that are made here about subjects concerning the holy faith.[20] Thus, if these stories are true, one would hope that it would be only profitable if after more than seventy years there were some implementation of the correction made to the said work.

The point is to find adequate and decorous advantages for doing so; I have no doubt that Your Paternity, being most knowledgeable about these matters, will be able to think of some that are appropriate to induce the right persons to permit the above-mentioned correction; to accomplish this, I have something to tell you. In regard to remedying or relaxing that prohibition, I know I heard from the late Lord Cardinal Leopold that he discussed the matter with some Lords of that Congregation and was told that there would have been ways of making everyone happy (of which I have no knowledge), and of doing so not only without damage but also with a gain in reputation. [155] In short, one frankly trusts the authority, esteem, knowledge, and skill of Your Most Reverend Paternity, together with the affection you have always shown toward our common fatherland and toward the doctrines of the author for whom you have thus far been so partial that, if the desired end is not achieved now through you, one does not hope that it will ever be achieved by anyone else. Thus, for you alone we reserve our applause, and from you alone we await the happy achievement of such a noble undertaking; in the meantime I look forward to your reply so as to be able to relate that already you have willingly and with no difficulty undertaken such a reasonable request in regard to something that is well known to be most arduous.[21]

The initial reference to a third party who would be very pleased if the anti-Copernican censures were removed is puzzling. It may be a reference to Leibniz, who (as we shall see presently), during his Italian trip of 1689–1690, engaged in a considerable effort to that effect, including making his own plea to Baldigiani.[22]

Viviani was requesting something very modest, namely that the ban on the *Dialogue* be made contingent on its being appropriately revised or "corrected" in a way analogous to the way Copernicus's *Revolutions* had been corrected. In other words, Viviani was asking that the 1620 correction of Copernicus's book be further implemented by applying it to Galileo's book. Thus the unexpurgated *Dialogue* would still be banned, and Galileo would still remain condemned. That is, Viviani was requesting neither the total repeal of the book's prohibition, nor the rehabilitation of the person.

In fact, we know from other sources that at about this time a manuscript copy of Galileo's *Dialogue* was prepared from the printed 1632 edition with some corrections and emendations. It has an appendix with Galileo's letter to Dini of 16 February 1615, Galileo's letter to Christina, Zúñiga's commentary on Job, the Inquisition's sentence, and Galileo's abjuration. And this was probably commissioned by Viviani as part of his projected edition of Galileo's works.[23] The publishing project was then a third reason, in addition to Leibniz's concern and Baldigiani's career advancement, why Viviani made the plea.

However, nothing came of Viviani's effort. Although the relevant correspondence has not survived, we know that Viviani never published his edition of Galileo's works. In fact, on 25 January 1693, Baldigiani wrote Viviani in Florence that all Rome was up in arms against mathematicians and physicists, and that very long lists of authors were being compiled for prohibition.[24] Thus, in the same year, Viviani decided to build a public monument to Galileo in the facade of his (Viviani's) house with an inscription full of praise.[25] He had obviously given up hope for the repeal of the condemnations.

5.2 The Ghost of Bellarmine: Adrien Auzout (1665)

In 1661, the French Jesuit Honoré Fabri (1607–1688) published a small book on Saturn and its rings.[26] The work was published in Rome, where Fabri had lived since 1646. It appeared under the name of Eustachio Divini (or Eustachius de Divinis), with whom Fabri had collaborated, because the book was part of an ongoing controversy with Christiaan Huygens; in fact, the previous year Fabri had published another book on the same topic and under the same collaborator's name, and Huygens had published a reply.[27] Despite the concealed authorship, there was no question about the identity of Huygens's real opponent because the controversy was being managed by Prince Leopold de' Medici, head of the Florentine Accademia del Cimento.

The details of the Saturn controversy are not pertinent here. What does concern us is a comment about the condemnation of Copernicanism that is found in the 1661 work. The book's author (nominally Divini) reported that he had consulted Fabri for advice on how to reply to Huygens's claims about the earth's motion, and that Fabri said one could reply as follows:

> It has been more than once asked of your leaders whether they had a demonstration for asserting the motion of the earth. They have never affirmed they had. Therefore, nothing hinders that the Church may understand those Scriptural passages that speak of this matter in a literal sense, and declare that they should be so understood as long as the contrary is not evinced by any demonstration; if perhaps it should be found out by you (which I can hardly believe it will), in this case the Church will not at all scruple to declare that those passages are to be understood in a figurative and improper sense, according to that of the poet's "the land and the cities slip backward."[28]

Today we know that such a judgment had been advanced by Cardinal Bellarmine in a letter to Foscarini dated 12 April 1615. One of the many points in this important letter reads as follows:

> I say that if there were a true demonstration that the sun is at the center of the world and the earth in the third heaven, and that the sun does not circle the earth but the earth circles the sun, then one would have to proceed with great care in explaining the Scriptures that appear contrary, and say rather that we do not understand them than that what is demonstrated is false. But I will not believe that there is such a demonstration, until it is shown to me. Nor is it the

same to demonstrate that by supposing the sun to be at the center and the earth in heaven one can save the appearances, and to demonstrate that in truth the sun is at the center and the earth in heaven; for I believe the first demonstration may be available, but I have very great doubts about the second, and in case of doubt, one must not abandon the Holy Scripture as interpreted by the Holy Fathers.[29]

While there are slight differences in the two formulations, there is great overlap in the key judgment.

Given the similarity of content, Fabri was certainly echoing Bellarmine's view. It is even possible that Fabri had had access to Bellarmine's letter. At any rate, the view seemed to be relatively common in Jesuit circles. In fact, in 1624 Mario Guiducci had written Galileo from Rome saying that he had had occasion to talk to the Jesuit Orazio Grassi; the latter reportedly asserted that if a demonstration of the earth's motion were found, then the scriptural passages on the earth's rest would have to be interpreted differently from the traditional interpretation, and that this had been Bellarmine's opinion.[30]

However, Bellarmine's letter to Foscarini was generally unknown for two and a half centuries; it was first published by Domenico Berti in 1876.[31] And the situation is similar for Guiducci's letter. Thus, for a long time the judgment in question was regarded as originating from Fabri, and as such it was widely quoted and discussed.[32] It was first taken up by Adrien Auzout in a book published in Paris in 1665 in the form of a letter written to one Monsieur l'Abbé Charles.[33] This work need not otherwise concern us, but Auzout's discussion contains an analysis and critique of Fabri's remark that are extremely insightful and important. It thus deserves extended quotation:

> [58] I thought I should quote here the words of the Reverend Father Fabri, so that one would know how to explain the prohibitions which the Inquisition formerly issued against supporting the earth's motion in regard to Galileo, perhaps because he was suspected of wanting to introduce novelties in religion as well as in philosophy; for he found much to criticize in that of Aristotle, which at that time almost the whole world followed as the only true philosophy, and upon [59] which one had grafted almost all the mysteries of theology. . . .[34]
>
> This passage has seemed strange to all who have examined it; for how can one say that there is nothing to prevent the Church from construing, and declaring that one must construe, literally the passages in question if she can later declare that one can construe them otherwise; and how will she declare that one can construe them in a figurative and improper sense if she has previously declared that one must construe them literally? It seems to me that from this one can conclude at least that Father Fabri believed that this question had not been decided absolutely, but only provisionally, "as long as the contrary is not evinced by any demonstration," to prevent the scandal which the novelty was causing or could cause. For it is not likely that he would have

explained his view on a subject so delicate in Rome without having explored the feelings that prevailed at the moment. If one believed that the question had been decided absolutely, one would be obliged to claim that one could not find a contrary demonstration, and not say that if a demonstration were found then the Church would declare, etc.

For in truth, these passages [60] must be construed either literally or not. If they must be construed literally and they teach the earth's immobility, they can never be construed in a figurative and improper sense (these are his terms), like the poet's words, "the land and the cities slip backward." If they can be construed figuratively in some way, one cannot at present declare that they must be construed literally, and one can only regard this decree as a disciplinary judgment aimed to prevent the scandal that this doctrine was causing. For it would be impossible to want to decide absolutely something for which one could fear or hope to have a contrary demonstration in the future; truth being eternal, one cannot say that at one time the words must be construed literally, and that at another time they can be construed figuratively.

Given this, and given that Father Fabri's argument assures us that the Inquisition has not declared absolutely that these scriptural passages must be construed literally because the Church can make a contrary declaration, I do not see that one should be afraid to follow the hypothesis of the earth's motion; the only thing with which one should perhaps comply would be to not support it publicly until the prohibitions are removed. It is desirable that this be done as soon as possible, so that learned astronomers who unfortunately do not follow the Roman Church will no longer blame us for being so slavish that we follow its decisions not only on matters of religion but also for what concerns physics and astronomy. It does not appear at all that God has wanted to teach us anything in particular about nature; on the contrary, almost all who have wanted to find the principles of their philosophy in Scripture have fallen into untenable errors; in it we should [61] only look for the maxims of religion and morality, and not for the principles of physics or astronomy, which are as useless for the other life as they are useful for this one.

It would also be desirable for Father Fabri to procure this liberty for all astronomers since, being in the position he is and being as learned as he is, he could perhaps testify more effectively than others that this hypothesis is neither absurd nor false in philosophy, as one believed at first; nor is it in any way prejudicial to the Faith, for the most subtle dialectician and the most troublesome sophist could not draw from it any conclusion that conflicts with the least article of our religion; and when one construes these scriptural passages in a figurative sense and in accordance with the appearances, one does nothing contrary to Scripture, since one would have to construe them in that manner if one were to find in the future the demonstration of which Father Fabri does not entirely despair. . . .[35]

[62] One can also show in a way different from Father Fabri's argument that the decree must have been provisional; for it was based on the common opinion of that time; that is, that hypothesis was also declared "absurd and false in philosophy."[36] But Father Fabri and all scholars on his side know well,

and must reflect accordingly, that it is neither absurd nor false in philosophy, and that it conflicts with neither physics nor astronomy. One can see what one should think on the matter from the replies explicitly given by Father Riccioli in his long treatise "On the System of the Moving Earth"[37] to the absurdities and falsehoods alleged by the Peripatetics; to be sure, he does say he has found no solid reply to two arguments which he advances, one taken from the percussion of heavy bodies that descend, and the other from bodies projected in different directions of the world; but it is good that his reasons are taken from mechanics, for one can demonstrate to him the incorrectness of his reasoning, as it is customary to do in mathematics, where the principles are assured. I would do that at length, if this were the proper place. But to just say a word in passing about them . . .[38]

[63] However, these are the only two things that have given trouble to Father Riccioli and Father Grimaldi in the hypothesis of the earth's motion, having dismissed or answered all the others. From this it is easy to see that he believed that that hypothesis implied neither absurdities nor falsehoods. While it is certain that at that time it was declared absurd and false in philosophy, as well as contrary to Scripture, one can even think that it was declared contrary to the sense of Scripture only because one believed it to be absurd and false. For there are many passages of Scripture that need not be construed literally; and in matters of physics, astronomy, etc., one knows well that it does not speak in order to instruct us, but that it speaks only in accordance with the appearances and ordinary human opinion and not in accordance with the truth of things. [64] Furthermore, even if the authors of the Sacred Books had known that the earth turns around the sun as the other planets do, one should not be surprised that they spoke as they did, that is, in accordance with what appears to us and what people believe; for they speak to persons most of whom are largely ignorant of astronomy and have no need to be instructed in these things; and that is how those who follow that opinion speak in ordinary language insofar as, outside the contexts where they treat professionally of heavenly motions, they speak of sunrise, sunset, its noon elevation, its approaching various stars, etc., as if it were in motion; for the same effects happen in experience, whether it or the earth is in motion, and this suffices as an explanation in ordinary life and whenever one does not want to teach astronomy.

This should persuade us that that decree was only issued provisionally, out of the fear one felt that this hypothesis would have bad consequences by reversing the philosophy that was accepted at that time; according to it, one was accustomed to construe the passages in question in accordance with what they seemed to mean, although there was not a single one that could be construed literally in all its parts, and although most of them had to be construed figuratively in all their parts. It would be easy to show this, if I had not already been long enough and if so many others had not already done it.

I wanted to dwell upon this matter in order to disabuse those who have not carefully examined the circumstances of that decree and have not explored the prevailing feelings about it (as Father Fabri has been able to do), and who thus condemn inappropriately those holding the earth's motion and speak as if the Church had decided this question absolutely; but this is far removed from the truth and even from the explicit or tacit acknowledgement of those who are most concerned with the matter.

[65] But one must wait and examine whether the motion of the last comet can convince us in any way of the earth's motion; this would not be a metaphysical or mathematical conviction, which involves the impossible (as one ordinarily says), for one need not expect such a kind; rather it would be a reasonable conviction, like the one that makes us judge that the sun with the other planets does not turn around Jupiter or Saturn, but these planets turn around it; for if one wanted to search for a demonstration of the first kind, I defy all astronomers in the world to prove to me that the sun and the earth do not turn around Jupiter, Saturn, or even the moon, although they all feel assured it is false that Jupiter or Saturn is the center and extravagant that the moon is. And yet if there were inhabitants on the moon, they would feel to be motionless, as we believe ourselves to be here when we base ourselves only on the appearances, and they would attribute to other bodies all motion that appeared to them, because they could not perceive the contrary; and we would mock them if they wanted to claim that the sun with its whole system and the stars were obliged to turn around them, instead of them turning with the earth around the sun; the inhabitants of the other planets, if one supposes there are any, would have the same reason to mock us for wanting to oblige them to turn around us every day together with the sun (which is the principle of their motion), rather than wanting to follow with them the motion of the vortex in which we as well as they find ourselves. And certainly Jupiter (which has four moons) and Saturn (which has one, as well as a ring that is such an extraordinary body) would have good reason to dispute that [centrality] to the earth (which does not have [66] a retinue as beautiful as they do and is perhaps a thousand times smaller).

Nevertheless, in all this I do not pretend to take sides stubbornly, and I am ready to submit to and follow all that the Church might order. But I thought it good to show that those who suppose the earth's motion can do it (it seems to me) without being disrespectful and without incurring the censure of those who have never carefully examined what happened; they do not know the intentions underlying the temporary prohibition to support that hypothesis, "as long as the contrary is not evinced by any demonstration" (as Father Fabri says), or rather until the fear that it might carry along some novelty pernicious to religion has passed; this should have arrived a long time ago. However, one must be satisfied with a reasonable demonstration, taking into account the subject, for it is impossible to advance any reason why the sun with its system should turn around the earth rather than around Saturn, Jupiter, Mars, Venus, or Mercury; and yet everyone feels certain that it does not turn around them.

Thus, since we are certain that if the earth turned we could not perceive it with our senses, and that if the sun with the earth turned around another planet we could not perceive that either, one cannot but be satisfied with reasonable evidences and analogies. They agree so well with that hypothesis that there is not any which can be imagined should exist but does not actually exist, and there is no effect that should occur on the assumption of the earth's motion but does not occur.[39]

This passage contains at least three arguments that are strikingly original, historically important and influential, and philosophically sophisticated and cogent. First, let us recall that Fabri's main contention was the biconditional claim that if a demonstration of the earth's motion is discovered then geostatic passages in Scripture will be interpreted figuratively, but that as long as the geokinetic demonstration is not available then those Scriptural passages must be interpreted literally. Auzout advanced the ingenious argument that if we accept the first part of Fabri's claim, it follows that the condemnation of the earth's motion was provisional, and not absolute or permanent. His reasoning is that truth is eternal, and this proposition should be applied to principles of biblical interpretation; hence if a figurative interpretation of a scriptural passage is legitimate after the demonstration of the earth's motion, it is also legitimate beforehand. Similarly, if a literal interpretation is mandatory before a demonstration is found, it is mandatory afterwards. Auzout seemed to be suggesting that the two parts of Fabri's biconditional were inconsistent with one another and that the proper way to resolve the inconsistency was to reject biblical literalism (in astronomy and physics).

Auzout also had a second argument to show the provisionality of the condemnation. This argument stressed the evaluation of Copernicanism as philosophically false and absurd that was explicitly contained in the Inquisition's sentence of 1633 and more cryptically in the anti-Copernican decree of 1616. He also insightfully suggested that it was probably this alleged philosophical untenability that led to the evaluation of Copernicanism as contrary to Scripture. Next Auzout easily showed that at his time (1665), the earth's motion was no longer absurd philosophically, a conclusion derivable even from Riccioli's work. It followed that the condemnation of the earth's motion should be withdrawn now that it was deprived of its natural-philosophical grounding. With these considerations Auzout was anticipating arguments that were later elaborated by Pietro Lazzari during the partial unbanning of Copernicanism between 1753 and 1758 and by Maurizio Olivieri during and in the aftermath of the Settele affair of 1820.[40]

A third argument involved the concept of demonstration. Auzout distinguished what he called a reasonable demonstration from mathematical and metaphysical proofs. He claimed that in the problem at hand, the relevant notion was that of reasonable demonstration, in part because otherwise astronomers would be unable to assert propositions which no one doubts in the least, such as that the sun does not circle Jupiter. Then he suggested that the earth's motion was at that time susceptible of a reasonable demonstration. It followed from Fabri's own (first) conditional claim that geostatic scriptural passages should be interpreted figuratively. In this connection it should be recalled that Bellarmine, in the passage of his letter to Foscarini quoted above, stressed the impossibility of a demonstration of the earth's motion (which Fabri only parenthetically questioned); in so doing, Bellarmine was using a very different concept of demonstration. Thus, Auzout's third argument would not have easily convinced the cardinal, although his first two would apply with equal force to Bellarmine as to Fabri.

Besides these well-reasoned and novel arguments, Auzout made other claims that were common at that time but no less important. He distinguished between the private pursuit and the public support of an idea; and he suggested that while he might be willing to accept the prohibition on public support, he did not think there was anything wrong with the private pursuit of Copernicanism; this had been a common response by Catholic astronomers and natural philosophers to the condemnations of 1616 and 1633. And he claimed that the geostatic interpretation of biblical passages usually presupposed a figurative interpretation of various parts of them, and so biblical literalism was generally untenable or internally inconsistent; this had been argued by Galileo at the end of his letters to Castelli and to Christina.[41]

In the same year (1665) as Auzout published the book that ended with these arguments, Fabri's remark and Auzout's argument received wider dissemination in the *Philosophical Transactions of the Royal Society of London*. The fourth issue contained several articles that discussed various parts of Auzout's book. One of these articles was an account of Auzout's critique of Giuseppe Campani's work on telescopes of great magnification and the observation of Saturn and Jupiter. This article contained a brief discussion of our topic. Fabri's remark was quoted from his 1661 book, and then Auzout's argument was described in these terms: "Whence this Author concludes, that the said *Jesuite* assuring us that the *inquisition* hath not *absolutely* declared, that those Scripture places are to be understood *literally*, seeing that the *Church* may make a contrary declaration, no man ought to scruple to follow the *Hypothesis* of the *Earths motion*, but only forbear to maintain it in *publick*, till the prohibition be called in."[42]

5.3 Diplomacy Fails: Leibniz (1679–1704)

Although Leibniz is best known for his coinvention of the calculus, his metaphysics of monads, and his contributions to dynamics, he also played a significant role as a cultural politician and diplomat. One of his dreams was the ecumenical project of reuniting the Catholic and Protestant Churches; his failure should not make us underestimate the amount of effort he devoted to it.[43] Another, even greater involvement was his work as a genealogical expert and dynastic historian of the ducal family of Hanover, the House of Brunswick, his employer; if this task strikes us moderns as unworthy of a genius such as Leibniz, we must not fail to appreciate that his genius

prevailed here too, and so in the process he managed to become a pioneer-
ing historian who understood the need to consult the original sources in
the archives, to criticize mythological accounts, and to evaluate alternative
interpretations; thus, since one particular issue was whether the House of
Brunswick was related to the House of D'Este in Modena, Leibniz traveled
to Italy for about a year in 1689–1690.[44] A third aspect of Leibniz's general
cultural activities was his work as a librarian and bibliographer, in which he
again excelled; for example, in 1692 he was able to track down a book
about which he had been consulted by Antonio Magliabecchi, the great
librarian to Grand Duke Cosimo III de' Medici, thus impressing both with
his "prodigious talent";[45] and in 1695 Leibniz was offered the position of
director of the Vatican Library if he converted to Catholicism,[46] an offer he
politely declined. A fourth cultural activity, and the one that concerns us
here, was Leibniz's effort to have the Church repeal its condemnation of
Copernicanism.[47]

Leibniz's earliest position on the matter was written between 1679 and
1686. It bears the revealing title "Apologia for the Catholic Faith Based on
Right Reason," and it appeared for the first time in 1948 in a collection of
previously unpublished manuscripts held at the library of Hanover.[48] The
most relevant part of that essay is this:

[329] Just as in physical inquiries everyone prefers to trust sense experience
rather than reason and any prudent astronomer includes no more in his
hypothesis than is contained in observation made according to the rules, so
when reason and revelation seem to conflict it is safer to distrust reason than
to accommodate revelation to reason by means of a strained interpretation.
However, I will not deny (and the most serious theologians agree with me)
that if something has been proved by very clear arguments taken from natural
reason, and if revelation can be easily reconciled with them by an interpretation that does not stray too far from the proper meaning of words either by
violence or by lack of example, it is more prudent to follow the meaning that
accords with reason rather than their literal meaning, so that we do not seem
to want to gouge out our eyes under the pretext of wanting to avoid a bright
light; this is especially so since nothing is more offensive to intelligent men
and alienates them more from true religion than their seeing some things
defended with absurdity and obstinacy by masters of religion. Thus, if nowadays someone were to absolutely condemn the antipodes (as Lactantius and
even Augustine did) or the earth's motion (as the theologians of recent years
have), although the most eminent astronomers judge it to be evident, he
would be offending the minds of the simple in a manner that would be stupefying in this enlightened century full of eminent minds, and he would be
prostituting the authority of the Church in the eyes of outsiders as much as it
was within his power. What would really happen if, in due course, one were to
find an irrefutable demonstration, given that those currently available are
very close to being conclusive? Here I recall what an eminent theologian of
the Society of Jesus correctly wrote: in that case the Church will readily admit

a figurative interpretation of the statements of Scripture, in the sense of taking it to speak in the same manner as the statement "the land and the cities slip backward" speaks for navigators. And another no less celebrated theologian, of the order of the Minims, judged that censures of this kind of doctrine made at one time or another are not always absolute condemnations of opinions but may be judgments of prudence, so that some things are not taught lightly or to the detriment of simple people. In the meantime, [330] in case of doubt, one must always hold that it is safer for a Christian to accept the proper meaning of divine words.[49]

It is certainly surprising to see Leibniz say that reason must bow not only to observation but also to revelation. So much for his alleged "rationalism." To be sure, after such a nonrationalist beginning, he went on to qualify the claim: if there is a conflict between a physical proposition supported by very clear reasons and a scriptural statement interpreted literally, and if the scriptural statement can be given a nonliteral interpretation that is not too strained or unprecedented and also reconciles it with the physical proposition, then the nonliteral interpretation is preferable. Then Leibniz applied this qualified principle to the earth's motion to conclude that in his time the continuation of the anti-Copernican decree was a perversion of Church authority. He ended by pointing out that his conclusion agreed with the judgments on the question expressed by two famous theologians.

The Jesuit was obviously Fabri, and the judgment in question was clearly the one I have quoted and discussed above.[50] The celebrated Minim was obviously Mersenne, but it is unclear what passage Leibniz was referring to.[51] On the other hand, we can see that the claim of prudential judgment attributed to Mersenne corresponds also to the claim of disciplinary judgment made by Auzout.[52]

Other Leibnizian discussions on this issue occur in his correspondence with Landgrave Ernst von Essen-Rheinfels. The landgrave had converted to Catholicism in 1652 and occasionally tried to convert Leibniz. However, they were both diplomatic enough not to let their religious differences prevent an open and frank dialogue. For example, in a letter dated 10/20 October 1684, among many other topics, Leibniz pointed out that Church authorities often embarrass the most well-intentioned persons; the best thing to do is to say sincerely what one believes, otherwise one might have trouble later; for example, an astronomer should explicitly declare that he thinks Copernicanism is correct, despite what the Roman Inquisition has prescribed to the contrary.[53] The landgrave replied confessing that he held many beliefs contrary to those of the Inquisition.[54] The most relevant letter is the one Leibniz wrote in summer 1688. It had a postscript that read as follows:

[200] P.S. When Your Most Serene Highness writes to Rome, it would be appropriate to sound out the Most Eminent Cardinals about whether there is any inclination to remove the interim censure formerly published against

Copernicus's opinion of the earth's motion. For this hypothesis is now confirmed by so many reasons, taken from new discoveries, that the greatest astronomers hardly doubt it any longer. Some very competent Jesuits (such as Father de Challes)[55] have publicly acknowledged that it would be very difficult to ever find another hypothesis that could explain everything so easily, so naturally, and so perfectly; and one sees clearly that nothing but the censure prevents him from openly yielding to it. Father Mersenne (a Minim friar) and Father Honoré Fabri (a Jesuit) have recognized and taught in their writings that the prohibition was merely provisional, until one had a better understanding, and that it was judged expedient at that time in order to prevent the scandal which that doctrine (then propagated by Galilei) seemed to engender in the minds of the weak. Nowadays one has recovered from that jolt, and all persons of good sense recognize easily that [201] even if Copernicus's hypothesis is true many times over, Holy Scripture will thereby receive no injury. If Joshua had been a disciple of Aristarchus or Copernicus, he would not have avoided speaking as he did, otherwise he would have shocked his assistants and good sense. All Copernicans, when they speak to ordinary people and even to one another in nonscientific discussions, will always say that the sun has risen or set and will never say this of the earth. These terms are appropriate for phenomena and not for causes. It matters to the Catholic Church that philosophers should have the reasonable liberty that belongs to them.

It is hard to believe how much harm is done by the censure of Copernicus. For the most learned men of England, Holland, and the whole North (to say nothing of France) are almost convinced of the truth of that hypothesis, and so they regard that censure as unjust slavery. And as they see that the greatest mathematicians among Catholics (and even among the Jesuits) are sufficiently informed of the incomparable advantages of that doctrine and yet continue to be obliged to reject it, they do not know what to say and are tempted to suspect them of insincerity; this gives them a bad idea of the Catholic Church. Moreover, nothing is more contrary to solid and generous minds than such a constraint. Others have already mentioned excellent passages from St. Augustine in which he shows that to hinder people about philosophical truths amounts to prostituting Holy Scripture and the Church and abusing their authority.

It should be possible to find some expedient in Rome for prescribing that all those who want to maintain that Copernicus's hypothesis is true must declare at the same time that Holy Scripture could not have properly spoken otherwise than it [202] did and that it did not diverge from proper expression. And if the Congregation should change or reduce the former censure, decreed suddenly when the facts were not sufficiently understood, that would not harm its authority and still less the authority of the Church, given that His Holiness had not intervened. There is no tribunal that does not sometimes revise its judgments; and some of the Church Fathers took the same step in regard to the antipodes; hence I do not see that one has to be so worried on a similar matter. I believe this forthrightness would have a more excellent and fruitful result than one thinks; for although this matter is not accessible to common people, it touches deeply the most learned men and the most excellent minds. And the authority and example of able persons, although small in number, has considerable power over the others.[56]

Here Leibniz was presenting to the landgrave arguments he might use with the Roman authorities for repealing the anti-Copernican censure. In stressing the provisional character of this censure, he was claiming to agree with Fabri, but was really agreeing with Auzout's interpretation of Fabri's remark. In saying that "His Holiness had not intervened," Leibniz (like Descartes) was showing his appreciation that the censure had not been formally endorsed by the pope speaking ex cathedra; this was another reason why the censure *could* be repealed. Leibniz ended with a third reason why the prohibition *should* be repealed, namely that revisability indicates intellectual honesty and so is actually a good thing.

Although Leibniz was too diplomatic to rely explicitly on the arguments of Kepler and Galileo, he did so implicitly. When he argued in the first paragraph that the truth of Copernicanism would not contradict Scripture because the latter speaks of phenomena and the former of causes, he was (without saying it) paraphrasing Kepler's introduction to the *Astronomia Nova*. When he mentioned at the end of the second paragraph that other people have argued on the basis of Saint Augustine's views that to limit the search for natural philosophical truth is a prostitution and abuse of biblical and ecclesiastical authority, he was referring to Galileo's Letter to Christina in the *Nov-antiqua* of 1636.[57]

When Leibniz suggested at the beginning of the third paragraph that allowing scholars to maintain the truth of the earth's motion should perhaps be conjoined with requiring them to declare that Scripture does not speak improperly on the topic, perhaps he was carrying the art of diplomacy too far, to an undiplomatic excess so to speak.

At any rate, Leibniz had the opportunity to test his diplomatic skills soon after he wrote these words, when he visited Italy for about a year in 1689–1690.[58] The official reason for the journey had to do with his duties as librarian and historian of the dukes of Hanover; Leibniz was charged with compiling a genealogical tree and history of this dynasty, allegedly going back to the year 1060.[59] Although Leibniz was disciplined enough that he never neglected the central purpose of his trip and could thereby justify almost all of his moves, he also exploited the opportunity to meet scholars, visit academies, and generally promote the republic of letters, freedom of thought, and ecumenism. He visited Venice, Modena, Florence, Rome, and Naples, among other cities.

Leibniz stayed for about six months in Rome, where fortuitous circumstances enabled him to witness several unusual and instructive events. In April 1689, there was the death and funeral of Queen Christina of Sweden, who had been living in Rome. Soon thereafter, there was the death and funeral of the old pope, and then a conclave that resulted in the election and inauguration of Alexander VIII in the summer. Leibniz's Roman stay also coincided with the return of the Jesuit missionaries from China and their departure for a second mission, headed by Father Grimaldi.

On the more strictly scientific and philosophical plane, it was in 1689 in Rome that Leibniz first read Newton's *Principia*,[60] published two years earlier. He got involved in the activities of the Accademia Fisicomatematica, which had been founded in 1677 by Cardinal Giovanni G. Ciampini and operated under his direction until his death in 1698; there he probably met such Italians as Francesco Bianchini, Giuseppe Campani, and Antonio Baldigiani, and Frenchmen such as Adrien Auzout, who was living in Rome at the time. Leibniz also found the time to write several essays on physical and astronomical subjects.[61]

Although Leibniz did not go to Rome with the intention of defending the Copernican cause, he naturally and quickly became involved with it, since freedom of thought was essential to the republic of letters, which he did want to advocate. He thought he could promote it on three levels: with the new Pope Alexander VIII; with the more active members of the Accademia Fisicomatematica; and with the Society of Jesus. For this purpose Leibniz wrote several overlapping texts that exist in several versions.[62] These were addressed and probably shown to the Jesuit Antonio Baldigiani,[63] who was a relatively authoritative official. Leibniz's most important and novel argument is contained in an untitled essay that begins with the words "Cum geometricis demonstrationibus."[64] The argument is based on a relativistic conception of motion and on an intriguing conception of truth that is harder to characterize.

Leibniz began the essay by elaborating the relativistic concept of motion in a way reminiscent of Descartes but less paradoxical and clearer; for instead of focusing on the legitimacy of asserting both that the earth moves and that it is motionless, Leibniz stressed the mathematical equivalence of alternative kinetic hypotheses. However, he also noted that this equivalence follows only if "the bodies are moving freely or colliding with one another."[65] Thus we could add that this necessary condition not only no longer applies today but had already been invalidated two years earlier by Newton's *Principia,* in which the bodies are neither moving freely nor colliding, because of the gravitational force (and inertia).

The essay went on to sketch a relatively original concept of truth. Leibniz's concept was partly rationalist insofar as it equated truth with intelligibility, but also partly pragmatist and contextualist insofar as it equated intelligibility with contextually appropriate usefulness. The concept implied that while Tycho (and the geostatic thesis) was wrong in the context of "theoretical astronomy," so was Copernicus (and the geokinetic thesis) in the context of spherical astronomy as well as in common speech, but that in its context Scripture spoke properly and truthfully.

Several surprising consequences followed. One the one hand, one did not have to accept the thesis (suggested by Fabri's remark and explicitly drawn by Auzout) that the anti-Copernican censure was provisional; nor was a retraction on the part of Church censors needed; thus ecclesiastical authority and dignity would be preserved. On the other hand, theoretical astronomers were justified in holding and defending the truth of Copernicanism because this claim was identical to the claim that the Copernican hypothesis was the best alternative in terms of intelligibility, intellectual adequacy, explanatory power, and simplicity, and even the censors agreed to the latter claim; thus, Leibniz wrote, "We can finally restore philosophical freedom."[66]

In short, Leibniz's 1689 argument was an attempt to make both sides happy;[67] and he thought he could accomplish this by means of a reinterpretation of the situation. This reinterpretation involved an explicit reaffirmation of the concept of motion as a relation and an elaboration of a concept of truth that was partly rationalist and partly context-dependent, and in that sense also "relativist."

Such an epistemological doctrine must have seemed unacceptable to Baldigiani and other officials. Perhaps Leibniz himself had doubts about the efficacy of such a radical argument, and this may be the reason why he composed another essay on the topic. It too is untitled, but begins with the words "Praeclarum Ciceronis dictum est."[68] There Leibniz in part reiterated the doctrine of the relativity of motion. In part he gave arguments analogous to those that had been advanced in 1665 by Auzout.[69] However, he also made an assertion that could be interpreted as an endorsement of Galileo's condemnation and justification of the Inquisition:

> Censorship has been rightly applied to the audacity of those who seemed to judge Holy Scripture less reverently, that is, as if it has not spoken accurately, with the pretext that its aim is to teach the way to salvation, not philosophy. In fact, it is more respectful and truer to acknowledge that in the holy texts are also hidden all the recondite treasures of science, and that absolutely correct things are said not less about astronomical matters than about all the others. This can be stated without damage to the new system. In fact the holy authors could not express the thoughts of their minds in a different way without absurdity, even if the new system were regarded as true a thousand times."[70]

Here, Leibniz seemed to be asserting that Scripture is a scientific, astronomical, and philosophical authority, and so those who denied its authority in this domain were audacious, irreverent, disrespectful, and deserved the censure they got. And of course, Galileo was the main proponent of that denial.

Again, the ecclesiastical silence and inaction that followed might be taken as an indication that Baldigiani and the other officials did not want to

go as far as Leibniz seemed willing to go. Moreover, in the light of this statement of apparent biblical literalism and traditionalism, one cannot help
wondering whether Leibniz was being consistent and what his real aim was.
Was he really defending Galileo, the Copernican cause, or even philosophic freedom? Perhaps he was pursuing ecumenism first and foremost.
Perhaps he simply suffered the fate of diplomats and arbitrators, namely to
end up leaving both sides unhappy.

At any rate, Leibniz's efforts continued. In 1694, he tried to convince a
young German studying at the Jesuit Roman College to take up the cause
of lifting the Church's anti-Copernican ban.[71] In 1699, he wrote to the
Medicis' librarian, Magliabecchi, recalling with pride his decade-old efforts,
and appearing to be still hopeful.[72] Leibniz's diplomatic attitude also continued, as we can see, for example, from a 1702 letter to one Reuschenberg.
There, at one point Leibniz made a statement that can be paraphrased thus:
"If the whole Church were to rise against Copernicus and Galileo, she would
be wrong, and a man who is all alone but right would be greater than she."[73]
On the other hand, Leibniz also asserted that "if someone replies that the
question of Copernicus's system is outside the jurisdiction of the Church, I
would answer that an infinity of other questions which one wants the Church
to decide are not at all less philosophical . . . or historical, and consequently
not subject to such decisions."[74]

Finally, we must examine a passage which Leibniz wrote in 1704 in the
New Essays on Human Understanding. He planned to publish this book during his lifetime, although it did not appear until 1765.[75] The passage is
important in light of the mature age of the author, the public nature of the
composition, and the intellectual context of the work; it treats of the general theory of knowledge and critically examines the views of John Locke.
Chapter 20 of book 4 of the *Essays* is titled "Of Error" and discusses four
main reasons for errors: want of proofs, inability to use them, unwillingness
to use them, and wrong measures of probability.[76] The wrong measures of
probability are subdivided into four types that involve "doubtful propositions taken for principles[,] . . . received hypotheses[,] . . . predominant
passions or inclinations[,] . . . authority."[77] In his discussion of received
hypotheses, Leibniz gives the example of the Ptolemaic theory in the
Copernican controversy.

Here Leibniz seemed to lean more toward the Galilean and against the
ecclesiastical side. For he reendorsed the Fabri-related provisionality thesis
and reiterated the claim that the condemnation of Copernicanism had
harmed progress. And he made a novel and extremely important point:
that for all their talk of hypotheses, the anti-Copernicans had the tendency
to regard only their opponents' view as hypothetical but their own as factually and categorically true. This attitude was an example of the error of
treating a received hypothesis as nonhypothetical.

This judgment of Leibniz is extremely important because a cause célèbre in the subsequent Galileo affair has been the criticism of Galileo as rash and overzealous in his commitment to the geokinetic hypothesis. This issue had already germinated in the 1678 correspondence between Baldigiani and Viviani. Now Leibniz was turning the tables on Galileo's opponents by claiming that they committed the classic error of imprudently overestimating the epistemological solidity of their own received hypothesis.

In conclusion, in his 1654 biography of Galileo, Viviani formulated an account of the trial that was gently critical and mildly appreciative of Galileo, as well as essentially uncritical and largely favorable toward the Church. In his 1690 letter to Baldigiani, Viviani made a very explicit pro-Galilean plea, but his request was very modest (the unbanning of the *Dialogue* after suitable revisions), and his reasons were mostly appeals to ecclesiastical self-interest. In his 1665 commentary, Auzout tried to build upon two authoritative ecclesiastical judgments, namely the conditional prediction published by Fabri and stemming from Bellarmine, and the philosophical evaluation of Copernicanism mentioned in the trial documents as a key reason for the condemnations of 1616 and 1633; the brilliance and elegance of his argument lay in the fact that he derived reformist conclusions critical of the Church from those judgments. In Leibniz we find someone who was temperamentally and methodologically moderate, bipartisan, diplomatic, and ecumenically minded; he attempted several such moves, and it may have been their failure that led him to a relatively pro-Galilean position in his most mature and public pronouncement.

Chapter 6

Myth-making or Enlightenment?
Pascal, Voltaire, the Encyclopedia *(1657–1777)*

The subsequent Galileo affair is too interdisciplinary, international, long-standing, and far-reaching a controversy to be susceptible of any neat chronological periodization, monotonically progressive development, or analytically simple taxonomy of problems. Thus, we now need to examine a miscellany of texts and events that begin in the latter part of the period we have already examined but extend into the later eighteenth century. Although at first these may seem to reiterate or regurgitate old points, they are really digesting them and paving the way for new developments.

6.1 From Copernicanism to Jansenism:
Pascal (1657) and Arnauld (1691)

In 1657, in the midst of the Jansenist controversy, Blaise Pascal found it only natural to comment on Galileo's condemnation in his own polemic against the Jesuits. He could have hardly missed the opportunity, given the existence of the Jesuit conspiracy theory. Thus, in the last (no. 18) of the *Provincial Letters,* addressing one Father Annat, Pascal wrote the following:

> [314] It was as much in vain that you obtained that decree from Rome against Galileo which condemned his opinion touching the motion of the earth. This will never be an argument to prove that it stands still, and if men had sure observations which proved[1] it is the earth which turns, not all mankind together would prevent its[2] turning, nor prevent their own turning with it. Nor would I have you imagine that the letters of Pope Zachary for the excommunication of St. Virgil, because he asserted that there were antipodes to us, have annihilated that new world;[3] and that notwithstanding he had declared this opinion to be a dangerous error, that therefore the King of Spain has not found great benefit by believing in Christopher Columbus, who came from

thence, rather than the judgment of this Pope who had never been there;[4] or, that the Church has not received great advantage since this discovery has procured the knowledge of the Gospel[5] to so great a multitude of people, who had otherwise perished in their infidelity.

[315] You see from hence, Father, what is the nature of facts, and by what principles we ought to judge of them: from whence it is easy to conclude as to our subject, that if these five propositions are not of Jansenius, it is impossible they should be extracted from his book, and that the only means of judging rightly of them, and to persuade the world, is to examine this book in a regular conference, as we have desired you to do for so long a time. Till that is done you have no right to call your adversaries obstinate: for they will be blameless as to this point of fact, as they are without errors on the points of faith; Catholics with regard to right, reasonable with regard to fact, and innocent in both cases.[6]

Here Pascal was making an analogy between the geokinetic question and both the old issue of the antipodes and the new Jansenist controversy. The comparison between the questions of the earth's motion and of the antipodes was a natural one to make. It had occurred to Copernicus when, in the preface to the *Revolutions,* he mentioned with disdain Lactantius's view of a flat earth.[7] Galileo had picked up on it when, in the Letter to Christina, he quoted in full that passage from Copernicus's preface.[8] And Descartes, in his April 1634 letter to Mersenne, had made the comparison with the problem of the antipodes, without naming Lactantius.[9] Pascal was adding new details to the story by mentioning the names of the censuring pope and the censured person. Pascal's version was widely repeated for a long time. For example, as we have seen (in chapter 5.3), Leibniz also made the comparison, both in the essay of "Apologia for the Catholic Faith Based on Right Reason" and in his summer 1688 letter to Landgrave Ernst,[10] although it is possible of course that Leibniz discovered the case without the influence of Pascal's *Provincial Letters*. And, like almost everything regarding Galileo's trial, the comparison eventually generated a controversy, which I shall discuss later in this chapter when I examine the reappearance of Pascal's comparison in the French *Encyclopedia*.

The similarity between the Copernican and Jansenist controversies was presumably the following. For Pascal, the Church had the right to condemn as heretical such propositions as the assertion that human beings are unable to resist or earn divine grace; that their actions do not give them merit or credit for their salvation; and that they are saved by faith alone.[11] However, to condemn Jansenius in this regard presupposes that such propositions are contained in his works. But to say that such propositions are contained there is a factual or historical claim and is not and cannot be an article of faith; so its acceptance cannot be dictated nor its denial condemned by the Church. This situation is like the question of the earth's motion and cannot be decided by ecclesiastical decree any more than the

earth could be made motionless by the condemnation of Galileo. Therefore, there is nothing wrong with accepting the Church's condemnation of
those propositions on grace and merit and denying that Jansenius in fact
held them.

In light of the classic status which the *Provincial Letters* were destined to
acquire, Pascal contributed considerably to spreading the Jesuit conspiracy
theory. It also appears that Pascal was suggesting an empiricist interpretation of the Copernican revolution, or at least of the geokinetic issue. For
when Pascal said that "if men had sure observations which proved it is the
earth which turns,"[12] he was suggesting that the issue could be decided by
simple observation, just as the spherical shape of the earth had been empirically proved by the voyages of Columbus (and others). This empiricist
interpretation was of course destined to become another cause célèbre of
the subsequent Galileo affair.

Another distinguished French Jansenist soon found the occasion to criticize the Church for its handling of the Galileo affair. In 1691, Antoine
Arnauld (1612–1694) published an explicit criticism of anti-Copernican
book censorship[13] in a work of more than seven hundred pages defending
the Oratorians of the Belgian town of Mons. The extremely complex details
of this controversy need not be related here.[14] It was sparked in February
1690 when one Baron de Surlet obtained approval to establish at Mons a
house of retreat under the supervision of the Oratorians, after which the
Jesuits became extremely upset and began a memorandum war. As a result,
the archbishop of Cambrai appointed a three-member commission to
investigate the affair. The commission was headed by Martinus Steyaert,
professor of theology at Louvain, who wrote a report in July 1690. Soon
Arnauld got involved; he wrote a lengthy criticism, which he published the
following year in Cologne in a book titled *Difficulties Proposed to Mr. Steyaert
on the Memorandum Given by Him to Monsignor the Archbishop of Cambrai to
Report on His Charge to Investigate the Rumors Spread against the Doctrine and the
Conduct of the Priests of the Oratory at Mons.*[15]

Arnauld discussed one hundred difficulties, the ninety-fourth of which
dealt with the problem that "there can be prohibitions that are unjust."[16] A
total of sixteen examples of this problem were given, the fifteenth of which
was an "example from philosophy, that is, the condemnation of Galileo's
books for having taught Copernicus."[17] Arnauld seemed to have more than
a superficial understanding of Galileo's trial, for he stated that Galileo's
troubles began with the publication of the *Sunspot Letters* (in 1613) and with
his advocating that the "sun turns on its own axis."[18] Furthermore, Arnauld
was clear that a key issue had been whether "one could defend as probable
an opinion that had been declared contrary to Scripture."[19] He understood
that the condemnation was also based on the claim that the earth's motion
was "false and absurd in philosophy."[20] He claimed that what was a hypoth-

esis in 1613–1616 had become "a certain truth . . . nowadays"[21] and went on to give a brief justification of this claim.

Next Arnauld discussed the Scriptural question, and with references to Saint Augustine and Saint Thomas Aquinas he argued in favor of the "very judicious rule"[22] that "in regard to natural things, whose truth can be known from convincing proofs or from manifest experiences, one must not rashly decide to give Scripture meanings that could be contrary to things so well proved."[23] He gave the example of the biblical account of Solomon's building a container that was allegedly ten cubits in diameter and thirty in circumference.[24] Arnauld also adopted the accommodationist principle that the sacred authors spoke in accordance with common opinion.

Finally, in an apparent reference to the case of Martin E. van Velden (whom he did not name),[25] Arnauld criticized Steyaert's university (Louvain) for censuring one of its philosophy professors by opposing to him the Inquisition decrees of 1616 and 1633. For Arnauld, it would have done more honor to the Inquisition to distinguish the situations in the earlier and the later part of the century and to say that although the censure had had some justification earlier, it no longer did. Arnauld felt he could find such a distinction and such a differential judgment implicit even in the behavior of the Inquisition, which after Galileo's trial had not condemned any of the numerous works that taught the earth's motion.

Such views echo those of Galileo's *Nov-antiqua* (1636) and of Auzout's *Letter to Abbé Charles* (1665). Whether or not these influenced Arnauld directly, the similarity reflects the attitude of a growing segment of the Catholic world. Arnauld's views, especially his point about the change in epistemological status of Copernicanism and the absence of specific post-trial condemnations of Copernican books, presaged similar arguments to come.

6.2 From Prison to Biblical Satire:
Un-Enlightened Myths (1709–1773)

Despite all its enlightenment about other matters, the eighteenth century was almost a golden age for the invention and diffusion of myths about Galileo's trial. In 1709, in a work on the history of heresy, Domenico Bernini asserted that Galileo was held in an Inquisition prison for five years.[26] This assertion is not true, but it was widely repeated for a long time;[27] thus Bernini may be said to have started one of the minor myths about Galileo's trial. On the other hand, the myth was not a pure fabrication because the Inquisition's sentence (which had by then been broadly publicized and printed in several works) did speak of formal imprisonment at the pleasure of the Holy Office, whereas information about the various

locations and forms of such imprisonment was harder to come by; even
Peiresc, a few years after the condemnation, was not sure about the precise
nature of Galileo's arrest. After his death, information about his imprisonment became even scarcer. The issue was not resolved until at least 1774,
when it was widely debated in Tuscany and the relevant documents were
accessed.[28]

In 1733, Iacopo Panzanini, the inheritor of Viviani's estate, died, and by
the terms of Viviani's will his books and papers were given to the library
of the hospital of Santa Maria Nuova, and his real estate was given in trust to
the family of Giovanni Battista Clemente de' Nelli for the purpose of building an appropriate mausoleum for Galileo when the opportunity arose.[29]
This change offered a pretext to Grand Duke Gian Gastone and the Florentine elites for asserting their independence from ecclesiastical authorities,[30] and so the Florentine inquisitor was asked whether there was any
Inquisition decision preventing the construction of such a mausoleum. The
inquisitor found no record of such a prohibition, and so on 8 June 1734 he
wrote to Rome for instructions.[31] The Roman Inquisition handled the matter in a routine manner. First its consultants were asked to study the matter
and make a recommendation. They recommended that there was no obstacle to erecting such a mausoleum for Galileo, and on 16 June 1734 the
cardinal-inquisitors endorsed this recommendation.[32] It took three years
for the arrangements to be made and the work to be completed. Finally, on
12 March 1737, in a ceremony attended by all the Florentine intelligentsia
and nobility, Galileo's body was exhumed from its original modest tomb in
Santa Croce's bell tower and moved to a sumptuous mausoleum on the
north side of the church's main aisle, across from Michelangelo's tomb.[33]

In this connection, it is interesting to read a summary of Galileo's trial
found in the Vatican file of trial documents, which appears to be a digest of
the trial prepared in 1734 by the Inquisition's consultants for the benefit
of the cardinal-inquisitors. It gives us a glimpse of how relatively inaccurately the affair was viewed in the mid-eighteenth century, even by people
who had access to all the proceedings and documents. It reads as follows:

> Galileo Galilei, a Florentine mathematician, was tried in the Holy Office of
> Florence for the following propositions: that the sun is at the center of the
> world and consequently immobile in regard to local motion; and that the
> earth is not the center of the world and is not immobile, but moves as a whole
> and also with diurnal motion.
>
> Summoned to Rome, he was imprisoned in this Holy Office, where the
> case was presented to the pope; on 16 June 1633, His Holiness decreed that
> the said Galileo should be interrogated about his intention, even by threaten-
> ing him with torture; that if he sustained it, after a vehement abjuration
> conducted at a meeting of the full Holy Office, he should be condemned
> to imprisonment at the pleasure of the Sacred Congregation; that he be
> enjoined that in the future he should not treat of the earth's mobility and

sun's stability in any way whatever, orally or in writing, on pain of relapse; that the book written by him and titled *Dialogue of Galileo Galilei Linceo* should be prohibited; and further that copies of the sentence, written as indicated above, should be transmitted to all apostolic nuncios and inquisitors, especially the one in Florence, who should read the said sentence publicly at a plenary meeting to include in particular professors of mathematics; all this was indeed done.

On 23 June of the said year, His Holiness commuted his detention from the prison of the Holy Office to the palace of the grand duke at Trinità dei Monti in lieu of prison; and on the first of December of the same year it was commuted to his villa, to live in solitude and without receiving anyone with whom to converse, as long as it so pleased His Holiness.[34]

The attitude here seemed to be to describe Galileo's penalties in the 1633 trial to determine whether they included any prohibition against the proposed reburial. And of course they did not. This prohibition, as we saw in chapter 4.3, had been an additional ad hoc stipulation requested by Pope Urban VIII when Galileo died in 1642.

However, the summary stated puzzlingly and incorrectly that Galileo was tried by the Florentine Inquisition. Moreover, the summary made it sound as if Galileo had been held in the Inquisition's prison throughout the 1633 trial, until the location was commuted on the day after the sentence (23 June). This was a misleading impression because, as was shown after the relevant diplomatic correspondence was published in 1774,[35] for the duration of the trial Galileo was under house arrest at the Tuscan embassy, except for the period between the first deposition (12 April) and the second deposition (April 30), when he was held in the prosecutor's apartment at the Inquisition palace. This error is significant because, if it could be committed by an Inquisition official, it could be committed more readily by someone like Bernini and the many others who followed him in believing and spreading the prison myth.

This document is important, too, for its stress on the role attributed to the pope in the conclusion of the trial; for the contents of the papal decision of 16 June 1633; and for its reference to a possible future relapse and what this reference implies about the existence of a special injunction in 1616.

In 1755, in a book on the history of astronomy published in Paris, Pierre Estève stated that Galileo had his eyes gouged out as part of his punishment.[36] This is not true. However, this claim acquired some following, although its diffusion was never large, not even as large as for the prison myth. The claim also had some basis in fact, although a weaker basis than the prison myth. What may have happened for the eye-gouging myth is that one started with the fact that toward the end of his life Galileo was tried, condemned, and punished by the Inquisition; then one considered the fact that at the end of his life he was blind; and then the imagination invented a connection between these two facts. Thus, Estève may be credited (or blamed) for having started another one of the minor myths about Galileo's trial.

In 1757, the legend of Galileo's uttering the phrase *e pur si muove* made its first appearance in print, in an English-language book by Giuseppe Baretti published in London and titled *The Italian Library*. Baretti (1719–1789) was a literary critic who was born in Turin but lived most of his life in London.[37] This work was essentially an annotated bibliography of books in Italian. In a section dealing with "natural philosophers"[38] there was an entry on the *Dialogo di Galileo Galilei*.[39] After briefly describing this book, Baretti identified the author as follows: "This is the celebrated *Galileo*, who was put in the inquisition for six years, and put to the torture, for saying, that *the earth moved*. The moment he was set at liberty, he looked up to the sky and down to the ground, and, stamping with his foot, in a contemplative mood, said, *Eppur si muove;* that is, *still it moves*, meaning the earth."[40]

Baretti had the sense to attribute these words to Galileo when "he was set at liberty," whereas other versions of the myth speak of the utterance having been made at the trial after the abjuration.[41] Such an utterance would have been proof of a relapse into heresy and thus grounds for immediate execution. Thus, not only did Galileo know better, but the fact that he was not burned should be proof enough that he did not utter those words.

It is unclear what Baretti was referring to when he spoke of Galileo "being set at liberty." Since he seemed to accept a version of the prison myth, perhaps Baretti was thinking literally of the time after Galileo had completed his alleged six-year prison term.

The myth of *e pur si muove* was in part based on the judgment that Galileo did not really change his mind about the earth's motion as a result of the trial.[42] He did abjure, of course, but the abjuration amounted to an external verbal compliance. What he thought did not necessarily correspond to the words he was willing to recite. This judgment is a respectable, and indeed plausible, historical interpretation, although it has become the subject of controversy; indeed it has become a key part of the cause célèbre.

The myth may have in part originated from a painting by the Spanish artist B. E. Murillo, dating from 1643 or 1645; it depicted Galileo in prison, pointing to a diagram of the solar system on the prison wall that bears the words *e pur si muove*. It was painted in Madrid, probably on commission from Ottavio Piccolomini, a general in the service of the king of Spain and brother of Galileo's friend Ascanio Piccolomini, archbishop of Siena.[43]

Even so, Baretti's passage remains apparently the first *statement* of the myth. Similarly, that passage contains one of the earliest explicit statements, if not indeed the earliest, of the torture thesis; and we saw in chapter 1 that the basis of the torture thesis is the reference in the Inquisition's sentence to the process of "rigorous examination." But with Baretti we are still in the prehistory of the torture controversy.

One final minor myth was created during this period. In 1773, Angelo Fabroni published in Florence a collection of previously unknown correspondence pertaining to Galileo's telescopic discoveries of 1610, the anti-Copernican decree of 1616, and the activities of Prince (and later Cardinal) Leopold de' Medici between 1639 and 1671. In 1775, Fabroni published a second volume of correspondence involving the trial of 1633, as well as Leopold's activities from the 1650s to the 1670s. Fabroni (1732–1803) was a clergyman who since 1769 had been head of the University of Pisa, and who since 1771 had been editor of the important *Giornale de' letterati*.[44] He had gathered the published documents mostly from the Medici archives, and this work was meant as a companion to a monumental twenty-volume biographical encyclopedia titled *Lives of Italians*, which he edited between 1778 and 1803, with the last two volumes appearing posthumously in 1804 and 1805.

In a footnote to the first volume of correspondence,[45] Fabroni claimed that the Dominican friar Caccini, in his sermon of 21 December 1614, discussed the suggestive biblical verse "Ye men of Galilee, why stand ye gazing up into heaven?" (Acts 1:11).[46] It is indeed true that on that date, at the church of Santa Maria Novella in Florence, Caccini preached a sermon against mathematicians in general and Galileo in particular because their beliefs and practices allegedly contradicted the Bible and thus deviated from the Catholic faith.[47] But before Fabroni there is no documentary evidence that Caccini focused on this passage. On the other hand, it would have been brilliant to exploit the ambiguity of the biblical verse to use it against Galileo. So here we have a case where someone's aesthetic or rhetorical imagination suggests something that should or could have happened, even if it did not. And so the story has been subsequently repeated many times, sometimes with and sometimes without a degree of skepticism.[48]

6.3 Whose Ignorance and Prejudice? Voltaire (1728–1770)

Few persons can be taken to represent the eighteenth century and to embody the spirit of the Enlightenment as well as Voltaire (1694–1778). It is perhaps not surprising that he found many occasions to comment on Galileo's trial, given his multidisciplinary interests and his involvement in the critique of religion and the struggle for freedom of expression.

Voltaire's earliest comment was included in letter 14 of the *Philosophical Letters*, published in 1734.[49] This letter had been written in 1728,[50] during Voltaire's exile in London (from 1726 until 1729),[51] and it was a critical comparison of Descartes and Newton.[52] Voltaire was an admirer of Newton, but he was reacting to the anti-Cartesian English criticism of Fontenelle's eloge of Newton. The result was a surprisingly balanced interpretation and assessment. On the one hand, Voltaire approved of the fact that Fontenelle had included some criticism of Newton, for Voltaire found that "he is in this country a second fabulous Hercules, to whose single valor the ignorant have ascribed the exploits of all the others."[53] On the other hand, Voltaire was "far from asserting that there are not [an] abundance of mistakes in Descartes."[54] Furthermore, Voltaire remarked that Newton was fortunate to live in a free country and in tranquillity, whereas Descartes had to emigrate to Holland in search of liberty but even there was persecuted for atheism. And yet: "His name, at length, made such a noise that it was proposed to engage him to return into France, by rewarding him according to his merit. He was offered a pension of a thousand crowns. Trusting in this, he actually returned, paid the charge of the patent, was disappointed of his pension, and went back to philosophize in the solitudes of North Holland; while the great Galileo, at the age of fourscore, groaned away his days in the dungeons of the Inquisition, because he had demonstrated[55] the motion of the earth."[56]

This was all that the letter said about Galileo. Voltaire had fallen under the spell of the prison myth. Moreover, he explained the condemnation as caused by Galileo's *demonstration* of the earth's motion; and by describing the Galilean accomplishment in this manner, Voltaire was contributing to the creation of another anticlerical myth.[57] For it is indeed true that Galileo was condemned partly for *supporting* the earth's motion, and that his geokinetic arguments were convincing and did convince almost all open-minded or progressive thinkers; but to speak of demonstration is an exaggeration and oversimplification, and to propose it as the cause of the condemnation makes the Church's conduct seem more incoherent and irrational than it really was or needed to be. At any rate, the question of the degree of cogency of Galileo's arguments for the earth's motion later became a key issue.

A second revealing passage is found in *The Age of Louis XIV,* first published in 1751. Voltaire held that there had been four great ages in history: ancient Greece, especially Athens; ancient Rome, especially the time of Caesar Augustus, Cicero, and Virgil; Renaissance Italy, especially Florence under the Medici; and seventeenth-century Europe, especially France.[58] For Voltaire, one of the reasons for the greatness of the seventeenth century was the scientific revolution, in regard to which he admitted that the contributions of England were so great that they "might give occasion to the calling of this age the age of the English as well as that of Louis XIV."[59] He began his discussion of science[60] by explaining how and why the century had started rather inauspiciously, and there made the following comment about the Galileo affair:

> This happy age, which has seen a revolution produced in the human mind, did not seem destined to it. To begin with philosophy, there was no appearance in the time of Louis XIII[61] that it would have emerged out of the chaos into which it was plunged. The Inquisition in Italy, Spain, and Portugal had

linked the errors of philosophy to the tenets of religion; the civil wars in France, and the disputes of Calvinism were not more adapted to cultivate human reason than was the fanaticism of Cromwell's time in England. Though a canon of Thorn[62] renewed the ancient planetary system of the Chaldaeans, which had been exploded for so long a time, this truth was condemned at Rome; and the congregation of the holy office, composed of seven cardinals, having declared not only heretical but absurd the motion of the earth, without which there is no true astronomy—the great Galileo having asked pardon at the age of seventy for being in the right—there was no appearance that the truth would be received in the world.[63]

Voltaire's reference to seven cardinals indicates that he was acquainted with the text of the Inquisition's 1633 sentence, which as we have seen (in chapter 1.1) had been signed by seven cardinal-inquisitors, although the same document told us that the full congregation numbered ten cardinals. If Voltaire had read the sentence, he was being careless in attributing to the Inquisition the assessment of the earth's motion as heretical and absurd, for as we have seen that the same sentence attributes this assessment to the Inquisition's consultants in 1616 and endorses only the assessment of "false and contrary to Scripture." However, this carelessness was no greater than that displayed in Nuncio Carafa's summary of the sentence published in Liège in September 1633, which we have seen (in chapter 2.3) committed the same error. And Voltaire may have been acquainted only with that summary.

A slightly longer but still capsule account of the Galileo affair was given by Voltaire in 1753 in a book titled *Essay on the Customs and Spirit of Nations and on the Principal Facts of History after Charlemagne and until Louis XIII.*[64] The topic was treated in a chapter on "Customs of the Fifteenth and Sixteenth Century and the State of the Fine Arts."[65] Voltaire wrote:

> True philosophy did not begin to enlighten men until the end of the sixteenth century. Galileo was the first who made physics speak the language of truth and reason. It was a little before him that Copernicus, at the frontiers of Poland, had discovered the true system of the world. Galileo was not only the first good physicist, but he wrote as elegantly as Plato, and he had over the Greek philosopher the incomparable advantage of saying only things that are certain and intelligible. The manner in which this great man was treated by the Inquisition toward the end of his days would bring eternal disgrace on Italy, were not this disgrace erased by the very glory of Galileo. In a decree issued in 1616, a congregation of theologians declared Copernicus's opinion, so well brought to light by the Florentine philosopher, "not only heretical in the faith, but also absurd in philosophy." This judgment against a truth later proved in so many ways is clear testimony of the force of prejudice. It should teach those who have nothing but power to be silent when philosophy speaks and not to interfere by deciding what is not within their jurisdiction. Then in 1633, Galileo was condemned by the same tribunal to prison and to do penance, and he was obliged to recant on his knees. In truth, his sentence was

milder than that of Socrates; but it was no less disgraceful to the reason of the judges of Rome than the condemnation of Socrates was to the enlightenment of the judges of Athens. Mankind seems to have this destiny: that truth should be persecuted after it begins to make its appearance. Always hindered, philosophy could not in the sixteenth century make as much progress as the fine arts.[66]

Voltaire penned an elegant image and expression in saying that the very glory of Galileo redeemed Italy from the disgrace of his condemnation; however, Galileo was not only an Italian but also a Catholic, indeed a sincere Catholic and more of a Catholic than an Italian (since Italy did not yet exist as a nation-state), and so one could modify and extend Voltaire's image and say that Galileo redeemed the Catholic Church from the stigma of having condemned him. Voltaire appeared to quote the Index's 1616 decree regarding the religious heresy and philosophical absurdity of Copernicanism, but he was really *paraphrasing* a statement in the Inquisition's 1633 sentence that was itself a *paraphrase* of the 1616 *consultants' report*. He explained that decree as the result of prejudice, and his explanation was destined to acquire a great following and generate controversy. The comparison to Socrates was an obvious one, and Peiresc had made it earlier in private, in his letter to the Vatican secretary of state pleading for Galileo's freedom;[67] Voltaire's statement may very well be the first such statement in print.

Voltaire found occasion to elaborate the prejudice explanation of Galileo's condemnation and the comparison with Socrates in a fragment written in 1756 and usually printed as section 2 of the entry "Newton and Descartes" in Voltaire's *Philosophical Dictionary*.[68]

[113] When we reflect that Newton, Locke, Clarke, and Leibniz,[69] would have been persecuted in France, imprisoned in Rome, and burnt in Lisbon, what are we to think of human reason? Happily, she was by this time born in England. In the time of Queen Mary, an active and bitter persecution had been carried on respecting the manner of pronouncing Greek; and the persecutors were the party that happened to be mistaken. Those who enjoined penance upon Galileo were more mistaken still. Every inquisitor ought to be overwhelmed by a feeling of shame in the deepest recesses of his soul at the very sight of one of the spheres of Copernicus. Yet if Newton had been born in Portugal, and any Dominican[70] had discovered a heresy in his inverse ratio of the squares of the distances, he would without hesitation have been clothed in a "san-benito," and burnt as a sacrifice acceptable to God at an "auto-da-fé."[71]

It has frequently been asked, how it happens that those who by their profession are bound to obtain knowledge and show indulgence, have so frequently been, on the contrary, ignorant and unrelenting. They have been ignorant, because they had long studied; and they have been cruel, because they perceived that their ill-chosen studies were objects of contempt to the truly discerning and wise. The inquisitors who had [114] the effrontery[72] to condemn the system of Copernicus not merely as heretical, but as absurd, had certainly nothing to fear[73] from that system. The earth, as well as the other planets, could move round the sun without their sustaining any loss of revenue or of honour. A dogma is always secure enough when it is assailed only by philosophers;[74] all the academies in the world will produce no change in the creed of the common people. What then is the foundation of that rage which has so often exasperated an Anytus[75] against a Socrates? It is that Anytus in the bottom of his heart says—Socrates despises me.[76]

Here Voltaire added ignorance and hatred to prejudice as the root causes of Galileo's condemnation; Voltaire apparently reasoned that the ignorant tend to feel themselves to be the object of contempt by the wise and so develop hatred for them. He was thereby rejecting the idea that Galileo was condemned in order to prevent his ideas from scandalizing common people, which was an explanation commonly advanced by Church officials (as Auzout and Leibniz had noted); Voltaire was objecting that the beliefs of common people cannot be changed by intellectuals and scholars.

As time went on, Voltaire's attitude apparently became more strident, as we can see from another fragment first published in 1770 and usually included as the entry "Authority" in his *Philosophical Dictionary*.[77]

[501] Miserable human beings, whether in green robes, turbans, black robes or surplices, mantles and bands, do not try to use authority when it is a question only of reason; otherwise consent to be mocked in all ages as the most impertinent of all men and to endure public hatred as the most unjust.

You have been told a hundred times of the insolent absurdity with which you condemned Galileo, and I myself am going to tell you for the hundred and first. I wish you would always celebrate its anniversary. I also wish you would have this inscription above the door to your Holy Office: "Here seven cardinals, assisted by some minor friars, decided to throw into prison the leading thinker of Italy at the age of seventy and to make him fast on bread and water, because he instructed the human race and they were ignoramuses."

[502] There they rendered a judgment in favor of the categories of Aristotle, and they decreed wisely and equitably the penalty of the galleys against anyone who was sufficiently daring to disagree with the Stagirite, whose books had been formerly condemned by two councils.[78]

It is unclear where Voltaire got the misinformation that Galileo's penalties included fasting on bread and water. Perhaps he was imaginatively embellishing the prison myth. On the other hand, Voltaire was mentioning Galileo's trial under the topic of authority, and he was making it clear that he had in mind various kinds of authorities, religious and civil, Catholic and non-Catholic. Thus the main issue was presumably the struggle between authority and reason, and this was a relatively novel and important interpretation of the affair; this view would also be elaborated in the twentieth century by the playwright Bertolt Brecht (chapter 15.1).

6.4 "Theology's War on Science":
D'Alembert and the French *Encyclopedia* (1751–1777)

Strangely enough, there was no entry on Galileo in the original edition of the French *Encyclopedia, or Reasoned Dictionary of Sciences, Arts, and Crafts,* edited by Denis Diderot and Jean D'Alembert and published in seventeen volumes from 1751 to 1765.[79] The reasons for this omission are unclear, although some have blamed it on the editors' excessively high opinion of Descartes and his rationalism on the one hand and of Francis Bacon and his empiricism on the other. The oversight was remedied when the four volumes of *Supplement to the Encyclopedia* were published in 1776–1777; they included an article titled "Galileo, Philosophy of" by the Italian scientist and Barnabite priest Paolo Frisi.[80] This article had originally been published in Italian in 1766 and then translated into French in 1767.[81] It was a comprehensive and reasonable account of Galileo's scientific work, if somewhat apologetic and excessive in praise; but it said nothing about the trial.

The original *Encyclopedia* nevertheless contained several revealing discussions of the trial. The first occurred in the general introduction included in volume 1 (1751) and titled "Preliminary Discourse."[82] It was authored by D'Alembert (1717–1783), an important physicist and thinker in his own right. The "Discourse" has been called, with some justice, "the most representative work of its age,"[83] as well as "the Manifesto of the Enlightenment."[84] It contained a philosophical history of the mind and a history of intellectual progress since the Renaissance, besides a prospectus for the whole *Encyclopedia* and a list of contributors to the first volume.[85] About halfway through the essay,[86] D'Alembert discussed various reasons why the fine arts and belles lettres were somewhat out of phase with philosophy and natural science, and why the former were able to progress earlier than the latter disciplines; one alleged reason was the excessive deference among natural philosophers toward the ancients in general and Aristotle in particular, especially as practiced by Scholasticism. Then came a passage concerning Galileo's condemnation.[87]

D'Alembert claimed that an additional factor retarding philosophic and scientific progress had been an attitude by a minority of theologians that involved several elements: an "abuse" of power over the people;[88] the fear of "blind reason";[89] the zealotry of "enthusiasts";[90] and the failure to appreciate that the best defense against the attacks of reason is to fight reason with reason. Another factor, he observed, was what might be called theological expansionism or imperialism: the tendency to expand the articles of faith necessary for salvation by indiscriminately adding opinions that theologians or clergymen happen to hold. Then there was a cause which D'Alembert himself called "theological despotism,"[91] that is, the desire to dictate belief beyond the spiritual domain of faith and morals, into matters of fact and natural philosophy; as examples of such despotism, D'Alembert gave the condemnation of Galileo and Pope Zachary's condemnation of the antipodes. D'Alembert concluded by advancing the warfare metaphor and portraying the image of theology (or religion) having "made open war"[92] against philosophy (or science).

Here D'Alembert was creating or promulgating a conflictual image of the relationship between science and religion in general, as well as interpreting Galileo's trial as epitomizing the conflict between science and religion. This view was destined to generate considerable heat but less light in the centuries to come. Yet for the most part D'Alembert exhibited a respectful, balanced, and nuanced tone. In alleging the abuse of power, he stressed that such abusive theologians were few in number and warned against hastily overgeneralizing from some to all. In talking of the absurdity of religions, he explicitly excluded Catholicism (labeling it "ours"[93]). He maintained the same tone when he clarified that the threat to religion came not from philosophic reason as such, which can actually help religion, but from the zealotry of those who fear it. And in advancing his warfare generalization, he qualified it by restricting it to the enemies that are "poorly instructed or badly intentioned."[94]

Finally, regarding the role of Scripture,[95] D'Alembert expressed the accommodationist principle that Scripture adapts itself to the common language and beliefs about physical reality. This was a widely accepted principle that had been advanced at least as early as Kepler's introduction to the *Astronomia Nova*. But D'Alembert seemed to go further by limiting scriptural authority to faith and morals but denying it a role in natural philosophy, giving as a reason that "the All-Powerful has expressly left to our disputations"[96] matters about the world systems. These views reflect the Baronio principle advanced in Galileo's *Nov-antiqua* (1636).

A briefer reference to Galileo occurred in the article on astronomy, which also appeared in the first volume of the *Encyclopedia* and was also written by D'Alembert. He devoted a short paragraph to Galileo's contributions to astronomy, adding this judgment: "Galileo's opinions attracted the censures of the Roman Inquisition. But these censures did not prevent his being regarded as one of the greatest geniuses that have appeared in a long time."[97] In that first volume, D'Alembert also had an article on the "Antipodes," in which he did not mention Galileo but advanced a very anticlerical account of Pope Zachary's condemnation of Bishop Virgil's belief in the antipodes.[98]

For this and other obvious reasons, the first volume of the *Encyclopedia* became the subject of a long, critical review in the October 1751 issue of the Jesuit *Journal de Trévoux,* published in Paris.[99] The review attacked the *Encyclopedia* on many points, only one of which was the condemnation of the antipodes. On that topic, the criticism included a devastating exposé of mistaken dates, names, titles, and other such facts.[100] Most important, it argued that the crime with which Virgil was being charged was not the belief in the existence of the physical antipodes, but rather the theological claim that there were on the other side of the earth inhabitants who did not descend from Adam and so had not been redeemed by Jesus; hence the crime was one involving theological, not philosophical or physical, belief, and so the Church was acting within its proper jurisdiction and its behavior was justified. This is a move often made in such controversies, and in chapter 8.3 we will see it emerging for the case of Galileo at the end of the eighteenth century. The argument was also made with great fanfare for the case of Giordano Bruno in 1942, when the summary of his trial was rediscovered and first published.[101] At any rate, D'Alembert was unconvinced, and so he was to repeat the story in the article on Copernicus in 1754, as we shall see presently.

D'Alembert also expanded this kind of comparison, for example in the article on Aristarchus of Samos in the fourteenth volume (under the entry "Samos").[102] After mentioning that Aristarchus was accused of impiety by Cleanthes, it added: "The accusation against Aristarchus should be less surprising than the treatment given in the last century to the celebrated Galileo. This respectable man, to whom astronomy, physics, and geometry are so much obliged, was forced to publicly certify that the opinion of the earth's motion was a heresy. He was even condemned to prison for an undetermined period. This fact is one of those that show us that, as the world becomes older, it does not become any wiser."[103]

However, the main discussion of Galileo is in the entry on Copernicus, contributed by D'Alembert and published in the fourth volume in 1754. It deserves extended quotation and careful examination. After a few short paragraphs describing the content of the Copernican system and its history, it continued:

[173] Nowadays this system is generally followed in France and England, especially after Descartes and Newton each tried to confirm it by means of physical explanations. The latter of these philosophers, in particular, has developed the main point of the Copernican system with admirable clarity and surprising precision. In regard to Descartes, the manner in which he tried to elaborate it, although ingenious, was too vague to have followers for long; thus hardly any of them remain today among true scholars.

In Italy they prohibit Copernicus's system, [174] which they regard as contrary to Scripture on account of the earth's motion that is supposed by this system. *(See "System.")* The great Galileo was formerly tried by the Inquisition, and his opinion of the earth's motion was condemned as heretical; in the decree which the inquisitors issued against him, they did not spare the name of Copernicus, who had updated it after the Cardinal of Cusa, nor that of Diego de Zúñiga, who had taught it in his *Commentaries on Job,* nor that of the Italian Carmelite Father Foscarini, who in a learned letter addressed to his general had just proved that that opinion was not at all contrary to Scripture. Despite this censure, Galileo continued to dogmatize on the earth's motion, and so he was condemned again and obliged to publicly retract it and to abjure his alleged error both orally and in writing; this he did on 22 June 1633.[104] Having promised on his knees and with his hand on the gospels that he would never say or do anything contrary to this sentence, he was sent back to the Inquisition's prison, from which he was soon released. This event frightened so much Descartes (who was very obedient to the Holy See) that he stopped the publication of his treatise on the world, which was ready to appear. *(See all these details in the biography of Descartes by Mr. Baillet.)*

Since that time, the most enlightened philosophers and astronomers in Italy have not dared support the Copernican system. If by chance they could seem to adopt it, they need to give notice that they regard it merely as a hypothesis and additionally that they are very obedient to the decrees of the supreme pontiffs on this subject.

It would be desirable that a country as full of intelligence and learning as Italy recognize an error so harmful to scientific progress and that she think of this subject as we do in France! Such a change would be worthy of the enlightened pontiff who governs the Church nowadays. Friend of the sciences and himself a scholar, he ought to legislate to the inquisitors on this subject, as he has already done for more important subjects. There is no inquisitor who should not blush upon seeing a Copernican sphere, said a celebrated author.[105] This furor of the Inquisition against the earth's motion harms religion itself.[106] In fact, what will the weak and the simple think of the real dogmas which faith obliges us to believe, if these dogmas are mixed with doubtful or false opinions? Is it not better to say that in matters of faith Scripture speaks in accordance with the Holy Spirit, but that in matters of physics it must speak in accordance with people's beliefs and in their language, in order to make itself accessible to them?[107] By means of this distinction one takes care of everything; physics and faith are equally safe. One of the principal causes of the discredit in which the Copernican system finds itself in Spain and Italy is that one is persuaded some supreme pontiffs have decided the earth does not turn and that one believes the judgment of the pope is infallible even in matters that do not concern Christianity at all. In France one recognizes only the Church as infallible, and one is in a better position than elsewhere to believe, in regard to the system of the world, astronomical observations rather than the Inquisition's decrees; for the same reason, the king of Spain, says Pascal,[108] preferred to believe, in regard to the existence of the antipodes, Christopher Columbus (who returned from there) rather than Pope Zachary (who had never been there). *(See "Antipodes" and "Cosmography.")*

In his biography of Descartes, which we just cited, Mr. Baillet accuses the Jesuit Father Scheiner of having denounced Galileo to the Inquisition for his opinion of the earth's motion. In fact, this priest was jealous or displeased toward Galileo in regard to the discovery of sunspots, which Galileo contested to him. But if it is true that Father Scheiner perpetrated this vengeance on his opponent, such an action would bring more discredit to his memory than the real or presumed discovery of sunspots would bring him honor.[109] (*See "Spots."*)

In France one supports the Copernican system without fear, and one is persuaded for the reasons we have mentioned that this system is not at all contrary to Faith, even though Joshua said "Sun, stand thou still."[110] So it is that

one responds in a solid and satisfactory manner to all the difficulties of unbe-
lievers in regard to certain passages of Scripture where they claim, without
reason, to find gross physical or astronomical errors.[111]

D'Alembert was repeating the common errors of attributing the anti-
Copernican decree of 1616 to the Inquisition (rather than the Index) and
interpreting it to have condemned the earth's motion as heretical (rather
than merely contrary to Scripture); by now we know that he was in good
company in these oversights.

D'Alembert's point about the alleged infallibility of the pope and/or the
Church raised some relatively novel issues. He was right that many people
believed that there were papal decrees against the earth's motion and/or
against Galileo himself. And this was not merely a belief of the uneducated
masses. For example, when the Minim friars Thomas Le Seur and François
Jacquier published their widely acclaimed annotated edition of Newton's
Principia in 1739–1742, their qualification about the earth's motion explicitly spoke of *papal* decrees: "Newton in this third book supposes the motion
of the earth. We could not explain the author's propositions otherwise than
by making the same supposition. We are therefore forced to sustain a character which is not our own; but we profess to pay the obsequious reverence
which is due to the decrees pronounced by the supreme Pontiffs against the
motion of the earth."[112] D'Alembert was perhaps rejecting this attribution
to the pope; for he contrasted the situation in France with that in Spain and
Italy, and he went on to speak of the *Inquisition's* decrees, thus implicitly distinguishing between papal and Inquisition decrees, although, as we have
seen, he did not seem to distinguish between the Inquisition's decrees and
those of the Index. But he was not too clear or explicit about this matter.

However, D'Alembert was explicitly distinguishing between decrees of
the pope and decrees of Church councils, for he went on to say that even if
Copernicus and Galileo had been condemned by papal decrees, these were
not necessarily infallible, as decrees of Church councils were. In thus denying papal infallibility, D'Alembert was relying partly on the familiar
Galilean distinction between questions of faith and morals on the one hand
and questions of physics and astronomy on the other. His denial was also
based partly on the French policy of asserting national autonomy from the
pope; this position went back at least to the time of Richelieu, when, as we
saw in chapter 4, the condemnation of Copernicanism would have needed
to be ratified by the Sorbonne and the Paris Parliament to be valid in
France (but it was not). Since then such nationalism had strengthened; for
example, in 1682 the French clergy had issued a declaration known as the
"Four Articles" that reduced papal infallibility to insignificance; and
although in 1690 Pope Alexander VIII had annulled the declaration, he
did not brand it heretical, and so the declaration continued to be tolerated
by Rome and to spread in France.[113]

D'Alembert's most intriguing point was his reference to Pope Benedict XIV. D'Alembert was sincere in his respectful and positive expressions about him, and indeed this pope was generally liked and respected. Here D'Alembert was making an indirect plea to the pope. In 1757 Benedict XIV did act to relax the anti-Copernican censures. Thus, D'Alembert's plea may have been a causal factor in his decision, as some scholars have suggested.[114] On the other hand, to regard that plea as the only cause or as sufficient would be an oversimplification and might commit the fallacy of *post hoc ergo propter hoc*. However, the details of that story are the subject of the next chapter.

Chapter 7

Incompetence or Enlightenment?

Pope Benedict XIV (1740–1758)

In 1740 Prospero Lambertini, from Bologna, was elected pope Benedict XIV; he reigned until 1758. He was widely respected and liked by Catholic, non-Catholic, and non-Christian rulers, scholars, and common people.[1] For example, Voltaire exchanged letters, compliments, and gifts with Benedict, claiming in 1745 that in his study he had an engraving of the pope with the caption: "Here is Lambertini, fittingly the father of Rome and of all the earth, / Who teaches the world by his writings and beautifies it by his goodness."[2] And we have already seen (in chapter 6.4) that D'Alembert spoke highly of him in the article on Copernicus in the French *Encyclopedia* and acted on his belief by indirectly addressing to the pope a plea for the relaxation of the anti-Copernican censures.

Benedict's reputation for enlightenment was generally well deserved. For example, Lodovico Muratori interacted with him with mutual respect; and as Muratori's *De Ingeniorum Moderatione* (1714) continued to be a favorite target of dogmatists, in order "to stop the unending sniping at it, the pope had the work examined by the Committee of the Index which reported that 'no censure was needed, as the author had attacked only abuses and popular opinions which had no official sanction.'"[3] Another example was the liberal-minded letter he wrote in 1748 to the grand inquisitor of Spain, advising caution and tolerance with regard to the works of one Cardinal Noris, which had been placed on the *Index* by the Spanish Inquisition.[4] And in 1754 Benedict intervened to defend a professor at the Jesuit Roman College, Carlo Benvenuti, who had sponsored a student thesis and public discussion advocating an approach to natural philosophy that was Newtonian and independent from metaphysics; the Jesuit authorities removed him from his position and were planning to transfer him to the provinces, but the pope had him simply transferred to another department at the College.[5]

More to the point, it was during the papacy of Benedict XIV that two important events in the subsequent Galileo affair occurred: in 1744 Galileo's *Dialogue* was republished for the first time with the Church's approval as the fourth volume of the Padua edition of his collected works; and in 1758 the new edition of the *Index* dropped the prohibition against "all books teaching the earth's motion and the sun's immobility."[6] These episodes deserve careful examination.

7.1 Galileo's *Dialogue* Unbanned, Sort Of:
Toaldo's Edition and Calmet's Introduction (1741–1744)

On 29 September 1741, Padua's inquisitor, Paolo A. Ambrogi, wrote to the Inquisition in Rome to ask its opinion on a projected publication by the press at Padua's seminary of Galileo's complete works, including the *Dialogue*. The editors had promised to revise the book to make it "hypothetical"; to have the revision done by persons who were both learned and of proven Catholic faith; and to include Galileo's abjuration and any other declaration required by the Inquisition.[7] On 9 October, the Inquisition decided to approve the project as described.[8]

However, on 10 February 1742 Inquisitor Ambrogi wrote again to Rome describing delays and difficulties encountered by the project, as well as some changes.[9] The editors were now planning to leave the text of the *Dialogue* unchanged but to add an apologetic editorial preface and the Inquisition's 1633 sentence, besides Galileo's abjuration; moreover, they were planning to include the Letter to Christina. In light of these changes, the Paduan inquisitor was asking the Roman Inquisition to reexamine the matter. In his letter he enclosed a copy of the proposed preface, which was the following:

> [136] O learned Christian reader, here is a beautiful example of humility and submission to the decisions of the Holy Roman Church. What I present to you is Galileo Galilei's famous *Dialogue on the Two Chief Systems, Ptolemaic and Copernican*. In this *Dialogue,* he showed too much fondness for the second, which is not compatible with Holy Writ; thus, he later repented and performed a solemn abjuration and retraction. Having decided to publish the whole body of the works of this author, which are avidly sought after, I did not think I could omit this one, which reveals not only the loftiness of his mind but also the docility and rectitude of his heart. Indeed, I have wanted the remedy to precede the disease in print, by prefacing to the dialogue itself the sentence pronounced against him and the ready mortification he showed toward the venerable decisions of the Holy Office; for he declared that what he had written on the subject, impulsively and out of intellectual vanity, was not only false but also improbable, because it was contrary to the divine scriptures. Given, [137] then, that the Copernican hypothesis is false and untenable, and that I

also condemn and detest it in the clearest manner and for the same reason, you can make use of the other admirable doctrines that are coincidentally found scattered on almost every page.[10]

This preface, without criticizing the Church, attempted to portray Galileo in a favorable light. Such a bipartisan account is clearly reminiscent of the one advanced in Viviani's biography[11] of Galileo and was probably influenced by it.

On 14 March 1742 Rome replied that, as long as the latest stipulations were observed, the Paduan inquisitor could grant the usual imprimatur.[12] This reply was apparently based on the following memorandum, which reads like a pair of consultant reports and is found in the file of Index documents for this episode:

[137] [I][13] Last [September][14] the Father Inquisitor [of Padua] informed this Supreme Congregation of the petition made to him for permission to reprint all of Galilei's works. To obtain it, the printer obliged himself to print all declarations that might be prescribed by this Supreme Congregation; to include in the fourth volume the abjuration made by the author; to do everything possible to change the exposition to a hypothetical one, as it had been done there [in Padua] for the reprinting of Pourchot;[15] and finally to have the correction done with the assistance of men who are learned and of proven Catholic religion.

When the petition was considered by the Consultants at their meeting of 9 October, they reached the following decision [138] (the Most Eminent Lord Cardinals being absent [on autumn recess],[16] and the meeting not having been held at the Minerva): "The committee of Consultants specially appointed by His Holiness decided that one should reply to the Father Inquisitor of Padua to permit the printing of the works in question, but only on the conditions described by the Father Inquisitor."

Now the Father Inquisitor informs us that the previous agreement has been modified, insofar as they want the imprimatur to be contingent only on a declaration to be printed at the beginning of the book and on the omission of the things mentioned in the same declaration.[17] I have the honor of reading this declaration to Your Eminences. And there was a reading of the flyer "O learned Christian reader, here is . . ."

[II] Note that the needed searches have been made in the archives and the chancery of this Supreme Tribunal in regard to Galileo's works. It has been found that according to the sentence emanated before the pope on 16 June 1633, Galileo (who was imprisoned in this Holy Office) was condemned to a vehement abjuration and to prison at its pleasure; that he was given the injunction not to treat any longer either orally or in writing of the earth's mobility or sun's stability, on pain of relapse; and that his book entitled *Dialogue of Galileo Galilei Lincean* was likewise prohibited. And on 9 September of the same year 1633, the pope ordered that the Inquisitor of Florence be severely reprimanded for having allowed the publication of the works of the said Galileo, but this reprimand does not indicate whether these works were different from the said *Dialogue,* which had appeared in 1632; this work is however listed among the prohibited books. It is true that in 1718 Galileo's collected works were published in Florence in three volumes, together with various treatises not previously published and accompanied by many [139] other extras; but there is no indication in the proceedings of the Holy Office or in the Roman *Index* whether that edition was prohibited.[18]

The first part of this memorandum was a summary of recent developments regarding the projected edition of Galileo's collected works. Its accuracy left something to be desired, since it spoke as if the present question had begun the previous November (rather than September), and as if it were the editorial preface (rather that the separate project description) that indicated what hermeneutical essays were being printed with the *Dialogue.*

The second part of the memorandum was a summary of relevant facts regarding the condemnation of Galileo and his works. It is puzzling in several respects. It claimed to have searched the archives but focused on the papal decision of 16 June 1633 made at the Inquisition meeting presided over by the pope on that day, rather than, for example, on the text of the sentence of June 22; this suggests at least that in the archives of the Inquisition the minutes were more easily accessible than the records of trial proceedings or final sentences, or perhaps even that the sentence was already missing from the files, as it is today and was discovered in the course of the nineteenth century. In this regard, this summary was analogous to the one produced by the Inquisition in 1734 in considering the reburial of Galileo (as we saw in chapter 6.2). Like that earlier summary, this one also spoke of an injunction administered to Galileo at the conclusion of the 1633 trial, without mentioning that (according to the sentence of June 22) that injunction had supposedly already been given to Galileo in 1616; this suggested that the author of this summary had not read the final sentence or, if he had, that he was inclined to dismiss the 1616 special injunction as invalid. Similarly, just as the 1734 summary seemed to be centered on the issue of whether Galileo's penalties included any stipulations about his burial, now the specific issue was the prohibition of the *Dialogue,* and the 1742 summary was trying to recount the facts relevant to this particular issue.[19]

Other changes were in store before the edition finally appeared. On 20 May 1742, the Paduan inquisitor again wrote to Rome with a slightly different proposal.[20] The editors still did not think it feasible or appropriate to change the text of the *Dialogue,* but they suggested that they were ready and willing to make deletions and changes in the marginal postils that dotted the pages of the book and that read like a running interpretive commentary by the author about the topics discussed by the three interlocutors. Moreover, the editors had dropped the idea of including Galileo's Letter to Christina; instead they were planning to include an essay already published

in Italian by the French Benedictine friar[21] and biblical scholar Augustin
Calmet that presumably defended the geostatic worldview on the basis of
Scripture. They were still thinking of reprinting the text of both the Inquisition's sentence and Galileo's abjuration. Finally, they were willing to
rewrite their editorial preface.

On June 6, the cardinal-inquisitors considered the matter, but before
making a decision they wanted more information about when and how the
Church had decided that the Copernican system could be admitted as a
hypothesis.[22] A consultant, Friar Luigi Maria Giovasco, was then commissioned to research the matter and make a recommendation. After this
report was submitted, on 13 June 1742 the cardinal-inquisitors approved
the publication as described in the latest prospectus.[23] The opinion written
by Giovasco is extremely important:

[146] At the beginning of the last century there appeared in print a work *On
the Revolutions of the Heavenly Spheres* by Nicolaus Copernicus (a celebrated
author on astronomical subjects) and a work by Diego de Zúñiga (a commen-
tator on the books [147] of Job). They supported the ancient opinion of
Pythagoras, who taught that the Sun was the motionless center of the world
and that the terraqueous globe of the Earth turned around it with perpetual
motion. The Carmelite Father Paolo Antonio Foscarini adopted such a system and defended it against the censure of theologians, who judged it false
and contrary to Sacred Scripture.

This system, which is commonly called Copernican for having been
reawakened by Copernicus from the ashes of the ancient philosophy of
Pythagoras, was denounced to the Sacred Congregation of the Index. On 5
March 1616, this Congregation published a decree prohibiting the system as
a false Pythagorean doctrine contrary to Sacred Scripture and prejudicial to
Catholic truth. But there was this difference: that Father Foscarini's *Letter* was
prohibited absolutely, whereas Copernicus's book and Diego de Zúñiga's
Commentaries on Job were merely suspended, until corrected.

Then some publishers approached the same Sacred Congregation of the
Index to have the corrections of the above-mentioned works and to be able to
publish them, exempt from the announced suspension. Father Master Capiferreo, secretary of that Congregation at that time, was commissioned with
this correction. So another decree appeared declaring that the system should
be understood as condemned only when it was expounded as an absolute thesis, but not when it was expounded as a hypothesis to better know the revolutions of the heavenly spheres.

These corrections appeared in a decree of the Sacred Congregation of the
Index of the year 1620. They emend the chapters of Copernicus's work in
such a way that [148] the printed text is left intact where it speaks problematically, and it is changed to mere hypothesis where it speaks in the manner of a
doctrinal and absolute thesis. Corrected in this way, Copernicus's work is even
today free of any condemnation.

Indeed, all astronomers study the moon by following Copernicus and tell
us that they follow such a system in the manner of a hypothesis and not in the

manner of a thesis, for they think it is more useful for contemplating the oppositions and phenomena of the stars.

In the year 1633, there appeared the *Dialogue of Galileo Galilei Lincean on the Two Chief World Systems, Ptolemaic and Copernican,* in which he established the Pythagorean system in the manner of a thesis. So it was prohibited by the same Sacred Congregation of the Index on 23 August 1634 because it defended and advocated such a system in the manner of a thesis and not in the manner of an imagined hypothesis. Moreover, since this author showed an obstinate unwillingness to submit to the above-mentioned censure, he had the misfortune that the Holy Office conducted proceedings against him, imprisoning him and obliging him to a public retraction.

One can read this whole story in *Philosophia Neo-Palaea,* by Father Master Agnani, librarian of the Casanatense Library, printed in Rome by Majnardi in the year 1734, from p. 159 (paragraph "Secondly, I answer") to p. 165.

This author gives an excellent and learned defense of the Roman condemnation of such a system when expounded in the manner of a thesis and not in the manner of a hypothesis, against modern ultramontane philosophers; they claim that this system is not contrary to Scripture but is a matter of opinion in regard to which philosophers, who fight one another for and against [149] such a system, may be mistaken, as St. Augustine was mistaken in maintaining the impossibility of the antipodes.

Thus it seems that by reprinting in Padua the works of Galileo Galilei, among which there is the prohibited *Dialogue,* in accordance with what has been said above; by including the decrees and Galileo's retraction, as the printer promises; with the marginal notes referring to the prohibition to speak of the subject in the manner of a thesis and to the fact that one may discuss it only in the manner of a hypothesis; with the addition of Father Calmet's "Dissertation," which for its part confutes such a system if taken in the manner of a thesis; by all these means one remedies very well the [potential] damage of this printing, and one corrects the daring of the modern philosophers who accuse of injustice the Roman condemnation and censure of such a system.

Fra Luigi Maria Giovasco,[24]
Consultant to the Holy Office.[25]

In dating the original publication of Copernicus's *Revolutions* to the beginning of the seventeenth century, this consultant was spreading misinformation and advertising his own carelessness. He was similarly inaccurate about Zúñiga's *Commentaries on Job,* which first appeared in Toledo in 1584 and then in Rome in 1591. More important, he made it sound as if the earth's motion was as central to Zúñiga's work as it was to that of Copernicus or Foscarini; but in so doing, he was following the suggestion implied by the 1616 anti-Copernican decree of the Index.

The report suggested that the revision of Copernicus's book had been made by the secretary of the Index, the Dominican friar Francesco Capiferreo. As secretary, he certainly had a role, and indeed his signature appeared at the end of the 1620 decree (as well as the 1616 decree). However, today we know that Francesco Ingoli was mainly responsible for those revisions.[26] Thus Giovasco was probably merely extrapolating from that signature.

More important, the report was explicitly drawing the thesis-hypothesis distinction implied by the 1620 decree and generally adopted by Catholics afterwards, and applying the distinction in order to claim that Copernicanism was prohibited and condemned if treated as a thesis, but allowed if treated as a hypothesis. However, the report did not distinguish between the instrumentalist notion of hypothesis and the probabilist conception, which, as we saw in chapter 1.4, was crucial for understanding, to say nothing of evaluating, the Galileo affair. On the other hand, the consultant was attempting to clarify the distinction the report did make, for he did not merely use the labels of thesis and hypothesis; he also associated a thesis with a family of notions such as "absolute" and "doctrinal"; and he correlated a hypothesis with the cluster of "to better know the revolutions of the heavenly spheres," "more useful for contemplating . . . phenomena," "imagined," and speaking "problematically." The last notion is especially interesting because it is reminiscent of a concept used by Peiresc in his plea for Galileo's liberation, when he defended the *Dialogue* as a *scherzo problematico* (see chapter 3.3).

With regard to the 1633 trial, the consultant appeared to give 1633 as the date of publication of the *Dialogue* and to accept some version of the prison myth. Thus, not only must he not have consulted the trial proceedings, but he probably did not even read the text of the Inquisition's 1633 sentence, which made it clear that the book had been published the previous year. However, he did apparently get a hold of the 1634 Index's decree prohibiting Galileo's book.

Toward the end, the report cited G. D. Agnani's *Philosophia Neo-Palaea* both as a source for the story of the affair and a justification of the condemnation based on the thesis-hypothesis distinction. Besides being a useful and responsible reference, this citation may also be a confession by Giovasco that he largely relied on Agnani's account rather than examining the primary sources and documents himself. In any case, here Agnani and Giovasco were providing a new kind of clerical apology that would later acquire something of a following and cause further controversy.

However, another consequence of the thesis-hypothesis distinction was the legitimacy of correcting a prohibited thesis-oriented work to make it hypothetical. In the last paragraph the report drew this conclusion with regard to the *Dialogue*. That had been the main point of Viviani's modest but unsuccessful plea in his 1690 letter to Baldigiani (as we saw in chapter 5.1). After more than half a century, one of Viviani's dreams was coming true.

Without further significant complications, the edition, in four volumes, was published by Padua's seminary in 1744. It was edited by Giuseppe

Toaldo, who had conceived, nurtured, and executed the project.[27] The *Dialogue*, with some related material, was in volume 4. The edition bore the usual ecclesiastical imprimatur by the local Paduan officials, with one noteworthy difference. There were two, rather than just one, sets of imprimaturs: one set in volume 1, applying to the published works in general and dated June and July 1742; and another set in volume 4, applying specifically to the *Dialogue* and dated May and June 1743.[28]

The text in the body of the *Dialogue* had indeed been left intact. Only the marginal postils had been "corrected": sixteen of them were deleted and forty-six edited to qualify the earth's motion as "hypothetical."[29] The Inquisition's sentence of 1633 and Galileo's abjuration preceded the text; they were printed in Latin, having been taken from Riccioli's *Almagestum Novum*. Also preceding the text, in accordance with the latest approved plan, were Calmet's essay on biblical exegesis and an editorial preface by Toaldo.

The editorial preface was not the one originally proposed and quoted above, although it was equally brief. The change is easy to understand in light of the changes made in the project since that original proposal. In the new preface actually published,[30] Toaldo mostly echoed Galileo himself rather than Viviani (as the earlier proposed preface had done). Thus, the published preface stated that it endorsed Galileo's own "retraction and qualification." It declared that the earth's motion was nothing but a "pure mathematical hypothesis," which was Galileo's own wording in the preface to the *Dialogue*.[31] It mentioned the removal or emendation of marginal postils that were not "indeterminate," which was an oblique reference to the book's full title: *Dialogue by Galileo Galilei, Lincean Academician, Extraordinary Mathematician at the University of Pisa, and Philosopher and Chief Mathematician to the Most Serene Grand Duke of Tuscany; where in meetings over the course of four days one discusses the Two Chief World Systems, Ptolemaic and Copernican, proposing indeterminately the philosophical and natural reasons for the one as well as for the other side.*[32] This title had been paraphrased (while still including the crucial word) in the latest version of the project description approved by Rome.[33] However, in the actual edition, the title was reduced to *Dialogue*.[34] Apparently the full title was being censored as insufficiently hypothetical, like so many of the marginal postils.[35]

Toaldo's preface also called attention to Calmet's biblical introduction and described it in a manner which I would regard as noncommittal. His description said nothing about the anti-Copernican or anti-Galilean flavor of that introduction. By displaying such descriptive moderation, caution, and sensitivity, Toaldo turned out to make an accurate statement, for, as we shall see presently, Calmet's introduction was not only not critical of Galileo but was really pro-Galilean.

To begin with, we should note the full title of Calmet's essay: "Dissertation on the World System of the Ancient Jews."[36] It was long (occupying twenty pages of small print), scholarly (with about ten citations per page), and erudite (with many biblical verses quoted in Hebrew). Its prologue and epilogue are worth quoting:

[1] It is really astonishing that the world is so little known. After so many centuries that the universe has been the subject of human investigations and disputes ("He surrendered the world to their disputations"),[37] one barely knows the shape and structure of the earth we inhabit, and we know only its surface and a very small part at that. Regarding all the rest of the universe, we are limited to devising systems and constructing mere hypotheses, without hope of ever arriving at a precise demonstrative knowledge of the things we study. Everything that the ancients had invented on this subject, all the discoveries they believed to have made, all their world systems have been either refuted or reformed in the last few centuries. Who doubts that one day we in turn will be refuted and abandoned by those who will be born after us? On this subject, there will always be obscurities and insuperable difficulties. It seems that God, being jealous (so to speak) of the beauty and magnificence of his work, has reserved only for himself the perfect knowledge of its structure and the secrets of its motions and revolutions. He lets us see enough to oblige us to recognize the wisdom and to make us admire the [2] infinite power of the Maker, but not to satisfy our curiosity and our desire. The study of the world and its parts is one of those laborious occupations which the Lord has given to men so that they may keep trying: "A thankless task God has appointed for men to be busied about."[38] However great is the progress made in this study, there will always be much more to be known: "Beyond these, many things lie hid; only a few of his works have we seen."[39]

One should never require or pretend that the sacred writers explain themselves with philosophical rigor or with the precision which the professors of the human sciences expect from their disciples. The Holy Spirit speaks to everybody and wants to be understood by the ignorant as well as by the learned. The latter understand popular expressions as well as the people do; but the people could not understand philosophical and technical expressions. Thus, so that no one would lose anything and all would profit, the wisdom of God decided to adapt itself to the simple ones in regard to the manner of speaking, and to give scholars something to struggle with in regard to the greatness and majesty of the things that face them. So one must have a most profound respect for conduct so full of gracious condescension and goodness.

Commentators who became involved in clarifying the occult meanings of the Holy Books and to explain their obscure terminology did not always abide by this principle. When they came across passages in which a Sacred Author expresses himself in the popular manner, instead of studying the beliefs which he presupposed in the mind of those to whom he spoke, they tried to show the truth of what they wanted to say and to reform his expressions on the basis of the ideas supplied to them by religion and philosophy. When, for example, Scripture attributes intelligence to animals, a body to God, and a soul to sensible things, interpreters do not neglect to note that these are popular and imprecise manners of speaking. That is all very well. But it would also be necessary to tell us what the people thought about the subject; what their ideas were, regardless of whether they were true or false; and then refute them, if those ideas deserved a refutation. Instead all commentators have wanted to squeeze out of the Sacred Author their own opinion, making him say what they wanted him to say; and so they made Moses or Solomon speak as Ptolemy, Galileo, Copernicus, or Descartes would have spoken. In the first chapter of Genesis, which treats of the creation of the world, they found all the systems they had in their mind. This is so true that a few years ago a book was published entitled *Descartes as Moses*,[40] in which the author undertakes to show that the worldview of Moses is the very same as that of Descartes.

Here we do not claim to impose laws on others, or to make it sound as if we had greater enlightenment than those who have preceded us. On the contrary, we confess that very often we have followed the current, and that (forewarned by the opinions of the schools) we have supposed that the Sacred Author wanted to say what we believe. However, by comparing the various expressions of Scripture on the arrangement of the parts of the universe, we have found that the world system of the ancient Jews was most different from ours, and that we often do unreasonable violence to the text when we want to adjust it to our preconceptions. One thing has been very instrumental in undeceiving us and in motivating our doubts on this subject: that is, reading the ancient philosophers and the Church Fathers. The former, whether by tradition or in some other way, had almost the same opinions as the Israelites on the structure of the world. The latter, imbued with respect for the Divine Scriptures and less prone than we are to take the liberty of making them fit [3] their opinions, nevertheless took them literally and followed the first idea that came to mind, and from that they constructed a viable system that was most similar to that of the ancient Jews. After exhibiting the expressions of the Sacred Writers, we will compare their hypothesis to that of the ancient philosophers and the Church Fathers. That is the method we propose to use in the present dissertation. . . .[41]

[19] From what has been said so far, it appears that the world system of the Jews, as we have expounded it, has a very great similarity to that of the ancient philosophers. This hypothesis is simple, easy, and intelligible; proportionate to the ability of the people; and appropriate to give them a good idea of the wisdom and power of God and to inspire in them vivid feelings of their own weakness and their total dependence. Thus, it is the most useful one [20] for the intention of the Holy Spirit, which is to lead us to God by means of fear and love; for the aim of all Scriptures is this: "The last word, when all is heard: Fear God and keep his commandments, for this is man's all."[42] Error in this sort of thing is of no consequence at all in regard to eternity. There is no doubt, says St. Augustine, that our Sacred Authors knew with certainty the whole truth about the world system; but the Most Divine Spirit, who spoke through their mouths, did not judge it appropriate to teach it to men, for these are things that do not pertain at all to salvation or lead us at all to become more just and better: "It should be said that our authors did know the truth about the shape of heaven, but the Spirit of God, which was speaking through them, did not want to teach men these things which are of no use to salvation."[43]

Now let no one tell us that since what they teach us about this is contrary to truth and to experience, one cannot give any weight to the rest of their statements, for they did not ascertain whether things were as they said. They simply assumed those things; and they did not expound their own beliefs, but the opinion of the people. In all of Scripture, there is not a single chapter aimed to instruct us with precision about those subjects that are indifferent in regard to our ultimate purpose. Are philosophers and theologians obliged by any chance, when they speak to the people, to use the same expressions used in schools and in scholarly books to explain the secrets of nature or the mysteries of religion? And if this is permitted daily to scholars and philosophers, why should it have not been permitted to authors who wanted to be useful to the many and to express themselves in such a way as to be understood by the simplest persons?[44]

Calmet began by elaborating the theme of epistemological modesty and partial revelation by God through his work, which Galileo had mentioned on the last page of the *Dialogue;* although the Benedictine friar was expressing more skepticism and pessimism than Galileo would be likely to endorse, Galileo too had made the point. Then Calmet went on to formulate and discuss the principle of accommodation, which Galileo had also espoused, although it was not at all original with him; by then it had become a widely accepted idea that facilitated acceptance of Copernicanism vis-à-vis Scripture. In the third paragraph, Calmet was criticizing with clarity and forcefulness the common abuse of reading one's own preconceptions in Scripture and was proposing a relatively novel approach that would pay more attention to the historical and intellectual context of the writers and audience.

Next Calmet described in more detail his contextual, historical, and comparative approach, which led him to conclude that the biblical worldview was very different from the contemporary one. This conclusion was then elaborated in meticulous detail in the body of the dissertation (which I have not quoted). It claimed that the biblical worldview was that of a flat earth capped by a tentlike heavenly vault. It followed, of course, that Aristotelian cosmology and the Ptolemaic system were as much contrary to Scripture as Copernicanism was. Thus, if the choice was between Ptolemy (or Tycho) and Copernicus, Scripture did not favor the former any more than the latter, or conversely did not undermine the latter any more than the former. Calmet's point was that the biblical worldview was scientifically (philosophically) untenable, and so one had better not regard Scripture as a philosophical authority. This final conclusion had been, of course, Galileo's own view of the matter (immortalized in Baronio's aphorism).

Calmet stated as much in the epilogue of his essay. It is revealing that there (in the penultimate paragraph) Calmet quoted a passage from Saint Augustine that had also been quoted and capitalized upon by Galileo in his

Nov-antiqua.[45] The same passage would be quoted and stressed by Pope Leo XIII in his encyclical *Providentissimus Deus* (1893).[46]

In his last paragraph, Calmet may be taken to reject and criticize Riccioli's thesis that scriptural authority had to be upheld in astronomy because otherwise its authority would dissolve in other, more spiritually relevant subjects (see chapter 4.3).

In light of all this, it is difficult to think of a more pro-Galilean introduction to the *Dialogue*. Even if Toaldo had printed Galileo's own Letter to Christina, it might have not been as effective; for Calmet was a highly respected biblical scholar.

However, it is unclear whether this pro-Galilean statement was deliberate. For during the negotiations for the imprimatur, Calmet's dissertation was described as progeostatic and anti-Copernican. It had been first mentioned in the prospectus of May 1742, in which the earlier idea of including Galileo's Letter to Christina had been dropped;[47] that prospectus, in the version sent by the Paduan inquisitor to Rome, stated that "instead we will reprint Father Calmet's dissertation in Italian, published in Lucca, which supports the earth's stability and sun's motion in accordance with common belief."[48] Then Giovasco recommended its inclusion on the grounds that "Father Calmet's 'Dissertation' . . . for its part confutes such a [Copernican] system if taken in the manner of a thesis."[49] Finally, when in June 1742 the Inquisition's assessor, Giuseppe Maria Ferrone, wrote the Paduan inquisitor to inform him that the cardinal-inquisitors had approved the latest prospectus, he made sure to repeat in his own letter which items were being included and which previously mentioned essays were being omitted; and he described the essay in question as "Father Calmet's dissertation in Italian, published in Lucca, in which the said Copernican system is refuted."[50]

In other words, Calmet's essay was supposed to be one of several means to neutralize the text of the *Dialogue* (together with the revision of the marginal postils and the reprinting of the 1633 sentence and abjuration). In reality, the essay neutralized these latter documents. Thus, the question arises of whether the 1744 edition of the *Dialogue* was a sign of incompetence or enlightenment. Pierre-Noël Mayaud, who for the first time has published the relevant documents and analyzed them with great acumen, stresses on several occasions the "incompetence" of the officials involved;[51] and certainly it would be proper to question whether the Paduan inquisitor, or Consultant Giovasco, or Assessor Ferrone, or the cardinal-inquisitors had read Calmet's essay carefully, or at all, or if they still remembered its content, on the assumption that they had read it once. Mayaud also suggests the possibility of what might be called bureaucratic inefficiency or overwork: he points out, for example, that at the Inquisition's meeting of 13 June 1742, when the cardinal-inquisitors gave their final approval, this project was the *tenth* case they had deliberated on in the *first part* of that meeting;[52] and they met several times a week.

On the other hand, perhaps the tolerant and liberal climate created by Pope Benedict XIV was indirectly responsible, although there is no documentary evidence that he was directly involved in this particular episode. Thus, his enlightenment may have encouraged church officials to adopt toward the issue of the scientific (philosophical) authority of Scripture an attitude similar to that adopted toward the issue of the earth's motion. That is, one would pay lip service to the hypothetical character of Copernicanism but then elaborate the Copernican system in any way allowed by the observational evidence and the physical theorizing; similarly, one would pay lip service to biblical literalism but in reality develop new methods of biblical interpretation and new exegeses.

7.2 Copernicanism Unbanned, Sort Of: Lazzari's Consultant Report (1757)

In July 1753, Pope Benedict XIV issued a bull titled *Sollicita ac Provida* on the reform of the criteria for the censure and prohibition of books in the *Index*. In January 1754, the secretary of the Congregation of the Index, Agostino Ricchini, proposed to the pope some additional reforms of the *Index*, involving the restructuring of its contents and the possibility of lifting the prohibition of some books after proper correction; among these were the works of Descartes, Copernicus, and Galileo. On 12 February 1754, the pope approved the secretary's proposal. In March and April 1754, the Congregation began working on the publication of a restructured and reformed *Index*.[53]

On 18 December 1755, the proceedings of Galileo's trial were removed from volume 1181 of the Inquisition archives (where they had been kept since the trial) and placed into a self-contained, freestanding file, which soon began to acquire a life of its own. This relocation was probably required by the fact that the Congregation of the Index was in the process of implementing the just-mentioned reform and deciding what to do about Copernican books in general and those of Copernicus and of Galileo in particular.[54]

Soon thereafter, the Index commissioned one of its consultants (the Jesuit Pietro Lazzari, professor of church history at the Roman College) to make a recommendation specifically about the general prohibition of "all books teaching the earth's motion and the sun's immobility." This clause had been included in all editions of the *Index* since the anti-Copernican decree of 1616. Lazzari wrote a lengthy memorandum full of arguments in favor of removing this clause from the *Index*,[55] we shall examine it presently. On 16 April 1757, with the approval of Pope Benedict XIV, the Congregation of the Index decided to drop from the forthcoming edition of the *Index* the prohibition of "all books teaching the earth's motion and the sun's immobility."[56] And so the 1758 edition of the *Index* no longer included this general prohibition, although it still included the five prohibited books by Copernicus, Foscarini, Zúñiga, Kepler, and Galileo.[57]

The main puzzle in this episode is why the repeal of the anti-Copernican censure was so incomplete. Mayaud finds this decision illogical,[58] in the sense of self-contradictory; that is, retaining the five mentioned geokinetic books seems incompatible with dropping the general anti-geokinetic clause. This judgment also corresponds to his account of the operation of the Congregation of the Index during this period, which he describes as characterized by inadequate documentation, sloppy record keeping, careless communication among the various bureaucratic levels, and general inattention to detail.[59] However, although the partial repeal is problematic and internally incoherent, there is really no strict inconsistency. The point is that the step taken was a small one; although it suggested other steps, it did not "logically" necessitate them. In a sense, this situation was similar to that of the "correction" of Galileo's *Dialogue*.

In contrast to that previous case, there is some documentary evidence that Benedict XIV was directly involved.[60] Moreover, although documentation about the activities of the Congregation of the Index is scarce, one crucial piece of documentation has recently been discovered, transcribed, and published by Ugo Baldini: the opinion of the Index's consultant recommending the withdrawal of the general anti-Copernican clause.[61] Consultant Lazzari's report reads as follows:[62]

[486r][63] Reflections on the clause "all books teaching the earth's motion and the sun's immobility" (Decree of 5 March[64] 1616).

There are three reflections which I plan to make about this clause: (I) that at that time it was prescribed prudently and with good reasons; (II) that these reasons no longer exist for the purpose of retaining it; (III) that in the present situation it is expedient to remove it.

[I.][65] Thus, one can say that at that time there were good reasons or motives for its being prudently prescribed. I consider three of these reasons for prescribing it. Firstly, this opinion of the earth's motion was new and was rejected and branded with serious objections by most excellent astronomers and physicists. Secondly, it was deemed to be contrary to Scripture when taken in the proper and literal sense; and this was conceded even by the defenders of that opinion. Thirdly, no strong reason or demonstration was advanced to oblige or counsel us to so disregard Scripture and support this opinion. One cannot deny that the collection of these reasons was a good and strong motive for adding that clause to the *Index*. And although decrees usually do not include the reasons, still in the one issued then, one has enough to understand that these were the motives of the prohibition; for it said: "This Holy [486v] Congregation has also learned about the spreading and acceptance by many of the false Pythagorean doctrine, altogether contrary to the

Holy Scripture, that the earth moves and the sun is motionless, . . . Therefore, in order that this opinion may not creep any further to the prejudice of Catholic truth, the Congregation has decided."[66] I shall briefly explain these motives.

First, then, Copernicus's opinion was indeed making news and was gladly listened to, especially by the young, which is what the decree indicates; this even happened when Copernicus himself explained it in Rome "to a large audience of listeners" (as one reads in Gassendi's biography of him),[67] and there are reports that sometimes the audience reached more than a thousand; nevertheless, mathematicians deemed it a novel opinion, and the most serious professors and experts rejected and criticized it. Kepler attests to this, although he was one of its principal supporters; in chapter 3 of book 5 of *Harmonies*, a work which he published in 1619 (namely, three years after our decree), he says: "That the earth is one of the planets and moves among the stars around the motionless sun, is still a new thing for the masses of scholars and the most absurd doctrine ever heard by most of them."[68] What he, with a kind of disdainful contempt, calls "the masses of scholars" were the astronomers and the philosophers of that time, as it appears from the writings of those who lived in those times. Thus, the Fathers of the College of Coimbra, in their commentaries on Aristotle's books *On the Heavens* of 1608 (chapter 14, question 5, article 1), wrote: "One must assert with Aristotle and with the common schools of physicists as well as mathematicians that the earth stands still at the center of the world."[69] In 1617 in *Sphaera Mundi* (book 4, chapter 2), Giuseppe Biancani said: "Finally, the common authority of almost all philosophers and mathematicians agrees [487r] in placing it completely motionless at the center of the world."[70] Read Charpentier's *Descriptionis Universae Naturae*, chapter 7.[71] Read the philosophers and mathematicians of those times; you will find that they all speak in this manner. Not only do they speak in this manner of their own or others' opinion, but also with serious objections they criticize the opinion of the earth's motion. In the book *De Coelestibus Globis ac Motibus*, Giovanni Antonio Delfino called it "foolish, absurd, and long ago exploded by the schools." Jean Morin, Royal Professor of Mathematics in Paris, called it "most absurd and erroneous"; and when Gassendi replied to him in regard to other things they disagreed about, Gassendi openly declared himself in favor of the motionless earth and unable to approve the contrary, although when Gassendi wrote to Galileo he praised his ideas (as it is usually done), and he praised Copernicus in his biography. In book 4, question 25 of *Various Reflections*, Alessandro Tassoni called it an opinion "against nature, against the senses, against physical reasons, against astronomy and mathematics, and against religion."[72] It was branded explicitly contrary to faith, to religion, and to Scripture by prudent and learned men such as Justus Lipsius (*Physiologia*, book 2, dissertation 19),[73] Marin Mersenne, Nicolaus Serarius, etc. This originated from the second thing I have proposed, namely that such an opinion was deemed to be contrary to Sacred Scripture taken in its proper and literal sense.

This was clear, both to the authors who opposed it and to those who supported it. Even before our decree, the much-praised Fathers of the College of Coimbra said: "It is legitimate to gather this immobility from some testimony of the Sacred Pages, such as the verse in Psalm 74,[74] 'I have set firm its pillars.'" Around 1584, in chapter 62 of the book *De Sacra Philosophia*, François Valles explained the passage of Ecclesiastes "the sun rises and the sun goes down"[75] and then added: "Therefore, [487v] now one readily removes the opinion of some ancients and the subtleties of Copernicus; in fact, the earth stands still, and the sun revolves."[76] Copernicus's defenders replied only by saying that Scripture did not use a proper manner of speaking that had to be taken with all the rigor with which other locutions are taken, but an improper manner that was figurative and adapted for common people. Galileo used this reply in the long letter which he wrote on this subject and which was translated into Latin and published in Germany. He prefaced as a foundation that the meaning of Sacred Scripture is "frequently recondite and very different from what appears to be the literal meaning of the words."[77] He gives the example of speaking of God as if he had feet, hands, eyes, bodily sensations, and human feelings such as anger, regret, hate, and sometimes even the forgetfulness of things past and the ignorance of future ones. In his words, "since these propositions dictated by the Holy Spirit were expressed by the sacred writers in such a way as to accommodate the capacities of the very unrefined and undisciplined masses, for those who deserve to rise above the common people it is therefore necessary that wise interpreters formulate the true meaning and indicate the specific reasons why it is expressed by such words."[78] Others have followed Galileo. To just quote one of them, in chapter 5 of book 1 of *Geographia*, the celebrated Varenius wrote: "One replies that in physical matters Sacred Scripture speaks in accordance with appearances and with what is understood by the people, as for example when the moon and the sun are called great luminaries, etc. Thus, Scripture says that the sun comes from one extremity and returns to another extremity, when in reality no such extremities exist. Thus, in the book of Job the earth is attributed a flat and square shape, under which are placed columns that support it. . . . In fact, Holy Writ is aimed not at philosophizing, but at improving piety."[79] These two propositions displeased theologians very much: that is, to claim that from Scripture one should gather only what pertains to Christian [488r] dogmas and morals, and to explain Scripture by attributing an improper meaning to such words. This displeasure made them utter serious censures. Thus Riccioli (and with Riccioli others) reaches this conclusion against Copernicans: "The propositions of Sacred Scripture that ascribe motion to the sun and rest to the earth are to be taken literally, in accordance with the proper meaning of words."[80] This must be done in general, when we do not have on the other side either other propositions of Scripture that are clear and certain, or definitions of the Church, or an argument that is certain and evident; we do have this in that case of the earth being called flat and supported by columns.

And this is the third thing I said, namely that Copernicans did not adduce any strong reason or demonstration why one had to explain Scripture in that manner and maintain their opinion against the proper and explicit meaning of words. The truth of this is shown by the authors of that time with the reasons they advance. Even much later, when Varenius advanced all the arguments in favor of the earth's motion, he concluded that they were not at all demonstrative and that they succeeded merely in rendering the opinion probable, "not apodictic," in his words.[81] And whoever will bother to read them and evaluate them will perhaps not concede even as much as Varenius would like. For they are refuted by Riccioli,[82] by de Chales,[83] and by many others. One should add that not only was there no demonstration in favor, but that one judged that there were strong arguments against it. In chapter 1 of book 4 of *De Augmentis Scientiarum,* Bacon of Verulam said: "Similarly it is manifest that Copernicus's opinion of the earth's motion (which recently became stronger) cannot be evinced by astronomical principles just because it does not contradict phenomena, but it can be correctly assumed from principles of natural philosophy."[84] Those arguments can be seen discussed by the cited authors and by Charpentier. [488v] Taking all these reasons together, one cannot deny that they constituted a very prudent motive for proscribing the Copernican opinion. Therefore, it is true that this clause was at that time placed on the *Index* with good reasons and prudently.

II. I now come to the second point and reflection: that not one of these reasons, and still less the whole set, remains nowadays to retain the clause. In saying this I say less that one could, for some of these reasons are such that their opposite is the case. First, then, the opinion of the earth's motion is prevalent in the principal academies, even in Italy, and among the most celebrated and competent physicists and mathematicians. Second, they explain Scripture in a sense that is proper and most literal. Third, they advance a kind of demonstration in their favor.

Now then, I say first that the opinion of the earth's motion is today a common opinion in the principal academies and among the most celebrated philosophers and mathematicians. Soon after our decree or thereabouts, this opinion began to get established, mostly through the work of Kepler, as he himself tells us in the *Epitome of Copernican Astronomy.* Bacon of Verulam also said, as we have seen, that in his time the opinion was beginning to spread and expand. In book 1 of *Kosmotheoros,* Christiaan Huygens asserted: "Nowadays all astronomers, except those who are of a retarded mind or whose beliefs are subject to the will of men, accept without doubt the motion of the earth and its location among the planets."[85] [489r] This is even more true today after the discoveries of Newton or those made with the benefit of his system. It is enough to read the proceedings and journals of academies, even Catholic ones, and the works of the most celebrated philosophers and mathematicians, or even dictionaries and similar books that report on the most widely accepted opinions. And indeed, in the article on Copernicus in the *Encyclopedia, or Reasoned Dictionary of the Sciences,* the famous mathematician D'Alembert writes: "Nowadays this system is generally followed in France and England, especially after Descartes and Newton each tried to confirm it by means of physical explanations. . . . It would be desirable that a country as full of intelligence and learning as Italy recognize an error so harmful to scientific progress and that she think of this subject as we do in France! Such a change would be worthy of the enlightened pontiff who governs the Church nowadays. Friend of the sciences and himself a scholar, he ought to legislate to the inquisitors on this subject, as he has already done for more important subjects. . . . In France one supports the Copernican system without fear. . . ."[86] In *Chambers's Universal Dictionary,* according to the translation and edition of Venice in the year 1749, the entry "Sun" says: "According to the Copernican hypothesis, which now seems generally accepted and even has a demonstration, the sun is the center of the system of planets and comets; and all planets, the comets, and our earth among them revolve around it in different periods according to their different distances from the sun."[87] We read approximately the same thing in chapter 2 of the *Grammar of the Sciences,* published in the year 1750, also in Venice. It should be noted, as I said, that these are elementary books compiled for the use of young persons or of those who have an average education, in which [489v] a particular author does not teach his own sometimes peculiar opinions; instead, to be well made these books must report what is most prevalent in the republic of letters and the world of scholars, so to speak. And as I also mentioned, some of these books were published in Italy, and so one can see how well established such a system is even in Italy, given that it is now published without any reservation or protest. I have also seen a book published last year, 1756, in Pisa by Barnabite Father Paolo Frisi under the title *Dissertation on the Diurnal Motion of the Earth;* the first words of the Preface are precisely these: "The phenomena that have become known everywhere in our century through the studies of the most distinguished men in astronomy, mechanics, and physics, not only confirm the most elegant and celebrated opinion of the great Galileo, but also . . ." And this book was published not only with the usual approvals, but also with the imprimatur of the general of his order; and it was signed "Rome, at the ex college of Saints Blaise and Charles, 24 January 1756" and was based on the reports and endorsement of two of his theologians. There is no need to speak of other books since it is clear and known to anyone of average education that nowadays the prevalent opinion among the most competent astronomers and physicists is that the earth moves around the sun. Here in Rome itself we can find that this is true. I have frequently had occasion to speak with the two celebrated mathematicians of the order of St. Francis of Paola,[88] with Fathers Boscovich and Maire, and with Monsignor Archpriest Stay.[89] I can attest that this is also their opinion. And the said Father Boscovich, who has tried to reconcile the modern discoveries with the earth's rest, has told me several times that he regards his reconciliation and the earth's rest most improbable from the point of view of pure natural reason, and that to believe this it is necessary to bind the intellect in deference to Faith.

[490r] Secondly, on the other hand the defenders of the Copernican system deny such a necessity; they deny it not (as they did in the past) because they claim that the scriptural passages should be taken in a sense that is improper and far from the natural meaning of the words, but rather because they believe that while defending such a system they can keep a sense that is more proper and natural than any other.

One must distinguish two kinds of persons, and correspondingly also two ways of speaking. The first kind consists of common people, those who are uneducated as well as those who are educated when they speak to the uneducated or to one another about current events and everything pertaining to human and civic life. The second kind of persons are the philosophers, who examine with subtlety things in themselves as well as the words used to describe them, the correspondence between the two, and also the concepts formed in the mind for this purpose; and so they have their own special and philosophical way of speaking. For example, without thinking common people speak of "a healthy man, a healthy color, and a healthy food," and do not make any distinctions; similarly for "cutting hair" and "cutting a text," or "doing one's nails" and "doing the laundry."[90] Here philosophers explain different ways of speaking and with subtlety distinguish and differentiate one from the other.

Furthermore, we must distinguish two kinds of motion and rest. The first is *absolute;* involves what is called imaginary space; and is not subject to any sensation. The other is *relative* to the bodies that are involved and that determine location, which is also called relative. Thus, when a ship is in motion, whoever is sitting astern moves with absolute motion and stands still at rest relative to the ship. Now, absolute motion is the one that is the subject of the reflection of philosophers since it is not possible to apprehend it with any sensation; relative motion is the only one that is the subject of [490v] common sense. Thus, civil society has coined the words "motion" and "rest" to express, in accordance with the common usage of words, relative motion and relative rest. And in accordance with this common manner of speaking, this meaning is not improper but really most proper.

This may be seen by reflecting on what happens in those situations in which one mixes absolute motion with relative rest, or vice versa. One says with all propriety, "I was standing still at the stern of the ship that was sailing." And if someone from the bow comes toward me at the speed with which the ship is advancing and hits me with his hand, one does not relate the fact thus: "While I was running fast, that man by means of constant effort prevented his being transported by the ship as he had been earlier, and he arranged for his hand to always remain immobile at the same place, so that I with that fast motion of mine ended up colliding with that hand." Everyone would laugh and would perhaps be disgusted by such a manner of storytelling. Nevertheless, this was how the incident happened in relation to imaginary space; and whoever had watched from the shore with binoculars would have seen this. But this description should be reserved for the lecture podium and the classroom, and not for social clubs and common conversation and dealings; in the latter contexts one says, "While I was standing still astern, he was running fast and stuck me with his hand." Thus, the proper manner of speaking in common social situations implies that motion and rest are considered and meant to be relative to us or to the things that are near us, and not relative to ideal fixed points and to imaginary space.

Thus, if Sacred Scripture is construed in this manner when it speaks of the motion of the sun and the rest of the earth, namely as meaning relative motion and rest, in relation to us and the place where we are, exactly as in that ship, [491r] then I am construing it in a sense that is proper, obvious, natural, and in harmony with the common definition of words. Moreover, whether what is true relative motion of the sun and true relative rest of the earth is also absolute motion of the former and absolute rest of the latter, or the reverse, that is a question in which the words motion and rest are considered in the way that philosophers use them in school when they use a language that is philosophical and not common.

We have a proof of all this from philosophers themselves. For even when they attribute the diurnal and annual motion to the earth and not to the sun, still they say, "The sun has moved so many degrees; has risen above the horizon; etc."; since the motion which they are affirming of the sun here is relative motion, it follows that by the word "motion" they mean relative motion.

I believe this discussion is straightforward, easy, true, and solid. Nevertheless, one should not neglect to note that nowadays we have another advantage over the times of Copernicus and Galileo, in regard to that most troublesome proposition that the sun does not move. For in truth modern philosophers and astronomers do not regard it as immobile at all, as they did; that is, they supposed its center to be immobile, and at most supposed it only moving around its own axis. After Newton, the moderns generally regard as immobile only the common center of gravity of the sun and all planets and comets; and they think that the sun as well as the earth and the planets turn around this center, although the sun has such a greater mass and is so much closer to the said center that it moves much less than all the other planets.

But there is no need to linger on this, and I do not make much of it. [491v] I stress only what I had proposed earlier. That is, nowadays the principal foundation of the prohibition no longer subsists, in particular the point which was so often made against the Copernicans and regarding which they appeared to give up in defeat: that according to their system Sacred Scripture could be explained only in a manner that was improper and far from the natural and commonly understood meaning of words; instead, as I said, they explain it in the most proper and natural sense.

Thirdly, there remains the point that, whereas formerly the Copernicans advanced in favor of their opinion reasons that were not very convincing, now they advance some that may be called demonstrations; thus, these would suffice to persuade us even if Scripture had to be taken in an uncommon sense and meaning.

There is no doubt that if at the time of Galileo some demonstration had been adduced, one would not have proceeded to the prohibition of his book and of the Copernican system. There is about this an explicit testimony and a letter by Father Honoré Fabri published in the *Transactions* of June 1665.[91] Wolff reprinted it in his *Elements of Astronomy:* "Rewriting what Father Fabri once publicly asserted, the Church does not operate against the evidence, but will declare that the system of a moving earth is not contrary to Scripture as soon as a demonstration of it is put forth."[92]

There are two ways of proving [492r] that there is this kind of demonstration. The first and the more effective way is to adduce the demonstrations themselves. But this is not for someone who wants to propose only some brief, straightforward, and easy reflections; and perhaps it would not benefit our purpose much. The second way is thus to show it by the authority of the most reputable and competent persons in such subjects.

This is extremely easy, for that is the common opinion among such persons. A short while ago I reported the words of Chambers, who said that the

Copernican system now seems generally accepted and "even has a demonstration." The *Philosophical Grammar* mentioned earlier says the same thing while speaking of the belief that the sun moves from east to west: "We have no reason to believe it; instead we have some demonstrations to the contrary"; and it gives a few of them. In the *Institutions of Physics,* translated and published in Venice in the year 1743, on p. 63 while speaking of the Ptolemaic system, Madame du Châtelet says: "The insuperable difficulties of the consequences drawn from it induced Copernicus to abandon it entirely and adopt the contrary hypothesis, which corresponds so well to the phenomena that now its certainty is not far from a demonstration."[93] Then there is Keill's *Introduction to True Physics and Astronomy,* lesson 4, in which after reporting some reasons he concludes: "Induced by these indubitable reasons, we brought the earth into heaven, placed it among the planets, and thrust the sun down to the center. And so from indubitable principles and invincible arguments we proclaimed the true system of the world, its arrangement, and the motion of the bodies of the universe."[94] Add to this Bradley's letter to Halley on the aberration of fixed stars[95] and chapter 3 of book 3 of MacLaurin's [492v] *Account of Sir Isaac Newton's Philosophical Discoveries.*[96] And there is a great multitude of others who speak in a similar or more striking vein. And note that they are either authors of introductions, dictionaries, and similar works, which present common opinions and speak with a common voice, as is the case for the first two I mentioned, which have been also reprinted in Italy; or they are celebrated and profound mathematicians (as especially the last one is regarded), who insist on penetrating right reasoning and singling out false arguments. Do we want to believe that those reasons which they recognize to be extremely forceful and strong are not such? Were they unable to distinguish them from the grossest paralogisms? How is it that after so many years they still continue to write and express themselves in this manner? It is certain that we shall never convince of this the community of scholars, and that they will always regard as demonstrations those reasons which we contradict with such a prohibition. This is especially true because, as I mentioned, for a long time the most competent persons in physical and mathematical subjects have spoken in this manner, and the world has believed them; on the other hand, those men who now deny the system of the moving earth with the most fervor and commitment are either strange in their other opinions, or barely educated in the basic elements of geometry and mechanics; I could easily prove that, if this were the place.

But why did I speak of "this kind of demonstration" and of "reasons that may be called demonstrations"? In order to conform to the manner of speaking of some, for whom however it was not necessary to show such regard. These persons want to recognize as a demonstration only a geometrical proof or what resembles it, which renders something as evident to everyone as saying that three plus three equals six. I should like to remind these people that we also call demonstrations [493r] the arguments for the existence of God, for his unity, and for the immortality of the soul; and yet it is certain that none of these reach that degree of evidence. Thus we must admit several kinds or perhaps degrees of demonstrations, and one of them is the one that proceeds from an assortment of things taken together.

I shall explain this by means of a very ordinary example, which we experienced some time ago in Rome and which I saw and thought about a great deal. It involves a beautiful altar of fine marble that went through the stages of being planned, started, put together, erected, and shown for public viewing; then it was taken apart and the various marble pieces were packed in order to be shipped and transported to Lisbon. The point is: assume that one builds an object composed of many, many parts of different size and quality; then let the object be taken apart and its parts be mixed together and rearranged; then assume that after some time one wants to rebuild the object and manages to match all the parts, so that they fit very well and the object is remade; I ask: would we not have the certainty and, so to speak, a certain demonstration that the parts are the same and the whole is the same as before? It is true that, speaking absolutely, it is possible that there is another combination (or hypothesis) such that the parts correspond to one another and all fit equally well, and yet are not placed in their previous position; but when the number of unequal and dissimilar parts is large, these possibilities are not taken into account as such, but are rather regarded as true impossibilities; and no one would doubt but rather would feel certain, very certain, that the matter was like that. We are in a similar situation. Copernicus revived the system of the moving earth [493v] to better accommodate some parts of the world system; and so did Galileo and Kepler to better accommodate other parts. But many others remained. As this system was elaborated by Newton, it was discovered that they also had their place in it and that the phenomena corresponded very well to the system. Then after Newton, one continued to make other discoveries. All these particular things that have been newly observed and discovered have also been found to correspond wonderfully to what had already been established.

There is more that is truly surprising. For once this system was established, one sought out all the most minute and individual consequences that derived from it; these consequences embrace a prodigious quantity of things which are so complex that by means of pure observation it had not been possible to infer their laws and to calculate and predict them; only in that system were they seen to come about as a legitimate consequence. Having seen this, one started to carefully observe phenomena and to examine whether they corresponded to the consequences that had been legitimately deduced; careful observation revealed that this indeed happened, and that in nature and in the world occurred the things that by sitting at a table with pen and paper one had shown from the adopted system. Such observations are innumerable, so to speak; each one of them sufficed to belie the assumed system, if it had not been in accordance with the consequences. To name a few items, such are the laws of the aberrations of the moon, which are many; the motion of fixed stars, called aberration of starlight; the nutation of the equatorial axis; the laws of the tides; the motions of comets; etc. One has found that in this [494r] system everything is explained simply and wonderfully, and everything corresponds to everything else. And so as one gradually makes similar discoveries and researches their most minute and particular details, everything is found to fit such a system with the greatest ease and clarity.

And is this fit not a kind of certainty and demonstration that this is the true system of the world? This demonstration is exactly like the one available in astronomy regarding how eclipses occur; they can be predicted so far in advance that as a result no one any longer doubts that they depend from those causes from which they are calculated. It is a demonstration like the one I have for thinking that the sun will rise tomorrow, although one cannot advance a geometrical demonstration.

By contrast, in the system of a motionless earth, either one cannot explain all these things (and in fact no one has explained them so far), or to explain them one must adopt hypotheses that are very convoluted, purely arbitrary, or in short such as to appear inherently improbable to pure common sense. Thus the defenders of the earth's motion have a kind of certain and very plausible demonstration in their favor.

III. Now I must speak briefly of the third point: that it is expedient in the present situation for the Index to remove that clause (1) because to retain it does no good but rather great harm; and (2) because to remove it does no considerable harm but engenders much good.

To retain it does no good. What good can this clause bring? To ensure that similar books are not read, are not spread, and are not reprinted? Having the clause [494v] has not so far provided this benefit. Who among young people studying mathematics does not read Wolff's *Elements?* Varenius's *Geography?* The *Introduction* of Keill, of Musschenbroek, and of Madame du Châtelet? Who does not consult Chambers's *Dictionary?* All these books mentioned so far have been republished in Italy; all are found in every bookshop of average stock; all are sold, bought, and lent. Who does not want to be informed about Newton's system or does not have available the book of some Newtonian?

Thus let us go on to the other reason, namely that to retain this explicit prohibition causes great harm: that is, contempt, disregard, ridicule, and mockery for the decrees contained in the *Index;* this will increase more and more as one sees that they are disobeyed openly and with impunity through the republication of similar books in Italy. What shall we do? Shall we renew those decrees? Shall we insist that they be obeyed? To bring this about in most areas of Italy, and especially in the universities with a higher reputation, is completely impossible; and it would be an imprudent and unfounded hope. Shall we ensure that some qualification be inserted every few pages, using that single word "hypothesis" as a panacea? I have occasionally seen and read such books, and I always thought that that remedy was almost worse than the disease and created a ridiculous situation. For everyone knows, everyone sees, and everyone perceives that the author supports that opinion with all his might, that he believes it, and that he wants to convince others of it; and yet one inserts and reads "hypothesis."

There is another very considerable harm. That is, Protestants are very deeply convinced of the falsity of the system of the motionless earth and of the existence of demonstrations to the contrary, and this prohibition is always in their mouths and in their writings with the intention of showing that in Rome there is the greatest ignorance of the most well known things or the blindest obstinacy. And so they exploit it (most falsely, to be sure, but with a specious appearance of plausibility) in connection with other points regarding either the interpretation of Scripture, or the definition of dogmas, or the understanding of Church Fathers. How prejudicial this is to religion, is known to anyone who has come across one of these books. I repeat that theirs is a bad argument, and they are completely wrong. But it is certain that they advance it. Thus, why should we not prevent them from doing so, and take away from them such a powerful weapon?

Let us now see what harm might result; this is the second point I have proposed here. At first it seems somewhat undignified for the Tribunal to make something of a retraction on such a famous question, and surrender to the opponents, who will celebrate their triumph. But this is a shadow without body and without substance.

First, to retract a judgment when the whole world, or most of it and the most learned persons, have done so is to now follow as before their footsteps and their indications; it must not and cannot be ascribed to some flaw and be the source of blame if a tribunal finally does not claim to be infallible in its judgments. Thus, I think it would earn respect by showing that it yields whenever it can, and that if it does not do so for other things, the reason is not obstinacy but constancy. We see how often this is practiced by other tribunals. Are there books that are more famous and more available to all than the missal and the breviary? How many things in them have been changed by accepting the recommendations of erudite critics! Not long ago one was thinking of making other changes and very considerable ones. Sometimes the heretics have been the first or the most precise in giving us some insights [495v] about ecclesiastical matters; they have not been rejected.

And that is right, for that is implied by the love and custody of truth. Thus, in regard to the letters of the first popes that are found only in Isidorus's collection, no one better than the Calvinist Blondel has studied them and taught us to evaluate them; and Catholics have generally accepted and approved his efforts. Have we perhaps suffered in reputation? What is more famous than the Donation of Constantine, which was once believed, mentioned, and promoted on every occasion? Why do we now agree to reject it and to surrender to its critics? Has our reputation suffered? Our reputation would have suffered if we had stubbornly held onto that opinion, and if we had decided not to abandon it unless forced by a geometric demonstration; we would have waited for it in vain there, as we are here.

Nor is it relevant to say that here one is dealing with the interpretation of Scripture and an opinion considered to be against the Faith. It would be unfortunate if, whenever there has been a consensus in the past, we try now to maintain the old shared opinions. Once it was a common opinion, which was supported by citing Scripture, that the heavens were moved by intelligent beings. Thus at about the same time, in paragraph 4 of book 2 of his *Philosophical Course*, Cardinal Sfondrati said: "It was and is the opinion of almost all philosophers and theologians that the heavens are moved by intelligent beings; in question 6 of article 3 of *De Potentia*, St. Thomas says that it belongs to the Faith."[97] Who among the more erudite and enlightened philosophers or theologians holds it now? Nor do I think that a book denying it would represent a criticism of the Sacred Congregation.

Second, one should examine whether, whatever harm might result, it is greater than the harm that would be produced [496r] by retaining the clause in the new edition, or that has been produced by having it there; for everybody says that this derives from ignorance, inflexibility, and obstinacy, and everybody proclaims our opposition and enmity to the most refined arts and sciences and our alleged tyranny over the most advanced intellects. Earlier I quoted Huygens saying that "nowadays all astronomers, except those who are of a retarded mind or whose beliefs are subject to the will of men, accept without doubt the motion of the earth and its location among the planets." In the *Encyclopedia*, D'Alembert asserted: "There is no inquisitor who should not blush upon seeing a Copernican sphere, said a celebrated author. This furor of the Inquisition against the earth's motion harms religion itself. In fact, what will the weak and the simple think of the real dogmas which faith obliges us to believe, if these dogmas are mixed with doubtful or false opinions?"[98]

Even if in the past there had been some harm in removing this clause, today there cannot be because it is de facto removed. One has started to publish with impunity on this subject in Italy, as I have shown; and that law is tacitly considered to be antiquated. Indeed this sentiment is not only tacit but has been written in print in famous works that pass through the hands of readers: that is, that this prohibition is no longer subject to the former rigor; that now one thinks differently; and that at most one needs to hide the matter by throwing around the word "hypothesis." So writes Serry in *Theological Lectures* (vol., 1, part 2, lecture 1): [496v] "After Copernicus and the old Pythagoreans, a great part (and perhaps the best part) of recent astronomers shows the motion of the earth around the immobile sun by reasons that are not to be disparaged at all. In a sense, the Church does not prohibit or condemn it; 'in a sense,' let me repeat. Although it is true that Galileo, who established and championed this system, did not at first find the Roman sky propitious, nevertheless after that time and in our own age, many men who excel in piety and religion (including several cardinals) are allowed to think freely with Galileo and Copernicus."[99] Then he adds, apparently out of fear of being censured: "In the sense that they cannot say that the system is certain and indubitable, but must say that it serves as a convenient hypothesis for the use of astronomy."[100] In the *Dissertation on the Qualities of Bodies* (paragraph 159), Father Fortunato da Brescia reports with approval the same thing. Thus, at the present time we are not really in a state in which we can suffer this harm, although by renewing and confirming the decree we place ourselves in that state, which is not at all desirable.

There remains to prove that great benefit would follow the removal of this clause. But this is evident from what been said so far. So I shall only mention some of the points explained above: (1) We will remove the obvious contempt (which we can no longer hold back) and the disregard of the decrees of the Congregation (disregard that is visible to all but is not condemned and is deemed to be well founded); this contempt and this disregard are the result of the printing and reading of such books. (2) We will allay the worries of many people who want to be [497r] informed of the best physical and astronomical discoveries and do not know how to proceed; for when they find the earth's motion in all such books, they either stop and abandon the lesson, or are filled with perplexities, scruples, and doubts. (3) We will move out of the way of the best authors and of those books regarded as classics; and the old instructions and decrees have always had great respect for them. (4) We will belie the bad reputation of the authors of these decrees and prohibitions: questionable competence in physical and astronomical subjects; rash severity in issuing prohibitions; and obstinacy in upholding previous decisions even against manifest reason. (5) We will confer greater authority to the new *Index*, or at least we will avoid greater contempt.

I end with two more reflections. First, we find ourselves forced to either confirm solemnly this decree (in which case, what authority will the new *Index* carry among philosophers and mathematicians, especially foreign ones?), or take advantage of a favorable and desirable situation (that will not come again for a very long time) and remove it with minimum damage. For it has already been decided to make other changes and to remove other general prohibitions; moreover, those that remain will be moved from their old place in the middle of the *Index* where they are now (namely, in the entry "Libri omnes") and transferred to the beginning, after the general rules and in front [of the list of particular books]; if we place it among these, it will be read and noticed more than before, whereas if we remove it, there will be no problem that it is no longer there and was deleted along with others. There is no [497v] middle course between these two alternatives; for to publish the qualification that the earth's motion is prohibited as a thesis but not as a hypothesis would provide no benefit and would produce the same damage as before, or worse; it would clearly appear as a frivolous subterfuge which opponents would deride because they know very well that no one nowadays thinks or writes like that.

The second reflection is that if someone thought that the opposite of what I have said so far is true, and that all those books should remain prohibited because the Copernican system is contrary to the authority of Scripture, repugnant to the Faith, etc., even then this clause would be unnecessary because the general rules would suffice; in fact, they prohibit any book that contains things repugnant to Scripture and Faith.

Father Pietro Lazzari, S.J.[101]

This memorandum strikes me as well argued, impressively nuanced, and often insightful, although of course it is not beyond criticism.[102] For example, in arguing that the prohibition was originally justified (in part 1), Lazzari assumed that a scriptural passage must be interpreted literally unless its literal interpretation contradicts other clear scriptural passages, or officially defined Church doctrine, or propositions that have been conclusively demonstrated; and it is a sign of sophistication that he called attention to this assumption and formulated it with some nuance. However, the principle had not gone unchallenged, and in fact in the Letter to Christina Galileo had argued that it should be modified to say that Scripture must be interpreted literally unless the literal interpretation contradicts propositions that *are capable* of being conclusively demonstrated.[103] Lazzari missed this key point, and his discussion of Galileo's Letter focused merely on the principle of accommodation.

Similarly, of Lazzari's several arguments as to why the general prohibition was no longer justified (in part 2), the one claiming that the earth's motion was now generally regarded as consistent with Scripture literally interpreted had several merits: it exploited the well-known distinction between absolute and relative motion to derive the relatively novel conclusion that the earth's motion was *not* contrary to the *literal* meaning of Scripture; in the process Lazzari was also using a plausible theory of meaning that made a distinction between technical and everyday contexts. But I am not sure he got the context right at the crucial point of the argument, when he compared what Scripture says with what one says in regard to motions *within* a ship; it seems to me that the more proper context would be what a traveler on a ship would say about the motion or rest of other ships or of the land.

Again, there was an insightful analysis of the concept of demonstration contained in Lazzari's argument that the prohibition was no longer justified because the earth's motion was now generally regarded as demonstrated. He was clear that one was not dealing with mathematical proof. The relevant concept was, rather, multifaceted (it "proceeds from an assortment of things taken together");[104] indirect ("by means of pure observation it had not been possible to infer their laws and to calculate and predict them");[105] predictive ("in the world occurred the things that by sitting at a table with pen and paper one had shown from the adopted system");[106] and systemic and explanatory ("everything is explained simply and wonderfully, and everything corresponds to everything else. . . . And is this fit not a kind of certainty and demonstration that this is the true system of he world?").[107] We saw in chapter 5.2 that a century earlier Auzout had started to move in this direction. And we can add that Galileo had begun to struggle with and move toward such a concept; he did this theoretically by starting to articulate the concept in his "Considerations on the Copernican Opinion" (1615) and practically by constructing in the *Dialogue* an argument that possessed such features to a considerable degree.[108] Thus, if one applied Lazzari's own concept of demonstration to Galileo himself, one might have to make drastic revisions to his condemnation; but of course, this is one of the issues in the controversy.

In the third part of his memorandum, Lazzari argued that it was expedient to abolish the general prohibition. This argument, as befitted its practical nature, was a cost-benefit analysis, so to speak. To retain the prohibition was harmful because, among other reasons, it encouraged non-Catholics to extend their rejection of Catholic ideas from questions of natural philosophy to questions of faith and morals—a point stemming from Saint Augustine and Galileo. And to remove the prohibition was beneficial because, among other reasons, it is a virtue to admit one's errors and revise one's ideas—a point that had been made in Leibniz's summer 1688 letter to Landgrave Ernst (chapter 5.3).

It is clear that Lazzari was addressing not the issue of the condemnation of Galileo but that of the condemnation of Copernicanism; and because of the way Lazzari approached the latter, his memorandum had no obvious implications about the former, or at least no reformist implications. He was arguing that although the prohibition of Copernicanism should now be repealed, it was justified in 1616; and this original justification would tend to support the 1633 condemnation of Galileo, at least superficially, that is, if we do not probe too deeply into the *actual* reasons for his trial and do not appreciate the possibility that the status of Copernicanism may have been significantly improved between 1616 and 1632, not least by the *Dialogue.*

Similarly, Lazzari's argument might be taken to have implications about the five particular books included in the *Index* (by Copernicus, Zúñiga, Foscarini, Kepler, and Galileo). That is, the memorandum was directly discussing only the general anti-Copernican entry in the *Index,* and recommending in clear terms its abrogation; and it might seem that its arguments to that effect applied with equal force (at least) to these five books. If we accept this further implication, then we have the puzzle of why the Church did not take the further step of dropping those five books from the 1758 *Index;* as we saw earlier in this chapter, some scholars (e.g., Mayaud)[109] go so far as to regard such refusal as an instance of "illogicality."

However, in light of Lazzari's memorandum (which was not utilized and perhaps not known by Mayaud), it becomes easier to understand why the prohibition of the five specific Copernican books was not repealed. For, as we have seen, the consultant was as clear and forceful in arguing that the general prohibition was no longer justified, as he was in maintaining that originally it was reasonable and prudent. To remove those five books from the *Index* might have tended to suggest that they had not deserved to be prohibited earlier, whereas to keep them was a reminder that the original prohibition was justified. This point would later be explicitly made by Maurizio Olivieri in 1820, during the Settele affair.[110]

Chapter 8

New Lies, Documents, Myths, Apologies (1758–1797)

The partial unbanning of Copernicanism embodied in the 1758 *Index* was noticed by a few people. For example, in 1765, while visiting Rome, the French astronomer Joseph Lalande[1] attempted to have Galileo's *Dialogue* taken off the *Index* by exploiting the fact that the 1758 edition had withdrawn the general ban on Copernican books. But he was told by the head of the Congregation of the Index that Galileo's case was different because it involved a trial, and so one would first have to revoke the sentence pronounced against him; he was also told that the just-deceased Pope Clement XIII had been inclined to move in that direction. Lalande did not have the opportunity to pursue the matter.

However, although that partial unbanning was the cause (as well as the effect) of an ongoing liberalization in the Catholic world, by and large it went unnoticed and had few direct repercussions.[2] In the next half century other developments enriched our story. In 1767, the king of Spain, Charles III, expelled the Jesuits from Spain and Spanish colonies; about ten thousand clergymen were affected, and most of them went to Italy.[3] One of these, Juán Andrés, ended up in Mantua, where he came to be known as Giovanni Andres; he kept himself busy researching and writing one of the first books on Galileo's thought; it was a useful and balanced account.[4] In 1773, the Catholic Church abolished the Society of Jesus, which had been in existence since 1540, and it would not be revived until 1814. Another ex-Jesuit, Girolamo Tiraboschi, developed a scholarly interest in Galileo and wrote one of the first essays dealing explicitly with the Galileo affair, of which more presently. On 5 July 1782, Grand Duke Pietro Leopoldo issued a decree abolishing the Inquisition in Tuscany.[5]

The most important developments of this period involve four texts on which I plan to focus in this chapter. They are significant from the point of view of novelty vis-à-vis what came before and of their influence by way of their repercussions. Besides chronological proximity, historical novelty, and influence, they also possess a thematic affinity: the first is a deliberate lie; the second lies by exaggerating the truth; the third lies by setting one extreme against another; but the fourth combats lies by an attempt to display balanced moderation, sound documentation, cogent argument, and elegant prose—a display that turns out to be partially but not entirely successful.

8.1 Dishonorable "Onorato": Gaetani's Forged Letter (1770–1785)

In the latter part of the eighteenth century, a document about Galileo's trial began circulating, purporting to be a letter written by Galileo in December 1633 to his disciple Vincenzo Renieri. The letter read like an account of the trial by the victim himself and contained other interesting information and reflections about his life. As early as 1770, the Florentine senator Giovanni B. C. Nelli had obtained a copy, as he tells us in his *Life and Correspondence of Galileo Galilei*.[6] It was first published in 1785 by Girolamo Tiraboschi in the eighth volume of the Rome edition of his *History of Italian Literature*.[7] And for about a century it was widely read, quoted, and interpreted by those seeking to come to grips with the Galileo affair.[8]

However, the letter turned out to be a forgery by the Roman clergyman Onorato Gaetani, although it took some time for its apocryphal character to be established.[9] An Italian man of letters named Pietro Giordani (1774–1848), who lived in Florence from 1824 until 1830, became suspicious of the letter primarily because the style seemed un-Galilean. When he expressed this doubt to the grand duke of Tuscany, the latter ordered an investigation. This centered on the library in Rome of the Gaetani[10] family, who held the title of dukes of Sermoneta. In fact, editors of the letter usually claimed that the original manuscript was kept there. When the original manuscript was located (in the library), it was discovered not only that the handwriting was not Galileo's but also that there was a note at the end of the letter stating that it had been forged by one of the Gaetani dukes explicitly to deceive Tiraboschi.

This note did not specify which member of the Gaetani family perpetrated the forgery. But there were only two brothers in the relevant period, the older layman Francesco (1738–1810) and the younger clergyman Onorato (1742–1797), and Nelli implies that Onorato Gaetani was responsible. In fact, in his 1793 biography of Galileo, Nelli quoted[11] a brief passage of the letter to support his claim that Galileo taught Italian to Gustavus Adolphus (1594–1632), king of Sweden from 1611; Nelli noted that he had received a copy of the letter in March 1770, and that the original was in Rome in possession of Monsignor Onorato Gaetani, duke of Sermoneta.

Nelli was apparently unaware that the letter was apocryphal, for with this quotation he thought he was confirming Viviani's claim[12] that Galileo had taught the royal prince Gustavus. However, this claim about king Gustavus Adolphus is one of the letter's many inaccuracies; for later scholars have determined that King Gustavus was never in Italy, and that the person referred to was probably the Gustavus who was the son of Erik XIV of Sweden and who lived all his life in exile.[13]

The apocryphal character of the letter had been established by 1848, when Eugenio Albèri edited the seventh of sixteen volumes of his critical edition of Galileo's collected works; Albèri reprinted the letter for the record but recounted the story of its forgery.[14] Antonio Favaro agreed that the letter was apocryphal and decided not to include it the National Edition; and it is interesting to note that Favaro apparently did not find any letter by Renieri to Galileo, dated 17 June 1633, mentioned in the apocryphal text.[15]

Despite its apocryphal character, the letter is historically important, and its content deserves some discussion. Its text need not be reproduced here, since an English translation has recently been published and is easily available.[16] To begin with, both the salutation and the closing seem uncharacteristic of Galileo; the former reads, "O most esteemed Father Vincenzo,"[17] the latter, "Be well."[18] Equally uncharacteristic is Galileo's remark about his being admired by Pope Urban VIII despite his (Galileo's) inability "to compose epigrams and love sonnets";[19] this appears to be sarcasm toward Urban's pretension and reputation as a poet. On the other hand, the forger coined a number of memorable phrases that do sound Galilean, such as a reference to "a crime which I do not know to have committed";[20] the observation that "because I was reasonable, I was regarded as little less than a heretic";[21] the remark that "perhaps people will turn me from a professional philosopher into a professional historian of the Inquisition!";[22] and the comment that the writer had "little inclination to be a theologian and still less to be a criminologist."[23]

The letter contains many inaccuracies. Some of these involve a type of error that could have easily been made by Galileo and which was not and probably could not have been known to be erroneous at that time: for example, Galileo apparently remembered a Swedish prince named Gustavus having attended his lectures in Padua, and when Gustavus Adolphus astonished Europe with his victories during the Thirty Years' War in 1631–1632, this must have jolted Galileo's memory into identifying the two men. However, some inaccuracies are of a type which it would have been unlikely or impossible for Galileo to commit: for example, the letter's statement[24] that Galileo's ideas on the earth's motion first became known in Rome through his "Discourse on the Tides," written on 8 January 1616 at the request of Cardinal Orsini; these ideas in fact were disseminated mainly through Galileo's Letter to Castelli (21 December 1613) and to a lesser extent through the *Sidereal Messenger* (1610) and *Sunspot Letters* (1613). The letter also displays factual knowledge which Galileo probably never had, such as the fact that during the relevant period of the trial in 1632 and 1633, the Inquisition replaced both its commissary and its assessor; despite some incoherence in what the letter says in this regard, it attributed knowledge of the change[25] to the accused, even though it occurred before February 1633, when Galileo arrived in Rome; the new commissary, Vincenzo Maculano, was appointed on 22 December 1632,[26] and the new assessor, Pietro Paolo Febei, on 26 January 1633;[27] and the point about the change in commissary was so arcane that even Favaro did not seem to be aware of it.[28]

At least two things suggest that the counterfeiter may have used Viviani's biography as one of his sources: the story of Gustavus Adolphus and the description of Galileo's house arrest at Arcetri as "the freedom of the countryside."[29] The latter phrase is particularly striking because it is *identical* to that used by Viviani.[30] Another striking remark suggests another possible source. It is hard to miss the reference to Bellosguardo as the first place to which Galileo returned after his Siena exile, and before moving to Arcetri; this is not true, since Galileo moved directly to Arcetri;[31] but the same erroneous assertion was made by Targioni Tozzetti in his 1780 book on the history of Tuscan science.[32] Although this publication date is too late for this work to have been a source for this error in the apocryphal letter, perhaps both were using the same faulty sources.

More important than any other feature is the fact that the letter advanced an interpretation of the 1633 trial. It portrayed the inquisitors at the 1633 trial as unconcerned about scientific arguments or the precedents of 1616 and instead concerned exclusively with biblical interpretation. It depicted Galileo at the trial as engaged in biblical exegesis and theological disputation. The reported argument is indeed a Galilean one: if we take Scripture as a source of astronomical science, then a passage such as Job 37:18 commits us to saying that the heavens are hard, which has been refuted by recent discoveries; thus we ought not to take Scripture as an authority in astronomy or natural philosophy. Such an exclusive concentration on the theological question conveys the impression that Galileo was condemned only for some kind of theological transgression. On the basis of this letter and other evidence, some people soon drew this conclusion explicitly, adding a special twist to it. The original creator of this new interpretation was a Frenchman named Jacques Mallet du Pan, of whom more presently. In fact, Mallet also exploited another letter that had been written in 1616 by the Tuscan ambassador Piero Guicciardini and that was first published in the 1770s; so before discussing Mallet's account, it is useful to examine Guicciardini's portrayal of Galileo.

8.2 Undiplomatic Diplomat:
Guicciardini's 1616 Report Published (1773)

In 1773, Angelo Fabroni published the first of two volumes of correspondence pertaining directly or indirectly to Galileo. As we saw in chapter 6.2, Fabroni managed to invent the myth that the biblical verse used against Galileo by the Dominican preacher Caccini in 1614 was, "Ye men of Galilee, why stand ye gazing up into heaven?" (Acts 1:11). However, Fabroni's documents were extremely informative and revealing, and they immediately started being used by other writers on Galileo, for example Frisi (1775) and Andres (1776). One letter was especially important because it shed new light on the Galileo affair and would later be widely read, quoted, misquoted, interpreted, and misinterpreted.

The letter was a report to Grand Duke Cosimo II written by Piero Guicciardini (1560–1626), Tuscan ambassador in Rome from 1611 to 1621.[33] It was dated 4 March 1616, in the midst of the first phase of the affair, while Galileo was in Rome trying to prevent the condemnation of Copernicanism. We know that he did not succeed, and in fact the Index's decree was issued the following day. Galileo had been lodging at the duke's Villa Medici since his arrival in December 1615 and would remain there until his departure in June.

The letter is filled with overcharged language and makes it clear that the ambassador did not like Galileo. Guicciardini portrayed Galileo as unwilling to keep quiet, taking the matter too personally, passionate (or zealous), imprudent, vehement (or intense), fixated (or obsessed), aggressive, argumentative, dangerous, and a troublemaker; in short as anything but diplomatic. This image, which was destined to become widely accepted, may be labeled the imprudence thesis.

Guicciardini appeared to have some inside information about the proceedings, as would be expected, since his position of ambassador gave him direct access to the pope himself as well as to cardinals and other well-connected diplomats. Some of this information seemed to be essentially correct; for example, he correctly predicted that Copernicus and other Copernican authors would be prohibited or suspended until corrected and that Galileo would not be publicly censured; this prediction seemed to reflect prior knowledge of the anti-Copernican decree. Similarly, when Guicciardini stated that Copernicanism had been declared erroneous and heretical at a meeting a few days earlier, he may be presumed to be referring to the Inquisition's consultants' recommendation of February 1616 that provided the basis for the (temporary) resolution of that earlier phase of the affair. In the eighteenth century, one could even independently confirm the accuracy of such information from a careful reading of the text of the 1616 Index's decree and 1633 Inquisition's sentence.

As to other points in the letter, at the time of its original publication no one was in a position to check their accuracy. For example, nobody could verify the role attributed to Cardinal Alessandro Orsini (1593–1626), who was a member of a powerful Roman family and related to the Medici through his mother and who had been appointed cardinal (in December 1615) shortly before Guicciardini wrote his letter; Guicciardini stated that Orsini spoke aggressively in favor of Galileo at a consistory and thus alienated the pope, who ordered the Inquisition to take action. However, soon after that original publication, an aspect of that story (Orsini's pro-Galilean disposition in 1616) could be confirmed when Targioni Tozzetti first published Galileo's "Discourse on the Tides" in 1780; for this discourse is written in the form of a letter addressed to Cardinal Orsini.[34]

The letter observes that Pope Paul V and Cardinal Bellarmine agreed that Copernicanism was erroneous and heretical. This was and remains precious information.

At a higher and more interpretive level, Guicciardini did not seem to understand very well the purpose of Galileo's trip to Rome; and insofar as he could understand it, he did not approve. His letter suggested that Galileo was trying to convince the cardinals that Copernicus's opinion (the earth's motion) was true; namely, that he was attempting to persuade them to accept the earth's motion, as if he wanted the Church's endorsement of Copernicanism. Guicciardini showed no awareness that Galileo might be trying to prevent its condemnation; that is, to convince the Church that Copernicus's opinion should not be prohibited or condemned. Most likely Guicciardini had not read either Galileo's Letter to Castelli or his Letter to Christina. On the other hand, in 1773 some readers of Guicciardini's letter could have had some such awareness, for both Galilean letters had been published,[35] however limited their circulation may have been. Guicciardini's vagueness on this point would soon mislead some people into thinking that in 1615 Galileo went to Rome to get the Church's endorsement of Copernicanism and that he sought it on the basis of scriptural arguments. One of these people was Mallet du Pan.

8.3 From One Extreme to Another: Mallet du Pan's Formative Myth (1784–1797)

In 1784, Jacques Mallet du Pan published in the *Mercure de France* a ten-page article entitled "Lies Printed on the Subject of the Persecution of Galileo."[36] His explicit aim was to excuse or absolve the Inquisition of the alleged crime of having condemned Galileo.[37] Mallet claimed that according to the popular view the condemnation was the result of barbarism, ineptitude, fear, and ignorance, but that this view was fiction. We have seen that this view had been popularized by Voltaire, although Mallet did not name him. Mallet argued that it was Galileo's meddling with the interpretation of the Bible, together with his petulance, that caused his misfortunes. In Mallet's own memorable words, "Galileo was persecuted not at all insofar as he was a good astronomer, but insofar as he was a bad theologian."[38] His theological error was to support an astronomical thesis by means of biblical passages.

The documentary basis of Mallet's account appeared to be Galileo's apocryphal letter to Renieri of December 1633 and the correspondence published in Fabroni's collection of 1773–1775, especially Guicciardini's letter of 4 March 1616. Of course, Mallet did not know that the 1633 Renieri letter was a forgery. For example, Mallet asserted that when Galileo was in Rome in 1615 and 1616, he wanted the pope to declare that the earth's motion was sanctioned by the Bible; that Bellarmine's warning to Galileo was to stop trying to reconcile Scripture and the geokinetic idea, not to stop supporting this idea in other ways; that in 1616 Galileo "tried to make the question of the earth's rotation on its axis degenerate into a question of dogma";[39] and that his defense in the 1633 proceedings consisted of biblical interpretations. Throughout his essay, Mallet stressed the unusually kind treatment received by Galileo, such as the pope's appointment of the special commission to investigate the *Dialogue* before forwarding the case to the Inquisition; the advance notice of major developments conveyed by the pope to the Tuscan ambassador Niccolini; the fact that Galileo was never put in an actual prison; and the fact that when he was detained between the first and second depositions, he was allowed to lodge in the apartment of the Inquisition's prosecuting attorney.

The author of this article is as interesting as its content.[40] Mallet was a Swiss Protestant journalist well known for moderate views that displeased both radicals and reactionaries. Born in Geneva in 1749, he was educated in the republican and Calvinist environment of that city. In 1770 he published a pamphlet of radical political ideas, which was condemned and burned in Geneva. Voltaire recommended him for a professorship of history and belles lettres to Landgrave Hesse-Cassel in Germany, but Mallet resigned this position within a few months and started working for S. N. H. Linguet's *Annales Politiques, Civiles et Littéraires*. In 1783 he moved to Paris and became director of the *Mercure de France*, which was to become one of the most widely read magazines in France. When the revolution came, from the pages of this newspaper he was one of the very few who advanced reasonable views critical of both sides. In 1792, Louis XVI sent him on a diplomatic mission to the French émigrés, but when this failed Mallet was unable to return to France. He thus took up an itinerant life, continuing his journalistic efforts, and he ended up in London, where he died in 1800.

Mallet's article did not go unnoticed, and a critique of it appeared within six months. It is of some significance that the critical response was published in the same magazine. The author was Girolamo Ferri, an Italian writing from Rome, and the article was titled "Apologia for Galileo."

Ferri began with the wise and insightful observation that "the exaggeration of truth is a lie"[41] and went on to argue that Mallet had indeed exaggerated the truth. Ferri admitted that several stories about Galileo were false, namely that he was tortured,[42] that he had his eyes gouged out for punishment,[43] and that he was kept for several years in an actual prison.[44] He refuted Mallet's claim that Galileo wanted to make a religious dogma out of Copernicanism, by summarizing (correctly) the main point of the Letter to Christina.[45] Ferri advanced an interesting interpretation of the Galileo-Bellarmine meeting in February 1616: "The philosopher listened patiently to the Controversialist, and, since he was not convinced, the Inquisition commissary in the presence of several witnesses forbade him to support orally or in writing the system of the earth's motion."[46] Against Mallet's assertion that the publication of the *Dialogue* was "ridiculous disobedience" on Galileo's part, Ferri mentioned all the precautions Galileo had taken. Ferri also pointed out that the Inquisition's respectful treatment of Galileo was partly due to the fact that the grand duke was treating the matter as an affair of state. In regard to Mallet's claim that at the 1633 trial Galileo discussed not astronomical evidence but biblical interpretation, Ferri (with no knowledge of the proceedings) responded that Galileo had had little choice. Mallet had based this particular claim on the apocryphal letter. Ferri did not question the authenticity of this letter.

Ferri's response showed that the untenability and falsity of Mallet's bad-theologian thesis were apparent to perceptive readers and thinkers of the time. We will soon see that Tiraboschi was another such person, although he was engaged in his own defense of the Inquisition. However, I regard Mallet's bad-theologian thesis as worse than untenable and false; it is perverse insofar as it does not merely fall short of the truth but inverts and subverts the truth. I would argue that Galileo preached and practiced the principle that scriptural passages should *not* be used in astronomical investigation but only when dealing with questions of faith and morals; whereas the Inquisition found this principle intolerable and abominably erroneous and held and wanted to uphold the opposite principle, that Scripture *is* a scientific authority, as well as a moral and religious one. Despite such flaws, Mallet's bad-theologian thesis proved to be long-lasting.[47] Such longevity, together with its reversal of the truth, justifies our labeling it a myth. There is no point in relating the full details of its subsequent history, but it is useful to discuss its most immediate adoption, which amounted to a kind of taking root of Mallet's interpretation.

One of the first places where Mallet's thesis took root was in Antoine Bérault-Bercastel's history of the Church. This was a monumental work of twenty-four volumes published in Paris between 1778 and 1790. Volume 21, which was published in 1790, dealt with the period from the beginning of Jansenism in 1630 to the Treaty of Westphalia in 1648. On pp. 140–46 of this volume was a section titled "The Affair of Galileo with the Inquisition." The section began: "At about the time when the Holy Office was endorsed by Urban VIII, this institution rendered (in the name of this pope) a judgment on which all the efforts of a crowd of historians or orators have propagated nothing but the densest shadows. After almost two centuries during which, regarding the subject of the celebrated Galileo, one has uttered shouts of barbarism and ignorance against the Inquisition, one has almost annihilated the memory of what really happened in the course of this affair. Thus it will not be useless to relate it. Here it is."[48] Then the author gave an account that reads like an abridged reprint of Mallet's article. What saves Bérault-Bercastel from a charge of plagiarism is the fact that he gave a precise reference to Mallet's article and expressed his reliance on Mallet. For example, he ended his account by saying: "There you have the truth of the story of Galileo and his judges, which had been so strangely deformed. We owe its discovery to the sound criticism and fair-mindedness of a citizen of Geneva, who in such a matter is a guarantor beyond suspicion."[49]

Although Bérault-Bercastel's appropriation of Mallet's account was quite uncritical, it was more succinct, less polemical, and more focused on the biblical issue than the original, and so it gained somewhat in rhetorical effectiveness. That is, certain aspects of the Mallet account emerge more clearly and incisively from it. For example, whereas the following contrast between Copernicus and Galileo does not appear until the fifth paragraph of Mallet's article, in Bérault-Bercastel it appeared in the initial paragraph, as the first substantive thesis of the account, immediately following the introductory remarks quoted in the previous paragraph. Here is Bérault-Bercastel's version: "Copernicus was the first who promulgated that the earth turned round the sun, but did so in a manner purely as an opinion in relation to physics, and had never been reproved by any tribunal. Galileo was not satisfied with adopting that theory and publishing it extensively; he undertook to confirm it by the Holy Scriptures. He converted a subject of a speculative kind in natural philosophy into a dogmatic controversy, and dared even to attempt to oblige the Inquisition to make a decision in favor of his views."[50]

Similarly, Bérault-Bercastel's abridged account made it easier to detect the (real or invented) documentary basis of the account. He mostly retained Mallet's references to and quotations from the Renieri apocryphal letter, which thus constitutes a proportionately larger part of the abridged account. Moreover, a different problem is aired, involving Guicciardini's 1616 letter. Mallet had referred to, paraphrased, and quoted extensively from this letter. By contrast, Bérault-Bercastel quoted only a brief sentence from Guicciardini, namely that while in Rome in 1616 Galileo "demanded that the pope and the Holy Office declare the system of Copernicus founded on the Bible."[51] A comparison with Mallet's corresponding passage and with the text of Guicciardini's letter reveals that this sentence did not appear in the original letter; in fact, it is the only portion of Mallet's own quotation which must be regarded as a fabrication.[52] The error or swindle was easy to overlook when one read Mallet but became more obvious in Bérault-Bercastel. Thus, while even the anti-Galilean Guicciardini had not blamed Galileo for wanting Copernicanism endorsed by the Church on biblical grounds, in the hands of Mallet and Bérault-Bercastel, Galileo became guilty of this fault as well.

It is not difficult to understand and sympathize with Bérault-Bercastel's uncritical appropriation of Mallet's account. The Catholic historian could have hardly failed to be impressed by the fact that a Protestant, apparently motivated by the desire for historical truth, had come to the defense of the Inquisition. Others, too, were impressed in this manner. At about the same time, Mallet's article was being turned into encyclopedia entries. One of these involved Nicolas Bergier's *Dictionnaire de Théologie* (1788–1790). In its entry on Galileo, it stated that the usual view of Galileo's trial "is a calumny which we will refute . . . in the entry SCIENCE."[53] It formulated the usual view as the thesis that "this scientist was persecuted and imprisoned by the Inquisition for having taught with Copernicus that the earth turns around the sun."[54] The relevant entry was in fact titled "Sciences Humaines";[55] the last third of it dealt with Galileo's trial and was essentially a digest of Mallet's article, with an explicit acknowledgment to him. Two things are noteworthy in Bergier's account. He explicitly connected the interpretation of Galileo's trial with the question of the relationship between science and religion by saying at the outset that Galileo's trial is one of the main examples cited to prove that Christianity is the enemy of science. And he explicitly repeated Mallet's most characteristic slogan in his assertion that this incompatibility thesis is untenable because Galileo was not persecuted for being a good astronomer but for being a bad theologian.[56]

Another, similar encyclopedia entry with wide repercussions appeared in François Feller's *Dictionnaire Historique* (1797).[57] Although Feller discussed Galileo's life in general and his scientific work in particular, and although he referred to many primary and secondary works, a large part of the entry dealt with the trial and relied on Mallet's article and the Renieri apocryphal letter. Feller was even acquainted with, and mentioned, Ferri's criticism of Mallet, but he dismissed it. The only noteworthy item to add here is that Feller included several other myths besides that of Galileo as a bad theologian. That is, he spoke of Galileo's illegitimate birth, even though it was about fifty years since that claim had been conclusively refuted, and various scholars had explicitly retracted and corrected the claim since then;[58] and Feller also presented, as if it were true, the legendary claim that immediately after the abjuration at the trial, Galileo uttered the words "E pur si muove" (see chapter 6.2).

8.4 "Spots in the Sun":
Tiraboschi's Brilliant Apology (1792–1793)

Even when Mallet's interpretation was not adopted, it probably exerted an indirect influence by encouraging the articulation of explicitly proclerical apologies. One example is the case of Girolamo Tiraboschi, who advanced his account of the Galileo affair in two lectures which he delivered in 1792 and 1793 to the Accademia de' Dissonanti in Modena, and which were published in 1797 in an Appendix to the tenth and last volume of the Rome edition of his *History of Italian Literature*.[59] However, Tiraboschi's own earlier account of Galileo's life and work, in the first edition of his *History of Italian Literature* (1780), although it included the story of the trial, was relatively matter-of-fact and stayed away from controversial and sensitive issues.[60]

Tiraboschi (1731–1794) had become a Jesuit in 1746 and the librarian of the duke of Modena, Francesco III D'Este, in 1770.[61] In 1772 he began publishing a monumental history of Italian literature that ran to thirteen volumes and went through several revised editions and countless reprintings. Literature at that time included all branches of learning, and so his history contains accounts of the development of medicine and science. At the time of his Modena lectures, and indeed for most of his literary career, Tiraboschi was no longer a member of the Society of Jesus because in 1773 the Catholic Church had abolished this order.

The first of the two lectures in question was titled "On the First Proponents of the Copernican System" and argued that before Galileo the Church was supportive of the new theory. Tiraboschi discussed several examples: the German Nicholas of Cusa (1401?–1464), who was even appointed cardinal by Pope Nicholas V in 1448; Copernicus himself, who received encouragement from Cardinal Nicholas von Schönberg (1472–1537), from Bishop Tiedemann Giese (1480–1550), and from Pope Paul III; the German Johann A. Widmanstadt (1500–1577), who in 1533 presented and explained Copernicus's views to the Roman court; and the Italian Celio Calcagnini (1479–1541), who became a follower of Copernicus while under the patronage of Cardinal Ippolito D'Este and in 1525 wrote a Copernican essay that was posthumously published in 1544.[62]

In the second lecture, titled "On the Condemnation of Galileo and of the Copernican System," Tiraboschi articulated and defended the thesis that Galileo was condemned primarily because he was too aggressive, zealous, and rash in advocating Copernicanism, and that if he had not defended the earth's motion in that manner, he would not have gotten into trouble with the Church. This account was also suggested in Guicciardini's 1616 letter, and a germ of it appeared in Baldigiani's 1678 letter to Viviani.[63] But Tiraboschi was the first to elaborate it and to advocate it publicly and explicitly.

Tiraboschi's apology was original and well argued, and it became influential; it thus deserves to be reproduced in full:

[373] Gentlemen, it seems to be a constant law of nature that nothing made by art or genius is [374] ever so perfect as to be incapable of being corrected or improved, and similarly that there is no man who does not lack something or is free of all flaws or is beyond criticism. Indeed sometimes we see those men who seem to rise above all others for the power and acuteness of their minds, and who fly so high that they almost escape being seen by the surprised observers, go down and tumble with such a ruinous fall that the admiration and envy one had for them almost change to derision and contempt. It is as if in this manner nature wanted to give flattering comfort to those who are unable to undertake great deeds and perhaps would become discouraged too much unless they saw that even the greatest men sometimes fall down to their level and humbly crawl on the ground. Who would have ever suspected that the greatest philosophical genius who ever lived and to whom calculus, optics, astronomy, and physics owe so much, I mean the immortal Newton, would later turn to commenting on the Apocalypse[64] and write in all seriousness that the seven-horned animal is none other than the Roman Pontiff? Who would have believed that the man whose acuity of mind and breadth of erudition made him the most capable interpreter of ancient history, namely Father Harduino,[65] would view the *Aeneid* as St. Peter's trip to Rome described by a Benedictine monk and would believe Horace's *Odes* to be the work of a Dominican of the thirteenth century and Dante's *Divine Comedy* to be the offspring of a Wycliffite[66] of the fifteenth century? And how many more could I name whose rare mind was always constant and equal to itself but was eclipsed by moral flaws that prevented them from fully obtaining the honors and the praise that would have been due to them! In this regard, the condition of great men seems more unhappy than that of mediocre ones, for the very mediocrity of the latter removes them from the gaze of envious critics and prevents one from seeing their flaws, just as one does not see any great virtues in them. For the former, on the contrary, one's admiration for their individual talents awakens envy and renders it ingenious in finding faults; the more light those men shed, the greater the curiosity with which one searches for their stains; and unfortunately it is rare not to find some. Thus, gentlemen, it is a very serious thing for me tonight to have to discharge the odious task of being a rigorous critic of the character and conduct of one of the greatest men of whom Italy can boast and who will always be immortal in the annals of philosophy and mathematics. But in a sense you are yourself forcing me to do this. You have not forgotten, gentlemen, that last year, when our Academy [375] began to rise toward more sublime goals and I had the honor of lecturing to you from this podium, I tried to show you that the Copernican system subsequently condemned in Galileo had been favored and promoted by the Roman pontiffs and by illustrious cardinals and prelates for almost two centuries before Galileo's time; and from this I inferred that if Galileo had supported his opinion less fervently, and if various other circumstances had not conspired to render him suspect and odious to the Roman Tribunals, he would not have been subjected to the harassment which he suffered on

account of that system. This proposition of mine, which I uttered in front of you, gives you the right, gentlemen, to ask me for the proofs in its favor. I cannot do this without showing that Galileo was guilty of some fault on account of which, more than on account of the system itself, he underwent suffering and difficulties. Nevertheless, I shall endeavor to do it with that moderation and with that respect that is due to the greatest men. If I am allowed to use an expression that would be more appropriate in the last than in the present century, I should not be blamed for daring to find some spots in a man who dared to find so many in the sun.

It is very well known to all, and no one can legitimately doubt it, that Galileo was tried before the Tribunal of the Roman Inquisition; that he was detained there for some time; that he was condemned there; and that the opinion he taught was prohibited by the same Tribunal as heretical. But the circumstances that preceded and accompanied these facts are not equally well known; only from those circumstances can one determine whether Galileo was guilty in some way and what motives led that Tribunal to such a rigorous condemnation. Galileo had gone to Rome for the first time in 1611,[67] but in that first journey there was no talk of the Copernican system, either because he had not yet studied it sufficiently or because he had not yet made his opinion public. At that time, the satellites of Jupiter, recently discovered by him and named Medicean Planets, were the principal topic of the discussions he held with the Roman philosophers and mathematicians. He himself wrote to Secretary Vinta[68] that he had found Father Clavius[69] and two other very learned Jesuit astronomers working on confirming his discoveries with new observations, and that they laughed at a certain Francesco Sizzi, who had fought those discoveries.[70] We also see from the documents published by Dr. Giovanni Targioni Tozzetti [376] that Cardinal Bellarmine himself had wanted to observe the phenomena discovered by Galileo in heaven, and that he had requested the opinion of Father Clavius and three other Jesuit mathematicians, who reported to him confirming their truth; as we shall see, Bellarmine took part in the first prohibition of the Copernican system.[71] However, it was at the Lincean Academy, then recently established by Prince Federico Cesi, that Galileo was heard most frequently and with the greatest appreciation holding public discussions about his discoveries, which were then the subject of the books and the arguments of all scholars, especially in Italy and Germany. Thus, that first trip was for Galileo only a source of admiration and glory.

After returning to Tuscany, he began to develop and to communicate to others his ideas on the Copernican system; and as it often happens for anything that appears novel, although he found many admirers and followers, he also found many and perhaps more opponents and enemies; this happened either because they did not properly understand the foundations of such a system, or because the old professors of these disciplines were ashamed to confess that they had been in error until then, or because many felt that the Copernican system could not be reconciled with Sacred Scripture, which seems to suppose the motion of the sun and the immobility of the earth. This last reason was the one that resounded most loudly against Galileo because it was the only one that could be adduced without having to enter into astronomical questions, in regard to which one had too much to fear from Galileo's mind. So complaints began to spread against the reformer of the Copernican system, and there were even sermons against him from sacred pulpits; and someone flattered himself to have found in the Acts of the Apostles a prediction and a denigration of Galileo's opinion in these words: "Ye men of Galilee, why stand ye gazing up into heaven?"[72] The uproar arrived even in Rome, and Galileo was informed that his doctrine generated serious scandal with some people. Thus, toward the end of 1615 he went to Rome, either because he spontaneously decided to do so (as he writes in one of his letters),[73] or because he was summoned to give an account of his opinions (as Antonio Querengo writes in one of his letters).[74]

There, now at one house, now at another, he began to spread the system [377] he embraced and to respond to the difficulties which many people presented to him, and from such disputes he usually emerged victorious in the midst of applause and astonishment by the audience. But he was unable to use the moderation that is so necessary to great men, all the more so inasmuch as others fear being outdone and oppressed. "Galileo," writes ambassador Piero Guicciardini to Grand Duke Ferdinand[75] on 4 March 1616,[76] "gave more weight to his own opinion than to that of his friends: Lord Cardinal del Monte and I (to the small extent I was able to), as well as several cardinals of the Holy Office, tried to persuade him to keep quiet and not to stir up this business; they pointed out that if he wanted to hold this opinion, he should hold it quietly, without making such an effort to encourage and draw others to hold it too." And a little after that: "He gets inflamed with his opinions and feels very passionate about them; he does not have the fortitude and the prudence to be capable of prevailing."[77] Thus you see, gentlemen, that if Galileo had moderated somewhat his expressions, perhaps he would not have exposed himself to the difficulties which he later endured; and that if he had supported his opinion only in private, the cardinals mentioned by Guicciardini would not have done him any harm.

However, before proceeding, one must lay down some principles that are necessary for clarifying and justifying their conduct toward Galileo. It is certain among all Catholics that the original text of Sacred Scripture does not contain any falsehood, even for things that do not pertain to dogma, and that everything it affirms must he held to be true. It is also certain that it is not legitimate to move away from the literal sense of Sacred Scripture except when we are driven to do so by some evident proof which demonstrates that the literal sense would contain some falsehood or manifest error. Given this, one cannot deny that various passages of Sacred Scripture seem to indicate that the earth is immobile and the sun turns around it. These were the passages that were opposed to Galileo; if he had limited himself to responding that he was speaking merely as a philosopher and that when scriptural passages could not be explained otherwise he did not intend to oppose such a respectable authority, then the question probably would not have gone any further. But it appears that Galileo was not satisfied with this. There is a letter written by him at that time to his disciple Father Benedetto Castelli, which is probably the same one that is mentioned in the decree of condemnation and which has been [378] published by the above-mentioned Targioni,[78] although not in its entirety; and another letter written by him to the Grand Duchess of Tuscany; they show that Galileo would have liked to persuade us that the literal sense of Scripture should not be taken into account except for things that pertain to dogma. Now, this proposition may in a sense be acknowledged to be true; nevertheless, it was deemed to be and was in fact dangerous, especially at that time, when the memory was still fresh of the painful losses which the Roman Church had suffered in the North, and which originated in large measure from the innovators' freedom to interpret Sacred Scripture as they wished and to give it the sense that suited them best. Certainly the Roman theologians did not ignore that for things that are indifferent to the faith, and even for those that are not, if there is an evident reason that requires it, it is legitimate and sometimes necessary to move away from the literal sense. But they also knew that all theologians and philosophers of past ages had believed that the earth's immobility is clearly established in Sacred Scripture; that those who had supported the Copernican system before Galileo had spoken as philosophers and had not attempted to reconcile their opinion with the Sacred Text; and that only Copernicus had made some mention of the issue, but that his work was known only to a few scholars. They saw that Galileo endeavored to establish the Copernican system by means of ingenious demonstrations; but at the same time they saw that most philosophers of that time did not appear to be convinced. Thus, Galileo's opinion did not appear certain and evident so as to allow, at least publicly, to give to the words of the Sacred Text a sense different from the one they seemed to manifest. On the other hand, Galileo was publicly making so much noise about his opinion that it was already in everybody's mouth; nor could one conceal that many of the most learned philosophers and theologians were scandalized by it and viewed Galileo as a dangerous innovator; for he dared to be the first and almost the only one to oppose in such a solemn manner the literal sense of Scared Scripture. Therefore, it appeared to them that the will of a single person should not be allowed to give the Sacred Text an explanation different from the one that had been given until then.

These were the reasons that led the Roman consultants to the first condemnation of the Copernican system, which is reported in the second decree issued sixteen years later. No trial against Galileo was held then, and no penalty was imposed against [379] him. Two of Galileo's propositions were prohibited: the proposition that the sun was at the center of the world and did not possess local motion was condemned as heretical because contrary to Sacred Scripture; and the proposition that the earth was not the center of the world and moved with diurnal motion was condemned as erroneous in regard to the Faith. Then Cardinal Bellarmine exhorted him in a friendly way, and the commissary of the Roman Inquisition forbade him strictly, to maintain such propositions, and even to discuss them; and he was threatened with imprisonment if he dared to violate the prohibition. And it was simultaneously ordered that Copernicus's work and some other book that adopted his system be purged and corrected by removing those passages saying that Sacred Scripture was not contrary to that system.

Here one cannot conceal the fact that Galileo began then to act in bad faith. In two letters written by him to Secretary Vinta[79] on that occasion, he makes no mention of the prohibition enjoined on him, but discusses only the books whose correction had been ordered. Nor did he ever mention it in his works. When he was accused of having violated it, he chose to excuse himself by saying that he had only been forbidden to defend and maintain the Copernican system, and not to merely treat of it, as he claimed to have merely done in his celebrated *Dialogue*.[80] Thus, it seems certain that he was determined not to obey the command he had received from the Roman Tribunal, and that he deceived himself into thinking that if he was silent about it, no one else would remember it. After his return from Rome, he was occupied with writing the *Dialogue* on the system of the world, divided into four days; it was completed in 1630.[81] He knew very well that its publication would be risky after the Roman Inquisition's decree condemning the Copernican system as contrary to the authority of Scripture. Thus, he went to Rome and presented the *Dialogue* to the Master of the Sacred Palace, who examined it and, perhaps to Galileo's surprise, did not find in it anything deserving criticism or censure; so he gave the imprimatur. Galileo returned to Florence with the intention of putting the finishing touches on the work and then sending it to Rome for publication there. The plague that was then beginning to rage in Italy did not allow this to happen. So he obtained permission from the Master of the Sacred Palace to have the book published in Florence after a new review of the work by a consultant of the inquisitor of that same city; [380] in this manner it appeared in Florence in the year 1632. This is the substance of what happened, and it seems that there is nothing to rebuke Galileo for. But oftentimes an action that seems to be innocent when described simply is revealed culpable upon examination of the circumstances. Let us see whether this turns out to be true for the case of Galileo.

The Preface with which he began the *Dialogue* does not allow us to entirely justify him. Here is how he begins, in a way which no one could have conceived more appropriate for the purpose of deceiving the reviewers: "Some years ago there was published in Rome a salutary edict which, to prevent the dangerous scandals of the present age, imposed opportune silence upon the Pythagorean opinion of the earth's motion. There were some who rashly asserted that that decree was the offspring of extremely ill-informed passion and not of judicious examination; one also heard complaints that consultants who are totally ignorant of astronomical observations should not cut the wings of speculative intellects by means of an immediate prohibition. Upon noticing the audacity of such complaints, my zeal could not remain silent. Being fully informed about that most prudent decision, I thought it appropriate to appear publicly on the world scene as a sincere witness of the truth."[82] Could a declared apologist for the ancient world system, or even the most zealous inquisitor, have spoken differently if he had undertaken to refute the Copernican system? But there is more. Not only does Galileo pretend to venerate that decree, but he almost wants to make us believe that it was promulgated on his advice: "For at that time," he continues, "I had been present in Rome; I had had not only audiences but also endorsements by the most eminent prelates of that Court; nor did the publication of that decree follow without some prior knowledge on my part."[83] And here is how he gives us an idea of the book he is publishing: "Thus it is my intention in the present work to show to foreign nations that we in Italy, and especially in Rome, know as much about this subject as transalpine diligence can have ever imagined. Furthermore, by collecting together all my speculations on the Copernican system, I intend to make it known that an awareness of them all preceded the Roman censorship, and that from these parts emerge not only dogmas for the salvation of the soul, but also ingenious discoveries for the delight of the mind."[84] And a little after that, he says that with this he wants to make it known that "to assert the earth's rest and take the contrary solely as a mathematical whim does not derive from ignorance of others' thinking but, among other things, from those reasons provided by piety, religion, acknowledgment of divine omnipotence, and awareness of the weakness of the human mind."[85]

[381] After this debut, who would have believed that Galileo's *Dialogue* might be the most ingenious demonstration of the Copernican system that could be constructed at that time? It is true that now and then, and especially at the end, he mentions that this is a mere hypothesis. But it is also true that he has the interlocutor Simplicio, to whom he assigns the part of defending the ancient system, say so many simple-minded things and defend his opinion so feebly that the suspicion arises that under the name of Simplicio Galileo wanted to portray and mock some of his censors; and there were those who suspected, although in my opinion without reason, that under that name he meant to portray Pope Urban VIII himself. And I am not far from believing that the reviewers to whom Galileo's work was given for examination, after reading such a modest and religious preface, and being unable to properly understand the ingenious arguments discussed in the *Dialogue,* judged the interior of that edifice from its exterior facade and believed it to be built exactly in accordance with their expectations. We know for a fact that Urban VIII complained several times that they had not been sufficiently discerning and that they had let themselves be deceived by Galileo; and the pope was especially angry with Monsignor Ciampoli,[86] a prelate of great authority in Rome, because Urban had inquired several times about the matter and Ciampoli had always assured him about Galileo's religious submission and sound doctrine.[87]

Now tell me what you think, gentlemen: if the most moderate and just tribunal that ever existed in the world saw one of its rigorous prohibitions publicly violated and knew that the violator of the command, not satisfied with this, had also wanted to mock it, to cheat it with skill, and to cleverly extract from it a permission that with full knowledge of the situation would have been denied, would not such a tribunal feel indignation and would it not deem the transgressor to deserve severe punishment? Thus, should one be surprised that the Roman Inquisition proceeded against Galileo with extraordinary rigor?

Nevertheless, such was the respect he had gained with his profound knowledge and with his ingenious discoveries, that that severe Tribunal treated him with unusual kindness. The toughest aspect of the trial to which he was subject was its beginning. That is, in February 1633 a man of seventy years old had to journey to Rome, and all requests made even in the name of the grand duke to delay the trial to a better season were good for nothing.

[382] But for the rest of the whole process he was treated in a manner very different from the usual. At first, for two months he stayed at the house of the grand duke's ambassador, without any explicit prohibition to speak to others but only with the friendly advice not to have frequent visits and to keep himself as much as possible out of sight and in solitude.[88] When according to the laws of that Tribunal it was time for him to be locked in prison and undergo the trial, he was called to the Tribunal's palace, but the place assigned to him was the prosecutor's three rooms; he could leave them at will to go into the courtyard; he could keep his own servant; he could receive the grand ducal ambassador's servants, who brought him meals; and he could write to, and receive letters from, whomever he wanted. Even before the interrogations had ended, after about two weeks he was sent back to the ambassador's house, and he had permission to leave it to go by carriage to the nearby gardens.[89] Finally, in June of the same year there was the sentence for which one had waited so long. In it, the system was condemned; Galileo's book was prohibited; he was forced to a solemn retraction; and because since 1616 he had been threatened with prison if he again discussed publicly or wrote about such a topic, imprisonment was imposed. But before he was taken there, by order of the pope it was immediately commuted to detention at the garden of Trinità de' Monti, which was the grand duke's villa; from there he was occasionally allowed to go to Castel Gandolfo.[90] This detention also ended in a few days, and he was allowed to move to Siena at the residence of his friend the archbishop; from there later at the end of the year he could move to his villa at Arcetri.

This whole series of facts which I have briefly mentioned is not taken from the works of some papal writer or some apologist of the Roman Inquisition, but from the letters with which ambassador Niccolini[91] informed the grand duke of everything that was happening to Galileo from day to day. Thus you see, gentlemen, how much we should trust certain writers of anecdotes and dictionaries, who portray Galileo locked and buried for a long time in a dark prison and subjected to cruel torture as one would do with a criminal. These pictures are delineated by the imagination of poisonous writers who angrily bite or insult anything that pertains to religion.

[383] What shall we then say of those Protestants who flatter themselves to find in the condemnation of Galileo an invincible argument against the infallibility of the Church? The Copernican system, they say, was condemned and proscribed as heretical. Nevertheless, it is now known as the only one that conforms to experience and reason, and among Catholics themselves there is no one who has any scruple to maintain it. But they do not perceive, or to be more precise, they pretend not to perceive the weakness of their argument. The Church has never declared the supporters of the Copernican system to be heretics; this excessively rigorous censure came only from the Tribunal of the Roman Inquisition, to which no one even among the most zealous Catholics has ever attributed the trait of infallibility. On the contrary, even in this one must admire the providence of God toward the Church; for at a time when the greater part of theologians firmly believed that the Copernican system was contrary to the authority of Holy Writ, yet it did not allow that the Church proclaim such a solemn judgment.

Nor do I thereby mean to say that the behavior exhibited toward Galileo was in every way praiseworthy. On that occasion, one put too much trust in Peripatetic philosophers, who did not know how to respond to Galileo's arguments, and so used the authority of Sacred Scripture as a shield. One did not sufficiently examine whether Galileo's arguments had the degree of strength that would have legitimated the abandonment of the literal meaning; one simply assumed, as already mentioned, that the Sacred Text could not have any other meaning. I gladly grant all this. But you must grant me, gentlemen, that Galileo himself was in no small part responsible for his condemnation; if he had been at least a more cautious transgressor (not necessarily a more scrupulous follower) of the injunction issued to him, and if he had exasperated his enemies and his censors less than he did and had not shown that he wanted to ridicule them, his opinion would have been left in the tranquillity which it has now enjoyed for a long time.

May this example be a lesson to learned men: to be more cautious about clashing too much against not only the commonly received opinions of other scholars, but also against the prejudices of the people; and to become persuaded that the more the defenders of truth stay away from using violence or deception in pursuing their end, the more easily does truth manage to become known and to triumph over error.[92]

The opening paragraph may be regarded as a twist on Viviani's thesis that Galileo was led to err by divine providence to show that, as great as he was, he was still fallibly human.[93] For Tiraboschi was stressing that great men often make great errors, as if to comfort ordinary mortals. He was also creating a brilliant and inspired image when he compared his attempt to criticize Galileo to Galileo's own discovery of dark spots in the sun.

When Tiraboschi summarized the basic, well-known facts, he included the alleged fact that Galileo's opinion "was prohibited by the same Tribunal as heretical" (375). Thus, he committed the error (which we have encountered many times) of confusing the concept of heretical with that of contrary to Scripture, and the decisions of the Inquisition with the recommendations of its consultants.

Quoting from Guicciardini's letter, Tiraboschi went on adopt the imprudence interpretation. But he was careful to limit this criticism of Galileo to his response to the scriptural objection in 1613 to 1616, and not to his attitude toward Copernicanism in general. With regard to the latter, Tiraboschi wisely stressed that Galileo had been cautious, for he had not advocated it publicly after his telescopic discoveries of 1610, and he had been able to enjoy a triumphant reception in Rome in 1611 free of Copernican controversy.

With regard to the scriptural controversy and the events of 1613 to 1616, Tiraboschi saw a clash between two principles: on the one hand, the traditional principle that on all subjects, whether pertaining to faith and morals or astronomy and natural philosophy, Scripture must be interpreted literally unless there is a conclusive proof that the literal interpretation implies an obvious falsehood; on the other hand, Galileo's principle that the literal interpretation of Scripture should not be taken into account unless the subject pertains to faith and morals. Tiraboschi was even willing to admit that the new Galilean principle was essentially correct. But it was perceived to be, and was in fact, dangerous in light of the controversy between Catholics and Protestants about the interpretation of Scripture. Moreover, in terms of the traditional principle, there was no conclusive proof that the literal interpretation of relevant scriptural passages implied a falsehood, because the consensus was that Galileo's arguments for the earth's motion were not conclusive. To these two reasons, Tiraboschi added the scandal caused by Galileo's imprudent advocacy while in Rome in 1615 and 1616. These three reasons provided for Tiraboschi a sufficient *explanation* of the 1616 condemnation; he was also implicitly suggesting that they were plausible enough reasons so as to yield a *justification* of that condemnation.

Tiraboschi then discussed the second phase of the affair, focusing on the *Dialogue* and charging Galileo with acting in bad faith. For he disobeyed the special injunction and indeed never had any intention of complying with it; for example, when he wrote to the Tuscan secretary of state (Curzio Picchena) at the end of the 1616 episode, he did not even mention the special injunction and instead talked about the book censures, which did not directly affect him. And he deceptively violated the prohibition against defending Copernicanism, for the book actually defended this doctrine, although the preface deceptively tried to give the impression that the work was a sort of vindication of the 1616 anti-Copernican decree, and the text gave lip service to the hypothetical character of the discussion. Finally, Galileo even mocked the ecclesiastical authorities by deceptively extorting their official imprimatur. Despite Tiraboschi's toughness here, he did *not* add a fourth element of bad faith to these three, namely that Galileo insulted the pope by impersonating him with the character of Simplicio; Tiraboschi mentioned the allegation but dismissed it.

With regard to the 1633 proceedings, Tiraboschi gave a balanced and documented refutation of the prison myth, showing that Galileo was not kept in a real prison either during the trial (despite Inquisition practice and precedent) or afterward (despite the formal inclusion of prison among the penalties announced in the sentence). Instead the diplomatic correspondence of the Tuscan ambassador proved beyond any reasonable doubt that during the 1633 trial Galileo had been treated with a consideration and kindness unprecedented in the annals of the Inquisition. This fact suggested a radical revision of the prevailing ideas about the trial, and we have seen that some people (i.e., Mallet du Pan) went to the opposite extreme of constructing an anti-Galilean myth. But Tiraboschi refrained from drawing any sweeping inferences.

Tiraboschi's summary of the sentence is revealing. Besides mentioning the prohibition of the *Dialogue* and the abjuration, he said that the system was condemned. That is, Tiraboschi judged that the Inquisition's sentence of 1633 not only represented the condemnation of a person and his book but also reaffirmed the condemnation of the Copernican doctrine. He did not elaborate, but, as we saw in chapter 1.1, in effect the reaffirmation in the new context strengthened the censure against the doctrine. With regard to the penalty of imprisonment, Tiraboschi interpreted it as the activation of the clause of the special injunction that stated that if Galileo refused to comply, he would be imprisoned.

Clearly this account of the affair was an apologia for the Inquisition; but it avoided the crudities and distortions of Mallet du Pan's account and explicitly contradicted his bad-theologian myth. On the contrary, Tiraboschi's account is eloquent, subtle, insightful, and well argued. This is not to say that it is entirely acceptable, but its criticism is beyond the scope of this work. Let me instead mention one other merit.

Tiraboschi gives us the first formulation and resolution of the infallibility problem. This anticlerical objection claimed that now that everybody agreed that Galileo was right about the earth's motion, his trial showed that the Catholic Church is not infallible and thus refuted the Catholic doctrine of infallibility. Tiraboschi replied that this doctrine attributes infallibility to the Church (as a whole) and not to the Inquisition, which is a particular institution within the Church and is fallible. And he counterargued that the case of Galileo instead proved the Church is under the protection of divine providence, for, at a time when the Inquisition and most theologians regarded Copernicanism as heretical, the Church refrained from making an official declaration to that effect.

Tiraboschi ended with an explicit attempt to show his impartiality by mentioning two points against the ecclesiastical side: the Church was too uncritically favorable toward the arguments of Aristotelian philosophers and too critically unfavorable toward Galileo's arguments. But these were mere qualifications in what was otherwise an anti-Galilean and proclerical account. For, in conclusion, he reiterated that Galileo was largely responsible for his own condemnation; that he was incautious and imprudent in defending his cause in Rome in 1615–1616 and in disobeying the 1616 prescriptions with his *Dialogue* of 1632; and that he defended truth by means of violence and deception. Thus, while Tiraboschi's valiant effort was on a higher plane than anything preceding it, and while its merits remain, it too ultimately may not have escaped the predicament, so aptly described in Ferri's criticism of Mallet du Pan, that "the exaggeration of truth is a lie."[94]

Chapter 9

Napoleonic Wars and Trials
(1810–1821)

The French Revolution affected the Galileo affair not only in the general and indirect ways that might be expected,[1] but also in a very specific and concrete way. For in 1810 Emperor Napoleon Bonaparte decided to transfer to Paris all Church archives in Rome, paying special attention to certain documents such as the Vatican file of the Inquisition's proceedings of Galileo's trial. Napoleon started the process of publishing these documents, but the project remained incomplete. As a result, however, a few documents came to light, the most important being Galileo's confession at the second deposition of 30 April 1633 and the Inquisition's executive summary of the proceedings, compiled in May or June 1633 to serve as a basis for bringing the trial to its conclusion.

9.1 The Trial Proceedings to Paris:
Napoleon's Publication Plan (1810–1814)

In 1798 a French army occupied Rome, abolished the papal government, and established a Roman Republic. Pope Pius VI was deported to Florence, and the Inquisition palace in Rome was "plundered to some extent by a French military rabble, and a part of the archives burned."[2] In 1800 a new pope, Pius VII, was elected in Venice, and in 1806 he was allowed to return to Rome with limited powers of government. But in 1807 Pius VII refused to cooperate with Napoleon's plan for the economic isolation of England. Thus in 1808, Napoleon decreed the separation of the pope's spiritual and temporal powers and reclaimed for France the thousand-year-old territorial gifts of Charlemagne. In 1809, Napoleon again abolished papal government in Rome; the pope responded by excommunicating him. As a result, the pope was arrested and deported to France, and on 2 February 1810 everything in Rome pertaining to papal government was ordered moved to France. This situation did not change until 1814, when Napoleon freed the pope, restored the papal state, and began returning Church records and archives to Rome.[3]

Transferring the papal archives to France required a monumental effort and an astronomical sum: the first convoy (in February) used 3,239 cases and cost 179,320 Italian lire; the second (April) and third (July) convoys transported the Inquisition archives.[4] A few documents were shipped separately because of their special importance; among these were the papal bull of excommunication against Napoleon and the file of the original proceedings of Galileo's trial.[5]

In January 1811 in Paris, the file of Galileo's trial was handed over by the Ministry of Religions to Napoleon's personal librarian, Antoine A. Barbier. The transmittal document noted that the file was labeled volume 1,181; that the manuscript folios were numbered, starting with 337 and ending with 556; that folio 556 was followed by five unnumbered folios; and that to these the numbers 557 through 661 were then added.[6] These are important facts in the history of the Galilean file that had been created in 1755 and continues to exist today, for that pagination is still visible today[7] and thus helps us to check the integrity of what is available.

On 12 March 1811, Barbier filed a report to Napoleon recommending publication of the original proceedings of Galileo's trial side by side with a French translation, and presenting an itemized budget of anticipated expenses for the publication of 1,000 copies. The report reads as follows:

> Sire: Nothing is more famous in the history of science and of the Inquisition than the trial of Galileo. One of the most distinguished scientists of the seventeenth century was forced to retract his opinion on the motion of the earth around the sun, an opinion that has been taught for a long time in all schools, even ecclesiastical ones. Being as pious as he was enlightened, Galileo proved that this opinion was not contrary to holy scripture[8] when it is properly understood. But theologians found it contrary to their interests and so pursued his condemnation with obstinacy; and what is still more astonishing is that these theologians have found some defenders until near the end of the eighteenth century, the famous Mallet du Pan among others.
>
> Your Majesty's victories, which have made so many men and so many countries your subjects, have also given you possession of the documents that make up the famous trial. They demonstrate the good faith and the enlightenment of the learned astronomer at the same time that they reveal the perfidy and the ignorance of his accusers.
>
> The publication of these documents is worthy of Your Majesty's rule. Some are in Latin, but most are in Italian. It would be appropriate to place a French translation on the opposite pages from these various texts.
>
> The printing of the original texts and the translation will form a volume of about 400 quarto pages. The manuscript is full of abbreviations and contains many difficult passages; so one cannot avoid having it copied. Adding 2,000

francs for the cost of the copy and the translation, the total expense will reach only about 7,000 francs; by printing 1,000 copies of the book, one could place 800 copies for sale; sold at the price of 8 francs each, these would produce at least 6,000 francs.

I have the honor of proposing to Your Majesty to have the documents comprising Galileo's trial copied, translated, and printed.[9]

It is striking but not surprising that Barbier would find Galileo to be "as pious as he was enlightened." Barbier obviously knew more about the case than the content of that special file, as shown by his remark that Galileo had proved that the earth's motion is not contrary to Scripture; for this claim was not only a secondary conclusion in the Letter to Christina, but was also implicit in his primary conclusion that Scripture is not an astronomical authority; in fact, if Scripture is not an astronomical authority then its statements about the sun and the earth need not be taken as descriptions of physical reality, and so they do not contradict Copernican claims about physical reality.

In talking of the "interests" of the theologians opposing Galileo, Barbier was hinting at what might be called a political interpretation of the trial. The mention of Mallet du Pan[10] in a document of this sort attests to how widely his myth had spread; clearly Barbier had not been impressed by Mallet's manner of combating lies. Similarly, in claiming that the trial proceedings demonstrated Galileo's "good faith," Barbier indicated that he had been unimpressed by Tiraboschi's[11] attribution of bad faith to Galileo. And in saying that those proceedings also demonstrated the treachery and ignorance of Galileo's accusers, Barbier was apparently endorsing and echoing a type of account that had been stressed by Voltaire.[12]

The emperor apparently approved the project, for on 16 October 1812, in reply to an inquiry from the Ministry of Religions regarding the file of Galileo's trial documents, Barbier stated that he had recommended to the emperor the publication of the original together with a French translation, and that the emperor had approved the project.[13] The first twenty-five folios of the manuscripts, containing nine documents, were translated:[14] (1) the summary of the proceedings from 1615 to May 1633, prepared at the end of May or the beginning of June 1633 by Inquisition officials for the benefit of the pope and the cardinal-inquisitors, and amounting to about eight folio pages;[15] (2) the letter of complaint by Dominican friar Niccolò Lorini to the Roman Inquisition, dated 7 February 1615, which had started the earlier proceedings;[16] (3) Galileo's Letter to Castelli, dated 21 December 1613, which had been attached by Lorini to his complaint;[17] (4) an Inquisition consultant's report on Galileo's Letter to Castelli;[18] (5) a letter dated 8 March 1615 by the archbishop of Pisa answering a request by the Roman Inquisition;[19] (6) a letter dated 7 March 1615 by the inquisitor of Pisa answering a request by the Inquisition;[20] (7) a legal deposition by Dominican friar Tommaso Caccini, dated 20 March 1615;[21] (8) a clerical note that Caccini's deposition had been sent to the inquisitors of Florence and Milan to interrogate other witnesses there;[22] and (9) a letter dated 28 March 1615 by the archbishop of Pisa to the Roman Inquisition.[23]

Why the translation was not completed is unknown. Perhaps the stumbling block was the difficulty of the undertaking. One aspect of this difficulty had been implied in Barbier's report to Napoleon, namely the many abbreviations in the text of the manuscripts. Another difficulty may have been the translator's incompetence. This is suggested by Favaro who, in 1902, after examining the translation, stated that it was terrible and gave the following example.[24] The initial sentence in Caccini's complaint reads: "Besides the common duty of every good Christian, there is a limitless obligation that binds all Dominican friars, since they were designated by the Holy Father the black and white hounds of the Holy Office."[25] The translator was uncertain whether the Italian word *cani* (which literally means *dogs* and which I have rendered *hounds*) should be translated as *canons* (meaning clergymen having minor duties in a cathedral) or *canon lawyers*.

Since the Napoleonic edition of the trial proceedings was never published, various conjectures were later advanced to explain why not. For example, in 1850 Marino Marini wrote that the Italian historian Carlo Denina, who was living in Paris between 1810 and 1814, had advised Napoleon that the proceedings contained nothing worthy of being republicized.[26] Marini was one of several Vatican officials who had accompanied the papal archives to France to keep track of them and help keep them in order. After the fall of Napoleon, Marini was one of the officials charged with the retrieval of those archives. To this aspect of the story we now turn.

9.2 Lost and Found: Marini's Efforts (1814–1817)

The retrieval of the papal archives from the Napoleonic plunder was only partially successful. Some were retrieved and sent back to Rome, such as minutes of Inquisition meetings; book censures commissioned by the Inquisition; and the complete files of the Index.[27] Other archives were found, but, because of their bulk and the cost of transporting them, they were destroyed or sold to cardboard manufacturers; such was the fate of 3,600 volumes of Inquisition trial proceedings and 300 volumes of Inquisition sentences, amounting to about two-thirds of all the archives originally transferred to Paris.[28] This destruction must rank as one of the saddest losses and most ironical twists in European history. Still other documents were not found; the file of Galileo's trial was among these.

With the fall of Napoleon and the restoration of the French monarchy, in April 1814 the new French government decreed that all papal archives be returned to the pope.[29] On 6 November 1814, Marini was in Paris and wrote to the minister of domestic affairs specifically requesting the return of the Galileo file.[30] On 11 November 1814, the ministry of domestic affairs informed Marini that he should contact the ministry of the royal house.[31] On 20 November, Marini wrote to the minister of the royal house, Count Luis Casimir de Blacas d'Aulps. On 5 December, in reply to an inquiry from Blacas, Barbier (now director of the French royal library) stated that the Vatican file with the manuscript proceedings of Galileo's trial was in his possession, as it had been for the last three years.[32] On 15 December, Blacas wrote to Marini saying that the Galilean file had been found and he was ready to hand it over to Marini in person.[33] On 16 December, Barbier handed over the Galileo file to Blacas.[34]

Marini wasted no time in going to the count's residence to get the file, but they were unable to meet. For one reason or another the same thing happened several times. Finally, on 2 February 1815, Blacas wrote Marini that King Louis XVIII wanted to read the Galileo file, and so the file was now in the king's private study; Marini could have it as soon as the king was finished.[35] The next several months witnessed the upheaval of Napoleon's hundred-day return, and everybody had more urgent things to think about.

After the final defeat of Napoleon and the re-restoration of the monarchy, on 22 October 1815 Marini was back in Paris and wrote the new minister of the royal house, Count de Pradel, to get the Galileo file back.[36] On 6 November, Pradel wrote Marini that after searching for the file, they had been unable to find it; perhaps Marini should contact Blacas.[37] At this point the documentary trail is interrupted. However, we know that in 1816, while Marini was back in Rome, another Vatican representative, Count Giulio di Ginnosi, approached the minister of the royal house for the return of the Galileo file, but with no success.[38]

In 1817, the pope sent Marini back to Paris to make further attempts to retrieve the file. Marini followed all possible leads, but to no avail: he contacted the minister of foreign affairs (on 23 July and 13 September 1817); the minister of police (on 1 August); the minister of the royal house (on 2 August); and the Louvre, under the jurisdiction of the minister of justice (on 4 September).[39] A good indication of Marini's indefatigable efforts and of the ceaseless shenanigans he had to deal with is provided by the following letter he wrote to the minister of foreign affairs, the Duke of Richelieu, on 23 July 1817:

[212] Your Excellency: As a result of the decree of 19 April 1814 issued by His Royal Highness in the name of His Majesty, the archives were given back to the Holy See. The Home Office's commissary, Mr. Beugnot, ordered that they be given to Father De Gregorio and to the two Marinis (uncle and nephew), who immediately directed their efforts to bring them back to the Vatican. If justice and religious piety had induced His Royal Highness to the prompt restitution of those documents, no smaller devotion toward the Roman Church moved the King to alleviate the transportation costs. By competing in religiousness and generosity, those two princes encouraged the papal com-

missaries to display the good faith that has always marked them. Because of this, they deemed it excessive to seal the archives when they took possession of them, persuaded that what had not happened under the previous government would not happen after Providence had restored religion and justice to the throne of France. But what occurred was very different, not without some consent by the Monarch. Abusing the good faith of the papal commissaries, various people removed with a sacrilegious hand many precious documents. Among the most important of these one can include some volumes of papal bulls, especially those of Julius II; the same misfortune befell the proceedings of the trial of the Knights Templar and the original letters of Bousset and of the many bishops who participated in the debates of 1682. Under the pretext of collating these letters with those that had been improperly published in France by the Jansenists, the enemies of religion, who are always enemies of the Roman doctrine, removed all of them artfully. The papal commissaries, whose good faith did not make them suspect others of fraud, allowed such a collation, especially when they saw that the archivist Daunou approved it and even begged that they have access to those letters.

But after the archives arrived in Rome, the deception was discovered and the Roman Church saw itself stripped of documents that were one of the many apologias of its doctrine. I cannot specify the persons who can be accused of such thefts. It is certain that Messrs. Valéry, Delespine, and Raynouard were dealing very frequently with the documents of the papal archives; also involved many times was pseudo-Bishop Grégoire, who I believe sent the people who collated the above-mentioned letters. With regard to the file of Galilei's trial, I have been unable to even get a hold of it, despite the fact that I have made many requests to Mr. Count de Blacas and he has made many promises to me. However, I inform Your Excellency that His Holiness is no less eager to recover this file than the others mentioned above. I cannot [213] doubt in any way that the religious propensity of His Most Christian Majesty as well as of the most pious Count d'Artois to want to return these extremely important documents to the Holy See will excite in Your Excellency a religious zeal to ensure that they be all returned to me, for they have been improperly removed and yet were included in the above-mentioned decree of restitution. To what other minister could I speak with greater trust than to Your Excellency, who has the honor of counting among your ancestors a most celebrated cardinal of the Holy Church? And with courteous regards I take my leave.[40]

The fact here mentioned, that the restored monarchy helped to pay some of the cost of shipping the files back to Rome, together with the fact that two-thirds of the archives were turned into cardboard, gives us a better idea of how massive Napoleon's translocation had been. Although the file of Galileo's trial was not the only set of documents that had not been returned, it clearly was one of those uppermost in the mind of the pope. And the context of the situation suggests that, whether or not it was actually true that King Louis XVIII had ever examined the file himself, Count de Blacas may have had ulterior motives for not giving the file back to Marini.

Soon thereafter, Marini returned to Rome thinking that the file had been irretrievably lost. The Vatican, however, did not forget the matter. There is evidence that in 1835 it made a further attempt to retrieve the file, but to no avail.[41] Unexpectedly, however, in 1843 it was returned to the Holy See by the nuncio to Vienna, to whom it had been given by Blacas's widow. The count, who had been the French minister of the royal house, had followed King Charles X in exile as a result of the revolution of July 1830 and had died in Austria in 1839.[42]

9.3 The Napoleonic Translations: Delambre's Finding (1820)

Although the file of Galileo's trial was not published as Napoleon had envisaged, and although it was lost soon after his fall, this minor Napoleonic feat did contribute to the diffusion of more accurate information and better understanding. This came about by way of the few persons who had had the opportunity to read the manuscript documents between 1811 and 1814 and the few documents that had already been translated into French.

The climax of these developments occurred between 1818 and 1821 and primarily involved Jean Delambre and Giambattista Venturi. Delambre, who was perpetual secretary of the French Royal Academy of Sciences, was working on a history of astronomy that appeared in 1821. Venturi (1746–1822) was a retired professor of physics who had taught at the universities of Modena and Pavia; he was publishing a two-volume documentary history of Galileo's work and its reception, the first volume of which had appeared in 1818 and the second of which would appear in 1821.

As Venturi was putting his finishing touches on his second volume and trying to articulate an account of the trial based on all available documents, he made one last attempt to get hold of the Napoleonic documents. On 25 April 1820 he wrote this letter to Delambre:

[222] I send you many thanks for the trouble you are willing to go through to let me have the file of Galileo's trial documents. I shall write to Rome about it to see whether Mr. [223] Count de Blacas has any information in this regard; but I do not expect so and believe that the manuscripts disappeared when Napoleon returned to Paris.

Ultimately I have no great need of these manuscripts because I have: (1) the regular correspondence by which the Tuscan ambassador to Rome with each postal delivery informed the grand duke of the development of the Galileo affair, for the grand duke was very interested; (2) three accounts of the affair itself written, first, by Galileo himself after his condemnation, second, by a Tuscan who was in Rome at the time, etc.; (3) the sentence in its Italian original, for Riccioli translated it into Latin; and (4) manuscripts of several other letters that relate to the same trial.

Father Altieri,[43] who was in Paris in 1814 charged by Austria to recover its documents, read there at that time the file of Galileo's trial documents; he reported its content to me in general terms, and told me that it was in order. For in 1616 Cardinal Bellarmine had in the name of the pope forbidden Galileo to write any longer for the Copernican system; but Galileo wrote about it and wanted to justify himself by saying that his 1632 *Dialogue* had received the imprimatur of the reviewers; then he was asked whether he had informed the reviewers about the injunction issued to him not to write any longer for this purpose; this he had not done; and there you have his fault. When he was before the judges, he often resumed talking about the interpretation of biblical texts to prove that they were not contrary to the earth's motion; but thereupon the judges ridiculed him because the question then was not that, but rather whether or not he had disobeyed the orders received in 1616.

The judges did have an apparent reason to condemn him. However, ultimately the persecution did not come directly from them, but from Urban VIII; until then he had treated Galileo with great friendship, but afterwards he was displeased for what Galileo in the *Dialogue* had put in the mouth of Simplicio, mocking the arguments which the pope himself had earlier in conversation advanced against the Copernican system. There you have the true cause of Galileo's misfortunes; I have in my hands the authentic proofs, and I shall publish them with the rest.

Kepler justified himself in a letter to Italian booksellers written with great skill;[44] it is published among his letters, and I have inserted it in the second volume of my book on Galilei. At the moment they are printing precisely the documents I mentioned above. I expect that the rest will also be printed by about four months from now; it will be an honor and a duty for me to send it to you.

Please accept my thanks, my respectful regards, and the assurance of my high esteem.[45]

It is unclear why Venturi was saying that "ultimately" he did not really need the trial proceedings, for the other documents he mentioned (except for the sentence) constituted indirect evidence, whereas the file contained primary and direct documentation. With regard to the sentence, Venturi was aware that the original text was in Italian, in accordance with the Inquisition practice of favoring the native language of the accused person; although the Italian text had been published by Giorgio Polacco in Venice in 1644, that edition had been overshadowed by Riccioli's superb collection of documents in Latin.[46] The first group of documents Venturi mentioned referred to the diplomatic correspondence published mostly by Fabroni in 1773–1775 (see chapter 8.2). Of the three accounts of the trial, the first was Galileo's apocryphal letter to Renieri of December 1633, although of course Venturi did not know it was a forgery;[47] the second was Buonamici's account of July 1633;[48] for the third, it is hard to fathom what Venturi was referring to. By the fourth group of available documents, Venturi was probably referring to the following letters by Galileo, which he published in his book. Three of them had been published before, but certainly deserved republication; that is, the Letter to Castelli, the Letter to Dini (23 March

1615), and the Letter to Christina.[49] A fourth letter was Galileo's reply to Francesco Ingoli, written in 1624 but published for the first time (in 1813–1814)[50] only a few years before Venturi's writing; it was really a short treatise, a first draft of the *Dialogue,* containing a detailed analysis and criticism of the anti-Copernican argument advanced in a 1616 book by Ingoli.[51]

What Venturi said about Father Altieri's reading of the Galilean file sounds like Altieri's summary of that file. Altieri's stress on Galileo's alleged violation of the special injunction and deceptiveness about it did not go beyond what one could read in the text of the sentence.[52] And Altieri's report that at the trial Galileo defended himself by trying to reconcile the earth's motion with Scripture did correspond to the picture in the apocryphal letter to Renieri;[53] however, it did *not* correspond at all to the proceedings in the file, as we can see today.[54] Of course, Venturi was in no position to thus discredit this part of Altieri's account, but we must conclude that Altieri had not really read the file, or at least not completely or carefully.

Venturi's own interpretation seemed to be that Galileo's violation of the 1616 injunction was only the apparent, formal, or external cause of the 1633 condemnation, but that the true, real, and ultimate cause was the pope's displeasure at having been caricatured by the character Simplicio in the *Dialogue.* In this letter, however, Venturi was perhaps echoing Altieri, for he described the 1616 injunction as the prohibition "to write any longer for the Copernican system," which is cleverly ambiguous; for it could mean writing *in favor of* the Copernican system or writing *about* it, and the difference between these two is the difference between the Inquisition commissary's special injunction and Bellarmine's certificate. Finally, to understand what Venturi was referring to in boasting that he had the evidence to prove the Urban-insult thesis, we would have to read his book, or at least his own synthetic account, as we shall see presently.

Delambre did not reply immediately, but eventually, on 30 June 1820[55] he did write to Venturi. He had finally found the partial French translation of the trial proceedings commissioned by Napoleon and sent it to Venturi attached to his letter.[56]

In this letter,[57] Delambre gave an account of the details of the Napoleonic project that corresponds to what librarian Barbier had reported to the emperor. The attached French extracts prove that the project had been begun. And Delambre suggested that the project had not been completed merely because of the fall of Napoleon.

Delambre was clear that the translations were incomplete, but he judged that one now had adequate knowledge of the essence of the trial based on these extracts together with Riccioli's documents and those published by Venturi in the first volume of his book (which covered the period through 1616). Delambre's judgment was somewhat superficial and excessively optimistic, as the subsequent history of the Galileo affair shows. However, his

view was not without justification because one of the Napoleonic translations happened to be the summary of the proceedings. This was the first
document to be translated for the simple reason that it was placed at the
beginning of the file of trial proceedings; and as we shall see presently, this
summary was not only a detailed report of the proceedings from 1615 to
May 1633 but also contained precise references to the other parts of the file
(most of which were missing); so from the summary one could even construct the structure of the missing file.

Like so many other things in the Galileo affair, these Napoleonic trial
proceedings had a subsequent history. In 1828, they were sold to someone
named Villenave; on 18 July 1841 they were purchased by Guglielmo Libri;
and eventually they found their way to the Mediceo-Laurenziana Library in
Florence.[58]

9.4 Primary versus Accessory Causes: Venturi's Explanation (1820)

Before receiving Delambre's letter of 30 June 1820, Venturi had composed
the best account he could of the 1633 trial and delivered it as a lecture at a
meeting of the Institute of Sciences in Milan on 8 June 1820.[59] Because
Venturi had searched, collected, and studied all that was available about
Galileo in general and the trial in particular, his account may be regarded
as an excellent gauge of knowledge about the trial at that time. Moreover, it
contains some valuable insights. And it is not surprising that even after he
read the Napoleonic French translations, he felt they did not necessitate
any revisions of his account but rather confirmed and enriched it.[60] Venturi's account is thus as important a milestone in the interpretation of the
trial as his book was for its documentation. He published the account in the
second volume, which was then in press. Venturi wrote:

> [192] After the French occupied Rome in 1809, they transported many doc-
> uments pertaining to the arts and the sciences from there to Paris; among
> these was the file of Galileo's trial documents. Then when the monarch ruling
> today first returned to Paris, he had it brought to his apartment to read it. But
> Napoleon returned from Elba, and later had to flee once again. In the midst
> of these governmental revolutions, the Galilean file must have been lost. For
> on the one hand, monsignor Testa (secretary of His Holiness) assures me that
> it has not been returned to Rome; on the other hand, Mr. Delambre (perpet-
> ual secretary of the Academy of Sciences in Paris), who is generously trying to
> get me [193] a copy, has been unable to find it.[61] In the meantime, an edu-
> cated person[62] who held an office in 1814 in Paris, told me that he had read
> it and related to me a large part of its content; he assured me that it did not
> indicate that Galileo had been subjected to any painful corporal examination;

in short, he declared that the proceedings were in order, and that printing them would bring no dishonor to the judges.

In the absence of the original proceedings, on the basis of the documents available so far and of the above-mentioned account, it will be useful to put together and briefly present the genuine history of the trial.

The substance reduces to this. Galileo was examined and during these interrogations he confessed that in 1616 he had been prohibited by Cardinal Bellarmine, on superior orders from the Congregation of the Holy Office and the pope, from speaking any longer in favor of the Copernican system.— But then how did you dare to publish the *Dialogue?*—I had the inquisitor's permission.—Did you inform him of the prohibition enjoined on you?—I deemed it pointless to do so. There, by rigorous justice, you have his fault; for the permission obtained from a lower authority does not justify the infraction of the prohibition issued earlier by a higher authority; instead there is also the suspicion and presumption of malice for not having revealed to the lower authority the prohibition received from the higher authority. To strengthen this suspicion in our case one added the tenor of the *Dialogue,* which was wholly favorable to Copernicus and thus in a way evasive toward the received prohibition. This was the chief point and the foundation of the condemnation. Father Riccardi,[63] who had fallen out of favor with the pope for having permitted the publication of the *Dialogue,* said in confidence to the Tuscan ambassador that the above-mentioned circumstance (also noted by the judges) was enough to ruin Galileo (pp. 150 and 159 above).[64] Nor did the pope adduce to Niccolini[65] any other reason for the condemnation (p. 166). And this was also the point stressed by the other members of the Congregation to the same Niccolini (pp. 165, 167, 168). If each individual is allowed to violate with impunity the laws of whose intrinsic reasonableness he is not persuaded, then social order will be left without support.

One could have perhaps succeeded in concealing this Galilean transgression by imposing silence about the matter, as some hoped (p. 165), if other circumstances that were accessory and extrinsic to the question had not converged to ensure that he would be persecuted as rigorously as possible. I mentioned the first and perhaps the strongest at the bottom of p. 146, namely the indignation and the spite felt by Urban VIII against Galileo; for even though they shared the same city as motherland, and even though Urban had favored Galileo in every way both before and after becoming pope, yet in his *Dialogue* he had dared place in Simplicio's mouth the arguments which the pope had presented to him against the Copernican system. Not that before the publication of the *Dialogue* the pope pressed his own commitment to the point of pretending that the earth's rest and sun's motion were articles of faith (pp. 88, 113, 178). But when he saw in the *Dialogue* his arguments advanced by Simplicio and scorned, he exploded into extreme anger; for he was very "ambitious to legislate to all the sciences; and for this ambition no amount of the most refined prudence is enough, and one is doomed to ruin." So said about him a cardinal "who through literature rendered more illustrious the distinction of his birth"[66] (perhaps Cardinal Bentivoglio).[67] This was the principal reason for the aversion and rage with which Urban spoke of Galileo to Niccolini and for the grave dedication with which he more than anyone else persecuted him for the whole period of the trial (pp. 147, 152, 156, 161, 164). And even three years [194] later he still held a grudge, persisting in the idea that in his *Dialogue* Galileo had wanted to make fun of him (pp. 191, 192).[68] With good reason, therefore, in the Magliabecchiana Library, on the box holding a copy of Galileo's 1638[69] booklet *Nov-antiqua* in quarto, there is the following note in handwriting that is not modern: "Pope Urban became spiteful toward Galileo because he had incidentally discussed with him some points of his system of the earth's motion (before it was published), and in disclosing them the latter put them in the mouth of Simplicio. The pope became irritated and made him abjure, and the poor man had to appear wearing a tattered piece of cloth that moved one to pity." In his anger the pope himself told Niccolini that he had already presented to Galileo all the difficulties that existed against the Copernican system (p. 147), and it is probable that they were the common difficulties of the Peripatetics of that time. He then mentioned (p. 161) a particular argument of his; that is, one must not impose on God the necessity to make the world one way rather than another; at the end of the *Dialogue* Simplicio advances this same argument and says that he once heard it "from a most learned and most eminent person."

The other extrinsic reason that had very great influence on the condemnation of Galileo was the hatred which the Peripatetics and the friars harbored toward him and the doctrines he maintained. When he went to Rome in 1623[70] and was favored by the pope, his court, and the Linceans, Father Caccini, who had persecuted him in 1614 (vol. 1, p. 219), was going around Rome saying that if Galileo were not protected he should be put in prison and tried by the Inquisition (see Castelli's letter to Galileo found in Nelli's library). In part 3 of this volume, we have seen how many authors rose up to fight the *Dialogue* and with what animosity they did so. And it is credible that one of them instilled the poison in the pope's mind and persuaded him that in the said *Dialogue* Galileo had made fun of him. Father Campanella warned Galileo that meetings were being held of theologians who were angry with him and that they had started their complaints with many blows against the new philosophers (p. 144). It may well be that the Jesuits participated in this conspiracy (p. 188), and there is no lack of writers who claim that Father Scheiner himself denounced Galileo's *Dialogue* to the Inquisition; about them in general the poet Menzini once said: "They were the ones who stung Galileo / With a needle from the papal insignia." But to tell the truth, I have not found in the documents of the time any that clearly reveals this (except for the vague indication on p. 188). On the contrary, Father Riccardi, who had a vested interest in defending the *Dialogue* since he had signed the imprimatur, assured the Tuscan ambassador that the Jesuit who was a member of the special commission charged to examine the book "was proposed by himself, is a confidant of his, and has good intentions" (p. 178).[71] At that time Mr. Buonamici (p. 178) attributed the persecution of Galileo to the hatred which Father Firenzuola, who was one of the Inquisition's commissaries, harbored principally against Father Riccardi, who had approved the *Dialogue* (p. 117);[72] and perhaps Holste's letter (p. 182) refers to the same Father Firenzuola, who was a mathematician.[73] Galileo remembers that at his interrogation there were only commissary Lanci and assessor monsignor Vitrici, together with two Dominican friars; in vain did he try [195] to present to them a reinterpretation of biblical passages that seem contrary to Copernicus, for the only response he received was a shrugging of the shoulders (p. 181).[74] In fact, the question then was not to decide whether one should reform the 1616 decree and the consequent prohibition to Galileo; rather, given these, one sought to determine whether or not Galileo had transgressed them.

Taking together the pope's determination to hound and humiliate Galileo and the theologians' determination to uphold the 1616 decree of the Congregation of the Index, it should not be surprising that Galileo was condemned for having acted against that decree, as well as that he was found gravely suspected of still believing the earth's motion not as a mere philosophical whim but as a thesis, even though it had already been declared heretical or at least erroneous in faith. The Florentine philosopher had already long before prepared himself to avoid this accusation. In fact, when in 1618 he sent to the archduke Leopold of Austria his discourse on the tides (which he explained on the basis of the earth's motion), he claimed that he regarded it as a poem or dream after the "celestial choice of the ecclesiastical superiors, who have access to higher knowledge, has reawakened me and dissolved like a fog all my confused and contorted images" (p. 80).[75] Similarly in 1624, just before defending the Copernican system against Ingoli, he declared that in so doing he had no intention of "supporting as true a proposition which has already been declared suspect and repugnant to a doctrine higher than physical and astronomical disciplines in dignity and authority";[76] but that he did it to show that there was at least one Catholic who knew the physical evidence for that system, although "he gives priority to the reverence and trust which is due to the sacred authors over all the arguments and observations of astronomers and philosophers put together."[77] Finally, in the preface to the *Dialogue* he says he wrote it to show that "it was not without prior knowledge of all speculations pertaining to the Copernican system that Rome had promulgated the salutary edict which imposed opportune silence about the Pythagorean opinion of the earth's motion."[78] After him, Gassendi and Angeli[79] said more simply that they respected the decree of 1616, but that the physical arguments by which some claimed to support it were not valid.

Then when the trial began, the Tuscan ambassador went around declaring to everybody in the name of Galileo that he was ready to obey, to abandon, and to retract all that would be commanded to him (pp. 154, 158, 159, 161). In light of his prompt obedience[80] regarding his coming to Rome despite his old age and his ailments, and in view of his sincere and loyal declarations, there was no need to subject him to torture. After staying for some time at the residence of the Tuscan ambassador, he was obliged to go to the Inquisition's palace. Here, not only was he not tortured, but he was not even placed in a prison cell; for lodging he had three rooms that belonged to the prosecutor, with the freedom of going also into the courtyard; they allowed his servant to attend him, to sleep in those rooms, and to go out whenever he wanted; and the ambassador's servants brought him food to his rooms every day (pp. 163, 188). Although he remained there about two weeks, he was interrogated only once or at most twice; and the cause of the delay in sending him back to the ambassador's residence was the fact that the pope was then in Castel Gandolfo (p. 164). The friars took advantage of his docility and resignation to extract from him a very clear and precise disapproval of the Copernican system; in this way he purged himself of any suspicion to the contrary; at the same time the theologians had the pleasure of seeing confirmed the decree prohibiting the Copernican system; and the pope was glad to see punished someone who he believed had shown no respect for him. The punishment was indefinite imprisonment, as a penalty for transgressing the injunction of 1616, as mentioned above; the sentence (pp. 171–72) indicates no other reason; nor, after the sentence was pronounced, was another reason adduced by the pope to Niccolini (pp. 166, 168).

[196] Mr. Lalande relates that Cardinal Bentivoglio, who was a member of the Congregation charged to judge Galileo, did whatever he could to save him. It is very true that, since at the beginning of the trial the grand duke had written to him and to Cardinal Scaglia a letter of support for Galileo (p. 183), they both showed themselves well disposed to favor the protégé (p. 158) and moved very much in unison to protect him (p. 162). But how could one avoid his condemnation in view of the pope's resentment and the rigor of justice? Thus, "with no one dissenting,"[81] the Congregation decided to punish him (p. 166). The Lord Cardinal Antonio Barberini also showed himself favorable to Galileo after he arrived in Rome and contributed to mitigating the exacerbated feelings of the pope (pp. 164, 191).

Father Riccardi, Master of the Sacred Palace, who had approved the publication of the *Dialogue,* at first hoped that the Congregation would not prohibit it but only correct and revise it (p. 148); and in fact he undertook to review the work again to try to fix it so that it could be tolerated. But he had his own troubles (pp. 156, 168). He excused himself by saying that he had approved the publication of the *Dialogue* because he had received an order from the pope to do so; the pope denied it saying that these were just words, not to be trusted; but finally the Father Master produced a note by monsignor Ciampoli, secretary to the pope, in which it was stated that His Holiness (in whose presence Ciampoli claimed to be writing) was ordering him to approve the book (p. 179). The pope said that this was an action typical of Ciampoli (p. 159), and that his secretary and Galileo had double-crossed him (p. 147); he had already removed Ciampoli, and he also dismissed Riccardi from his position.

Some have seen fit to accuse Ferdinand II of weakness, because he allowed his mathematician to be persecuted in this manner; they blame it on the pernicious influence of Cioli,[82] his prime minister, as if the latter for his own private reasons did not want to upset the pope. Those who talk in this manner may not have noticed how much authority the Roman court exercised at that time over the various Italian states, except the Venetian Republic. It was only at the end of the last century that Italian rulers regained in large measure the exercise of their rights; but the situation was very different in the seventeenth century. At that time Tuscany would not have dared fight against the ecclesiastic pretensions of the Roman court, especially in affairs of the Inquisition and especially at the time of Urban VIII, who was a born Florentine and showed the greatest possible regard for his native city; in view of the grand duke's warm and strong displays, the pope claimed to have used every regard toward Galileo, in an affair that belonged exclusively to the Holy Inquisition. Reading ambassador Niccolini's letters, one can see that he was also of the opinion that to oppose the pope would have been useless and would have served only to enrage him further.

In Paris the Copernican system almost suffered an affront similar to that experienced in Rome. Incited by Rome or by some Scholastic philosopher, Cardinal Richelieu proposed to the Sorbonne to decide whether to condemn that system. Already in a session of this assembly they were about to confirm the decree of the Roman Inquisition by a plurality of votes. But one of its members, who was a person of talent, advanced such reflections that they stopped the blow; and they allowed that the question of the rest or motion of the earth be handled by philosophers, despite the efforts of those who would have liked the intervention of ecclesiastical authority (Montucla, *Hist. des Mathematiques*, vol. 1, p. 527).[83]

The decree of the Roman Inquisition in 1633 based itself on the one [197] in 1616; thus, what I said earlier about this first one applies equally in regard to that second one. Read what I said in vol. 1, pp. 273–74.[84]

It was clear from the numerous references to the documents in his own collection that Venturi's account of the 1633 trial was very well documented. Besides the sources mentioned in his letter to Delambre, the documentation also included unpublished letters from the Nelli Library. More generally, in his work Venturi had tried to include all that was relevant and available.

However, Venturi was no mere document collector, and so he developed an insightful interpretive explanation. His insight was that the primary cause of Galileo's 1633 condemnation was his violation of the 1616 injunction, but that this cause was not sufficient and had to be combined with two other, secondary factors: the offense felt by pope Urban for Simplicio's caricature of him; and the hatred of Galileo held by Aristotelian philosophers, Dominicans, and Jesuits. Whatever one may think of the details, Venturi's multifaceted approach was a step in the right direction. Moreover, he had ideas about the relationship between the two kinds of causes: the primary cause was described as the "foundation," whereas the secondary factors were "circumstantial, accessory, and extrinsic"; in the April 1820 letter to Delambre, Venturi had described their difference as "apparent" versus "ultimate" or "true" cause; these two descriptions seem incompatible with each other, and in any case all these notions required further clarification; nevertheless, such remarks are highly suggestive and indicate an awareness about levels of causation.

On the other hand, Venturi's account of Galileo's violation of the 1616 injunction was somewhat superficial, for he failed to distinguish between the content of the special injunction, of Bellarmine's certificate, and of the Index's decree, thus failing to clarify exactly what Galileo had been forbidden to do in 1616. His account of the Simplicio-caricature factor was relatively uncritical toward the note in the Magliabecchiana Library; it also failed to notice that the key supporting evidence of Castelli's letters dates from a period (1635–1636) that is not previous to or contemporaneous with the trial, as Pieralisi argued in 1875.[85] And Venturi's remarks on the role played by Riccardi and Ciampoli were mostly a repetition of Buonamici's account of July 1633. By contrast, in his account of the Jesuit conspiracy (classified as part of the second extrinsic factor), Venturi seemed to show some appropriate nuances and qualifications.

In his concluding paragraph, Venturi was saying that Galileo's 1633 condemnation was based on the 1616 condemnation of Copernicanism, which raised the question of the status of the latter. For this he referred readers to his earlier volume, where he did briefly discuss it. He held that the Copernican condemnation was no longer justified because the earth's motion had been proved beyond any reasonable doubt; he also held that it had been set aside "after the wise Benedict XIV erased the above-mentioned decree from the *Index* of prohibited books, which is to say that he annulled it."[86]

9.5 Galileo's Confession: The Inquisition's Trial Summary Revealed (1821)

After receiving the Napoleonic documents attached to Delambre's letter, Venturi wrote an addendum to his account. He did not think he needed to revise it,[87] but he added supplementary judgments as well as some long and brand-new quotations from the Napoleonic extracts. Of course these were in French, and so Venturi had to translate them back into Italian. He published his addendum in 1821 in the second volume of his book.[88]

In the meantime, Delambre was doing something similar in his *History of Modern Astronomy*, which also appeared in 1821. In a long preface, he stated that he had written his account of Galileo and the trial in the body of the work[89] before reading the Napoleonic translations of the file; and although he did not think that he needed to make any drastic revisions, he had decided to add to that account by describing and quoting from those extracts.[90]

Both Venturi and Delambre focused on the Inquisition's executive summary of the proceedings from 1615 to May 1633, compiled at the end of May or beginning of June 1633. This was a crucial document because, in standard Inquisition practice, it was the legal basis on which the cardinal-inquisitors and the pope reached a final decision on the case being tried. However, neither Venturi's addendum nor Delambre's preface was clear about the nature of this document, and in fact their exposition confused three sets of documents: (1) the Inquisition's summary, whose French translation they had in their possession and which referred to and described other documents they did not possess; (2) the other trial documents pertaining to the year 1615 whose translations were in their possession; and (3) the other trial documents from 1615 to 1633 whose translations were not in their possession. The confusion was compounded by the fact that the summary included an almost verbatim transcript of Galileo's second deposition of 30 April 1633, at which he confessed to some wrongdoing.

Despite the confusion, however, both authors quoted Galileo's confession in full, and so its content and tenor emerged clearly.[91] It was a very touching, moving, and eloquent statement, comparable in pathos and poignancy to the abjuration made available two centuries earlier. At that second deposition, Galileo confessed that the first deposition had prompted him to reread his *Dialogue;* he was surprised to find that the book gave readers the impression that the author was defending the earth's motion, even though this had not been his intention; he attributed his error to wanting to appear clever by making the weaker side look stronger; he was sorry and ready to make amends. The second deposition ends with Galileo humbling and prostrating himself in an attempt to move his judges to have compassion for an infirm old man.

The quotation of Galileo's confession makes up the last third of the Inquisition's executive summary. Venturi summarized the first two-thirds of that summary, especially the part about the charges by Lorini and Caccini that started the Inquisition proceedings.[92] Delambre quoted also the middle third of the summary, focusing on the 1630 negotiations for the imprimatur of the *Dialogue.*[93]

Both mentioned but failed to appreciate the significance of Galileo's first deposition (12 April 1633). The summary of that deposition made clear that Galileo did not feel he had received a special injunction not to discuss the earth's motion in any way whatever, and that he felt Bellarmine's certificate confirmed his understanding. Galileo also claimed that his book did not violate Bellarmine's warning not to hold or defend the earth's motion. His stated reason was that the book showed the opposite of the earth's motion and the invalidity of the Copernican arguments in favor of it. This was, of course, an exaggeration, to say the least; and he changed his mind in his second deposition on 30 April. Nevertheless, his initial claim had some plausibility. For although the book did not show that the earth was at rest, it did show that the Copernican arguments, while they were very good, were not conclusive; from this it would follow that Galileo was in no position to *hold* the geokinetic thesis. And on the question of whether the book *defended* the earth's motion, the subsequent defenders of Galileo had pointed out and would continue to point out that the book was a discussion of all the arguments, and if by evaluating them one discovered that the arguments on one side were better than the ones on the other side, that was part of an objective discussion and not an objectionable defense. As Nicolas Peiresc had put it, the book was a "philosophical play."[94]

Although Venturi did not defend Galileo in this manner, he did end his addendum with the following syllogism:

> I believe no one should reproach Galileo for submitting to the will of the Roman tribunals. For either one admits that those tribunals exercised a legitimate and regular authority over our mathematician, or they did not. In the first case, by publishing the *Dialogue* Galileo had done something wrong (if we speak with all due rigor); but then he did his duty in submitting to the will of his judges and showing himself ready to retract. On the other hand, if there is someone who claims that the Inquisition abused its power and undertook against him a violent and illegal persecution, even in this case, like a man who is taken prisoner by a mutinous crowd, the Florentine philosopher was right to try to avoid death with those exterior submissions that were the only means of saving himself. Thus, in neither case should one find reprehensible the behavior of Galileo on the occasion of his disgusting trial.[95]

By contrast, Delambre, after quoting the same April 30 confession, observed, without argument, that he saw only "weakness and absolute lack of sincerity."[96] Thus was born another subcontroversy of the Galileo affair.

At any rate, as a result of the efforts of Napoleon, Venturi, and Delambre, by 1821 the European public had access to the text of Galileo's confession at the second deposition of the 1633 trial and to two-thirds of the Inquisition's executive summary. There was still a long way to go to publish the complete proceedings, but this was a giant leap.

Chapter 10

The Inquisition on Galileo's Side?

The Settele Affair (1820) and Beyond (1835)

In 1820 a controversy raged in Rome that came to be called the Settele affair.[1] The surface issue was whether to allow the publication of an astronomy textbook in which Giuseppe Settele treated the earth's motion as a fact. The Inquisition sided with him, but both were opposed by the chief censor in Rome. Settele won his case in 1820, but two other steps took longer: only in 1822 did the Inquisition rule that Catholics in general were free to accept the earth's motion as a fact in accordance with modern astronomy, and only in 1835 were Galileo's and Copernicus's books taken off the *Index*.

Settele (1770–1841) was a clergyman who worked as a canon in a Roman church and as a professor at the University "La Sapienza." Siding with him was Maurizio Benedetto Olivieri (1769–1845), a Dominican friar, professor of Old Testament at the same university, and Inquisition consultant; in July 1820 he became commissary general of the Inquisition and held that position until his death. Also on Settele's side was Fabrizio Turiozzi (1775–1826), who was also a Dominican and held the office of Inquisition's assessor, or chief legal adviser. Another important figure in the story was Antonio Maria Grandi (1760–1822), a Barnabite friar and Inquisition consultant, who wrote the main evaluations favoring Settele's case. The protagonist of the opposition was the Dominican friar Filippo Anfossi (1748–1825), who from 1815 held the position of Master of the Sacred Palace, which amounted to being the chief book censor for the city of Rome. Siding with Anfossi was the Vatican majordomo, whose role was minor but revealing because this position was that of chief of staff of the Vatican palace and was then occupied by someone (Antonio Frosini) who later (1823) was appointed cardinal.

Many documents pertaining to the Settele affair have survived, and almost all of them have been recently published. The three most important documents are Settele's diary, Olivieri's summary, and the attachments to this summary. Settele kept a diary from 1810 to 1836 which is a gold mine of information about the Roman history of the period; it was published in 1987, in its entirety for the year 1820 and in its relevant parts for the rest.[2] Olivieri, in his role as Inquisition commissary, in October–November 1820 wrote an account of the events and of the issues to be used by the cardinal-inquisitors and other officials to reach a decision; the compilation of such a summary was standard Inquisition practice, as we have seen (chapter 9.5) for Galileo's own trial (whose May-June 1633 summary was being discovered by Delambre and Venturi at about the same time that Olivieri was compiling the summary for the Settele affair); Olivieri's executive summary is now available in two editions (one facsimile and one critical) that complement each other.[3]

As regards Olivieri's attachments to his summary, these were the documents that had been accumulating in the Inquisition's files since the affair began, and they were also the documents to which (as was customary) he referred and from which he quoted. He took the unusual step of having them printed together with his summary for internal administrative circulation; these attachments occupy 130 of the book's 154 pages, and they too have now been published in a facsimile edition and in a critical edition.[4] These documents will be mentioned and described below, but it is useful to briefly introduce the principal ones now. There were Settele's two appeals to the pope and an "Insert" note to his book.[5] There were also Anfossi's three increasingly long explanations and justifications of his refusal: an "Appendix" to a book on another topic; an essay stating his justifying "Motives" sent to the pope; and a printed booklet titled *Reasons,* also submitted to the pope.[6] As for Olivieri's writings, besides the summary already mentioned, there was his own lengthy and detailed essay of "Reflections" criticizing Anfossi's justifications.[7] Finally, there was Grandi's first consultant report.[8]

A few other important documents were generated after Olivieri compiled his summary and collection of attachments, and they have survived and have been published.[9] They will be mentioned in passing in what follows, but would have to be examined further in a fuller account of the Settele affair.

10.1 More Unbanning of Copernicanism (1820–1835)

In March 1819, Settele published the first volume (dealing with optics) of a textbook on optics and astronomy. Bearing the imprint date of 1818, this book was a written version of his class lectures; the rector of the university had recently encouraged professors to write up and publish their lectures, and Settele was one of the first to do so. The second volume dealt with astronomy and caused a controversy that took up the whole of the year 1820; it was finally published in 1821 (but with an imprint date of 1819).[10]

On 3 January 1820, Master Anfossi refused to give his approval for the publication of the second volume, on the grounds that it held the thesis of the earth's motion.[11] Settele sought the advice of Consultant Olivieri and Assessor Turiozzi; with their encouragement, Settele wrote a formal appeal to Pope Pius VII.[12] In March 1820, the pope forwarded the case to the Inquisition.[13]

The Inquisition, unable to consult the file of Galilean trial documents that had gone missing after the Napoleonic transfer, did the next best thing; it requested the Congregation of the Index to provide the file on the 1758 edition of the *Index*, which contained the partial and silent retraction of the anti-Copernican ban of 1616. The Index delivered the file to the Inquisition on March 28.[14] In the meantime, newspapers in Germany, France, and Holland were publishing articles about this ecclesiastical censorship.[15]

In April, Assessor Turiozzi talked to the pope, who seemed to give his informal approval to Settele's appeal.[16] In May, the assessor conveyed to Master Anfossi the pope's wish to give the imprimatur to Settele's book.[17] Anfossi replied that he respected the pope's wishes but could not in good conscience give the imprimatur.[18] Instead Anfossi showed Turiozzi a book he had just published, to which he had added an "Appendix" explaining his reasons for his refusal.[19]

On June 3, several Inquisition officials, including Assessor Turiozzi, Consultant Olivieri, and the chief prosecutor (named Libert), met with Settele to see whether he was willing to make some changes of wording in his manuscript to try to appease Master Anfossi. Settele accepted Olivieri's suggestion to replace the crucial chapter's initial phrase, "The Earth being in motion around the sun," with the clause "Given the earth's motion around the sun," although Settele said that this did not change the substance and that the book still contained a demonstration of the earth's motion. The prosecutor added that when he was a student at the Jesuit Roman College, he participated in a public debate where he defended the earth's motion as a thesis.[20]

During the same month, Olivieri finished writing an essay with a detailed refutation of Anfossi's view and with arguments trying to convince him to grant the imprimatur.[21] On June 27, Settele had a long discussion with the Vatican majordomo, who was on Anfossi's side and tried to convince Settele to treat the earth's motion as a hypothesis rather than as a demonstrated fact.[22] In the meantime, as a result of widespread criticism, Anfossi withdrew from circulation his booklet with the "Appendix" against the earth's motion.[23]

In July 1820, the Dominican friar Olivieri, who had been a consultant until then, was appointed commissary general of the Inquisition.[24] This development was a turning point in the affair.

On August 1, Settele, with the advice and consent of Olivieri and Turiozzi, submitted to the pope a second appeal requesting a formal ruling by the Inquisition. He enclosed three documents as attachments: a copy of his first appeal to the pope; a copy of Anfossi's "Appendix" published in May; and a critical article from the periodical *Italian Library* published in Milan. The pope agreed.[25]

Thus the Inquisition started its usual proceedings. It asked one of its consultants for a formal written opinion on the matter.[26] The consultant, Antonio Grandi, wrote an opinion concluding that it was proper to defend the Copernican thesis that the earth moves in the way in which it was customarily defended by Catholic astronomers, and also that Settele should insert a note in his book explaining that the Copernican system no longer suffered from the difficulties from which it had suffered at the time of Copernicus and Galileo.[27] On August 16, the cardinal-inquisitors unanimously accepted the consultant's recommendation, and the pope ratified the decision.[28] In the meantime, Settele compiled his "Insert," and on August 23, the Inquisition approved it.[29]

Anfossi did not acquiesce. In the next few days, he sent the pope a memorandum explaining his "Motives" why he thought Settele's book should not be published.[30]

Then the Inquisition went on the counterattack, intellectually and bureaucratically. Olivieri wrote a reply to Anfossi's memorandum and gave it to Turiozzi.[31] On August 27, Turiozzi officially conveyed in writing to Anfossi the decision by the Inquisition and the pope on the Settele case.[32] Anfossi then started temporizing: on the following day he wrote Turiozzi that he would give the imprimatur, but suggested that the Inquisition get two more evaluations of Settele's book by consultants; he also mentioned that he had given the pope a written memorandum explaining his objections.[33] The day after that, Turiozzi agreed to request two additional reviews and appointed Pietro Ostini and Giuseppe Mazzetti for the task.[34]

In September, the two consultants both confirmed that Settele's book did not contradict Catholic faith and morals.[35] But Anfossi formally withdrew his imprimatur and printed in a booklet titled *Reasons* a lengthy explanation, including objections to Settele's "Insert" note that had been requested and approved by the Inquisition; and he maneuvered to appeal the Inquisition's ruling to the pope.[36] Seeking to avoid a disciplinary confrontation, on 20 September the pope approved the idea that if Anfossi did not want to sign his name, the book's imprimatur be granted by another authority: the pope's own deputy as bishop of Rome, the so-called vicar apostolic.[37]

The controversy still did not end. In September and October reports appeared in the popular European press that the Inquisition had ruled in favor of publishing Settele's book; other reports added that despite this approval, the book's publication was being opposed and delayed by Anfossi.[38] And indeed, on November 11 the Vatican majordomo was instructed by the pope to forward to the Inquisition Anfossi's latest (September) printed booklet against Settele and the earth's motion.[39]

However, Olivieri had already started his reply, and sometime in November he finished writing his criticism of Anfossi's latest effort and included it in his "Summary" of the whole affair; he attached all the available documents and had the whole thing printed for internal distribution and use.[40] Anfossi immediately wrote a brief reply to Olivieri's latest criticism.[41] Then Grandi was again consulted, and he in turn wrote a lengthy criticism of Anfossi's latest objections.[42]

On November 20, the Inquisition consultants met and approved the immediate publication of Settele's book.[43] On December 14, the Inquisition cardinals agreed that the imprimatur would be given by the vicar apostolic, and the pope approved the decision.[44]

Finally, on 2 January 1821, Settele's book appeared. It bore an imprint date of 1819 on the title page and a statement at the end, just preceding the table of contents, saying, "Printing completed on 3 December 1820." But this latter date was a typographical error and should have read 30 December 1820. The book carried imprimaturs and endorsements by the vicar apostolic, Candido Frattini, as well as by the consultants Mazzetti and Ostini and the rector of the University of Rome, Belisario Cristaldi.[45]

Although the publication of Settele's astronomical work settled the issue involving that particular book, the more general doctrinal question had not really been settled. Thus before long, new difficulties arose.

In April 1822, Anfossi refused to give his imprimatur to an extract of Settele's book on astronomy by a certain Dr. D. de Crollis scheduled to be published in a Roman periodical, *Arcadian Journal*.[46] This refusal was followed in July by Anfossi's anonymous publication of a booklet against the earth's motion,[47] despite the fact that Settele had acquired minor celebrity status; for example, that July he received a medal from the Emperor of Austria, to whom (among others) he had sent a copy of his book.[48] Thus, the Inquisition became involved once again, and in August Olivieri got back into action by writing an opinion criticizing Anfossi's latest refusal.[49]

On 11 September 1822, the Inquisition ruled that in the future the Master of the Sacred Palace should not refuse the imprimatur to publications teaching the earth's motion; but it postponed a decision about removing from the *Index* five particular Copernican books including Galileo's *Dialogue;* Olivieri was given the task of examining whether these five explicitly mentioned books were prohibited because of advocating the earth's motion or for some other reason.[50] On September 25 Pope Pius VII ratified the Inquisition's decision to permit works teaching the earth's motion.[51] Thus, after several months of controversy, the proposed extract of Settele's book was published in the October 1822 issue of the *Arcadian Journal;* the imprimatur was formally signed by Anfossi himself, who finally yielded to the Inquisition and the Vatican secretary of state.[52]

In June to October 1823, a former student of Settele, Sebastiano Purgotti, published a booklet refuting Anfossi's views as articulated in his anonymous 1822 work; the book was published in central Italy, outside Anfossi's jurisdiction, but with the proper imprimatur.[53] In October, Olivieri finished writing his account of the original reasons for prohibiting the five Copernican books explicitly mentioned in the *Index.*[54] On November 10, the Inquisition's consultants discussed these books and decided to ask the commissary (Olivieri) to write out answers to various questions that came up at the meeting.[55] On December 1, the Inquisition consultants discussed Olivieri's answers and decided to request the opinion of two other experts, B. Garofalo and Bartolomeo Capellari (who would later be elected Pope Gregory XVI).[56]

At this point the documentary trail is lost, but not the historical connection. For on 20 May 1833, while deliberating on a new proposed edition of the *Index,* Pope Gregory XVI decided that it would omit the five books by Galileo, Copernicus, Kepler, Foscarini, and Zúñiga, but that this omission would be made without explicit comment.[57] Thus the 1835 edition of the *Index* for the first time omitted from the list Galileo's *Dialogue,* as well as the four other books.

Until this decision, even after the 1820 and 1822 progressive steps, the situation had remained uncertain. For example, in 1827 Settele thought it newsworthy that the astronomer Giuseppe Piazzi had been reported as saying that the Copernican system was not as certain and well demonstrated as commonly believed.[58] In 1829, "when a statue to Copernicus was being unveiled at Warsaw, and a great convocation had met in the church for the celebration of the mass as part of the ceremony, at the last moment the clergy refused in a body to attend a service in honor of a man whose book was on the Index."[59] And the same year, a Spanish bishop consulted the Roman Inquisition about whether the Copernican system could be maintained, and instead of a definite answer he was sent the recent rulings stemming from the Settele episode.[60]

10.2 Anti-Copernican Insubordination:
Olivieri's Official Summary (1820)

As already mentioned, in October and early November 1820 the Inquisition's commissary, Olivieri, wrote a summary of the events and issues of the Settele affair to be used by cardinal-inquisitors and other officials in reaching a final decision on the case. The document also contained his recommendations. Olivieri wrote:

Most Eminent and Most Reverend Fathers:

1. Here is once again before Your Eminences the question of the publication of Prof. Settele's *Elements of Astronomy*, which maintains the earth's motion as a "thesis."

2. At the meeting on Wednesday, August 16, it was decided (Sum., doc. VIII, p. 92)[61] that "nothing ought to prevent defending Copernicus's opinion of the earth's motion in the way that nowadays Catholic authors usually defend it." It was also said that "the Most Rev. Fr. Master of the S.A.P.[62] is ordered not to prevent the publication of Canon Giuseppe Settele's *Elements*." This resolution was approved by His Holiness in the evening of the same day (ibid., p. 93); and on the 27th of the same month it was communicated with an official memorandum (Sum., doc. X, p. 95) by the Msgr. Assessor to the Most Rev. Fr. Master, who acknowledged receipt on the following day the 28th (Sum., doc. XI).

3. The Fr. Master suggested to the Monsignor that he get the approval of one or two reviewers and left the choice to him (ibid.); "then," he said, "I will give the imprimatur." But even before these approvals, he had been quick to give it to some sections of the book that had been shown to him and that contained the disputed doctrine, as well as to the cautionary "Insert"[63] requested from Settele by the above mentioned decree and accepted by the Sacred Congregation on Wednesday, August 23.

4. Nevertheless, in the above-mentioned acknowledgment the Master expressed a contrary opinion, referring to the essay he had presented to His Holiness (Sum., doc. XII, pp. 98–104)[64] and giving some reasons against the above-mentioned cautionary "Insert." Then, with a delay of only one or two days,[65] he recalled those book sections and the "Insert" and erased the imprimatur he had written earlier.

5. Then on September 14, the Monsignor Assessor (Sum., doc. XIII, p. 105) forwarded to him, as agreed, the approvals of the two reviewers and sent back to him the said book sections. At the same time the Assessor replied to his objections to the "Insert," adding that it had already been "accepted as sufficiently right and regular by the Supreme Congregation and by the Holy Father." But the Most Rev. Fr. was unwilling to yield. On the 17th he replied (Sum., doc. XIV, p. 106) that he had given the imprimatur "on the assumption that such was the desire of His Holiness"; these words imply that the Monsignor Assessor in his official memorandum had falsified His Holiness's intention. In regard to the "Insert," he said he had given his imprimatur to it based on a description of it by others; yet the "Insert" is very short (Sum., pp. 93–95), and in his August 28 letter he appeared not only to have already read it but also to want to criticize it.

6. On the 22nd the Master again wrote the Assessor a letter (Sum., doc. XV, p. 107), in which at first he reports having said to the printer "that he should do whatever Your Most Rev. and Illustrious Lordship ordered." But then he goes on to declare what he would like, which in essence was the "hypothesis" qualification; he also resumes speaking against the "Insert"; and

finally he makes this confession: "I do not want to have anything more to do with this affair, unless the professor obliges me to publish my reasons. Your Most Rev. and Illustrious Lordship should deal with it as you deem most appropriate."

7. These letters were answered the same day (the 22nd) by the Monsignor Assessor (Sum., doc. XVI, p. 108). His letter discusses the history of the affair; tells the Master that "he forgives the personal insult" (§5 above); points out that his proposals were in opposition to the decree of the Supreme Congregation and that they "tend . . . to suggest that the Most Rev. Master of the S.A.P. knows matters of faith better than two respectable reviewers, one distinguished consultant (who examined the issues), the entire body of consultants, and the Lord Cardinals General Inquisitors." The following day the Master replied (Sum., doc. XVII, p. 109), but only stated that he wanted to apologize, if he "might have written down some disrespectful expression."

8. In the meantime, we learned that the decision of the Supreme Congregation had become public and newspapers were talking about it (Sum., doc. XX, p. 111). And the Most Rev. Fr. did not conceal his plan to publish his reasons, which in fact happened with a booklet of fifteen pages, without date or place of publication and without any imprimatur (Sum., doc. II, pp. 2–12);[66] on Saturday, October 7, just passed, by way of the Monsignor Majordomo it was presented to His Holiness, who told him that "he had forwarded everything to the Holy Office" (Sum., doc. I, p. 1). Thus, with a memorandum dated the 11th, the Majordomo forwarded it to the Monsignor Assessor.

9. By publishing this booklet, the Most Rev. Fr. Master of the S.A.P. has wanted to justify his denial of the imprimatur to Settele's *Astronomy;* and yet the decree communicated to him by the Supreme Congregation and approved by His Holiness (Sum., p. 96) indicated that "in the name of His Holiness there should be perpetual silence about the affair of the controversial publication."

10. The Sacred Congregation had ordered that an "Insert" by canon Settele be added; so it was compiled, heard, and approved (Sum., doc. IX, p. 93). But the Fr. Master has fought against it and has erased the imprimatur which he had initially given to it (Sum., pp. 6ff., 96, 106, 107).[67]

11. The printed booklet does not bear anyone's license or approval, not by the vicar's office nor by the superior of the Master's order. But he had been reminded (p. 96) of "the rules prescribed by the Sacred Council of Trent for the publication of books by all regular clergymen, and thus by the Fr. Masters of the S.A.P.; that is, they cannot publish any work except after approval by reviewers and by the general of their order; furthermore, it is not proper that in his own works the Fr. Master of the S.A.P. print his own permission; it should be only the permission of the vicar's office."

12. The booklet lacks the printer's name and the place and year of publication. Is the Most Rev. Fr. perhaps exempt from the instructions of Clement VII added to the rules of the Trent *Index?* Here, under the heading "On the Printing of Books," §3, one reads: "Bishops and inquisitors . . . shall be especially watchful that in the printing of particular books, the name of the printer, the place of the printing, and the year when the book was printed be noted at the beginning or the end."

13. From these four points there follow four presumably criminal charges[68] against the Most Rev. Fr. Master. These suggest the question whether it is appropriate to take disciplinary action against him, and if so, what it should be. Before discussing something that may aggravate his situation, one should take into account something that may justify him in his own eyes.

14. An extremely important point in his thinking seems to be that he regards himself independent of the Supreme Sacred Congregation in the exercise of his duties, and so not subject to the discipline of its decrees. To be sure, he has never said so explicitly. But he has constantly said, repeated, and reiterated that his authority over the press derives from Leo X with the approval of the Fifth Lateran Council; he said so in his initial "Appendix"[69] (Sum., doc. VIB, pp. 86–87), in his booklet *Reasons*[70] (ibid., doc. II, pp. 2–13), in his essay "Motives"[71] (ibid., doc. XII, pp. 98–104), and in his latest letter (ibid., doc. XI, pp. 96–97); so it seems he has formed the opinion that his authority derives in a transcendent manner. Moreover, when faced with the orders in the decrees of the Sacred Congregation, he has never said he was obeying it; instead he has always spoken of obedience to the Pope and has always avoided even naming the Sacred Congregation.

15. However, if we reread the words of Leo X in the tenth session of that Council, nothing more is attributed to the Master of the Sacred Palace than to the inquisitor of the smallest diocese. The decree says: ". . . unless the books are approved . . . in Rome by our deputy and by the Master of the Sacred Palace, and in other cities and dioceses by the bishop . . . and by the inquisitor for heretical depravity. . . ." (Sum., doc. XXII, p. 112). Thus, in exercising their role in publishing, the inquisitors are entirely under the authority of the Supreme Sacred Congregation.

16. The Council of Trent, which enacted general laws that applied even to those countries in which there was no branch office of the Inquisition, limited itself to speaking of sacred subjects, of the approval by bishops, and of the approval by superiors for the case of members of religious orders (ibid., p. 113). But in the rules for the *Index* published by order of that Council, rule no. 10 does indeed refer to the decree of the Lateran Council, but with an addendum for the case of Rome: "If a book is to be published in the good city of Rome, it must be *first examined* by the deputy of the Supreme Pontiff and by the Master of the Sacred Palace, *or by persons appointed by His Holiness.*"[72]

17. The Supreme Sacred Congregation was instituted by Paul III twenty-seven years after the said Lateran Council, which took place in the year 1515;[73] that was on 21 July 1542, the eighth year of his pontificate, after the general council later held at Trent had been announced, but before it had convened yet; he did it, in his own words, "so that things do not turn for the worse while one waits for the day of the Council which we have just announced, for although we are ready to foresee everything, we are unable to carry it out ourselves, even with the other hard workers of the curia" *(Bullar.,* Mainardi, vol. 4, part 1, const. 41, p. 211).[74] It lost no time in exercising the fullest regulation of books, their prohibition, sale, commerce, and printing; among the attachments (Sum., doc. XXII) there is exhibited one of its edicts issued as early as 12 July 1543, namely less than a full year after its creation.

This edict shows that the Supreme Congregation exercises all such supervision either on its own or through delegates, and that it prohibits any publication of books in Rome "unless they have *our explicit license* or that of the Pope's deputy, and unless they have been examined *by us or by the Master of the S.A.P.*" Thus, one can say in passing that the Sacred Congregation did something in accordance with its old prerogatives when it said (Sum., p. 92) that "inasmuch as the Fr. Master of the S.A.P. refuses to grant the imprimatur to Prof. Settele's work of which we speak, *this imprimatur shall be granted by the Most Eminent and Reverend Cardinal Secretary of the Congregation of the Holy Office.*"

18. The attachments also exhibit (Sum., p. 117) another edict, by Paul V and the Supreme Congregation in 1620; it gives some particular duties in the province of Rome to the Master of the Sacred Palace, "especially commissioned for this purpose by the above mentioned Congregation of the Holy Office."

19. Finally, there is an exhibit (Sum., doc. XXIII, p. 119) of the executive order of St. Pius V, which is still in force, and which places all departments in Rome under the Supreme Congregation; thus, it has become an axiom that "all tribunals submit" to the Supreme Congregation. It also follows that the Most Rev. Fr. Master may properly aspire to be the first executor of the decrees of the Supreme Congregation with regard to publishing, and to have a special place among its consultants; but he can never consider himself to be "one of the first magistrates of the papacy," who could "act in opposition to the first Congregation of our holy religion," as the Monsignor Majordomo seems to represent him (Sum., p. 1).

20. At one time the Fr. Master of the S.A.P. was the only one in Rome who discharged the policies of the Holy Inquisition which the popes wanted to implement, and the decree of Leo X at the Fifth Lateran Council belongs to that epoch. But then one established the Sacred Congregations of the Holy Office and of the Index, the latter to assist the former in regard to the regulation of books; the Fr. Master of the S.A.P. was made consultant of the former and assessor of the latter, and so he became one of their officials and an executor of their decrees. His old office was thus divided into three, for another member of his order was charged with the job of commissary of the Supreme Congregation, and a third one with that of secretary of the Sacred Congregation of the Index. Some might think that with these new directives and subdivisions the position of Master of the S.A.P. has become much less than it once was. But I believe the opposite is true. For as long as those duties all belonged to a single clergyman they were never as dignified, thorough, or authoritative as they became after the permanent creation of the said sacred congregations of Most Eminent Lord Cardinals, which have produced countless benefits to the Catholic Church.

21. It seems to me that the Most Rev. Fr. can be used as a witness against himself. For in his "Appendix" (Sum., p. 86) he speaks of "the two decrees of the Sacred Congregation of the Index, of which the incumbent Fr. Master of the S.A.P. is permanent assistant, and *as such he must ensure their execution.*" Why then does he not say that he must now "ensure the execution" of the decree of the Supreme Congregation? I remember hearing his distinguished predecessor say that in complex cases the Fr. Masters of the S.A.P. were in the

habit of resorting to the Supreme Congregation for a decision; thus, besides doing their duty, they avoided personal responsibility. Why does the Most Rev. Fr. now want to do the reverse?

22. In the present case, there are also two aggravating circumstances. One is that His Holiness has explicitly forwarded the affair to the Sacred Congregation of the Holy Office (Sum., doc. VI, p. 85). The other is that one is dealing with proceedings that are rooted in the Sacred Congregation from the very beginning. Look at the sentence against Galileo (Sum., p. 77), and you will see that everything was handled by the Holy Office. The initial charges, the evaluations by the consultants, the injunction to Galileo, the very decrees of the Sacred Congregation of the Index, the new proceedings against Galileo, and finally his condemnation and abjuration and the prohibition of his *Dialogue* by public edict, everything was done in this Supreme Sacred Congregation. Thus the same congregation has the right to clarify its intentions and the import of its decrees, and the Fr. Master has the duty to respect and execute its judgments, those of 1615, 1616, and 1633 as well as those of 1820.

23. However, the Most Rev. Father perceives that the present judgments are infected with all the criminal errors he has expressed in his letters, "Appendix," essay, and booklet. Thus, he felt obliged to withdraw the imprimatur; to publish a booklet, despite the prohibition to do so, and to do it in the manner he did (§9–12 above); and finally to present it to His Holiness and to wait for his decision. It is proper, then, to listen to him and to propose a second question: should the Congregation withdraw its decree as a result of what the Most Rev. Father writes to His Holiness?

24. Here it is useful to first relate the present controversy from the beginning. Last March Professor Settele, by way of Monsignor Soglia (a fellow professor at the University "La Sapienza"), presented an appeal (Sum., doc. III, p. 13) to His Holiness; the latter kept it for a few days and then gave it to his assistant the Monsignor Archbishop of Edessa with the order of forwarding it to the Monsignor Assessor. This was done with a letter dated the 19th of the same month of March, which states simply: "Here are the papers which by order of the Holy Father I must send to Your Most Illustrious and Rev. Lordship, as I explained to you this morning." The Monsignor Assessor advised His Holiness that it was important to examine what had been done by the Sacred Congregation of the Index in publishing the new *Index* of 1758. The Holy Father approved, and the Monsignor was commissioned to obtain the file from the Most Eminent Prefect of the Index; the latter gave the appropriate orders to the Father Secretary,[75] who on the 28th of the same month signed the memorandum of transfer (Sum., doc. V, p. 83). The Monsignor reported on the subject to His Holiness at the usual Wednesday audience of the following April 12, and His Holiness charged him to let the Father Master of the Sacred Palace know that in the future, whenever books teaching the Copernican system are proposed for publication, he shall approve them. The pope did not think that anything more needed to be done. But this most prudent decision by the Successor of St. Peter was not received with the necessary submission by the Most Rev. Father Master. For when on May 13 the said Monsignor went to see the Most Rev. Father and informed him of the will of the sovereign, he showed the Monsignor the "Appendix" he had published

(Sum., p. 86) in which he justified his denial of the imprimatur to Settele's book; he spoke as if he had been obliged to do so by the newspaper reports he had seen, and he also appeared to be a little angry with Settele for having appealed to His Holiness; some information about the appeal had transpired to him. After various arguments back and forth between the two, the Monsignor was unable to persuade the Master that he was merely an executor and would not contradict himself if in view of new orders he gave an imprimatur which he had denied earlier based on his own interpretation of the laws; the Monsignor managed only to get the Most Rev. Father to agree that the book could be published as long as he was not involved and did not have to sign its imprimatur. Then in the evening the Most Rev. Father, by way of a highly placed person, tried to change the mind of His Holiness.

25. The Monsignor believed that the affair could end by arranging that the imprimatur be signed by the assistant of the Fr. Master of the Sacred Palace; he gave this task to Fra Olivieri, who at that time was the assistant of the now-deceased commissary of the Holy Office. The Monsignor also arranged for the Master to see Settele's appeal and to be informed of the rest. But all was useless because the Father Master of the Sacred Palace did not agree, although he did make a copy of the appeal and later criticized it in his "Motives" (Sum., p. 98) and in his *Reasons* (Sum., p. 2), which he presented to His Holiness.

26. In the meantime Fra Olivieri, on orders from the Monsignor Assessor, examined the "Appendix" of the Most Rev. Father and wrote some "Reflections"; now, in accordance with the Assessor's wishes, and after some changes and additions, these "Reflections" are reproduced here (Sum., doc. IV, pp. 21–74);[76] there Fra Olivieri has tried, as much as it was within his power, to discuss the subject in the appropriate light. Then the *Italian Library* of Milan reported (Sum., p. 87) on the "Appendix" of the Most Rev. Father, and it did so with some humor for his having signed his own imprimatur.

27. At the beginning of August, Settele again appealed[77] to His Holiness, who forwarded the papers (Sum., p. 85) to this Supreme Congregation; these papers, together with the consultants' recommendation dated Monday 7 August, were presented at the meeting of Wednesday the 9th when it was decided that the consultant Most Rev. Father Grandi be asked to quickly write an opinion about what to do; his wise and erudite opinion (Sum., p. 89) was read to the consultants on Monday the 14th and, together with their recommendation, was distributed to Your Eminences on the same Monday; the cardinals discussed the topic at their next meeting on Wednesday 16 August and issued (Sum., p. 92) their decree, which was approved by His Holiness on the evening of the same day.

28. It must be mentioned that there was no prior circulation of material among the consultants, and that the Most Rev. Father Master of the Sacred Palace was not present at the two said meetings of the consultants (I do not know for what reasons). As a result, he was not aware of the proposal or of the discussion; and this was certainly unfortunate for him. For he would have heard the difficulties which some advanced at first, the solutions which others gave, and the ideas which everyone presented, until at the second meeting everyone shared an admirable consensus and seemed animated by the most

sincere religiousness; then, it seems, he could not have remained in his contrary fixation that makes it hard for his pride to say: I went too far in my "Appendix"; I was wrong in many ways. No less uniform were the feelings of the Most Eminent Lord Cardinals; thus the decision had all the signs of having been dictated by the Holy Spirit.

29. After this historical sketch of the affair, I come immediately to the "Motives" and the *Reasons* given by the Most Rev. Father Master for his contrary view. First, in favor of his "Appendix" he says (Sum., p. 97): "I have done nothing but repeat what has been published by Father Salvatore Maria Roselli here in Rome with all the approvals." I believe I have shown very clearly (Sum., p. 97) how poor a guide Roselli was for the Most Rev. Father. Then it must be noted that Roselli plays the role of a debater, who tries to strengthen the doctrines he opposes, whereas the Most Rev. Father represents the figure of a judge, who cannot legitimately say whatever he wishes might be the case but rather only what one must believe and what is the case. Finally, whatever Roselli may have said, a Most Rev. Master of the Sacred Palace cannot be excused for ignoring the *Index* that has been in force since 1758 and declaring prohibited books that certainly are no longer such.

30. One error must be immediately noted in the text of the full title of his "Motives" (Sum., p. 98): "One must not allow Settele to teach as a thesis . . . the stability of the sun at the center of the world." Along with modern astronomers, Settele does not teach that the sun is at the center of the world: for it is not the center of the fixed stars; it is not the center of heavy bodies, which fall toward the center of our world, namely of the earth; nor is it the center of the planetary system because it does not lie in the middle, or center, but to one side at one of the foci of the elliptical orbits that all planets trace. Still less does he teach that the sun is motionless; on the contrary, it has a rotational motion around itself and also a translational motion which it performs while carrying along the outfit of all its planets.

31. He says in "Motives" (no. 1, p. 98)[78] and in *Reasons* (p. 2) that Settele's manuscript begins with the words, "As the earth moves around the sun." This is an error due to lack of reflection. Various chapters of the second volume, which had been reviewed earlier by the Most Rev. Father, precede the one in which one reads this proposition.

32. He claims (p. 2) that "he does not adjust his judgment with the help of the theories of philosophers and astronomers, which are subject to countless exceptions and errors, but with the doctrine of the Church, founded on the Scriptures, tradition, the Church Fathers, and the definitions of the Holy Apostolic See. With its help he has examined Settele's manuscript and his 'Insert.'" Excellent idea! But still it is necessary to know what the theories of philosophers and astronomers are before saying that they have been condemned. Settele's first appeal, which the Most Rev. Father wants to refute, is entirely aimed to prove that what was condemned by the 1616 decree is not the doctrine of modern astronomers. What does the Most Rev. Father advance to tear down those proofs? Nothing. He merely repeats that "the system as defended today is still the same" (ibid., p. 10). This point alone destroys all his "Motives" and his *Reasons*. With much common sense, the Monsignor Assessor had warned him about it (ibid., p. 105). See also the

excellent considerations of the Most Rev. Grandi (p. 89). And I have tried not to leave out anything that might persuade the most reluctant person (ibid., pp. 32f. and 45ff.).

33. Then it is very well that he should say that he goes by "the doctrine of the Church." But what research has he done to investigate it? He adds: "Founded on the Scriptures." But where are the scriptural passages and the corresponding interpretations that he might have advanced? "The Church Fathers." Again, excellent! But which ones, how, why? "The definitions of the Holy Apostolic See." He attempts to turn these to his purpose. One must listen to him.

34. In his "Motives" (nos. 1 and 2, ibid., pp. 98ff.), in his *Reasons* (ibid., pp. 2ff.), and also in his letters (p. 107), he insists that the doctrine of the earth's motion around the sun has been condemned "as formally heretical or at least erroneous in the Faith" with a censure and a condemnation that "could not have been more authentic or more solemn" (ibid., p. 3). But it is easy to see that such an assertion is refuted by just reading the documents with which the Most Rev. Father supports it. The Pope and the Cardinals of the Holy Office ordered the theologians to evaluate the two propositions of the sun's stability and earth's mobility.[79] Very well. So the evaluations of theologians constitute an authentic and solemn judgment of the Pope and of the Cardinals? Not at all, for it happens frequently that such recommended censures are not approved or are adopted only in part. But afterwards the Pope had Galileo ordered not to maintain this doctrine, on pain of imprisonment. Thus, "not only did he approve the censure of the evaluating theologians; but also in a sense he sanctioned it by means of the prescribed penalty of prison" (p. 110). I fear that with this unheard-of interpretation of dogmatic definitions, the Most Rev. Father exposes to ridicule by enemies the very serious discussions of articles of Faith. From this prescribed penalty one can deduce nothing but that the Pope judged that Galileo could not be allowed to teach this doctrine.

35. The censure adopted by the Sacred Congregations is merely that of "false and contrary to Sacred Scripture," as it is clear from the decrees and from the sentence against Galileo; in it he is attributed the crime of having defended, or at least having represented as probable, an opinion "after it had been declared and defined contrary to the Sacred Scripture." All the rest is folly.

36. In his "Motives" (nos. 3 and 4, p. 100f.) and in his *Reasons* (§4 and 8,[80] pp. 4 and 6), the Most Rev. Father stresses the decrees of 1616 and 1620. I must warn the Most Rev. Father about an error that seems to be contained in "Motives," no. 4, where he says: "Does he want to be authorized to teach 'principles that are repugnant to Sacred Scripture and its true and Catholic interpretation, which is not to be tolerated at all in a Christian man'[81] and especially in a canon, and to teach them not as hypotheses (for which there would be no difficulty) but as theses?" (p. 101). Now, this proposition seems to me to be infected with intolerable absurdity. The Most Rev. Father should consider the excess to which he has been led. Nor is this an accidental proposition, cursorily written down as an afterthought; rather it is a perfect expression of what he wants condemned, combined to what he wants permitted. The Most Rev. Grandi has said it very well (p. 89): "If this system had been

judged erroneous or heretical, the Church would never have allowed it to be maintained even as a hypothesis; the reason is that otherwise those who studied it would be placed at risk of sinning against the Faith, in case they judged the system to be manifestly demonstrated."

37. From the fact that the hypothesis was allowed, I have demonstrated (I hope incontrovertibly): (1) that the system had not been condemned as regards the astronomical motions of terrestrial rotation and translation, that is, in its foundation and per se; but (2) that it had been condemned as regards the terrestrial difficulties besetting the doctrine of its defenders.[82] Thus, now that the system is taught without such difficulties, it is no longer subject to the condemnation (Sum., pp. 32f., 44ff., 58ff.).

38. They were very real in the doctrine as it was proposed at that time, although nowadays it may seem ridiculous to some that such objections should have been made. For example, Msgr. Fabroni in §86 of his biography of Galileo,[83] says this: "The Roman theologians were stressing the great disturbances of which we spoke, that is, the confusion of things produced by the earth's motion. In fact, they were afraid that if Galileo won the dispute, then the waters of the sea, the flow of rivers, the waters of wells, the flight of birds, and all atmospheric phenomena would be completely disturbed and intermingled." Even the Most Rev. Father admits that because of such serious resulting disorder, that system had been called "philosophically false and absurd"; but he does not want to take this into account for the theological condemnation. Here are his words from §9 of his refutation (Sum., p. 7) of Settele's "Insert": "Considering that system as philosophers, given the serious difficulties that would follow, they called it 'philosophically false and absurd'; but the Holy See took no account of this. They considered it as theologians and declared it 'formally heretical or at least erroneous in the Faith' because contrary to the Divine Scriptures; and the Holy See condemned it. If in the judgment of the Holy See it was contrary to the Divine Scriptures in 1616, so it is still in 1820."[84] This is the great misconception which the Most Rev. Father has in his head, and on account of which he has not followed the path that would bring him to understand what everyone else has understood; the point had been clearly explained by Settele in his first appeal and was well repeated in his "Insert," but the Master did not perceive it there either.

39. Let us try, if possible, to free him from such a misconception. Most Rev. Father, you have evaluated so well the censures of the "eleven very competent, indeed extremely competent, theologians" (pp. 3, 99) that you transformed them into "a condemnation that could not have been more authentic or more solemn" (p. 3); so tell me: is it not true that the doctrine declared "heretical or at least erroneous in the Faith" was also "philosophically false and absurd"? You do not deny, nor could you or anyone else deny, that the theological consultants said so. The same identical doctrine was judged "philosophically false and absurd," and then "formally heretical or at least erroneous in the Faith." Implicit here is a syllogism,[85] with major and minor premises, and it is accepted by you, me, and all. But the doctrine of modern astronomers, Settele's corresponding proposition "as the earth moves around the sun" and others like it, can no longer be called "philosophically false and absurd" by anyone because it is most certain that philosophically and by natu-

ral reason they contain no "falsehoods or absurdities" and do not imply, to use your own words, "the serious difficulties that would follow." Here I do not see that you produce any proof to the contrary; neither your Roselli nor anyone else presents any (Sum., p. 55, etc.) that does not dissolve as soon as it is examined, and indeed that does not change into a proof in favor. Therefore, Most Rev. Father (and here is the necessary consequence), the doctrine of modern astronomers is not the one judged "heretical or at least erroneous in the Faith" by the eleven "very competent, indeed extremely competent, theologians."

40. You say: "They called it 'philosophically false and absurd'; but the Holy See took no account of this. They considered it as theologians and declared it 'formally heretical or at least erroneous in the Faith' because contrary to the Divine Scriptures; and the Holy See condemned it." In these words lies an equivocal exchange of ideas that must be revealed to you. It is most true that the Holy See, or more precisely the Sacred Congregations, did not condemn that system only because it was "philosophically false and absurd"; for this is a judgment that does not belong to matters of faith. But they did condemn it as a doctrine "contrary to the Divine Scriptures"; the words of the 1616 decree are: "that false doctrine, altogether contrary to Divine Scripture"; so tell me how you think they arrived at such a "contrariety to the Sacred Scriptures." On what foundation did they establish it? What exactly were they basing it on? Tell me also: why did it dawn on those eleven "very competent, indeed extremely competent, theologians" to precede the theological judgment with the philosophical judgment of "philosophically false and absurd"? Why did the Sacred Congregation in its decree say "false" before "altogether contrary to Divine Scripture"? You, Most Rev. Father, speak loudly of Scripture, the Church Fathers, the doctrine of the Church, the decrees of the Holy See, and the condemnations it proclaims; but the fact is that you say nothing with any perspicacity or with distinct clarity.

41. Please reflect that if philosophical absurdity (that is, falsity or absurdity recognized as such by the light of reason) is attributed to the words of Sacred Scripture, it becomes an interpretation which ecclesiastical authority can very well define as "contrary to Sacred Scripture";[86] and this is precisely our case. Such was the case of the devastating motion from which Copernicus and Galileo had been unable to free the motions of axial rotation and orbital revolution which they ascribed to the earth; such devastating motion was certainly contrary to Sacred Scripture, for example to the passage in Psalm 92: "He has made the world firm, not to be moved."[87] I believe I have sufficiently explained it (Sum., pp. 32f., 45ff.). This is also reiterated by Settele's "Insert," which sounded so bad to your ears, but so good to others. Thus, those "very competent, indeed extremely competent" theologians of yours wanted to support the theological evaluations they gave to the censured propositions with philosophical evaluations, and so they combined the two by placing the latter before the former. What then shall we say of your words that "they called it 'philosophically false and absurd'; *but the Holy See took no account of this.*" Does it not seem most certain to you now that the Holy See took it very much into account? Remove this support from the theological censure and then check whether you can support it with Scripture, the Church fathers, the doctrine of

the Church, and the definitions of the Holy See. You will certainly find in Scripture and in the Church Fathers assertions of terrestrial immobility that is opposed to the devastating mobility; but to properly understand the latter with its problematic characteristics, you will have to focus on what you perceive in experience and apprehend by reason, for here one is not dealing with a supernatural mystery but with something accessible to experience and observation; that is, you will need philosophy to make you perceive the falsity and absurdity, so that based on these you can understand the language of Scripture and of the Church Fathers, which uses experimental notions. This is the way it must be; and this is in fact shown by those theologians and by the Sacred Congregation, both of whom pronounced the doctrine false before calling it contrary to Sacred Scripture; by so doing they warned us to fix our attention on the philosophical falsity, and thus to not go astray in thinking of contrariety to Sacred Scripture, for mobility and immobility are not things which God has chosen to reveal to us; rather he has inspired the Sacred Writers to express to us what our senses perceive in the way they perceive it. Recall the statement of our Holy Teacher: "Moses describes what is obvious to sense, out of condescension to the ignorance of the people" (Sum., p. 35).[88]

42. I hope the Most Rev. Father can quietly accept that that system was not declared "heretical" or "erroneous in the Faith"; that due to their ignorance, Copernicus and Galileo were unable to remove the "serious difficulties" affecting our globe, and so their system was infected with a devastating motion; that therefore the condemnation was based on the philosophical absurdities on account of which the system had consequences implying that the doctrine (I mean *their* doctrine) could be called contrary to Sacred Scripture; and that all this does not harm in the least the respect due to the decrees of the Sacred Congregations.

43. Next, Settele did not say that the Holy See was "deceived in condemning a doctrine pertaining to the Faith." It is most true that when such a doctrine was condemned, it was contrary to Scripture. But now, after it has been rectified and corrected, it no longer is. The Most Rev. Fr. should be careful to note that one is talking about a doctrine ("that doctrine," as the 1616 decree says), not of things in themselves but of things as they are in human teaching. This teaching is susceptible of being corrected or perverted, and of turning from guilty to innocent or from innocent to guilty, whereas things remain as they are in themselves. Then the comment on the word "appeared" instead of "was" (p. 7, §10) is inept. Is it not enough that the system "appeared contrary to the literal meaning of Sacred Scripture and implied serious difficulties," as Settele expresses it in his "Insert" (Sum., p. 93), for it to be deservedly prohibited?

44. The Most Rev. Fr. (§10) attacks the "Insert" because it advances "the fundamental rule that excludes the proper literal meaning of the Sacred Scriptures whenever it leads to absurdities" (Sum., p. 94). You make so much noise against such a maxim, Most Rev. Fr., and yet it is familiar to theologians. For example, in the dissertation you cited in your "Appendix"[89] regarding the stopping of the sun at the command of Joshua, Natalis Alexander uses it in this formulation: "Certainly the words of Sacred Scripture are to be taken and explained in the proper literal meaning, when nothing follows from them

against the truth of the faith or the purpose of charity." Did you not use the same language, Most Rev. Fr., when in your "Appendix" (Sum., p. 86, no. 2) you said about the Church Fathers that "on this point they had construed scriptural texts *literally*"?

45. This principle does not lead at all to the personal interpretations of the heretics. The point is extremely simple. The "arm of God" is an expression that sounds absurd if understood *literally;* thus it is interpreted in a *figurative* sense, as a figure of speech. You are right that in this case the figurative interpretation technically coincides with the literal meaning. But we must not engage in verbal disputes by equivocating on the meaning of words, promiscuously using both their common meaning and their technical meaning and so misunderstanding what the people who use them are trying to say. Then, to put aside the fear that one might fall into the personal interpretations of the heretics, it is enough to reflect that Catholics learn from the Church and study in its theological schools when one should regard as absurd the meaning of scriptural words variously labeled literal, material, natural, etc. and adopt a meaning variously called translated, improper, and what not.

46. The Most Rev. Fr. claims (Sum., p. 8, §11) that the sun's stability and the earth's mobility, which do not belong directly to the Faith, have been declared "formally heretical" or at least "erroneous in the Faith" because they are contrary to divine revelation, which is part of the Faith. But since he does not prove in any way that what he understands by the sun's stability and the earth's mobility has been the subject of revelation, by claiming this he does not say anything.

47. The Most Rev. Fr. shouts "most harmful absurdity" when (ibid., p. 9, §12) he discusses the statement in Settele's "Insert" that "subsequent discoveries provide us with equally enlightening proofs of the truth of the controversial system," as if the correct understanding of Scriptures depended on these discoveries. But what difficulty is there if by subsequent discoveries men correct what they thought was contrary to the Sacred Scriptures? Or if those who are more knowledgeable in the sciences are in a better position to correctly understand what the Scriptures say about them? One should always remember that like the other sacred writers, "Moses describes what is obvious to sense, out of condescension to the ignorance of the people," and that according to the same St. Thomas (who quotes the great Augustine on this topic), "Therefore the words of Scripture are explained in many ways so that they are shielded from the mockery of the secular letters" (Sum., p. 57).

48. This is the true meaning of St. Augustine in regard to these physical subjects, and this is what the Most Rev. Fr. must use to explain the passage he quoted from the same Church Father in §17 (p. 12). And for greater confirmation he can consult what I have quoted on p. 37 and other similar passages from such a great Doctor.

49. The Most Rev. Fr. must be joking when (ibid., p. 9, §12) he says that "these gentlemen . . . try to tell us that what is stated many times by the Holy Spirit is false, but that what their stellar parallax and aberration tell them is true." Then he calls as a witness Fr. Jamin,[90] to persuade them of the incomprehensibility of God's works. He also dares say that "the best astronomers and philosophers do not agree among themselves in regard to these discover-

ies." But he does not mention anyone. However, the fact is, as I hear from those who are well informed, that although there is no universal consensus among the experts in the field about the annual parallax of fixed stars, the aberration of fixed stars and of the planets has been verified for at least a century and is regarded by all astronomers as a true physical demonstration of the earth's annual motion (Sum., p. 54). See also the opinion of the Most Rev. Grandi (Sum., pp. 90, 91). Thus, it is not surprising that the Most Rev. Fr., who has not had the patience of mastering these astronomical matters, should appear to be incredulous, and that so does the Monsignor Majordomo, who in his memorandum claims to be "convinced of the uncertainty and the great deceptiveness of astronomical science" (Sum., p. 1). For one always gets a reconfirmation of the assertion made by such a great man as Fr. Montfaucon to the effect that "thus finally it has always happened that what astronomers establish by *clear demonstrations,* others incompetent in such things reject with derision" (Sum., p. 37).[91]

50. The Most Rev. Fr. tries to strengthen his case with the names and the authority of some astronomers. He mentions Boscovich, Tycho, and Gassendi. Boscovich, of the Society of Jesus, seems to have impressed him to such an extent that, based on two passages from Boscovich (Sum., p. 107), he withdrew his imprimatur to Settele's book and "Insert." He praises Boscovich with Andres's words that "he was accustomed to live among the stars and deserved to penetrate their secrets" (Sum., p. 2). Since Andres has been quoted, I must mention that he was completely persuaded of the truth of the Copernican system, as one can see in vols. 1 and 4 of his *History of Literature,* where he discusses astronomy. I shall quote a few lines: "Nowadays the Copernican system is respected as a great discovery and an astronomical truth. . . . It laid the foundations of true modern astronomy" (Rome, 1812, vol. 4, p. 348f.). And on p. 359, one finds this language: "Unfortunately, in all nations and in all ages a misconceived zeal for religion has led one to commit violence and fall into error. Among philosophers, Galileo's fate is not new; nor is it a special fault of Rome to have condemned a philosophical opinion as contrary to religion."[92] But let us move on to Boscovich.

51. I have had occasion to speak with an old Roman scholar about the Most Rev. Fr.'s attempt to seek help in the authority of Boscovich; he told me that he had been on friendly terms with Boscovich and was more certain of his Copernicanism than of our existence there as we were talking. And indeed, the two passages quoted by the Most Rev. Fr. appear to demonstrate it well enough. What does the first one say? It speaks of the earth's motion, "which, having been condemned by the sacred authority, it is not legitimate for us here in Rome to embrace." Thus, he limits the "it is not legitimate to embrace" to "here in Rome," and he lets us understand that he did not follow this opinion because he was in Rome, where the sacred authority had formerly condemned it; so elsewhere he would have followed it, and in his heart he already held it. The other passage is this: "Telluris quies *ut* in sacris literis revelata admitti omnino debet." To me this means: "The earth's rest, *as* (or *in the way that)* it is revealed in the Holy Scriptures, must be accepted completely." Here too, by means of '*ut*', '*as*', or '*in the way that*', he lets us see his inclination toward the earth's motion, speaking absolutely or simpliciter (as

the Scholastics say), but toward its rest *secundum quid* or κατα τί; that is, limiting oneself to certain special meanings of the word 'rest', the special meanings that are encountered in Sacred Scripture, it must be accepted. These meanings are indeed found in it (Sum., pp. 30f., 60). But they let the earth be simultaneously in motion, in the sense understood by astronomers.

52. Here in passing, it is useful to observe that there are many other examples of scholars who (while they advanced the refutations, qualifications, and corrections commanded to them) at the same time were careful to let it be understood that they were being forced to do so by others, but that their own real inclination was for the earth's motion. Elsewhere (Sum., p. 55) I have observed that Fr. Schettini lets it transpire that he is in this situation. I fear that Roselli was in it too, for while he piles up so many ineffectual arguments against this system (which are impossible for any man of intellect to present in good faith), at the same time he admits that it satisfies all phenomena. And the Most Rev. Fr.? I find his internal incoherence stupefying (Sum., pp. 38ff.). I will also mention the editors of Galileo's *Works* (Padua, 1744). With the imprimatur from this Supreme Congregation (whose decree has been found in the archives), in vol. 4 they reprinted the famous *Dialogue,* which caused its author the condemnation of 1633; they reprinted the sentence and his abjuration before the text; they deleted or made hypothetical the marginal postils; and in the editorial preface they made the fullest declarations. But they also added a dissertation by Calmet, some passages of which I have quoted elsewhere (Sum., pp. 28ff.).[93] Does not this dissertation aim to tacitly persuade the reader of the contrary? The deformity, moral and otherwise, of these external submissions, which are in reality simulated, is for me a strong reason for removing the prohibition. Now I come to Tycho.

53. The Most. Rev. Fr. has been careless about chronology. Since I consider it to be necessary for seeing things in their proper place, I felt it was my duty to go and find the dates and assign them. So Tycho was a Danish astronomer very famous for his industriousness and indefatigable persistence in observing the heavens. He died in 1601, at the age of 55. Thus he lived before the invention of the telescope, which occurred around 1609 (Lalande, *Astron.,* vol. 2, p. 569), and before the wonderful celestial discoveries made with it by Galileo, before the trouble he experienced, and before all the knowledge acquired later. Does the Most Rev. Fr. think that today Tycho would declare himself against the earth's motion, against the universal persuasion acquired by astronomers more than two centuries after him, now that they believe the system of the earth's motion has been "proved as much as anything physical can be," as Lalande says (vol. 1, p. 421; cf. Sum., pp. 52f., 90f., 94). The Most Rev. Fr. has given no thought to any of this. Instead, in his "Motives" (Sum., p. 101, no. 5) he asks: "Will Canon Settele not show toward the assertions of Scripture the respect which a heretic had for them?" Allow us to ask him in turn: in astronomy Protestants have abandoned Tycho, despite his extremely great merit, and they believe his system is a monstrous absurdity; instead they have turned to following Copernicus, Kepler, and Galileo (all great Catholic men), who are believed to hold the truth, the physical evidence, and the increasingly stronger observational confirmations; why then do you want to

oblige a Catholic to follow a heretic who has been abandoned by his fellow heretics?

54. What shall we say now of Gassendi? He died in 1655, at the age of 64; so he was 29 in 1620, when the second decree of the Index was published and the one of 1616 republished (Sum., p. 22); and he was 42 in 1633, when Galileo's condemnation took place. Do you really think it is a point in your favor, Most Rev. Fr., that at that time Gassendi wrote, "although this opinion of the earth's motion is regarded as having been rendered probable, nevertheless there is no demonstration that proves it to be true," as you report (Sum., p. 11, §16)? This language assigns to such an opinion that degree of probability which Galileo's sentence called an extremely serious error; you admit this yourself when you say that "the Sacred Congregation has told us that it is an extremely serious error to maintain that the rotation of our globe is probable" (Sum., pp. 10–11, §15).[94] Since we know that Gassendi, who was a canon and a theologian from Digne in Provence and later professor of mathematics in Paris, was also celebrated for his supreme piety and rigorous probity, his authority creates more difficulties for you, Most Rev. Fr. But let us set aside all this and let us note that some of the most cogent proofs, such as nutation and the annual aberration of heavenly bodies, had not been discovered at the time of Gassendi.

55. Nor had they been discovered at the time of Nieuwentijd,[95] who I see died in 1718, whereas the discovery of the aberration and nutation is assigned to 1727. Since you (Sum., p. 11, §16) do not name "the most renowned astronomers mentioned" by him, I pass up researching the topic to find them; for they certainly belong to an earlier age. But let us go on to other things.

56. In his "Motives" (Sum., p. 102, no. 7), the Most Rev. Fr. puts forth "the unrevisability of pontifical decrees." But we have already proved that this is saved: the doctrine in question at that time was infected with a devastating motion, which is certainly contrary to the Sacred Scriptures, as it was declared.

57. In "Motives," nos. 8 and 9, he raises objections based on the decorum of the Holy See, the Congregations, and the popes of that time. Let us repeat that this is also safe. I think of it this way. Imagine those popes coming back to life: Paul V, under whom the proceedings from 1615 to 1620 occurred; and Urban VIII, who ruled in 1633 when Galileo was condemned, and about whom Fabroni[96] wrote in §95 of his biography of Galileo, "It must seem really strange that the principal author of this catastrophe should be Urban VIII, supreme pontiff," and then proceeded to relate how this happened. Imagine also coming back to life the cardinals and theologians who participated in these decisions. Would it not do an extremely great disservice to them to believe that they would want to persecute anyone for today's Copernican system, which is followed by all astronomers, as they did at that time when it was followed by few, who appeared as innovators? That they would want to condemn the doctrine of the earth's motion now that it has been rectified and freed from the absurdities envisaged then; now that it has been confirmed, and indeed demonstrated as much as a celestial fact can be, based on new

wonderful discoveries, as astronomers themselves assure us? Now that almost two centuries have elapsed; now that it roars throughout Christendom and has spread everywhere; now that the popes, by not giving any thought to stopping it, have in fact given it free rein? . . .[97] One shudders just to think of it. Thus, if one does not want to do them a disservice, one must do what they themselves would have done in our time. Finally, the Most Rev. Fr. mentions various passages of Sacred Scripture. But I have already shown elsewhere that in reality they do not say anything to the contrary (Sum., pp. 30ff.).

58. Among the things on which the Most Rev. Fr. seems to me to be most wrong in this affair is the obstinacy of never ceasing to try to evade (with bad reasons) the 1757 decree and its implementation in the 1758 Index. He says (Sum., p. 5, §7) that this Index "has confirmed the condemnation of such a doctrine by leaving in the prohibited list *the books of those who teach it,* and indeed they are still there." As it is expressed, this proposition is completely false. For it is most false that this Index still contains in general "the books of those who teach it." It is true that it contains *some books,* namely the particular books (Sum., pp. 22ff.) by Copernicus, Zúñiga, Foscarini, Kepler, and Galileo, the last of which was inserted in 1634; but it is not true that it contains other books (at least as far as I have been able to determine) and still less books in general that teach it. And there is a reason for leaving those books there (Sum., p. 58), for they belong to the age in which the earth's motion had not been freed from its implication of devastating mobility.[98]

59. It is no less blameworthy to interpret that episode as "an omission of certain edicts accomplished with the consent of Benedict XIV" and as "not publishing an edict for good reasons," that is "for reasons known to them" (Sum., p. 5, §7).[99] But, Most Rev. Fr., after the 1664 Index the edict was no longer being reprinted.[100] And why do you speak of "reasons known to them" and of "good reasons known to him" (Sum., p. 102), i.e., Benedict XIV? Here this jargon means wanting to say something without knowing what to say; instead of using such jargon, you, who are "perpetual assistant" of the Sacred Congregation of the Index, should have retrieved the proceedings of that episode in its archives. You would have found that it was preceded by an extensive and learned written opinion, which is still preserved;[101] that it contains the very clear reasons of the Sacred Congregation and the pope; and that they really wanted to remove the prohibition. Why, Most Rev. Father, instead of talking off the top of your head, did you not consider the arguments in Settele's appeal, which destroy the things you say? Why did you not take seriously the collection of general decrees of prohibition in the new Index and the existence of the exclusionary rule about whether or not to count as prohibited those books that are not specifically mentioned or generally described?[102] Why did you not notice that by talking of "good reasons known to them" you make it sound as if the Sacred Congregation and the pope were guilty of dishonesty? For to begin with, you claim that the condemnation was a solemn judgment; that it originated from the pope himself and the Holy See; that it was an unrevisable judgment; that it declared a doctrine "heretical or at least erroneous in the Faith"; and that it targeted the doctrine of the earth's motion as it is taught even today. Then you do not realize you are committed to the inexorable self-refuting argument that after 1634 popes

have been deceptive because they have no longer spoken against this doctrine despite the fact that it was constantly acquiring more and more embellishments from supporters and was becoming universally held. Finally, with your alleged omission of publication, especially in a situation when it was supremely necessary to bring it about, namely on the occasion of the renewal of the *Index* and the collection of decrees of prohibition, you come along and tell us that "for reasons known to them" they have neglected to acknowledge the truth of the faith. But, Most Rev. Father, this smells a little of the doctrine that some truths are being obscured in the Church, especially on the part of the Holy Apostolic See; and this doctrine is indeed heretical and was condemned as such in the bull *Auctorem Fidei*, in the first proposition, if I am not mistaken. You know that, for you have defended this bull. So you are in the position of judging yourself by your own principles.

60. I hope the Most Rev. Fr. has noticed that it was irrelevant for him to mention (Sum., pp. 5, 102) the omission of the annual reading on Holy Thursday of the bull *Coena Domini* (Sum., p. 26). Since he referred to the Tribunal of the Sacred Penitentiary (Sum., p. 102), I decided to consult one of its principal officials; he replied that the bull contains an indication that that publication was valid until the next publication. In any case, if this bull were no longer published and not included even in editions of the *Bullarium;* and if one compiled a collection of promulgated censures and said (as in the 1758 *Index*) that this was done so that in case of doubt about a particular case of censure, one could determine whether it should be regarded as subsumed under the censures; then clearly by such omission the validity of such censures would cease, especially if one knew (as in the present case one knows) that this omission was due not to oversight but to the explicit will of the pope.

61. The Most Rev. Father has also stressed the point that Galileo was judged "vehemently suspected of heresy" by the Holy Office. This is indeed true. But this heresy is not the doctrine of modern astronomers, but the maxim that one may defend as probable a doctrine after it has been defined false and contrary to Sacred Scripture; and the heresy also refers to the philosophically absurd and false propositions of the earth's mobility and sun's immobility that are indeed contrary to Sacred Scripture but also far from the doctrine of modern astronomers.

62. The Most Rev. Father makes a lot of noise about Pius IV's Profession of Faith (Sum., pp. 4, 12). But he is far from having given in support of his claim either the doctrine of the Church or the unanimous consent of the Church Fathers; thus, his words are irrelevant both for the doctrine of modern astronomers and for our concerns here.

63. He tries to elude the authority of Cardinal Gerdil. He gives everything away when he says that Gerdil "did not praise Copernicus's works any more than the Sacred Congregation did in 1620" (Sum., p. 12).[103] There is no doubt that the Sacred Congregation praised Copernicus, but at the same time it condemned the Copernican system "because he does not treat as hypotheses, but advances as completely true, principles about the location and the motion of the terrestrial globe that are repugnant to the true and Catholic interpretation of the Holy Scripture; this is hardly to be tolerated in a Christian" (Sum., p. 76).[104] By contrast, the Most Eminent Gerdil praises the system

as well as its author, saying that "he produced and perfected the famous system that is the basis of the most beautiful theories of the new philosophy" (Sum., p. 95).[105]

64. Finally, I am ashamed for him of what he says about other examples of publications already accomplished in Rome. Volume 10 of Tiraboschi was published in Rome in 1797 with the explicit imprimatur of the Master of the Sacred Palace of that time, Fr. Pani. Tiraboschi's letter of retraction to the Most Rev. Mamachi is a satirical joke at his expense. It is printed in that volume.[106] You should not say, "He had the courage of retracting himself in his letter to the Most Rev. Father Mamachi" (Sum., pp. 11–12).[107] Instead, if you read this letter, you too will say, "He had the courage of making fun of Mamachi by mockery that is sometimes more subtle and sometimes more blatant." Poor Father Mamachi, who was a very decent man, but an irascible kind of person!

65. Guglielmini's two booklets[108] were published one in Rome with the imprimatur of the Master of the S.A.P., the other in Bologna with the imprimatur of the archbishop and the inquisitor. Calandrelli's booklet[109] had the imprimatur of the Master of the S.A.P. and also was dedicated to the ruling pope; it systematically tries to demonstrate the annual motion of the earth, and indeed its opening words assert both the annual and the diurnal motions (Sum., p. 4); now it is evident that its teaching could not have been overlooked by either the Master of the S.A.P. or His Holiness, who accepted the dedication. The Most Rev. Father dares to compare these with the appalling books published in Rome during the republican period, when there was no impediment to obscenities and impieties.

66. Before stopping this modest writing of mine, I must not be silent about the Msgr. Majordomo's assertion that "one can maintain as a thesis only what is true or what is believed to be incontrovertibly true" (Sum., p. 1). Very well; let us suppose that this is so; let us take this as our "hypothesis," so to speak. But the fact is that nowadays astronomers really seem to be so convinced of the earth's motion that they "believe it to be incontrovertibly true." To Cardinal Nicholas of Cusa it appeared manifest for the elegance with which the apparent motions in the heavens would be explained (Sum., p. 52). To Copernicus it seemed to be an inescapable premise for the purpose of rendering planetary motions regular (Sum., p. 57). Others were struck by the analogy with the planets: like the earth, they were round, opaque, and illuminated by the sun; some had moons that accompanied them and simultaneously performed revolutions around them; and then the planets could be observed to possess axial rotation and orbital revolution, namely diurnal and annual motions; thus, given that the planets had all these properties, one did not see why the earth should not also have them (Sum., p. 71). However, it is certain that nutation, annual aberration, and other data that require more subtlety to be detected are believed to provide a new irresistible argument (Sum., p. 53f.).

67. At any rate, in order to be able to assert something, which is to say to call it a "thesis," probability is sufficient. In everyday life we are constantly affirming and denying based on probable data; and probability has an extremely large role not only in daily life but also in the disciplines. Moreover, in

our case the mere assertion that probability favors the earth's motion would be "a most serious error," as we were saying a short while ago (§54); to be more precise, so would also be the assertion that it was possibly true or even that it was uncertain, as is clear from the text of Galileo's sentence; if one were really talking of that kind of motion that had been declared "false and contrary to Sacred Scripture." But the good Monsignor should not be afraid, for such a condemned motion is kept at a distance. If he still wants to call that other motion false, let him speak in "hypothetical" terms, as when one supposes something that is false; for example, "if by hypothesis men did not eat, they would not cultivate the soil." But if for Settele his intellect, which is the required power, makes him perceive that motion to be not only possible but also probable, and indeed manifest, he should be allowed to speak of it accordingly and to use the language of "thesis," as is done by the experts (the astronomers) in consequence of what they perceive.

68. I believe I have demonstrated that nothing that has been produced by the Most Rev. Father has any validity; on the contrary, he has produced very many things that are wrong, which are not really his but stem from the fixation that has overtaken him. Therefore, Settele's "Insert," which had already been approved by the Sacred Congregation, should continue to be endorsed; and the resolution of August 16 of this year should be judged more than ever necessary, right, and dictated by the spirit of religiousness and prudence. Thus, my answer to the second question[110] is: the previous decision should remain standing. And I now dare suggest something else. After all the publicity which the affair has received (Sum., p. 111f.), after the equally publicized resistance of the Most Rev. Father Master of the S.A.P., it might be appropriate that the printed edition of Settele's *Elements of Astronomy* display a formal decree of the Supreme Congregation granting the imprimatur. I have in mind the one printed in Fr. Maracci's refutation of the Koran, published in Padua in the year 1698; the book contains its entire Arabic text and its complete Latin translation. I transcribe this decree here: "Wednesday, 7 November 1691. . . . The request by Fr. Ludovico Maracci, of the order of the Mother of God, was presented, and the Most Eminent and Most Rev. Lord Cardinals General Inquisitors mentioned above, after considering what needed to be considered, granted to the said Fr. Maracci the permission to publish the refutation of the Koran with the entire Arabic text and the Latin translation made by himself. When His Holiness was informed, he endorsed the decision." I express this opinion with trepidation. I do not dare describe the precise terms in which the decree should be expressed, but I defer to whatever might be decided in due course, if my recommendation is approved.

69. With regard to the first question[111] about the Most Rev. Fr. Master, I dare not propose any resolution other than whatever the assessor and His Holiness decide. The Supreme Congregation did not issue an injunction, but requested His Holiness to do it. Thus it seems that the matter is in the hands of His Holiness. I very much want that he be treated with compassion. He is someone toward whom I have shown veneration since my earliest youth. In many books he has energetically defended the Holy See and the cause of religion. He became involved in this affair without noticing that he was not sufficiently knowledgeable about it, in either its scientific, or its scholarly, or its

theological aspects. He feels he has acted with zeal. He has a real fixation and did not realize what he was doing when he opposed the decision of the Supreme Congregation, which was unanimously agreed upon by the Most Eminent Judges, after the unanimous opinion of the extremely distinguished consultants. He has been seduced by unknown persons who are incompetent, but who seemed to him (and I am sure were) "beyond any suspicion" (Sum., p. 106). Finally, although he had his booklet printed and he presented it to His Holiness, I do not think he actually published it.

This is my opinion; from the bottom of my heart I submit it in all its parts and in every way to the understanding of the consultants and to the wisdom and authority of Your Eminences.

Fra Maurizio Benedetto Olivieri, Commissary and Consultant.

P.S.[112] The passages from Boscovich quoted by the Most Rev. Fr. (§51 above) may be found in Pino, who actually gives them on p. 75 of vol. 3.[113] But a serious person took the trouble of consulting Boscovich's dissertation "On the Ebb and Flow of the Sea," both in the edition of 1747 and in his collected works, published in Bassano in 1785 (if I am not mistaken); he assures me that the passage on the earth's motion, "which, having been condemned by the sacred authority, it is not legitimate for us here in Rome to embrace," is not there. The other passage was: "the earth's rest, *as* (or *in the way that*) it is revealed in the Holy Scriptures, must be accepted completely"; this was being attributed to others, who so claimed and whose claim he was explaining, after he had presented another explanation that presupposed the earth's motion. Since the 1747 edition is the second one, Boscovich must have revised those passages after the first edition.[114]

10.3 Solomonic Injustice: 1820 versus 1616

Obviously this document is a gold mine of information as well as a priceless compendium of the issues that needs to be reread, studied, and analyzed to be digested, interpreted, and evaluated. Here I can focus only on the main theme.

The most immediate question (§1–22) was whether to take disciplinary measures against Anfossi, whose apparent insubordination raised fundamental legal issues about the precise role of the Master of the Sacred Palace in the Church hierarchy. But the most important question (§23–69) was whether the Inquisition should withdraw its approval of Settele's book in light of Anfossi's arguments. Olivieri's conclusion was clear and unambiguous: none of Anfossi's reasons were valid, and so the Inquisition should let its previous approval stand.

A key part of Anfossi's position was that although he could not allow the teaching of the earth's motion as a thesis, since it was heretical, there was no problem in allowing it as a hypothesis, and he had tried to convince Settele to make his claims hypothetical. This distinction, which was suggested by the 1620 decree correcting Copernicus's book, had become a semi-official position and had been generally accepted and used by Catholic scholars to give themselves a modicum of freedom. Following the consultant Grandi, Olivieri objected (§36–37) that it was inconsistent to allow the earth's motion as a hypothesis and to regard it as heretical or erroneous, for, if a doctrine were heretical, the Church would not place believers at risk, and so would not allow its acceptance even as a hypothesis; consequently, given that acceptance of the geokinetic hypothesis was legitimate, it followed that the earth's motion was problematic in some other way than heresy or religious erroneousness. Olivieri stressed that at the time of Copernicus and Galileo it had been problematic as regards the mechanical consequences for terrestrial bodies that seemed derivable from it. Olivieri seemed to have in mind such consequences as that on a moving earth falling bodies would follow a path slanted westward; that westward gunshots would range farther than eastward ones; that birds would not have the freedom of movement that allows them to fly; and that air would generate a constant wind blowing from the east. And he suggested that since in 1820 the earth's motion was no longer problematic in this regard, the condemnation no longer applied. In this document Olivieri did not say when and how the status of the earth's motion changed, but from other writings[115] we can gather that he thought the crucial development was the discovery that air has weight. He held that Galileo was not aware of this fact; that the mechanical difficulties with the earth's motion depended crucially on the assumption that air has no weight; that after Torricelli and Pascal demonstrated the weight of air, these difficulties could be resolved; and that Pope Benedict XIV's partial unbanning of Copernicanism indicated that the Church had gradually recognized the change. This position was historically untenable insofar as Galileo was clearly aware that air has weight;[116] and it was scientifically misconceived because most of the mechanical difficulties depended on such questions as conservation and composition of motion and the principle of inertia, and not on the weight of air.[117]

Then (§38–42) Olivieri reinforced his contrast between the situations in 1820 and in 1616 by arguing that since all relevant documents mentioned that the earth's motion was philosophically false and absurd, it followed that this philosophical evaluation was a reason for the theological evaluation of "contrary to Scripture." In this summary he did not actually spell out this argument,[118] but he used it to try to show Anfossi how something might be contrary to Scripture in 1616 but not in 1820. Olivieri's point was that given that the earth's motion was declared contrary to Scripture (in part) because it was philosophically false and absurd, once it was discovered that it was no longer such, then one was no longer forced to say that it was contrary to Scripture either. Here, Olivieri's thinking reflected an assessment we have encountered before, in Auzout (1665) and Lazzari (1757).

For Olivieri (§43), this change did not imply that the earlier condemnation had been wrong because the earlier condemnation obviously applied to the earlier doctrine, and the earlier doctrine referred to an earth's motion that was problematic, and that problematic earth's motion was indeed contrary to Scripture. The change did presuppose the hermeneutical principle that scriptural passages should not be interpreted literally if their literal interpretation implies absurdities; and he thought that this principle was sound. But he did not take it to imply the denial of the scientific (philosophical) authority of Scripture. Thus although Olivieri's hermeneutical reflections corresponded to some Galilean views, he did not seem to use or explicitly endorse Galileo's key principle (immortalized in Baronio's aphorism).

Another key objection by Anfossi had been that papal decrees were unrevisable, and since the earth's motion had been condemned once, there could not be another decree withdrawing or revising the first. Olivieri did not reply by denying that the condemnation of 1616 was a *papal* decree but rather by denying that the earlier decree needed revision (§56). He had an argument why the condemnation of the earth's motion as contrary to Scripture did not have to be revised: it did not refer to motion per se, or as it exists in itself; what had been condemned was the proposition that the earth moved in the sense of motion that implied all the mechanical difficulties that seemed derivable from it; and the earth's motion in this "devastating" sense was indeed contrary to Scripture. Correspondingly, the earth's motion theorized by the astronomy of Settele's time was a motion freed of such difficulties, and so it was not contrary to Scripture.

This reply is interesting. Insofar as it spoke of unrevisability rather than infallibility, it was dealing with a more manageable concept. Moreover, it seemed to presuppose that there was a papal decree against the earth's motion, and so Olivieri's criterion for a papal decree seems less stringent than those prevailing today. He seemed to regard a papal decree as one which the pope made while discharging his official functions, such as being president of the Congregation of the Holy Office; examples of such decrees would be Paul V's decision that the earth's motion was contrary to Scripture (endorsed at the Inquisition meetings of 25 February and 3 March 1616) and Urban VIII's decision that Galileo be condemned (reached at the Inquisition meeting of 16 June 1633). Although Olivieri's criterion was probably historically correct, it is also important to point out that the definition of a papal decree ex cathedra was undergoing some evolution; thus by the end of the nineteenth century such a decree had to contain an explicit self-referential description that the decree was being characterized as ex cathedra and infallible.[119] Finally, Olivieri seemed to presuppose a peculiar theory of meaning according to which the meaning of a proposition includes the consequences implied by it, or perhaps the consequences allegedly derivable from it; but this theory of meaning does not seem to be at all plausible.

One of Anfossi's greatest oversights, according to Olivieri (§58–60), was the failure to appreciate Benedict XIV's decision of 1757–1758. Anfossi tended to stress more the fact that the particular books of Galileo, Copernicus, and others were still prohibited than the repeal of the general prohibition on all books teaching the earth's motion; and the Master tended to interpret the latter repeal as an omission of a previous decree rather than a real decree abolishing the previous one. However, Olivieri argued that the documents in the archives showed that the decision was significant and deliberate because it was based on explicit discussion and on a written and reasoned recommendation, and that the repeal of the general prohibition was more important because it was relevant to the contemporary situation, whereas the five particular prohibitions reflected the historical situation at the earlier time.

Finally, in his reply to Anfossi's reminder that Galileo had been convicted of "vehement suspicion of heresy" (§61), Olivieri did not question its legitimacy or correctness; he only proposed a reinterpretation of the (suspected) heresy in question. He was careful enough to admit the twofold character of the heresy, and that one of them was the methodological and hermeneutical principle that denies philosophical authority to Scripture. But he took the other (suspected) heresy to be the theory of the earth's motion, including the old Aristotelian physics (that led to insuperable difficulties and mechanical absurdities for the simple reason that the combination was internally incoherent) and the thesis that the sun is completely motionless (which had been long refuted by modern astronomy).

All these critiques underscore a general feature of Olivieri's analysis of the issues, namely that while his argument justified a liberalization in 1820, it simultaneously justified the earlier condemnation at the time it occurred. In that sense, it was an update and reaffirmation of Lazzari's analysis, which had led to the partial unbanning of Copernicanism in 1757–1758. Thus, although some may admire Olivieri's balanced impartiality, his argument was Solomonic in more than one sense; it was a double-edged sword of questionable value to a friend of the historical Galileo.

Chapter 11

Varieties of Torture

Demythologizing Galileo's Trial? (1835–1867)

In the middle part of the nineteenth century, Galileo's trial started receiving an unprecedented amount of attention. Sustained discussion spread from Italy and France to England, Ireland, America, and Germany. Key issues started to be seriously debated with a critical dialogue of arguments and counterarguments. The controversy also grew more heated and bitter. The variety of topics and approaches seem to have coalesced around two themes: torture and demythologization.

The question of whether Galileo had been physically tortured became a cause célèbre. But physical torture was not the only kind that was argued about. Some authors who rejected the physical-torture thesis nevertheless claimed that the Church's treatment of Galileo amounted to *moral* torture. One critic of the physical-torture thesis engaged in such a manipulation of the documents that he may be described as having tortured the texts. On the other hand, another such critic was so earnest in his attempt to avoid one-sidedness that he ended up displaying a type of incoherence that could be labeled tortured thinking.

The other trend involved a move away from portraying Galileo as a martyr or hero and toward depicting him as a fallible and flawed human being, mostly a moral weakling and coward. It also involved moving away from interpreting the trial in terms of grand notions such as science versus religion, philosophy versus theology, and scriptural authority versus natural reason, and toward emphasizing contingent and accidental circumstances and petty human motives. By following such an antiheroic and circumstantialist approach, such authors may be said to have been trying to demythologize Galileo's trial; but one may question their success insofar as while some myths were being discarded, others were being retained or created.

The mid-nineteenth century was also the time when the superb collection of documents in the two volumes of Venturi (1818–1821) became available; when the first book-length biography of Galileo in English appeared (Drinkwater Bethune 1832); when the lost file of trial proceedings was found and returned to the Vatican (1843); and when an excellent critical edition of Galileo's works was published, in sixteen volumes, by Eugenio Albèri (1842–1856). Finally, this was the time when several previously developed accounts were exercising their influence and being assimilated: Mallet du Pan's thesis that Galileo was condemned for his bad theology and not for his good astronomy; Tiraboschi's thesis that in 1633 Galileo was condemned neither for his bad theology nor for his good astronomy, but for his disobedience of the 1616 stipulations; the imprudence thesis, holding that Galileo was condemned for being too rash in his attitude toward Copernicanism; and the papal personal-insult thesis, holding that Galileo was condemned because Pope Urban VIII felt insulted by being caricatured as Simplicio in the *Dialogue*.

11.1 "Martyr of Science"? Victim of Torture?
Brewster and Libri (1835–1841)

The Scottish physicist David Brewster (1781–1868) is best known for his discoveries pertaining to the polarization of light. However, he originally received a theological education and a preacher's license; he was an evangelical who later joined the Free Church of Scotland. Moreover, he was a notable intellectual who was the editor at various times of the *Edinburgh Magazine,* the *Edinburgh Encyclopedia,* and the *Edinburgh Philosophical Journal;* a founder of the British Association for the Advancement of Science in 1831; and the principal of the University of Edinburgh in the last decade of his life.[1] In 1835, Brewster published a biography of Galileo as part of the biographical volumes of *The Cabinet Cyclopedia;* later he included it, with very minor changes, in his *The Martyrs of Science, or the Lives of Galileo, Tycho Brahe, and Kepler* (1841). In it, far from portraying Galileo as a martyr, Brewster depicted him as having cowardly avoided martyrdom (in 1633), thus in effect harming the cause of science and benefiting that of the Church. Brewster also portrayed him as reckless and too bold in 1613–1616, for failing to appreciate the justifiable mental inertia of his opponents. And Brewster explicitly criticized the explanation of the trial as caused by Pope Urban's resentment for being caricatured as Simplicio.

For the first phase of Galileo's trial, the substance of Brewster's account is the familiar imprudence thesis, which, as we have seen, originated in Guicciardini's 1616 report and can be traced through Baldigiani's 1678 letter to Viviani and through Tiraboschi's 1793 lecture.[2] However, Brewster was the first to elaborate it in English, and his expression was inspired to a very high degree. For these reasons it deserves our attention:

[57] The ardour of Galileo's mind, the keenness of his temper, his clear perception of truth, and his inextinguishable love of it, combined to exasperate and prolong the hostility of his enemies. When argument failed to enlighten their judgment, and reason to dispel their prejudices, he wielded against them his powerful weapons of [58] ridicule and sarcasm; and in his unrelenting warfare, he seems to have forgotten that Providence had withheld from his enemies those very gifts which he had so liberally received. He who is allowed to take the start of his species, and to penetrate the veil which conceals from common minds the mysteries of nature, must not expect that the world will be patiently dragged at the chariot wheels of his philosophy. Mind has its inertia as well as matter; and its progress to truth can only be insured by the gradual and patient removal of the obstructions which surround it.

The boldness—may we not say the recklessness—with which Galileo insisted upon making proselytes of his enemies, served but to alienate them from the truth. Errors thus assailed speedily entrench themselves in general feelings, and become embalmed in the virulence of the passions. The various classes of his opponents marshaled themselves for their mutual defence. The Aristotelian professors, the temporizing Jesuits, the political churchmen, and that timid but respectable body who at all times dread innovation, whether it be in religion or in science, entered [59] into an alliance against the philosophical tyrant who threatened them with the penalties of knowledge.[3]

For the second phase of Galileo's trial (in 1633), Brewster's account again reached poetic heights but was also novel in any language:

[93] The account which we have now given of [94] the trial and sentence of Galileo, is pregnant with the deepest interest and instruction. Human nature is here drawn in its darkest colouring; and in surveying the melancholy picture, it is difficult to decide whether religion or philosophy has been most degraded. While we witness the presumptuous priest pronouncing infallible the decree of his own erring judgment, we see the high-minded philosopher abjuring the eternal and immutable truths which he had himself the glory of establishing. In the ignorance and prejudices of the age—in a too literal interpretation of the language of Scripture—in a mistaken respect for the errors that had become venerable from their antiquity—and in the peculiar position which Galileo had taken among the avowed enemies of the church, we may find the elements of an apology, poor though it be, for the conduct of the Inquisition. But what excuses can we devise for the humiliating confession and abjuration of Galileo? Why did this master-spirit of the age—this high priest of the stars—this hoary sage, whose career of glory was near its consummation—why did he reject the crown of martyrdom which he had [95] himself coveted, and which, plaited with immortal laurels, was about to descend upon his head? If, in place of disavowing the laws of Nature, and surrendering in his own person the intellectual dignity of his species, he had boldly asserted the truth of his opinions, and confided his character to posterity, and his cause to an all-ruling Providence, he would have strung up the hair-suspended sabre, and disarmed forever the hostility which threatened to overwhelm him. The philosopher, however, was supported only by philoso-

phy; and in the love of truth he found a miserable substitute for the hopes of the martyr. Galileo cowered under the fear of man, and his submission was the salvation of the church. The sword of the Inquisition descended on his prostrate neck; and though its stroke was not physical, yet it fell with a moral influence fatal to the character of its victim, and to the dignity of science.[4]

Certainly Brewster was right that "mind has its inertia as well as matter" (58). But the problem is how to overcome such mental inertia in the search for truth. If one opposes it, one may be judged "reckless." If one acquiesces and "rejects the crown of martyrdom" (94), one may be judged a coward. At different times, in light of different circumstances, Galileo did both, and so he earned a double dose of blame from Brewster. Far from being the "martyr" of science suggested in Brewster's title, Galileo allegedly harmed science first by alienating potential converts and antagonizing opponents, and later when he "cowered under the fear of man" (95) and let his enemies (figuratively) decapitate him. Although Brewster was apparently no friend of the Inquisition, clearly he was no friend of Galileo either. Brewster's account was primarily an evaluation of Galileo's trial, a negative evaluation—a new retrial, so to speak.

Brewster's account was criticized along different lines by Guglielmo Libri in a wide-ranging series of four review articles published in the *Journal des Savants* in 1840–1841. As the length of the review suggested, Libri was interested primarily in using Brewster's work to develop his own account of Galileo's life and work in general; in fact in 1841–1842, Libri published an account that circulated widely in French, Italian, and German.[5] Nevertheless, Libri criticized Brewster for inadequate knowledge of the relevant documents, which were available from Nelli's 1793 biography and from Venturi's collection of 1818–1821; and here Libri also mentioned some documents which he had presumably discovered himself, such as Peiresc's letters to Cardinal Barberini and Galileo's July 1634 letter to Diodati.[6] Libri also criticized Brewster for portraying the Inquisition's treatment of Galileo as excessively kind, countering with a brief argument to the effect that he had been tortured.[7] This torture argument was only a preview of the thesis Libri later elaborated in his own account, to which we now turn.

Libri was aware that the file of the original manuscripts of the Inquisition proceedings of Galileo's trial (transferred from Rome to France by order of Napoleon) was missing.[8] He also claimed that he was in possession of the original manuscripts of an Inquisition trial in Novara in 1705; that the procedure and terminology used there were the same as those used in Galileo's trial; and that in the Novara case, it was clear that the defendant was tortured.[9] Libri discussed explicitly and systematically whether Galileo was tortured, arguing that he was indeed tortured.[10] Libri's argument was based partly on the wording of the sentence, which uses a phrase ("rigorous examination") connoting torture; partly on the Novara manuscripts, in which the language was analogous to that in Galileo's sentence and which clearly involved torture; partly on the fact that according to the Inquisition manual of procedure,[11] torture was standard practice in cases like Galileo's, where the defendant's intention was in doubt; partly on the fact that after the trial, Galileo was afflicted with a hernia; and partly on the fact that Galileo's silence can be explained by the fact that the Inquisition's defendants were sworn to secrecy.

Although the torture question had been explicitly discussed in a subdued manner in some 1774 correspondence between Frisi and Fabroni,[12] with Libri in 1841 it became a cause célèbre. For besides being well documented and well argued, his account was embedded in a generally anticlerical position whose flavor is conveyed by the concluding paragraph in Libri's book:

[46] Scholastic philosophy was unable to ever recover from the blow Galileo gave to it, and the Church, which unfortunately became the instrument of the Peripatetics' hatred, shared their defeat. In fact, how can one dare claim infallibility after declaring "false, absurd, heretical, and contrary to Scripture" a fundamental truth of natural philosophy, a fact that is incontestable and now admitted by all scholars? The persecution of Galileo was odious and cruel, more odious and more cruel than if the victim had been made to perish during torture. For by nature all human individuals have the same rights, and there are no privileges as regards physical suffering; and when tortured, Galileo did not then deserve any greater compassion than [47] other less famous victims of the Inquisition. But they were not intent only on Galileo's body; they wanted to strike him morally; they forbade him to make discoveries. Enclosed in a circle of iron, blind, and isolated, he was left to be consumed by the anguish of a man who knows his strength but who is prevented from using it. This ill-fated vengeance, which Galileo had to endure for such a long time, had the aim of silencing him; it frightened his successors and retarded the progress of philosophy; it deprived humanity of the new truths which his sublime mind might have discovered. To restrain genius; to frighten thinkers; to hinder the progress of philosophy; that is what Galileo's persecutors tried to do. It is a stain which they will never wash away.[13]

Here in a nutshell we have the infallibility objection, the torture thesis, the accusation of retarding scientific progress, and one more anticlerical criticism: that after the trial Galileo was subjected to something that may be called moral torture. Soon the moral-torture thesis was widely adopted, especially by those authors who rejected the physical-torture thesis.

11.2 Immoral Disobedience?
Dublin's Cooper and Cincinnati's J. Q. Adams (1838–1844)

At about the same time that Libri was writing his anticlerical compendium, others were publishing an anti-Galilean one. This happened in 1838 with the publication in the *Dublin Review* of an article by Peter Cooper titled "Galileo—The Roman Inquisition." Moreover, Cooper's article deserves attention in its own right insofar as it may be the first lengthy apologetic account of Galileo's trial in the English language; and it is also interesting as an expression of the attitude of Irish (and English) Catholics.

Cooper's article was occasioned in part by the publication of John Elliot Drinkwater Bethune's *Life of Galileo* (1832) and of the brief account of the trial included in William Whewell's *History of the Inductive Sciences* (1837).[14] In fact, these works were listed at the beginning of the essay, in the style characteristic of a review article. However, the initial paragraph made it obvious that Cooper was also animated by the same motive as Mallet du Pan[15] (to whom he refers), namely to dispel the many falsehoods spread in regard to the trial of Galileo: for example, that he was held in prison for five years; that his eyes were gouged out as punishment; and that the case is indicative of religion's jealousy and mistrust toward science.[16] Moreover, the broad scope of Cooper's article is suggested by the fact that he showed direct acquaintance with such works as Fabroni's collection of letters (1773–1775), Giovanni Targioni Tozzetti's book on Tuscan science in the seventeenth century (1780), Nelli's biography (1793), Jean Biot's article (1816), Venturi's collection (1818–1821), and Jean Delambre's *History of Astronomy* (1821).

Cooper elaborated and defended the following theses: Copernicanism was never properly and officially declared a heresy; this, of course, was absolutely correct. Galileo was the one to begin interfering with biblical interpretation; this was a mythological distortion à la Mallet. In 1615 and 1616, Galileo acted rashly, and that was why the Church silenced him; this is the imprudence thesis. Galileo wanted the Church to sanction Copernicanism; again, this was an element of Mallet's myth. Galileo was condemned insofar as he was a bad theologian, not insofar as he was a good astronomer; this was the canonical Mallet thesis. He believed wrongly that Copernicanism had been conclusively proved and that the tidal argument provided a conclusive proof; this is an exaggeration. In 1633, he was condemned for disobedience, not for heresy; this was a version of Tiraboschi's explanation. The Church favored science by directly supporting Copernicanism before Galileo and by indirectly supporting first the Lincean Academy and later the Accademia Fisicomatematica; this echoed Tiraboschi's 1792 lecture.

In addition to a typical dose of inaccuracies, we find uncritical references to and uncritical acceptance of the Renieri apocryphal letter, Bergier's and Bérault-Bercastel's accounts, and the fabricated sentence from Guicciardini's letter of 4 March 1616.[17] Moreover, Cooper exploited the key points of Mallet's and Tiraboschi's theses to formulate a relatively novel and original apology:

The distinguished individual with whose story we have been all this while occupied, was never condemned—never indeed so much as arraigned—but once; and then not for his science, or his religion, or any other mere matter of opinion whatsoever, but for the *moral* fault of having in a most flagrant manner transgressed a solemn injunction placed on him by the highest tribunal in the land; a tribunal to which he had himself appealed,—whose decision he loudly and pertinaciously demanded, and at last succeeded in extorting. For the transgression of an injunction like this, aggravated, too, by circumstances of insult and contumely against the authority that awarded it, was he condemned for the first and last time, towards the close of his life, [in] 1633; in one word for a grievous contempt of court.[18]

Cooper was relatively clear about the chronology of the situation, and this explanation was meant to apply to the one and only time that Galileo was formally tried, that is, in 1633. Cooper was saying that Galileo was condemned neither for his Copernican astronomy nor for his biblical hermeneutics, but for his immoral disobedience of the special injunction of 1616. Galileo's disobedience was immoral because he himself believed that the Inquisition had the right to pass judgment on such matters; and the evidence that Galileo believed this was that in 1615–1616 he supposedly tried to have the Inquisition rule favorably on the earth's motion, and to do so on biblical grounds. But this last claim was essentially Mallet's myth. Mallet himself may have been somewhat unclear as to whether his interpretation applied to the first or the second phase of the trial, but once one distinguished the two phases, then Cooper could go on holding the bad-theologian thesis for the first phase and the immoral-disobedience claim for the second phase. That is how Cooper's account was an interesting attempt to elaborate and justify a relatively novel apology by exploiting Mallet's and Tiraboschi's theses.

Soon after Cooper's account was published in 1838, it spread to America under circumstances that deserve a brief recounting.[19]

During the U.S. Congress debates over the founding of the Smithsonian Institution between 1838 and 1846, the congressman and ex-president John Quincy Adams delivered several speeches in various cities in favor of astronomy in general and an astronomical observatory in particular. One of these lectures was given at the Cincinnati Astronomical Society on 10 November 1843. Some of Adams's assertions offended and upset Catholics, and within a few months a reply was published in the form of an anonymous book consisting of Cooper's 1838 article from the *Dublin Review*, preceded by a long introduction explicitly and severely critical of Adams's lecture. Although the book did not name the editor or author of the introduction, the title page was revealing enough: *Galileo—The Roman Inquisition: A Defence of the Catholic Church from the Charge of Having Persecuted Galileo for His Philosophical Opinions; from the Dublin Review, with an Introduction by an American Catholic,* Cincinnati: Published for the Catholic Book Society by Monfort and Conahans, 1844. This unnamed "American Catholic" was probably John B. Purcell, archbishop of Cincinnati.[20]

There is little question that Adams's account of Galileo was relatively superficial, contained factual inaccuracies, and expressed interpretations unflattering to the Catholic Church. Purcell tried to capitalize on the factual inaccuracies to discredit Adams in general, but he also tried to refute Adams's interpretations by using Cooper's account (and also Brewster's), including the claim that Galileo was not condemned for his scientific ideas but for theologically unsound hermeneutics. One irony of this exchange is that Purcell failed to see that Adams happened to share a key part of this claim, and so their disagreement involved deeper questions, rather than merely factual issues.

In fact, Adams had explained the 1633 condemnation in these terms: "At 70 years of age, Galileo was compelled by the sentence of these inquisitor cardinals, to crave pardon for having maintained the truth, and abjured it as absurdity, error and heresy, upon his knees, with his hands upon the gospel."[21] And then he had derived the following more general lesson: "In the lives of Copernicus, of Tycho Brahe, of Kepler, and of Galileo, we see the destiny of almost all the great benefactors of mankind. We see, too, the irrepressible energies of the human mind, in the pursuit of knowledge and of truth, in conflict with the prejudices, the envy, the jealousy, the hatred, and the lawless power of their contemporaries upon the earth."[22] To this Purcell replied:

[19] Galileo strove not for truth, but for victory! For the vindication of the Church from the odious charge of persecuting science in the person of Galileo, we do not choose to rest content with the palliative statement of Hallam that "for eighty years the theory of the Earth's motion had been maintained without censure; and it could only be the greater boldness of Galileo which drew down upon him the notice of the Church."[23] Nor with the admission of Sir David Brewster that "the Church party were not disposed to interfere with the prosecution of Science, however much they may have dreaded its influence."[24] Nor yet with the very candid and eloquent exposition of the Edinburgh Review (October 1837) which, more than anything we have seen from Protestant authority, presents this point in its true light. . . .[25] [20] The facts connected with Galileo's first attempt to force from Rome the concession that the Copernican doctrine was consistent with Scripture, set this important matter at rest. We will not here anticipate the admirable exposition of this point and its accompanying circumstances made by the writer in the Dublin Review.[26]

The penultimate sentence in this quotation embodies the Mallet myth. However, this was a thesis with which Adams seemed to agree. In fact, when describing the initial phase (1613–1616) of Galileo's troubles, Adams had asserted: "He was denounced, before the tribunal of the inquisition, and, in his own defence, wrote memoir upon memoir, to prevail upon the Pope, and the inquisitors, to declare the Copernican system, to be in strict conformity with the Holy Scriptures. As the Pope, and seven cardinals,[27] appointed by him to solve this knotty question, pronounced, that the doctrine of the earth's motion, was an absurdity in physics, and a damnable heresy in religion, Galileo was expressly forbidden, ever again to maintain, by word of mouth, or in writing, that the rotary motion of the earth was countenanced by the holy scriptures."[28]

The difference between Adams and Purcell seemed to be that for Adams the Inquisition had no right to silence Galileo for his opinions, be they the physical idea that the earth moves or the theological claim that the earth's motion is compatible with Scripture; hence he interpreted the condemnation as an abuse of power. On the other hand, for Purcell the Inquisition certainly had the right to limit Galileo's freedom of thought, especially regarding the question of the compatibility of the earth's motion with Scripture; and that was why portraying Galileo as playing the theologian (and a bad one at that) became crucial for justifying the Inquisition's condemnation.

11.3 Torturing People versus Torturing Texts: Marini's Semi-Official Apologia (1850)

The next important "retrial"[29] of Galileo occurred in 1850 with Marino Marini's *Galileo and the Inquisition*. At that time Marini was the prefect (director) of the Vatican Secret Archives, which held the file of trial proceedings. As we saw in chapter 9.2, he was the official who from 1814 to 1817 had been in Paris to retrieve the Church archives transferred by Napoleon, but had been unable to regain possession of the Galileo file; however, it had been found and returned to Rome in 1843. After this near-loss and the aborted Napoleonic project to publish the documents, the general expectation was that the Church would bring about publication. Instead it was decided to produce an updated apologia, for which Marini was the ideal author and whose semi-official status is suggested by the fact that it was published by the press of the Sacred Congregation for the Propagation of the Faith.

The book reached new heights of extremism: Marini ended his account of the trial by claiming to have shown that "to render due praise to the justice, wisdom, and moderation of the Inquisition, we must affirm that perhaps there has never been a judicial action as just and as wise as this one."[30] The work did contain an unusual number of quotations from the proceedings, which were not accessible to anyone else. It was obvious, however, that these quotations were highly selective and taken out of context.[31]

Marini gave a general justification of the Inquisition based on its historical role in saving Europe from heresy and on a comparison between its practices and the treatment of heretics by Protestant churches and of ordinary criminals by lay state courts. We need not agree with Marini here, but he certainly was raising an important issue.

More specifically, he argued at length that although Galileo was threatened with torture, he was not actually tortured, as Libri claimed. First,[32] Marini tried to show that the term "rigorous examination" (used in the text of the 1633 Inquisition sentence) was *not* synonymous with torture; he argued that a rigorous examination was an interrogation conducted with the verbal threat that if the defendant did not tell the truth he would be tortured; physical torture followed if and only if the replies were unsatisfactory. Marini could support his claim by referring to and quoting from one of the trial documents available only to himself, namely the fourth deposition held on 21 June 1633; this was the written record of Galileo's "rigorous examination."[33]

Marini also mentioned a fact that had been reported by Venturi in his account of the trial.[34] As we have seen, Venturi had consulted Father Carlo Altieri, who had read the file of trial proceedings in France, before it was lost. Altieri had reported that the proceedings contained no evidence that Galileo had been subjected to physical torture.

Another argument Marini gave[35] was that the manuals of Inquisition procedure stipulated that for torture to be administered, a prior vote and recommendation of the consultants were required, and if need be a vote by the cardinal-inquisitors.[36] And the prefect of the Vatican Secret Archives assured the reader that the trial documents did not contain any minutes of such decisions.

Marini also argued[37] that since the Tuscan ambassador Niccolini was having regular meetings with the pope and was informing the grand duke of all the details, the ambassador would presumably have been informed of this detail, and he would have reported it to Florence. Yet the correspondence contained no such report.

Finally, Marini gave a description[38] of the documents in the file after the May–June 1633 summary, to counter Libri's claim that the documents recording the torture might have been removed. Recall that when Delambre and Venturi accessed the French translation of the executive summary of the trial, compiled in May–June 1633, it contained page and folio references to the other documents since the beginning of the proceedings in 1615; but necessarily that summary did not contain references to subsequent documents. Thus Delambre and Venturi knew nothing about the post–May 1633 documents, and anticlericals like Libri advanced the suspicion that those documents might have been removed. So, if they could trust Marini, their suspicions would be allayed. Could they trust him? Not judging from the way he tortured other known texts in his apologetic effort.

In fact, Marini charged Galileo with all kinds of inconsistency, insincerity, and imprudence. He argued that the Inquisition acted justly and wisely in opposing, trying, and condemning Galileo, because (1) he supported a physical theory by means of biblical passages and expected the church to do the same and to officially endorse it; (2) he violated Bellarmine's injunction by publishing the *Dialogue;* and (3) he ridiculed Pope Urban VIII through the character of Simplicio.

Marini's discussion of the last allegation is a good example of excessive indulgence in portraying the situation in the worst possible light against Galileo.[39] In discussing the disobedience thesis,[40] Marini failed to mention that the special injunction document lacked Galileo's signature and that in the first deposition Galileo denied knowledge of the crucial clause in the special injunction. Of course, no one knew these facts at that time, for they were not revealed until later in the century, when the proceedings were published. But Marini had access to these documents; he simply acted as if he did not know anything about them.

As regards the first thesis, Marini relied uncritically on Mallet's 1784 article and on his repeaters, Feller, Bergier, and Cooper.[41] Marini explicitly referred to them with approval several times, and he was especially effusive about Mallet. In fact, Marini ended his book's central chapter, which dealt with Galileo's trial, as follows; after asserting, as quoted above, that the Inquisition's condemnation of Galileo was the wisest and most fair judicial act in recorded history, Marini added: "To remove any suspicion of partiality [on my part] toward the Inquisition, I refer the reader to the dissertation by the Genevan Mallet du Pan, included in Geneva's literary review of 1784, aimed against Voltaire and the French encyclopedists; in it are rebutted all the insults which some bad writers are accustomed to vomit against the Inquisition when the discussion turns to Galileo; in his affair, that Protestant affirms, he was wrong on all counts."[42] Marini also quoted the adulterated passage from Guicciardini's letter, containing the fabricated sentence that Galileo "wanted that the pope and the Holy Office declare the system of Copernicus founded on the Bible."[43]

However, Marini also did something admittedly ingenious in exploiting the documentation available only to him. This involved the Inquisition minutes of 16 June 1633, at which meeting the pope decided how to bring the trial to a conclusion; the minutes contain the ambiguous order to interrogate Galileo one more time under torture or threat of torture (later, after the publication of the proceedings, this document would cause much discussion). Another part of the minutes contains a sentence stating that Galileo "is to be enjoined that in the future he must no longer treat in any way (in writing or orally) of the earth's motion or sun's stability, *nor of the opposite,* on pain of relapse."[44] The crucial phrase is "nor of the opposite" and refers to the opposite of the Copernican hypothesis, namely to the geostatic thesis. Why was such a phrase included? The reason, Marini answered,[45] was that Galileo was to be told that it was wrong for him to argue in favor of the earth's rest, as well as in favor of the earth's motion.

Galileo's preferred manner of arguing was allegedly to use biblical texts in support of a physical claim. The Church thus was telling Galileo to stop this theologically unsound practice. Of course, this was the Mallet thesis. Marini's own formulation is worth quoting: "The Inquisition's indignation against Galileo was not elicited by his teaching a heretical doctrine, such as the Copernican one, regarding which the Church had been silent for so many years, but by his manner of elaborating it. For neither the Inquisition nor the Congregation of the Index prohibited the supposition that the sun stands still and the earth moves. Instead, what moved the Inquisition to indignation was to want to guarantee this supposition by means of scriptural texts, so as to show it to be in accordance with the meaning of the sacred word. Thus, Galileo's abuse of Scripture had brought against him the two Congregations, the Inquisition and the Index."[46] As if the abuse of wanting to support an astronomical hypothesis with biblical texts were not enough, Marini went on to claim that "Galileo wanted to have it proclaimed a dogma of the faith . . . but the Inquisition did not want to transform a scientific opinion into a dogma, and on account of this it rejected Galileo's absurd pretensions."[47]

So although Marini may be said to have advanced a new shred of evidence in favor of the bad-theologian thesis, it is questionable whether this ought to be counted to his credit; for such an effort was accompanied by a stunning blindness to the mountains of evidence against it. One of these mountains (though not the only one) is the whole *Dialogue on the Two Chief World Systems,* in which Galileo clearly stays away from biblically based arguments.

Finally, to see the extent to which Marini was willing to torture texts in his apologetic crusade, it must be mentioned that he quoted the passage from Galileo's Letter to Christina in which Galileo quoted Cardinal Baronio's aphorism; that was the principle that "the intention of the Holy Spirit is to teach us how one goes to heaven and not how heaven goes."[48] Marini made it sound as if Galileo was quoting a principle he rejected, instead of a principle he accepted![49]

11.4 "Moral Torture"? Antihero?
Biot's and Chasles's Circumstantialism (1858–1867)

The deeply flawed character of Marini's semi-official apologia for the Inquisition was apparent even to observers who were Catholic and well disposed toward the Church. For example, in 1858 the French physicist Jean Biot wrote about Marini that "in effect, his whole book is marked by such an incessant and bitter feeling of malevolence against the unfortunate Galileo that he really seems to have intended not to sincerely describe the circumstances of the trial, but rather to redo it worse than it had been the first time."[50]

In 1816 Biot had written the Galileo article for the *Biographie Universelle,* which treated Galileo's work in general but included a few pages on the trial.[51] However, in 1858 Biot published a new account of the trial[52] based on new primary and secondary sources published since then, such as Venturi's collection (1818–1821), Albèri's critical edition of Galileo's works and correspondence (1842–1856), and Marini's book. Biot claimed that "the crux"[53] of the trial was personal circumstances, such as the "envy"[54] of others and, primarily, Galileo's managing to sour his friendship with the pope.[55] Biot saw himself as giving both a revisionist and a bipartisan interpretation, in contrast to those accounts that blamed the whole episode on ecclesiastic obscurantism, Galilean errors (scientific and otherwise), or the alleged conflict between science and religion.[56] He explicitly argued that Galileo was not tortured physically but was subjected to "moral" torture:

> [42] No! Galileo was not physically tortured in his person. But what frightful moral torture must he have suffered when, under the terrible threat of corporal punishment and incarceration, he was miserably forced to perjure himself; that is, to deny the immortal consequences of his discoveries, [43] to declare true what he believed to be false, and to take an oath that he would no longer maintain what he believed to be the truth! Can one understand the anguish of this martyr, the grief with which this elite intelligence was filled? They did not forbid him only his earlier thoughts, but tried to chain them down forever. From this fateful year of 1633 until his death, on 8 January 1642, that is, during the last nine years of his life, the unfortunate Galileo remained in a state of callous suspicion and uneasy surveillance whose rigor followed him beyond the grave. Some fanatical theologians decided to contest the validity of his testament and his right to ecclesiastical burial, on the grounds that he had died while serving a sentence imposed by the Inquisition. But these odious efforts were judicially repelled, and his motherland Florence has nothing to be ashamed of for her attitude toward the memory of such a great genius, who had brought her so many honors.
>
> There you have the truth on the trial of Galileo. In its impartial justice, posterity has inverted the roles of persecuted and persecutor. But in honoring the genius and misfortune of the one without absolving the other, posterity must not exaggerate the violence that was in fact exercised, nor ignore or dissemble the unwise provocations that generated it. It is this double interplay of human emotions that I have wanted to stress in the affair I have just related.[57]

Libri had already spoken of moral torture, but in a context where it was added to physical torture, as well as to other ecclesiastical abuses. But with Biot we have the new thesis that explicitly affirmed moral torture, while denying physical torture.

Soon after Biot's, a similar account was elaborated by someone who approached the topic from a very different angle: Philarète Chasles, a distinguished literary critic, director of the Mazarin Library in Paris, and professor of foreign languages at the Collège de France. In 1862, he published a book titled *Galileo Galilei: His Life, His Trial, and His Contemporaries,* consisting largely of quotations from correspondence interspersed with commentary. Dismissing the idea that the trial was caused by scientific or theological controversies or the conflict between science and religion, Chasles claimed that it derived from ordinary human malice, especially envy: "In this affair, it is more of a question of personalities and small hatreds than of theology or doctrine."[58]

Chasles portrayed Galileo as riddled with conflicting tendencies but ultimately a coward and moral weakling; his enemies as motivated by envy and petty hatred, and engaging in what amounted to the moral assassination of an innocent man; and the society in which they lived as decadent and amoral and carrying to an excess the social art of living based on comfort, sensuality, accommodation, dissimulation, appearances, and face-saving. By way of conclusion, he described his own work as a case study of a "moral assassination" of a moral weakling in a "decadent society,"[59] and added:

> [278] What a sad spectacle! But posterity did not accept such a legacy. It invented another Galileo, the heroic Galileo. Forgetting that the Italian mores of the seventeenth century . . . could not produce heroes, it created a sublime myth, which it substituted for the real character. Thus in popular belief there was established the fiction of a martyr Galileo, indomitable philosopher with convictions; this fiction is a sweet and proud figure, a wonderful creation born from aspirations and desires; it is a daughter of the poetry of justice [279] and of the instinct of moral beauty that always live in the human soul to protest against the realities of history. . . .
>
> [280] I have wanted to disengage the true Galileo from the clouds of symbols. This Italian, half Greek, sublime discoverer of the secrets of the heavens; genius who preceded Newton, continued Bacon, announced Descartes, is not a hero of moral courage; he is a genius of enlightenment.[60]

Chasles's account is also noteworthy for three particular theses. Like Biot, he criticized the physical-torture thesis; but he elaborated the moral-torture thesis more extensively and even more eloquently. Like Biot, he stressed the Urban-Simplicio caricature factor; but he elaborated it in a plausible way by connecting it directly with the character traits and social mores on which he focused. And like Biot, he denied the *e pur si muove* legend; but he added that Galileo expressed such an attitude just before the beginning of the trial, in his letter to Diodati of January 1633. In other words, although it was not literally true that Galileo uttered this phrase just after his abjuration, it was an accurate representation of how he really felt before, during, and after the trial.

Regarding this last point, Chasles's thesis coincided with the one advanced at the same time in Geneva and expressed in a memorable formulation that deserves note. In 1862, Jules Barni, a professor at the Academy of Geneva, gave a series of ten public lectures on the martyrs of free thought, such as Socrates, Hypatia, Abelard, Ramus, Servetus, and Rousseau. The eighth lecture dealt with Bruno, Campanella, Vanini, and

Galileo. The account of Galileo ended with the claim that although the story of his *e pur si muove* "is not historically true, it is (if I may say so) philosophically true."[61]

Chasles's descriptions were generally inspired, vivid, interesting, and valuable. But ultimately his account was too one-sided, and this was due to his overstress on the correspondence and his neglect of the content of the *Dialogue* and of whatever trial proceedings were available. Moreover, his many quotations were really paraphrases, rather than verbatim translations, and they were never referenced; thus, from the point of view of literary genres, his work must be classified as halfway between a history and a novel. Finally, for all his antiheroic, circumstantialist, and demythologizing fervor, Chasles ended up merely substituting the old heroic portrayals with his own; certainly, to speak of Galileo as a "genius of enlightenment" is to speak of another kind of hero. More important, since to this talk there did not really correspond an actual analysis of Galileo as a hero of enlightenment, and given that Chasles did indeed stress portraying Galileo as a victim of moral assassination, such a victim turned into another kind of hero, as is often the case when victims are romanticized and made to wallow in their condition of victimhood.

Nevertheless, Chasles's account shows that Brewster's cowardice thesis, however untenable, had not been forgotten. It also shows that the moral-torture thesis was gaining popularity. And it strengthened an approach to Galileo's trial that may be called circumstantialist and demythologizing; this was an approach that had been advocated by Biot and that aimed to decouple the condemnation from larger issues of philosophy, science, theology, and conflict between them, and to root it in nothing more (and nothing less) that the vicissitudes of the human condition and social interaction.

During the same period (1860s) and in the same country (France), other accounts were published that were less one-sided and more scholarly than Chasles's. For example, the physician Maximien Parchappe (1800–1866) published a book in which he tried to give an impartial account, aware that impartiality should not require making illegitimate concessions out of an abstract duty to equity.[62] And Thomas Henri Martin published a work that must be regarded as a milestone in the Galileo affair, showing how far one could go and not go without those new documents whose publication marks the next phase.[63]

However, despite such works, for a while the circumstantialist and superficially antiheroic approach, à la Chasles, prevailed. In fact, this approach reached a climax with what was perhaps the first play written and performed about the Galileo affair. On 7 March 1867, the French playwright François Ponsard had a verse drama titled *Galileo* performed by the imperial company at the Comédie Française. Dedicated to Emperor Napoleon III, its performance had been delayed for two years to avoid disturbing French relations with the Holy See. It was now being staged by order of the emperor, overcoming the opposition of the clerical forces. The Paris newspaper *Le Figaro* published excerpts from the play on the first page, along with critical articles. This went on for about a month, while the play was being performed. But the public found it boring, and so it closed after a month. The play (inaccurately) portrayed Galileo as a married man and his motivation in abjuring as based largely on feelings of family and parental responsibility.[64]

11.5 Inquisition Right *and* Wrong?
Madden's Tortured Thinking (1863)

Many of the works discussed above (and others) were studied by Richard R. Madden (1798–1886), who attempted to take them into account and produce his own synthesis in a book published in London in 1863, *Galileo and the Inquisition*. Madden was an Irish Catholic physician who from 1833 to 1850 held various posts in the British colonial administration in the West Indies, Africa, and Australia, and then from 1850 to 1880 in Dublin. In his spare time he wrote about twenty books, mostly historical, including a two-volume work titled *The Life and Martyrdom of Savonarola*.

In *Galileo and the Inquisition*, Madden had chapters giving brief accounts and English translations of long passages of the works of Bérault-Bercastel (1790), Libri (1841), Marini (1850), Reumont (1853), Biot (1858), and Chasles (1862). He discussed the torture thesis, criticizing Libri's view and siding with Marini. And he had a chapter containing an instructive, revealing, and comprehensive account, of which more below.

His most novel contribution was a discussion of the origin and content of the Inquisition archives found in the library of Trinity College, Dublin.[65] In 1848–1849, a revolution in Rome overthrew the papal government and created a Roman Republic; the pope escaped to the fortress at Gaeta. But within a year a French army occupied the city and reestablished papal government. During the French occupation, seventy-seven volumes were removed from the Inquisition Archives; in 1850, a French army officer sold them to the English Duke of Manchester; then the latter sold them to the Irish Protestant Rev. Richard Gibbings; finally, they were purchased by a certain Dr. Wall, a Fellow of Trinity College, Dublin, who donated them to the library of that college. Around 1860 they were deposited there.

Based on these documents, Gibbings wrote three anti-Catholic books. Madden criticized Gibbings and in the process found some new evidence against the torture thesis. That is, the Trinity College proceedings sometimes explicitly speak of "torture" and sometimes only of "rigorous examination," thus proving that the two could not be equated. Thus, when the Inquisition's 1633 sentence states that Galileo was subjected to a "rigorous examination," this does not imply that he was tortured.

Madden tended to be careless in factual and scholarly details and incoherent in his reasoning. But he seemed reasonable in his stated goal of criticizing writers who made careless and untrue assertions about Galileo's trial. He also seemed reasonable in defending the Inquisition by means of a comparison with state criminal justice systems and of a historical contextualization of its original purpose and later development. In so doing, he was, of course, following Marini's general approach, but Madden made specific comparisons with the British situation. For example, in one appendix he quoted the text of "the preamble of the act of parliament of the twenty-second year of the reign of Henry VIII, 9th chap., ordaining that persons convicted of the crime of administering poison should be boiled to death."[66]

The climax of Madden's effort was a chapter titled "On the Justifiability of the Proceedings against Galileo." It advanced an account[67] whose most striking aspect is its tortuous and tortured style of exposition. This might seem an irrelevant detail were it not for the fact that it seems to reflect his attitude, his thinking, and the earnestness of his attempt to arrive at a satisfactory position.

Madden began by taking for granted that Galileo was neither a hero nor a martyr, based on his previous discussion and the recent works of others (148); then he went on to ask whether the Inquisition proceedings were justified. And his initial answer was that they were not advantageous in either the short or long run, but were rather lamentable.

As Madden proceeded to elaborate this answer, he stressed the 1616 decree that Copernicanism was contrary to Scripture and the injunction to Galileo to abstain from teaching that system; and he made the first of many statements of the bad-theologian thesis (stemming from Mallet du Pan), "that Galileo equally erroneously believed the same theory was sanctioned expressly by [Scripture]" (149). Madden then mentioned, lamented, and dismissed the Urban personal-insult thesis, adding that what worried the pope was not even Galileo's alleged heresy of Copernicanism; instead "he was arraigned for disobedience and obstinacy in promulgating an astronomical system [as] supported and confirmed by the Scriptures" (150), another twist on the Mallet myth.

Next Madden argued that in Galileo's time the Copernican theory was merely a hypothesis and not a demonstrated truth: his discoveries did indeed make it more worthy of serious consideration than it was in Copernicus's time, but Galileo knew nothing of Newton's universal gravitation, Bradley's aberration of starlight, and the earth's spheroidal shape or equatorial bulge; and these were the phenomena that proved the earth's motion (150–51). This was a relatively novel thesis; or to be more precise, to stress it as Madden did and to exploit its apologetic potential yielded a relatively novel account. An embryonic version of this account was contained in Olivieri's efforts to repeal the continued condemnation of the Copernican doctrine in 1820 while simultaneously justifying the original censure in

1616 (see chapter 10.2 and 10.3). Olivieri also explicitly elaborated such an account in some articles published anonymously in 1841 in French and German, and the same explicit account was contained in a posthumous work in Italian published in 1872.[68]

Given the nondemonstrative status of Galileo's pro-Copernican arguments, an immediate conclusion which Madden drew was that therefore the pope and the Inquisition were justified in refusing "the sanction of the Church for them, which Galileo so importunately solicited, or rather demanded" (151). At the same time, such refusal to sanction was no reason to condemn, especially on the part of an institution like the Catholic Church that had a long history of supporting astronomy, going back to Copernicus himself (152–53). On the other hand, Galileo had "a solemn engagement in 1616, to abstain from teaching the system of Copernicus as a system in obvious conformity with the Scriptures" (153). Thus, when "certain rivals and enemies of Galileo" managed to convince the pope that he had violated that injunction, Urban put him on trial (153).

Madden did not ask how Galileo's enemies could have convinced the pope in 1632 that Galileo was teaching Copernicanism "as a system in obvious conformity with the Scriptures" (153), given that the *Dialogue* completely avoided theological and scriptural considerations. I believe Madden was confused and careless about chronology, the content of the *Dialogue*, and the text of the 1633 sentence. A careful reading of the latter alone could have shown that the 1616 injunction prohibited Galileo not merely from supporting Copernicanism with scriptural passages, but rather from holding or defending it in general (according to the version remembered and documented by Galileo), or perhaps from discussing it at all (according to the version alleged by the Inquisition).

In an attempt to be judicious, Madden criticized one of Cooper's arguments (153–55). Cooper had argued that the Inquisition had never declared Copernicanism heretical but merely contrary to Scripture; and that claims to the contrary were based on the failure to understand that the consultants' evaluation of February 1616 was merely a recommendation (which was in fact not endorsed by the cardinal-inquisitors), and on the failure to appreciate that the Inquisition's talk of heresy reflected a broad concept according to which any transgression it prosecuted was a heresy (and so was a typical legal fiction). While avoiding such oversights, Madden nevertheless went on to suggest that the Inquisition had declared Copernicanism in 1616 and Galileo in 1633 heretical. He did not really explain his reasons, aside from advancing two related considerations: that such a declaration by the Inquisition should not be equated with a declaration by the Church, which in fact never happened (155); and that the Inquisition's declaration was erroneous, which should be admitted without difficulty because infallibility does not belong to the Inquisition but to the Church as a whole (155–57).

Madden was willing to admit "the blunder or the crime into which the Inquisition had fallen in its proceeding against Galileo" (155), as long as one did not exaggerate it. And as long as one did not forget certain "adventitious circumstances" (155), Madden was willing "to state that he had been charged indirectly with an offence which in point of fact was in itself no offence against God or man" (155), and that "it was a hard thing to compel that old man to retract an opinion on a scientific subject which was not at variance with any religious dogma or doctrine, and which he was firmly persuaded was true" (156). On the other hand, there was a crucially important adventitious circumstance, namely "Galileo's alleged attempts . . . to produce Scriptural authority in support of an astronomical theory, and to get those attempts sanctioned by the highest ecclesiastical authorities" (156); and in light of this circumstance, "it would be impossible . . . to find fault with such a condemnation" (156). This judgment was not a slip of the pen, for Madden felt the need to justify it by explaining that Scripture is not a scientific or philosophical authority, but one only for matters of faith and morals (156).

Unfortunately, this justification was groundless, since it was based on an adventitious circumstance that did not happen; in fact, it was Mallet's untenable bad-theologian thesis. Moreover, this justification seems to expose the incoherence of Madden's account; for apparently he was saying both that Galileo's trial was inexpedient, lamentable, a blunder, and a crime (all his words), and that it was impossible to fault. Thus, to his tortuous and tortured language there corresponded a logical self-contradiction. More charitably interpreted, Madden's position perhaps was that there were reasons for thinking that the trial was unjustified, and there were also reasons for thinking that it was justified; and there is no inconsistency in pointing this out, as it happens on many occasions in human affairs, which are usually characterized by complexity and multidimensionality.

Chapter 12

A Miscarriage of Justice?
The Documentation of Impropriety (1867–1879)

We have seen that in 1755 Church officials created a special file of proceedings of Galileo's trial by removing the relevant documents from one of the regular volumes of the Inquisition archives. We have also learned that Napoleon was the first to have made a serious plan (between 1810 and 1814) to publish that file of trial documents. After the dossier was returned to Rome in 1843, the prefect of the Vatican secret archives (Marini) was expected to publish it, but instead in 1850 published his own interpretive account. There is evidence that many other people tried unsuccessfully to publish it, or at least to consult it. Albèri tried in connection with his critical edition of Galileo's works (1842–1856); he was dealing with Marini, but when the latter died in 1855 Albèri lost the opportunity.[1] In 1864, the German scholar Moritz Cantor complained in print that he had been refused permission to examine the Galilean file.[2] In 1878, the Polish-Italian scholar Arturo Wolynski reported that he too had been refused access.[3]

However, from the late 1860s to the late 1870s, four scholars were allowed to consult and publish the file. The first was the Frenchman Henri de L'Epinois, who published a large selection in 1867; but besides being incomplete, this edition was full of errors and other imperfections, and so ten years later he published a complete and improved edition.[4] The second scholar was an Italian, Domenico Berti, who published his first edition in 1876 and his second, improved and complete edition in 1878.[5] The third one was an Italian priest, Sante Pieralisi, who was the director of the Barberini Library in Rome; he did not publish his own edition but compared the manuscripts with the editions of L'Epinois and Berti and published corrections to them.[6] The fourth scholar was the Austrian Karl von Gebler, who was granted permission after publishing an interpretive account based on L'Epinois's first edition; then from his own personal consultation and with the benefit of L'Epinois's second edition, Gebler published his own complete edition in 1877.[7] Thus, by 1878 there existed three essentially complete editions of the Galilean file: L'Epinois (1877), Gebler (1877), and Berti (1878).[8]

12.1 A Legal Impropriety:
Wohlwill's Radical Revisionism (1870)

L'Epinois's first edition contained a selection of the most important documents in the Galilean file (twenty-five pages of small print), and was preceded by a comprehensive account (seventy-eight pages of larger print).[9] The latter may be summarized as follows. According to L'Epinois, the conflict was primarily between science and Aristotelianism. Galileo's main opponents were Aristotelian philosophers, who at one point advanced the biblical objection. He defended himself by advancing novel interpretations of Scripture. On the physical doctrine (of the earth's motion), Galileo was right and the Inquisition wrong, but in his time most scientists thought the reverse (Inquisition right and Galileo wrong). *Both* sides were imprudent, Galileo for wanting to go too far too fast, the Inquisition for showing itself to be too circumspect. The Inquisition was right to be concerned about Galileo's novel interpretations of Scripture, but wrong to give its decisions an absolute value and an aura of finality, permanence, and definitiveness.

The emphasis on the split within natural philosophy was relatively novel; it had some plausibility, at least as long as the emphasis was not too excessive, too one-sided, too reductionist, or too exclusivist. In saying that Galileo answered the biblical objection by reinterpreting Scripture, L'Epinois was missing the main point, which was to deny the philosophical (i.e., scientific) authority of the Bible; to that extent, he was echoing Mallet's bad-theologian thesis. L'Epinois was giving an interesting twist to the imprudence thesis when he pointed out that the Church was being too conservative, too slow, and too cautious; it would follow that the Inquisition was also acting imprudently, recalling that prudence can mean wisdom or judiciousness as well as caution or carefulness (as L'Epinois presupposed).

However, although L'Epinois's account had some originality, plausibility, and appeal, it was the sort of thing one could have thought of without the newly published documents. In this regard, one could say that L'Epinois was merely confirming some of the allegations made by the file's earlier readers, namely that it contained nothing that was not already known and it did not deserve to be published.[10] Perhaps L'Epinois was too clerically minded, or perhaps he was too concerned with making the documents available to others so that they could reach their own conclusions,[11] to be able to notice the radical novelties the file contained. It was left to other, more open-minded readers to see them. In particular, a young German named Emil Wohlwill rose to the task.

Wohlwill (1835–1912) was born in Hamburg into a family of secular-minded Jews who had a keen appreciation for intellectual freedom.[12] He studied chemistry at the universities of Heidelberg, Berlin, and Göttingen and then returned to his native city to make a living as an industrial chemist. Working in a foundry that produced copper, he made important contributions to perfecting the process of the electrolysis of nonferrous materials. But his great love was the history of science, and he managed to become one of the leading Galilean scholars of his time. Although the first volume of his monumental biography of Galileo was published at the end of his life and the second posthumously, his most important contributions to the Galileo affair date from his middle age,[13] when the Vatican file of Galilean trial proceedings was being published.

Among the documents transcribed and published by L'Epinois, but unappreciated by him and to be exploited by Wohlwill, were the Inquisition minutes of 25 February 1616; the special-injunction transcript of 26 February 1616; Bellarmine's certificate of 26 May 1616; and Galileo's first deposition of 12 April 1633. Let us look at these documents before examining Wohlwill's interpretation.

The Inquisition minutes of February 25 were important because the meeting took place the day after the consultants had submitted their evaluation of Copernicanism; moreover, it was presided over by the pope, and so a decision was made about how to conclude the proceedings that had been going on for about a year:

Thursday, 25 February 1616.
The Most Illustrious Lord Cardinal Millini[14] notified the Reverend Fathers Lord Assessor and Lord Commissary of the Holy Office that, after the reporting of the judgment by the Father Theologians against the propositions of the mathematician Galileo (to the effect that the sun stands still at the center of the world and the earth moves even with the diurnal motion), His Holiness ordered the Most Illustrious Lord Cardinal Bellarmine to call Galileo before himself and warn him to abandon these opinions; and if he should refuse to obey, the Father Commissary, in the presence of a notary and witnesses, is to issue him an injunction to abstain completely from teaching or defending this doctrine and opinion or from discussing it; and further, if he should not acquiesce, he is to be imprisoned.[15]

Next, the importance of the special-injunction transcript was obvious in light of the stress placed on that 1616 prescription to Galileo in the 1633 sentence and the subsequent diffusion of the disobedience thesis:

[147] Friday, the 26th of the same month.
At the palace of the usual residence of the said Most Illustrious Lord Cardinal Bellarmine and in the chambers of His Most Illustrious Lordship, and fully in the presence of the Reverend Father Michelangelo Segizzi of Lodi, O.P. and Commissary General of the Holy Office, having summoned the above-

mentioned Galileo before himself, the same Most Illustrious Lord Cardinal warned Galileo that the above-mentioned opinion was erroneous and that he should abandon it; and thereafter, indeed immediately,[16] before me and witnesses, the Most Illustrious Lord Cardinal himself being also present still, the aforesaid Father Commissary, in the name of His Holiness the Pope and of the whole Congregation of the Holy Office, ordered and enjoined the said Galileo, who was himself still present, to abandon completely the above-mentioned opinion that the sun stands still at the center of the world and the earth moves, and henceforth not to hold, teach, or defend it in any way whatever, either orally or in writing; otherwise the Holy Office [148] would start proceedings against him. The same Galileo acquiesced in this injunction and promised to obey.

Done in Rome at the place mentioned above, in the presence, as witnesses, of the Reverend Badino Nores of Nicosia in the kingdom of Cyprus, and of Agostino Mongardo from the Abbey of Rose in the diocese of Montepulciano, both belonging to the household of the said Most Illustrious Lord Cardinal.[17]

As regards Bellarmine's certificate, it was important because, as the 1633 sentence explained, it was the key document presented by Galileo in his own defense at the trial. To be sure, the text of this document was already known; it had been published first by Nelli (1793) from some source (which he did not reveal) other than the Vatican file of trial proceedings; it had been reprinted by Venturi (1818–1821); and it had been quoted in full by Marini (1850).[18] However, no one had yet exploited its exculpatory potential. It reads as follows:

We, Robert Cardinal Bellarmine, have heard that Mr. Galileo Galilei is being slandered or alleged to have abjured in our hands and also to have been given salutary penances for this.[19] Having been sought about the truth of the matter, we say that the above-mentioned Galileo has not abjured in our hands, or in the hands of others here in Rome, or anywhere else that we know, any opinion or doctrine of his; nor has he received any penances, salutary or otherwise. On the contrary, he has only been notified of the declaration made by the Holy Father and published by the Sacred Congregation of the Index, whose content is that the doctrine attributed to Copernicus (that the earth moves around the sun and the sun stands at the center of the world without moving from east to west) is contrary to Holy Scripture and therefore cannot be defended or held. In witness whereof we have written and signed this with our own hands, on this 26th day of May 1616.
The same mentioned above,
Robert Cardinal Bellarmine.[20]

Finally, Galileo's first deposition (dated 12 April 1633) was important because it contained his initial answers before he decided to confess and admit some wrongdoing. As was known from the texts published by Delambre and Venturi in 1821, on April 30 Galileo admitted that in the *Dialogue* he had unintentionally violated Bellarmine's warning not to defend the earth's motion.[21] The full text of the first deposition took twelve manuscript folio pages and five printed pages in L'Epinois's edition.[22] Its gist is as follows.

Galileo was asked about the *Dialogue* and the events of 1616. He admitted receiving from Bellarmine the warning that the earth's motion could not be held or defended but only discussed hypothetically. He denied receiving a special injunction not to discuss the topic in any way whatever, and in his defense he introduced Bellarmine's certificate, which only mentioned the prohibition to hold or defend. Galileo also claimed that the book did not defend the earth's motion but rather suggested that the favorable arguments were inconclusive, and so did not violate Bellarmine's warning.

In 1870, three years after L'Epinois's publication of these documents, Wohlwill published a short book containing a radical revisionist reinterpretation of Galileo's trial. Wohlwill argued that the special injunction transcript was problematic because it conflicted with the papal decision of the day before, with Galileo's attitude and correspondence of that period, with the Index decree of 5 March 1616, with the Church's goodwill toward Galileo, and with the fact that the *Sunspot Letters* was not prohibited.[23] The special injunction also flatly contradicted Bellarmine's certificate, and so it was questionable whether it had really been served; Wohlwill was led to presume that the transcript was written improperly, perhaps after the date recorded on it.[24] An examination of Galileo's correspondence indicated that he never showed any awareness of the special injunction but rather tried to and did comply with Bellarmine's warning; if Bellarmine had been alive in 1632 and 1633, there would have been no trial.[25] Based on the 1633 depositions (especially the first one), Wohlwill argued for the existence of a mere warning (not to hold or defend) and for the nonexistence of a special injunction (not to discuss in any way).[26] Although the 1633 sentence *claimed* that Galileo also violated the requirements of Bellarmine's certificate, there was never any serious inquiry into what this certificate implied or into the discrepancy between the documents of 25 and 26 February 1616.[27] The only evidence for the existence of a special injunction was the February 26 transcript, and so no impartial judge could assign it the force of a proof.[28] Wohlwill concluded that Galileo's condemnation was based on a document that was legally inconclusive or invalid; he also conjectured that the document was forged in 1632 in order to justify a preordained verdict, which would have been insufficiently grounded on just Bellarmine's warning (not to defend).[29]

Wohlwill had advanced a revolutionary account. As suggested by my labeling one of his theses a conclusion and the other a conjecture, his claim of the legal impropriety of the special injunction was powerful, solid, and well argued, whereas his forgery thesis was highly speculative. And indeed during the next decade the impropriety thesis was reinforced, while the forgery thesis was disconfirmed.

12.2 Independent Evidence:
Gherardi's Inquisition Minutes (1870)

In the same year that Wohlwill published his radical revision of Galileo's trial, an equally significant contribution was made by Silvestro Gherardi (1802–1879). Gherardi was a professor of physics who taught at the universities of Bologna and Turin, as well as an Italian patriot who took an active part in the Risorgimento. His contribution to the understanding of Galileo's trial was occasioned by those activities.

We have already mentioned that during the Roman Republic that replaced the papacy in 1848–1849 and was suppressed by French military intervention, seventy-seven volumes of Inquisition archives were stolen by a French officer and eventually found a home at the Trinity College Library, Dublin.[30] During the revolutionary government, the Inquisition archives were moved from the Inquisition palace to the church of Saint Apollinaire, and the palace was used as a shelter for the poor; this decision prevailed, with the support of the revolutionary leader Giuseppe Mazzini, over the proposal by hotheads to raze the palace and in its place erect a commemorative column for the victims of the Inquisition.[31] Although the revolutionary government took possession of the building housing the Vatican Secret Archives, no documents were removed or tampered with.[32]

During the same period, however, two republican officials (Giacomo Manzoni and Gherardi) searched the Inquisition archives for material related to Galileo's trial. They found the archives organized into three main groups of material: files called *Decreta* ("Decrees"), containing minutes of meetings and summaries of resolutions; lengthy proceedings of trials consisting of charges, depositions, sentences, and the like, called *Processus* ("Trials"); and indexes to the decrees and trials, called *Rubricelle*. They did not find the proceedings of Galileo's trial because (unbeknownst to them) that dossier was kept in the Vatican Secret Archives rather than in the ordinary Inquisition archives. However, by examining the "Decrees," they found that many of the minutes contained information about Galileo's trial. So they copied relevant passages from thirty-two minutes, spanning the years 1611 to 1734, although most were from the crucial years 1616 and 1633. These documents remained unpublished for about twenty years, but eventually Gherardi published them with a critical commentary in 1870.

These documents were (and remain) invaluable because they constitute an independent and equally authoritative source of information about the trial; that is, independent of, and as authoritative as, the documents in the special file of trial proceedings. We have seen that this file was created in 1755, and we can now add that it was created by removing the appropriate documents from the relevant volume of "Trials." According to Inquisition practice, minutes of meetings were first written on loose sheets of paper during the meeting, and then they were copied into various volumes of "Decrees" or "Trials" or both.[33] Thus, many minutes of Inquisition meetings are found with identical wording in both the special Galilean file of trial proceedings and in Gherardi's set of minutes. A consequence of this coincidence was that the general accuracy and integrity of the file documents is reinforced. This was important because the vicissitudes of the special file (especially the Napoleonic removal and temporary loss) made the anticlericals suspect that the documents had been tampered with.

One of the best examples of the minutes copied and published by Gherardi, but also found in the special file, was the minutes of the all-important Inquisition meeting of 16 June 1633. The pope chaired that meeting and decided on the disposition of the case. It reads:

[282] *Thursday, 16 June 1633.*
The meeting of the Holy Office was held at the apostolic palace on the Quirinale hill, in the presence of His Holiness Urban VIII, pope by divine providence; of the most eminent and reverend Lord Cardinals Bentivoglio, di Cremona, Sant'Onofrio, Gessi, Verospi, Ginetti, [283] general inquisitors; and of the Rev. Lord Fathers the commissary general and the assessor of the Holy Office. Regarding the cases mentioned below, which the Lord Assessor presented in writing and conveyed to me (the notary), to wit . . . For the case of the Florentine Galileo Galilei, detained in this Holy Office and (because of old age and ill health) freed with the injunction not to leave the house where he has chosen to reside in Rome and to appear here whenever requested (on pain of some penalty to be determined by the Sacred Congregation); after the issues were presented, the proceedings were related, etc., and the various opinions were heard; His Holiness decided that the same Galileo is to be interrogated even with the threat of torture; and that if he holds up, after a vehement abjuration at a plenary meeting of the Holy Office, he is to be condemned to prison at the pleasure of the Sacred Congregation, and he is to be enjoined that in the future he must no longer treat in any way (in writing or orally) of the earth's motion or sun's stability, nor of the opposite, on pain of relapse; and that the book written by him and entitled *Dialogo di Galileo Galilei Linceo* is to be prohibited. Moreover, so that these things be known to all, he ordered that copies of the sentence containing the above be transmitted to all apostolic nuncios and to all inquisitors for heretical depravity, and especially to the inquisitor of Florence, who is to publicly read that sentence at a plenary meeting to which must be personally summoned most professors of the mathematical sciences.[34]

This document reignited the torture question, especially because its language is somewhat ambiguous. The ambiguity concerns the clause "Galileo is to be interrogated even with the threat of torture; and that if he holds up . . ." It is unclear whether this meant that "Galileo is to be interrogated, and tortured if need be, and if he endures or survives the torture, then . . ." or whether it meant merely that "Galileo is to be interrogated with the verbal threat of torture, and if he holds onto his previous confession denying any malicious intention, then . . ." This document was the one exploited by Marini to provide further evidence for the bad-theologian thesis.[35] And the document had also been often used by officials who wanted a brief summary of Galileo's condemnation without having to do too much research; for example, on the occasion of allowing Tuscany to build an honorific mausoleum for Galileo in 1734–1737 and the occasion of allowing Toaldo to publish an expurgated edition of the *Dialogue*.[36] At any rate, the key part of this document reappears in the special file, that is, the part detailing what the pope decided.[37]

However, due to bureaucratic sloppiness, the transcription of minutes from the loose sheets used at the actual meetings into the appropriate volumes did not happen with absolute regularity. Thus, some minutes were never recorded in the trial proceedings. A crucial example was the minutes of the meeting of 3 March 1616. This was an important meeting because Cardinal Bellarmine reported that he had done what he had been ordered to do at the previous Inquisition meeting presided over by the pope on 25 February. Gherardi had copied the March 3 minutes in 1849 and published the text along with the rest in 1870. The document reads as follows:

> The Most Illustrious Lord Cardinal Bellarmine having given the report that the mathematician Galileo Galilei had acquiesced when warned of the order by the Holy Congregation to abandon the opinion which he held till then, to the effect that the sun stands still at the center of the spheres but that the earth is in motion, and the Decree of the Congregation of the Index having been presented, in which were prohibited and suspended, respectively, the writings of Nicolaus Copernicus *On the Revolution of the Heavenly Spheres,* of Diego de Zúñiga *On Job,* and of the Carmelite Father Paolo Antonio Foscarini, His Holiness ordered that the edict of this suspension and prohibition, respectively, be published by the Master of the Sacred Palace.[38]

In his commentary,[39] Gherardi pointed out that Bellarmine's report provides further evidence against the special-injunction transcript. For the cardinal was reporting that he had warned Galileo to abandon his geokinetic opinion, that there had been no complications, and that Galileo had agreed. Bellarmine's warning was in accordance with the decision at the Inquisition meeting of 25 February 1616. That decision stipulated that Bellarmine should first warn Galileo to stop *defending* Copernicanism; that *if* Galileo refused, the Inquisition commissary should issue him a formal injunction to stop *discussing* the topic; and that if he did not acquiesce at this injunction, he should be imprisoned. Bellarmine's report stated that Galileo accepted the warning and said nothing of any refusal that would have triggered the stricter formal injunction. The report implied that, in accordance with the papal orders of February 25, there had been no need for a special injunction.

In short, the special injunction contradicted not only the minutes of 25 February 1616, Bellarmine's certificate, and Galileo's first deposition (12 April 1633), as Wohlwill had argued, but also the minutes of 3 March 1616, as suggested by Gherardi. Thus the legal validity or admissibility of the special injunction was further undermined, and correspondingly the legal-impropriety thesis was reinforced.

Also like Wohlwill, Gherardi was inclined to believe that the special-injunction transcript was a forgery, and was aware that he had no direct evidence. He only had the indirect argument that a forgery would explain the existence of the transcript document describing the special injunction, and that otherwise there was no way of explaining how it came about.

12.3 Plea Bargaining out of Court: Commissary Maculano's 1633 Letter Published (1875)

Although the new conclusions based on the new documents were accepted by many, they were rejected by others, including some who were well acquainted with these documents. We have already seen that L'Epinois, who was the first to publish an edition of the Vatican file, did not really appreciate the significance of its content. Nor did he change his mind much when he published his own comprehensive account a decade after his first edition.[40] Similar remarks apply to the other document editor, Berti, who explicitly argued against the new conclusions.[41] On the torture question, Berti argued plausibly that on the one hand torture was ordered by the papal decision of 16 June 1633 and was implied by the language of the sentence, but that on the other hand the rest of the proceedings (especially the deposition of 21 June 1633) implied that Galileo was merely threatened with torture, and that the Inquisition's commissary (Vincenzo Maculano da Firenzuola) had properly used his discretion and decided that Galileo was too old and ill to undergo actual torture.[42]

As regards the forgery thesis and the legal-impropriety thesis, Berti criticized them by criticizing their main basis, the contradiction between the special-injunction transcript and the other documents. Berti claimed that there was not much difference between the content of Bellarmine's informal warning and the content of the formal special injunction; and that the special injunction did not really contradict the papal orders of 25 February 1616, Bellarmine's Inquisition report of 3 March 1616, or Bellarmine's certificate to Galileo. But I do not see that Berti had any real argument or any plausible reason for his claim, and his discussion strikes me as exemplifying a failure to see the obvious.

Berti also tried to undermine the legal-impropriety thesis by arguing that Galileo could and would have been convicted of disobedience even without the special injunction, merely for violating the warning not to hold or defend the earth's motion; for he had never denied receiving a warning so phrased, and beginning with the second deposition (30 April 1633) he even admitted this violation (while insisting that it had been unintentional). Here, Berti was right that in 1632–1633 Galileo was open to the charge of having violated the warning not to hold or defend Copernicanism. However, whether he could and would have been convicted of this charge (without the special injunction) is questionable, as some of Berti's critics soon pointed out. Such criticism was partly based on a new document that was discovered at the time and first published by Sante Pieralisi in 1875.[43]

The new document was a letter dated 28 April 1633 by the Inquisition's commissary Maculano da Firenzuola to Cardinal Francesco Barberini, the Vatican secretary of state and a member of the Inquisition. This letter was written between Galileo's first deposition (April 12) and the second (April 30). We already know that at the first deposition he essentially denied receiving the special injunction (not to discuss Copernicanism in any way whatever, which the *Dialogue* would have necessarily and automatically violated), and that he admitted receiving but denied violating the warning not to hold or defend it. And we know from the second deposition, published by Venturi and Delambre in 1821 as part of the Napoleonic French translation of the trial proceedings, that on 30 April 1633 Galileo "confessed" having violated the warning not to defend Copernicanism, albeit unintentionally. One could not help wondering why Galileo changed his mind and conjecturing that it must have been in response to some kind of pressure from the Inquisition.

Maculano's letter to Barberini provided the evidence for this conjecture. The letter was not an official document, and so it was not in the Vatican file of Inquisition proceedings. It was found by Pieralisi in the Barberini family's archives in Rome. In this letter, Maculano conveyed the latest news to Francesco Barberini, who had not attended the last Inquisition meeting. Maculano said that, in light of Galileo's denials in the first deposition, he had obtained permission from the cardinal-inquisitors to engage in out-of-court negotiations with the defendant. During this meeting Maculano was able to convince Galileo to confess to some wrongdoing, and Galileo was now (April 28) writing a confession to present in court at the next deposition.

Partly exploiting the information contained in this letter, in 1878 J. A. Scartazzini articulated a position that took into account the other recent documents and interpretations. He tried to develop his position as being intermediate between two extremes.[44] On the one hand, there was the thesis (stemming from Wohlwill) that the condemnation of Galileo embodied a legal impropriety because it was based on the special injunction, and the document recording this injunction contradicted the other documents made available by L'Epinois and Gherardi. On the other hand, there was the view (stemming from Berti) that the special-injunction document was essentially consistent with the others, that they all proved that Galileo had indeed received an injunction not to defend Copernicanism, that the *Dialogue* was a defense of Copernicanism, and so he was guilty of disobeying this injunction, and therefore the condemnation was not improper in that sense. Scartazzini argued that the special injunction was crucial, at first, to get Galileo indicted and summoned to trial and, later, to get the trial started and the first interrogation defined; but after Galileo's denials in the first deposition and his introduction of Bellarmine's certificate, he was forced to confess having transgressed the prohibition to hold or defend Copernicanism. In other words, the special injunction transcript was important in the *dynamics* of the 1633 trial: Galileo was not convicted just on account of the special injunction; but he could not have been convicted (in the way he was actually convicted) without it; the special injunction was instrumental in extracting (or extorting) from him the confession of 30 April 1633, which paved the way for the conclusion of the trial. This account was a plausible and reasonable one.

12.4 Tampering with the Evidence:
Scartazzini on Paper Shuffling (1877–1878)

Scartazzini also articulated a more extreme position about the authenticity and completeness of the documents in the Vatican file.[45] He questioned not only the integrity of the special-injunction transcript, as Wohlwill had done, but also the integrity of several other documents, especially the fourth deposition (21 June 1633) describing Galileo's "rigorous examination." His arguments on these topics are noteworthy more for the quality of evidence and kinds of considerations used than for solidity of the conclusions they reach. They are based on considerations about where the various texts are found in the physical organization of the Vatican file. The points are subtle but intriguing.

The text of the fourth deposition[46] is the presumed record of an interrogation of Galileo under the verbal threat of torture. Galileo was repeatedly asked whether he held the heliocentric and geokinetic view; at one point he was threatened "that unless he decided to proffer the truth, one would have recourse to the remedies of the law and to appropriate steps against him";[47] later "he was told to tell the truth, otherwise one would have recourse to torture."[48] The defendant repeatedly denied holding this view;

he did have some inclination toward it before the condemnation of 1616, but not since; he wrote the *Dialogue* to discuss the evidence on both sides and to show that all arguments were inconclusive; "for the rest, here I am in your hands; do as you please."[49]

On the basis of this and other documents, in 1877 Wohlwill had raised anew the question of whether Galileo was tortured.[50] Of course, the content of this document seemed to suggest that although Galileo had been threatened with torture, he had not been actually tortured. However, Wohlwill argued that the document exhibited various irregularities: for example, there was no statement at the end of the interrogation, just before the signature, that the defendant had been sworn to secrecy, as was the norm and as may be seen in Galileo's first deposition of 12 April 1633. Moreover, some others of the newly published documents could be interpreted as suggesting torture; such was the case for the Inquisition minutes of June 16, also quoted above. More important, Wohlwill argued that the notion of "actual torture" was itself ambiguous because the label could refer either to the process of being taken to the torture chamber and being shown the instruments of torture or to the process of being put on the rack and being tormented; and he claimed that Galileo underwent at least the former.

However, critics objected[51] that Wohlwill had been insufficiently subtle: for the notion of "torture," as we have seen, included at least five stages or gradations: being verbally threatened with bodily torture; being taken to the torture chamber and being merely shown the instruments; being undressed, as if one were going to be tied to the instrument, without being actually tied; being tied to the instrument of torture, without torture being applied; and being tied to the instrument and having torture applied. Thus, whatever references to torture the documents contained, the process in question could be simply the first stage (verbal threat), which everyone acknowledged that Galileo had indeed undergone; one was not justified in claiming that it was the second stage (being shown the instruments), any more than Wohlwill himself felt justified in claiming that Galileo had undergone the third, fourth, or fifth stage. Wohlwill's book had certainly raised the discussion to a new level of subtlety and sophistication.

Scartazzini was inclined to agree with Wohlwill, but did not have much to add to his type of argument, whereas he had something original to contribute with regard to another category of evidence.[52] Scartazzini questioned the authenticity of the document of 21 June 1633 based on its physical location in the file. Where is it located? It is located on folios 452r, 452v, and 453r.[53] The problem is that folios 452 and 453 are (respectively) twins of folios 414 and 413, meaning that folios 452 and 414 are two parts of the same folded sheet, and so are folios 453 and 413; indeed, all these four folios and the two corresponding folded sheets belong to the same bundle of folded sheets comprising also folios 415–422, for a total of six folded sheets or 12 folios in this bundle; and on folios 413 and 414 (up to folio 419r) is written Galileo's first deposition (April 12), and in the rest of the bundle (folios 419r–422r) are the second and third depositions. To understand why this is a problem a very roundabout explanation is needed.

The Galilean file of trial proceedings consists not of loose leaves of paper collected or bound together, but rather of bundles of folded sheets of paper and single folded sheets; the number of sheets in each bundle varies but never exceeds about ten; and the folded sheets in a particular bundle are sometimes sewn together and sometimes not. In the terminology used here, a *folded sheet* generates two *folios*, each of which is the *twin* of the other; a folio yields two *pages* corresponding to its recto and verso sides; and several sheets folded together in half generate a *bundle*. In the seventeenth century, when a short document had to be written, one took a single folded sheet, started writing on the recto page of the first folio, continuing if need be on the verso page of the same folio, and then on the recto and verso sides of the second folio. For longer documents, one took several sheets folded in the middle and filled out the recto and verso pages of the various folios in sequence. As a rule, the text of a document took only part of the available space, and so whole pages and whole folios were often left blank at the end of a bundle. Such blank pages would be used only when writing up a new document on a closely related topic in close chronological proximity. When enough time had elapsed since a previous document, or a new topic or new type of document was being drawn, then one took a new bundle or a new folded sheet and started writing there. In any case, for documents (such as letters) of different origin, one received them already written on a folded sheet or a bundle, and then one added it to the file. In accordance with Inquisition practice, the order in which the various papers were organized in a file was largely chronological, but with some exceptions. The general chronological order was a natural one, since as new documents were received or generated they were simply added at the bottom of the collection. One deviation from the strict chronological order derived from the distinction between the order of receipt and date of composition; for example, Bellarmine's certificate, dated 26 May 1616, is found with the 1633 documents for the simple reason that it was presented by Galileo during the 1633 proceedings.

Such practices implied that one would normally not file the documents in a dossier by placing one bundle inside another, for the insertion would be likely to break the continuity of the text of the latter. If such insertion ever occurred, it was for clear and obvious reasons. In the whole file of Galileo's trial, there are only two exceptions to such norms, and they involve the special-injunction transcript of 26 February 1616 and the rigorous examination of 21 June 1633. The special injunction is written on folios 378v and 379r, and folio 379 is the twin of folio 357, on which is written Caccini's deposition of 20 March 1615; thus, the folded sheet consisting of folios 357 and 379 contains several bundles recording the proceedings between March 1615 and 26 February 1616.[54] And the June 21 deposition (as implied by the facts and figures mentioned earlier) is recorded on folios (452–453) that are same-bundle twins of folios 413–414 (on which is recorded part of the April 12 deposition); thus, between the two dates and the two depositions are eight other bundles containing all the documents and proceedings between the two dates.[55] Moreover, each of the problematic documents (the 1616 special injunction and the 1633 rigorous examination) is filed next to a bundle from which a folio has been cut out: immediately preceding the special injunction (folios 378v–379r) there is missing the twin of folio 376, which together with the folded sheet of folios 377 and 378 constitutes what is now an odd bundle of only three folios (or one and one-half folded sheets); and immediately following the June 21 deposition (folios 452r–453r) there is missing (having been cut out) the twin of folio 455, which together with the folded sheet of folio 453-duplicate and 454, makes up another odd bundle of three folios. In this case the irregularities are increased by the fact that there are two folios numbered 453.

On the basis of such considerations, consisting of the identification of sheets, twin folios, and bundles, and of the imaginary shuffling and reshuffling of the folios, sheets, and bundles making up the Galilean file, Scartazzini claimed that both documents as they exist today had been written (some time after the fact) on pages that were originally blank because they were the last pages of bundles that contained other documents whose text did not use up the entire bundle; at the same time, some adjoining folios (probably containing the original genuine wording of the corresponding documents) were cut out; and then those folios on which the forged text had been written were placed in the proper chronological sequence (next to later documents) by taking the appropriate number of other bundles and inserting them within the bundles being tampered with.

Taking these arguments, involving such paper shuffling, together with the arguments of Wohlwill, involving what might be called content analysis, Scartazzini felt sure that both documents were forgeries. He had apparently not been allowed to consult the actual manuscripts, and so he constructed such speculations on the basis of the information provided by Gebler in his "diplomatic" (exact replica) edition of the documents.[56] However, as Scartazzini suggested, his arguments can be made more comprehensible if one constructs for oneself a kind of replica of the file out of folded sheets of paper bundled as described by Gebler; in fact, making and using such physical models may be in this case the only way of making sense of what Scartazzini was trying to say.

12.5 Inaccurate but Not Forged Documents: Gebler's Balanced Synthesis (1879)

During the period under discussion, the publication, interpretation, and evaluation of the Galilean trial proceedings reached a climax with the work of Karl von Gebler. The son of an Austrian field marshal, he also became a soldier at first but soon left the army because of ill health. He studied Galileo for four years before publishing his first book in 1876. This gave him the opportunity to gain access to the Vatican file, which he consulted in mid-1877 for ten weeks, working fourteen hours a day; the result was a complete and diplomatically precise edition of the file, published in the same year. This archival and editorial work, together with new secondary literature, also led him to reassess parts of his earlier account. This reassessment (together with an update) is incorporated in the 1879 Italian and English translations of his 1876 German book. Born in 1850, Gebler died prematurely in 1878, just as the translations of his revised work were going to press.[57]

With the benefit of the earlier editions by L'Epinois and Berti and with his access to the Vatican archives, Gebler (1877) produced a superior edition of the Vatican file. Not only did he have unsurpassed knowledge of the relevant primary sources and the ability to appreciate the novel implications of the new documents, but he did so without going to the extremes to which Wohlwill and Scartazzini were led and with a willingness to change his mind when necessary. Thus, Gebler's revised account (1879a,b) is useful even today and could provide an excellent starting point. An appendix to his 1879 work contains a good summary of his account and a revealing indication of how that work is the climax of the period which may be called the golden age of the Galileo affair.

In that appendix,[58] Gebler began by stressing the important point that the Vatican file of trial documents was demonstrably incomplete. To the examples of missing documents he mentioned, one could add the most striking case of all, namely the sentence and abjuration; but this lacuna presented no problem, since copies of these documents had long been readily available as a result of the Church's unique effort to publicize Galileo's condemnation immediately after its occurrence.[59] Gebler was also right to stress that many documents in the file were standard copies of originals held elsewhere, while many others were simply clerical notes by Inquisition officials who wanted to keep track of relevant developments.

As regards Wohlwill's questioning the authenticity of the Inquisition's executive summary of the trial (composed in May–June 1633), Gebler correctly pointed out that such summaries were standard Inquisitorial practice. Moreover, he claimed that Wohlwill's suspicions about the integrity of various documents were the result of his lack of access to the Vatican manuscripts and his reliance on inadequate editions. This was a rhetorically devastating criticism but not entirely fair. For Scartazzini had relied on Gebler's own perfect edition and yet had managed to strengthen the various forgery theses; and as we shall see presently, Gebler's own answer to Wohlwill's conjecture about the falsification of the special injunction did not completely reject, but rather qualified, that conjecture.

In fact, after Gebler made his own personal examination of the Vatican file, what he found untenable was that the special-injunction document was a *later* falsification, but he remained more convinced than ever that it was "a downright untruth, exaggeration, or misinterpretation."[60] In short, it was not forged in 1632 in order to "get" Galileo but had been planted in 1616 as insurance. One of Gebler's arguments[61] was that the handwriting on the special-injunction document was the same as that of other 1616 (annotation) documents and different from the 1632–1633 documents; this was a pretty decisive reason.

Another one of Gebler's arguments[62] was based on the folio-bundle structure and on watermarks: (1) the special-injunction document is on folios 378v and 379r, which are respectively twins of folios (377 and 357) on which are written the consultants' report of 24 February 1616 and part of Caccini's deposition of March 1615; and (2) the watermarks on the folios of the special-injunction document are the same as those on the 1615–1616 documents and different from those on the 1632–1633 papers. But for these facts to prove what Gebler wanted them to prove, he had to also exclude the possibility of the kind of paper shuffling imagined by Scartazzini, which would have enabled the forger to use parts of existing and authentic documents to write inauthentic texts and then reshuffle those parts into inauthentic but apparently proper locations. This possibility could be excluded because of another feature of the documents which Scartazzini neglected or was not cognizant of: the 1615–1616 proceedings happen to bear a second set of folio numbers (949–992), which are the numbers assigned to them when they were originally produced and were part of a regular miscellaneous volume of Inquisition trial proceedings; this happened before the proceedings of 1632–1633 led the Inquisition to retrieve the earlier documents and combine them with the new, accumulating ones, of which more presently. Thus, the special-injunction document is written on folios that bear not only the numbers 378–379 mentioned previously, but also the older numbers 987–988, which are in proper sequence with the rest of the older numbers; a post-1616 forgery, using Scartazzini's otherwise powerful method, would have had to erase the older numbers and renumber the affected folios; and there is no trace of such alterations. Gebler was correct to conclude that "his theory therefore belongs to the realm of impossibilities."[63]

Gebler did not discuss Scartazzini's other forgery claim about the June 21 rigorous examination. But it could be refuted in a similar way. For the

1632–1633 proceedings also possess a second set of folio numbers, which were added by the official who wrote the May–June 1633 summary and who needed to make specific references to the documents in the course of his exposition. Thus, he numbered the folios of all the documents sequentially from 1 (the first page of Lorini's letter of complaint) to 103 (the last folio of the third consultant report on the *Dialogue*). Of course, the June 21 deposition was not numbered in this way since it was subsequent to the date of the report. However, according to Scartazzini, the June 21 rigorous examination as it exists today was written (in the nineteenth century) on folios which were originally blank; which belonged to the bundle of six folded sheets on which were recorded Galileo's first three depositions of April–May 1633; and which were twins of the first two folios of Galileo's first deposition (April 12). If that were the case, the two forged folios would bear numbers corresponding to the sequence created by the writer of the summary; and not only do they not, but there is no trace of erasure and renumbering. To be more specific, the forger decried by Scartazzini would have had to use folios numbered 79–80 (so numbered by the writer of the summary) because the bundle used for the forgery had six folded sheets with twelve folios, the first ten of which were numbered folios 69–78; but whereas the latter ten folios do display these numbers (along with numbers 413–422 in the standard numbering added later and available in Scartazzini's time and used by Favaro), the last two folios of this bundle (now displaying standard numbers 452–453) show no trace of numbers 79–80.

However, Gebler's refutation of the thesis of a *later* forgery did not affect his reaffirmation that the special-injunction transcript contained a factual misrepresentation of what happened on 26 February 1616 and was a legally worthless document. As we have seen, the reasons for thinking that it was factually inaccurate were that it deviated from the papal orders given at the Inquisition meeting of 25 February 1616; it contradicted Bellarmine's report at the Inquisition meeting of 3 March 1616; it contradicted Bellarmine's certificate written for Galileo on 26 May 1616; and, moreover, Galileo never showed or betrayed the slightest worry that he was under the obligation not to even discuss Copernicanism; instead he essentially denied it in his first deposition (12 April 1633). To these Gebler added a clinching proof: if the special injunction had really been issued and the annotation transcript were accurate, then there would have to be an original document with signatures, including Galileo's, because that was standard Inquisition practice, as shown by the case of the injunction dated 1 October 1633 summoning Galileo to go Rome to stand trial, whereas the available transcript is a mere clerical annotation in the file; if such an original document existed, then one should be able to find it in the Inquisition archives; but a search requested by Gebler and ordered by the Vatican secretary of state in 1877 yielded negative results. Moreover, if the original had ever existed, it would have been unearthed in 1633 after Galileo denied it at the first deposition;

but instead the issue was apparently dropped, and it resurfaced only by means of allegations in the final sentence. Besides proving the factual inaccuracy of the special injunction, such evidence also proved that the document was legally invalid and so inadmissible at the trial. Since that document was instrumental in convicting Galileo, his condemnation was a legal impropriety.

Gebler's account represented a convincing confirmation and important refinement of Wohlwill's radically revisionist thesis of legal impropriety, but also an elegant refutation of Wohlwill's and Scartazzini's forgery theses. It also amounted to a judicious assimilation of the trial proceedings finally made accessible by the Vatican and published in the previous decade by L'Epinois, Berti, and Gebler himself, as well as the independent documents published by Gherardi and Pieralisi. It turned out that despite the puzzling features of the Vatican file, its obvious incompleteness, and its other limitations, by an unplanned and almost miraculous set of coincidences it happened to provide within itself internal evidence of its own authenticity. Moreover, Gherardi's Inquisition minutes provided an external and independent check. Finally, I believe we may also speak of a third check on the authenticity of the file of proceedings, namely the incriminating evidence it contains; that is, the fact that it contains almost conclusive evidence that Galileo's condemnation embodied a judicial impropriety. If post-1633 apologists had really wanted to tamper with the evidence, they would have removed the evidence that conflicts with the special injunction, which was highly irregular within the Inquisition's own framework; they would not have removed the alleged evidence of actual torture (the presumed deposition of Galileo on the rack), which was standard procedure not only by the Inquisition's own rules but also by those of criminal justice systems throughout Europe and the whole world.

Chapter 13

Galileo Right Again, Wrong Again

Hermeneutics, Epistemology, "Heresy" (1866–1928)

During the last third of the nineteenth century, the most important developments in the Galileo affair were those discussed in chapter 12: the publication of the Vatican file of trial proceedings and other directly related documents; the interpretation and evaluation of these documents; and the discussion and refinement of various revisionist accounts that undermine the propriety of the Inquisition's having condemned Galileo for disobeying the special injunction. Overlapping with these developments there were others that deserve comment for a variety of other reasons.[1] Moreover, after a brief pause, around the turn of the century new and important developments once again entered the scene: among them were the theological rehabilitation of Galileo implicit in Pope Leo XIII's encyclical *Providentissimus Deus* (1893); the epistemological reconviction of the culprit in Duhem's *To Save the Phenomena* (1908); the collection and refinement of three centuries of anti-Galilean charges in Müller's compendium (1909); and a new justification of Galileo's condemnation based on Garzend's account of the concept of heresy (1912).

13.1 Cultural Penetration and Consolidation (1866–1928)

In 1866, William Ward published in London a book titled *The Authority of Doctrinal Decisions Which Are Not Definitions of Faith, Considered in a Short Series of Essays Reprinted from the Dublin Review.* Ward (1812–1882) was an Englishman who converted to Catholicism in 1845 and the editor of the *Dublin Review* from 1863.[2] In this book he argued that the condemnations of Copernicanism and Galileo were not papal decisions ex cathedra but rather decisions of papal congregations of cardinals; that congregational decisions did not have the character of infallibility, but nevertheless ought to be obeyed externally and internally by Catholics; and that those condemnations were not erroneous in any proper sense of the term because at that time the hypothesis of the earth's motion was less probable than the geostatic view.

Ward's arguments started a discussion that was soon joined by William Roberts, who in 1870 published in London a book titled *The Pontifical Decrees against the Motion of the Earth, Considered in their Bearing on Advanced Ultramontanism*. Roberts criticized Ward's theses and supported the view that the condemnations of Copernicanism and Galileo were papal decisions ex cathedra; that they were erroneous; that they thus showed that papal decisions ex cathedra are not infallible; and that they did not merit obedience by Catholics. Ward did not acquiesce, and so the following year from the pages of the *Dublin Review* he replied to Roberts.[3] The latter was not convinced, but he took no action until 1885, when he published another book containing an update of his earlier essay and a rebuttal of Ward's rebuttal.[4]

A few years after that, in *Galileo and His Judges*, F. R. Wegg-Prosser tried to steer a middle course between Ward and Roberts. On the one hand, he argued, the anti-Copernican decree and Galileo's sentence were not doctrinal and infallible, but disciplinary and fallible decrees. On the other hand, the congregations of the Index and Inquisition were wrong in claiming that the earth's motion was contrary to the Bible and scientifically improbable.[5]

By that time Ward was dead, but many others had since joined the discussion of infallibility, spurred in part by the Vatican Council of 1870, which (in session 3, chapter 4) had proclaimed the doctrine of papal infallibility.[6] In the process, they connected that particular issue to others. In the German-speaking countries of central Europe, the issue of infallibility became entangled with Wohlwill's forgery and torture theses, with the Jesuit conspiracy theory, and with the question of the autonomy of German Catholicism vis-à-vis central papal authority. Hartmann Grisar, an Austrian Jesuit who had had to leave Italy because of Italian state persecution of the Jesuits, defended the condemnation of Galileo, Roman authority, and papal infallibility; whereas Franz Reusch, a German priest who was excommunicated in 1872 because of his association with the nationalist German Catholic Church, defended Galileo and German autonomy and criticized papal infallibility.[7]

In Britain, St. George Jackson Mivart, a convert to Catholicism who was later excommunicated, tried to steer a middle course between uncritical acceptance and dogmatic rejection of Darwinism and argued for the compatibility of Christianity and evolution. But an Irish priest named Murphy objected to this view, claiming that the special creation of the human species was an article of faith, and so evolution was heretical and Mivart should retract his view. Mivart (1885) replied by arguing that Galileo's trial showed that the Church was fallible in scientific matters, and so modern

Catholics had complete freedom in scientific inquiry; but he argued that the Church's error on Copernicanism was a providential one, and so took his conclusion to be a positive lesson rather than a criticism of the Church.

In Spain, the general cultural relevance of Galileo's trial was illustrated in 1875 when a play titled *Galileo* by D. Eleuterio Llofriu y Sagrera was performed with great success at the Teatro Martin in Madrid.

In Italy, on 21 April 1887 a marble column commemorating Galileo was inaugurated in Rome, near Villa Medici. A similar monument had been proposed on 28 August 1872 in the form of a commemorative plaque to be affixed on the facade of the building; but such a plaque had been opposed by the French government, which owned the palace. The inscription, which had been written by Domenico Gnoli and approved in principle in 1872, was not installed until July 1887 and read: "The palace next to this spot, / which belonged formerly to the Medici, / was a prison for Galileo Galilei, / guilty of having seen / the earth turn around the sun. / SPQR / MDCCCLXXXVII."

The event was applauded by the anticlerical press and sharply criticized by the official Vatican newspaper, *L'Osservatore Romano.*[8] On 23 April 1887, in an unsigned article titled "Epigrafi ed offese" (i.e., "Epigraphs and Insults"), *L'Osservatore Romano* objected to the Villa Medici inscription by arguing, first, that Galileo was not "imprisoned" in Villa Medici but comfortably hosted there by the Tuscan ambassador. Furthermore, he was not found guilty of having "seen the earth turn around the sun" because this proposition was never formally declared a heresy by any pope. "What, then, was Galilei's fault . . . ? If you do not believe us, you should believe Sarpi himself, the impartial Balbo, and Guicciardini, who recognize that the great Tuscan philosopher had the fault and the recklessness of wanting to change the physical and astronomical question into a theological one. You should believe Galileo himself, who, in the last years of his life, regretted having engaged in arbitrary interpretations of the Bible based on private judgment, which were especially dangerous at that time when this was the practice of the heretics in many parts of Europe."[9] The last reference is unclear, but it probably referred to the Renieri apocryphal letter; on the other hand, the reference to Guicciardini was clearly a reference to his letter of 4 March 1616 (see chapter 8.2). Since I have previously mentioned that, although critical of Galileo, this letter does not attribute to him an attempt to base Copernicanism on the Bible, the article was probably relying on those accounts that misquoted the letter by fabricating a crucial sentence. And indeed in the previous paragraph, the unsigned article mentioned several authors, among whom were Marini and Mallet du Pan.[10]

To this period also belongs the conception and execution of the National Edition of Galileo's complete works. On 17 May 1883, the possibility of a new critical edition of Galileo's works began to be discussed among a group of Florentine linguistic purists associated with the Accademia della Crusca, chiefly Isidoro del Lungo and Cesare Guasti.[11] On 20 February 1887, a royal decree approved the project of publishing, at state expense, such an edition by a team of scholars under the general editorial supervision of Antonio Favaro.[12] The volumes began to appear in 1890 and continued at the rate of one a year, so that the twentieth and last volume came out in 1909. It was a monumental achievement and helped to set a new standard; in fact, it was part of a remarkable Europe-wide cultural movement that saw similar editions produced or started for the works of such figures as Kepler, Tycho Brahe, Huygens, and Descartes.

The trial documents were placed in volume 19, which appeared in 1908. But given their importance, Favaro printed them separately on an advance basis. For example, in 1902 a special edition of thirty copies was issued of the trial documents to be published in volume 19 of the National Edition; this was done for the convenience of specialists, given that volume 12 of that edition had just been published, containing the correspondence relevant to the first phase of the trial in 1615–1616.[13] And in 1907, when everything but the printing of volume 19 was completed, Favaro issued separately a book[14] with all the trial documents that were being included in the volume in press.

Favaro had been given complete access to the relevant Vatican secret archives, as well as the even more inaccessible archives of the Inquisition.[15] However, his access to the ecclesiastical archives in Florence encountered difficulties.[16] As early as 1882, he was shown the disordered state of the archives of the Florentine archdiocese, which had inherited the Florentine Inquisition's archives; at that time he was told that because of this situation he could not have permission to do research there. Later this permission was flatly denied. Still later, when he would have been granted permission, the Inquisition's archives had disappeared. But they had not been lost, and in fact in 1908 another scholar published a collection of Galilean documents from the archives of the Florentine Inquisition.[17]

The final development in this period subsumable under the present theme occurred between 1926 and 1928. With the cooperation of Vatican authorities, Rudolf Lämmel performed physical tests on the manuscripts contained in the Vatican file of trial proceedings.[18] The tests involved examining the manuscripts with X rays and ultraviolet light. The result was that there were no signs that previous writing had been erased and new words written over in any of the problematic documents, such as the special-injunction transcript of 26 February 1616 and the rigorous-examination deposition of 21 June 1633. This further confirmed that the file did not contain falsified documents of the kind conjectured by Wohlwill's and Scartazzini's forgery theses, as Gebler had already convincingly shown by more traditional arguments. At the same time, such tests did not affect in any way the legal-impropriety thesis and the claim that at least one of those documents (the special-injunction transcript) is false in the sense of being an inaccurate description of what happened.

13.2 Galileo Theologically Right:
Leo XIII's Encyclical *Providentissimus Deus* (1893)

One of the most novel developments in the Galileo affair at the end of the nineteenth century involved a Church action that was *not* taken directly with regard to Galileo but for other reasons, and yet it had consequences that affected a key issue of the Galileo affair. We saw in chapter 1.1 that the "suspected heresy" for which Galileo was condemned was twofold: one aspect was the physical proposition that the earth does not stand still at the center of the universe but revolves annually around the sun and daily on its own axis; the other was the methodological, theological, and hermeneutical principle that Scripture is not a philosophical or astronomical authority but an authority only on questions of faith and morals. And we have seen that with regard to the physical proposition, Galileo was proved right, and the Church gradually, if slowly and grudgingly, admitted its error with such actions as the retraction of the general prohibition of Copernican books in 1757, as well as the approval of Settele's textbook in 1820, the approval of Copernican books in general in 1822, and the silent withdrawal of the prohibition of Copernicus's and Galileo's books in 1835.

However, with regard to the hermeneutical principle, because of the nature of the topic, developments were not as discrete, clear-cut, or easy to identify and assess. Thus, I have not been following with equal attention the history of the problem of the nature of scriptural interpretation and its relationship to scientific or philosophical inquiry. Nevertheless a few points have emerged.[19] We have seen (in chapter 4.3) that a conservative reaction to Galileo's condemnation was the reaffirmation of biblical literalism, a leading example being Riccioli. We have also seen (in chapter 5.2) that some significant implications were derivable from the recurring discussions of Bellarmine's assertion, popularized by Fabri, to the effect that if a demonstration of the earth's motion were ever found, then the Church would revise its literal interpretation of biblical passages such as Joshua 10:12–13. And we have seen (in chapter 8.3) that a curious thing happened on the way to the consolidation of the novel principle that Scripture is not a scientific authority: Galileo was accused of having *violated* it, of having engaged in the practice of supporting astronomical propositions with biblical passages; so it was alleged that he was condemned for this reason, for being a bad theologian and not for being a good astronomer; this was Mallet du Pan's bad-theologian myth; while it illustrates the strange forms which the anti-Galilean animus can take and has taken, it also testifies to the fact that the denial of the scientific authority of Scripture was becoming an increasingly well-established principle.

A sort of climax of the hermeneutical aspect of the Galileo affair occurred in 1893 with Pope Leo XIII's encyclical letter *Providentissimus Deus,* for this document put forth a view of the relationship between biblical interpretation and scientific investigation that corresponded to the one advanced by Galileo in his letters to Castelli and Christina. To be sure, the encyclical did not even mention Galileo but rather was written in response to the controversy known as "the biblical question," which involved such issues as the nature, methods, and implications of the scientific study of the Bible and the validity of rationalist criticism of Scripture.

Pope Leo XIII ruled the Church from 1878 to 1903. This was a formative period for the modern papacy, which in 1870 had lost all temporal power when the just-unified Italian nation and the newly formed Kingdom of Italy had conquered the city of Rome. Leo's election soon led to a rapprochement between the papacy and Germany, which in 1873 had passed a series of laws abolishing papal jurisdiction over the Catholic Church in Germany, diluting the power of bishops, and giving the state bureaucratic control over priests; these laws were repealed in the 1880s.[20] In 1879, Leo took the unusual but significant step of declaring Thomism the official theology and philosophy of the Catholic Church.[21] In 1881, he opened the so-called Vatican Secret Archives to scholars for purposes of research, although these archives did not include those of the Inquisition; and he established an astronomical observatory, called the Specola Vaticana.[22] And it was during Leo's papacy that (on 10 February 1902) Favaro was granted by the director of the Vatican Library (Father Franz Ehrle) unprecedented access to all documents relevant to Galileo's trial found in the Vatican Secret Archives and in the Archives of the Roman Inquisition, so that they could be published as part of the National Edition of Galileo's works.[23]

On the other hand, when in 1886–1887 a detailed summary of Giordano Bruno's trial was discovered in the Vatican Secret Archives, Leo ordered that the discovery not be divulged and the document not be given to anyone;[24] the document had to be rediscovered in 1940 and was published in 1942.[25] When a statue to Bruno was being proposed, with the support of the Italian government of Prime Minister Francesco Crispi and an international committee of scholars, Leo strenuously opposed it; of course, eventually (on 9 June 1889) the statue was inaugurated in Rome at the Campo dei Fiori, the square where Bruno was burned at the stake in 1600.[26] And the Church's response was antiliberal when the American priest, scientist, and Notre Dame professor John Zahm (1851–1921) published *Evolution and Dogma* (1896), arguing that evolutionary theory was compatible with Scripture, with patristic and scholastic theology, and with the concept of God's creation (through the operation of natural laws); the Congregation of the Index (in 1898–1899) prohibited further publication and circulation of the book, and Zahm was forced to recant, although the Index decree was never formally published.[27]

This mixed record in Leo's papacy was reflected by the relatively ambivalent content of the encyclical *Providentissimus Deus*. Most of the encyclical stressed conservative views. For example, it asserted that the science of biblical studies needs the guidance of the Church[28] and needs to set aside "the arrogance of 'earthly' science."[29] It reiterated that Scripture was divinely inspired and that this divine inspiration meant it can contain no errors.[30] It clarified that the authority of Scripture on historical questions cannot be limited as it is in physical science.[31] And it warned that one must carefully observe "the rule so wisely laid down by St. Augustine—not to depart from the literal and obvious sense, except only where reason makes it untenable or necessity requires; a rule to which it is the more necessary to adhere strictly in these times, when the thirst for novelty and unrestrained freedom of thought make the danger of error most real and proximate."[32]

However, on the relationship between Scripture and physical science, the encyclical could be seen to advance Galilean views.[33] Just before discussing this topic, Leo discussed the problem of how to defend Scripture from the so-called "'higher criticism,' which pretends to judge of the origin, integrity and authority of each Book from internal indications alone."[34] He went on to formulate the problem of what might be called the "scientific criticism" of Scripture. This problem was the reverse of what Galileo had to deal with: he was trying to defend astronomical theory from objections based on scriptural assertions, whereas Leo was discussing how to defend Scripture from attempts "to vilify its contents"[35] based on physical science. However, their respective answers hinged on essentially the same point, the denial of the scientific authority of Scripture: in Galileo's case, one could see that biblical criticism of astronomy was an abuse and a misconception once one understood (with Cardinal Baronio) that "the intention of the Holy Spirit is to teach us how one goes to heaven and not how heaven goes";[36] in Leo's case, one could see that scientific criticism of physical assertions in Scripture was beside the point once one realized (with Saint Augustine) that "the Holy Ghost . . . did not intend to teach men these things (that is to say, the essential nature of the things of the visible universe), things in no way profitable unto salvation."[37]

Not only were both Galileo and Leo asserting the same principle that Scripture is not a scientific authority in answer to analogous problems involving questions of the relationship between Scripture and science (or natural philosophy), but they also shared some crucial aspects of the reasoning to justify this principle. The argument is this. Natural science and scriptural assertions cannot contradict each other, because both nature and Scripture derive from God. Hence, if there appears to be a contradiction, the conflict is only apparent, not real, and must be resolved. In particular, for Leo, "whatever they [physicists] can really demonstrate to be true of physical nature, we must show to be capable of reconciliation with our Scripture."[38] This is normally done by interpreting the biblical statement in a nonliteral fashion. Now, to see why this is done, "to understand how just is the rule here formulated,"[39] we have to remember that Scripture is not a scientific authority. Given this principle, the principle of accommodation also follows: "hence they [biblical authors] . . . described and dealt with things in more or less figurative language, or in terms which were commonly used at the time, and . . . what comes under the senses."[40]

In short, on the one hand Leo was supporting the principle denying the scientific authority of Scripture by means of a quotation from Saint Augustine. On the other hand, the encyclical also contained a supporting argument: this principle provides the explanation of the fact that demonstrated physical truth is given priority over literal biblical assertion, and the fact that sacred authors wrote by accommodating their writings to common language, belief, and observation; these facts are explained by the principle, which is in turn justified by them. This structure of reasoning was the same as that advanced by Galileo.[41]

Besides the formal similarity of problems, the substantive overlap of content, and the deep-structure correspondence of the reasoning, Leo's account was reminiscent of Galileo's even in its appearance, on the surface, and as a matter of initial impression. This parallelism involved the quotations from Saint Augustine and how they were interwoven with the rest of the argument. In fact, Leo's two main passages from Augustine had also been quoted by Galileo in his Letter to Christina: Augustine's statement of the priority of demonstrated physical truth ("whatever they can really demonstrate . . . , we must show to be capable of reconciliation with our Scripture")[42] and his statement of nonscientific authority of Scripture ("the Holy Ghost . . . did not intend to teach men . . . the things of the visible universe").[43]

It is not surprising that Leo's encyclical has been widely perceived as the Church's belated endorsement of the second fundamental belief for which Galileo had originally been condemned, namely that Scripture is not an authority in astronomy.[44] As we shall see in a later chapter, this interpretation was also endorsed by Pope John Paul II in 1979–1992.[45]

13.3 Blaming "Realism": Duhem's Epistemological Explanation (1908)

One might have thought that the implicit theological vindication of Galileo by an influential pope, coming soon after his judicial rehabilitation by the meticulous scholarship of the 1870s, on top of the older and more gradual scientific vindication provided by the proofs of the earth's motion from Bradley's stellar aberration (1729) to Foucault's pendulum (1851), would prevent or discourage further retrials of the victim. But to think so would be to underestimate the power of human ingenuity and the unique complexity of the Galileo affair. In fact, a novel apology was soon devised by a great scholar who combined knowledge of physics, history, and philosophy— Pierre Duhem (1861–1916).[46]

Duhem advanced his account of the Galileo affair in 1908 as part of a series of articles that were first published in the journal *Annales de Philosophie Chrétienne* and then (in the same year) appeared in book form under the title *To Save the Phenomena: An Essay on the Idea of Physical Theory from Plato to Galileo.* The journal of first publication gave a clue that the account was some kind of apologetic contribution;[47] but the matter is probably more complex, in light of the fact that that this journal was placed on the *Index* soon thereafter, the reason stemming largely from its opposition to the Church's antimodernism.[48] And the book's subtitle gave a clue that Duhem's view of Galileo's trial was some kind of epistemologically oriented account; in fact, Duhem's own self-conception was that that work was a "development"[49] of the epistemological thesis "that physical theories are by no means explanations, and that their hypotheses [are] not judgments about the nature of things, only premises intended to provide consequences conforming to experimental laws."[50] This was Duhem's version of an instrumentalist, antirealist epistemology.

It is difficult not to be struck by Duhem's memorable thesis that "logic was on the side of Osiander, Bellarmine, and Urban VIII, and not on the side of Kepler and Galileo; that the former had understood the exact import of the experimental method; and that, in this regard, the latter were mistaken."[51] The sense in which the former were right and the latter wrong is that the former understood and appreciated "that the hypotheses of physics are merely mathematical contrivances intended to save the phenomena,"[52] and not descriptions of reality capable of being true and of being false; whereas Galileo and Kepler did not have this understanding and appreciation.

However for Duhem, this clash of epistemological principles was only one of two key factors leading to Galileo's condemnation; the other factor was a substantive disagreement between realists, between the "impenitent realism of Galileo" and "the intransigent realism of the Peripatetics of the Holy Office."[53] Duhem saw little difference between the condemnation of 1616 and that of 1633: the former allegedly amounted to a prohibition to use the key Copernican hypotheses in any way whatever, "even for the sole purpose of 'saving the phenomena,'"[54] whereas "the condemnation of 1633 was a confirmation of the sentence of 1616."[55] The rationale underlying these condemnations was primarily a pair of realist principles which, according to Duhem, "both Copernicans and Ptolemaics by common agreement required of all acceptable astronomical hypotheses,"[56] namely that they be compatible with sound physics and with Scripture. The differ-

ence between the two sides concerned the application of these two principles, which yielded a negative answer for the Inquisition and a positive one
for Galileo.

By explaining the condemnation in terms of realism, Duhem was in a
sense blaming that tragic episode on realist epistemology, and thus further
discrediting realism. Moreover, the condemnation need not have occurred,
and indeed would not have occurred, if the realists on both sides had listened to the voices of reason, moderation, and prudence of people like
Cardinal Bellarmine and Pope Urban VIII. Their criticism of Galileo's realism was based on precepts stemming from an ancient tradition, for "these
precepts had been formulated by Posidonius, Ptolemy, Proclus, and Simpli-
cius, and an uninterrupted tradition had brought them to Osiander, Rein-
hold, and Melanchthon."[57] Duhem's counterfactual conditional proposition was obviously meant to strengthen the antirealist and instrumentalist
tradition of which he himself felt to be a part.[58]

This Duhemian interpretation of the Galileo affair has considerable
originality, simplicity, and elegance. Its simplicity lies primarily in the fact
that it explains both phases of the episode (1616 and 1633) by means of
the same epistemological interpretation, in terms of realism. Its elegance
lies in the fact that it makes sense of the key provision in the special injunction prohibiting Galileo to hold, defend, or teach the earth's motion in any
whatever; Duhem was suggesting that the phrase "in any way whatever"
referred to the discussion of the earth's motion even as a hypothesis meant
merely to save the appearances; the sentence of 1633 then became justified
as a literal application of this injunction to the *Dialogue*. Moreover, Duhem's
epistemological account has been and continues to be influential. But now
we must ask whether it is also correct. To evaluate its correctness, we must
critically examine what exactly Duhem meant by the "realism" he attributed
to Galileo and whether his attributions are really accurate.

I have elaborated the details of this critical analysis elsewhere,[59] and so
here it will suffice to summarize them. The realism attributed by Duhem to
Galileo consists in part of the idea that physical theories are not merely
instruments of calculation and prediction, but also descriptions of physical
reality; and this idea is indeed an important part of Galilean epistemology.
Duhem also attributes to Galileo other epistemological principles involving
metaphysics, the Bible, the ideal of certainty, and the methodology of crucial experiments. However, Galileo did not hold such principles, and it is
arbitrary to subsume them under the label of realism.

For Duhem, Galileo believed that astronomical (or physical) theories
should be confirmed or disconfirmed by means of metaphysical considerations. Duhem seems to suggest this by his assimilation of Galileo with
Kepler and with the methodology of the Inquisition consultants. But this
interpretation is refuted by Galilean practice and fails to properly distinguish natural philosophy and first philosophy.

Again, according to Duhem, Galileo thought that astronomical (or physical) theories should be confirmed or disconfirmed by means of biblical
texts. This interpretation results from a misunderstanding of the logic of
Galileo's critique of the biblical objection to Copernicanism and involves a
flawed reading of the Letter to Christina. Here Duhem succumbed to the
spell of Mallet du Pan's bad-theologian myth.

Third, Duhem's Galileo held that physical theories in general, and
Copernicanism in particular, are demonstrable with certainty. Duhem
makes such a questionable attribution largely on the basis of a passage from
Galileo's "Considerations on the Copernican Opinion." However, this passage happens to be an explicit and unambiguous statement that all we can
determine about the epistemological status of Copernicanism or of physical theory is that it "may be true." Duhem interprets "may be true" to mean
"must be true."

Finally, in Duhem's eyes, Galileo thought that Copernicanism had been
adequately established and that he could establish it (or a physical theory in
general) simply by refuting one inadequate alternative and saving a few
phenomena in Copernican terms. Duhem supports this interpretation by
assuming that Galileo regarded Copernicanism as established and reconstructing his supporting argument. Unfortunately, there is a vicious circu-
larity between this assumption and this reconstruction; the assumption is
unwarranted; and the reconstruction is not only logically inconclusive but
also internally incoherent insofar as it attributes to Galileo the very same
methodology of saving the phenomena whose neglect was allegedly his central error. At any rate, a proper examination of the issue would have to
examine Galileo's mature case in favor of Copernicanism (in the *Dialogue*
of 1632), and this would reveal both Galileo's reservations about the epistemological status of Copernicanism and the complexity and multifaceted
nature of the case in its favor. He did regard Copernicanism as capable of
being demonstrated, and as on its way toward being demonstrated, but not
as capable of being demonstrated with certainty and not as already demonstrated with adequacy.

13.4 Müller's Anti-Galilean Synthesis and Garzend's
Un-Apologetic Concept of Heresy (1909–1912)

In 1909, the German Jesuit Adolf Müller published a work in two volumes
criticizing Galileo and justifying the Church.[60] The book did not present
any really novel documentation, interpretation, or evaluation, but it is significant as a compendium of anti-Galilean criticism and proclerical justifications. Müller made systematic use of Favaro's recently completed
National Edition of Galileo's works and tried to take into account all
that had been written on the Galileo affair for the past three centuries

(although this aspect of his work was, understandably, more selective than systematic). Müller was an astronomer and was proficient in the Italian language. Thus he easily surpassed all previous examples of this genre of literature, such as Cooper 1838, Olivieri 1840, Marini 1850, Madden 1863, and L'Epinois 1878. Subsequent instances of this genre were unable even to approach Müller's accomplishment; what followed were mostly attempts to popularize this type of account. For example, one could read along these lines the account of Galileo in Arthur Koestler's *Sleepwalkers* (1959), of which more in chapter 15.2. In short, one might say that Müller gave a classic (scholarly) exposition of the anti-Galilean account and was a scholarly precursor of Koestler's popular libel against Galileo.

Although Müller admitted that Galileo made important contributions to kinematics, this point was not elaborated. Instead, Galileo was criticized as a poor methodologist, unaware that his geokinetic arguments were invalid or inconclusive; as an intrusive theologian, who meddled in questions outside his expertise; as a mischievous rebel bent on disobeying or circumventing ecclesiastical authority at every turn; as a megalomaniac with regard to his own accomplishments; and as a paranoiac with regard to the accomplishments and motives of others. Müller even criticized Galileo as a poor astronomer who did not keep up with the best recent work by men such as Clavius and Kepler; who was unable to properly compile ephemerides of Jupiter's satellites; and whose telescopic discoveries were made through mere luck and mostly duplicated simultaneously or earlier by other observers.

On the other hand, as regards the Church, Müller was willing to admit that it committed one error, but tried to justify it in every other way. The error was the judgment that the earth's motion was contrary to the Bible; but this error did not disprove papal infallibility and merely confirmed that the Inquisition, the Index, and individual officials are fallible; for it was not a formal pronouncement ex cathedra. Müller also claimed that the Index decree of 1616 was a judicious middle course between the extremes of declaring Copernicanism heretical and of taking no action;[61] that that decree did not eliminate freedom of research but rather wisely regulated it;[62] that in 1616 Galileo was properly served with the special injunction not to discuss the earth's motion in any manner whatsoever, and so his publication of the *Dialogue* clearly made him guilty; that the Church's intransigence with Galileo after the 1633 condemnation was an appropriate response to his own behavior, which showed no sign of repentance but rather indicated defiance;[63] that the condemnations of Copernicanism and of Galileo did not impede scientific progress;[64] and that forcing Galileo to abjure was not an abuse of power, nor a way of making Galileo perjure himself, but rather a reaffirmation of Church authority and of his submission.[65]

By way of general criticism, I would say that Müller's account is too one-sidedly anti-Galilean and proclerical; too uncritical regarding the factual accuracy or legal admissibility of the special injunction; and too superficial in its reading of key documents such as the *Dialogue* and the Letter to Christina.

Notably, Müller did not include in his account Mallet du Pan's bad-theologian thesis. This myth had run its course and performed its formative function in the development of the Galileo affair; but it could no longer survive critical scrutiny, even the scrutiny of such an extreme anti-Galilean critic as Müller.[66] However, Müller did speak as if Galileo interfered improperly in theological matters, and he regarded this interference as a factor contributing to Galileo's condemnation. This thesis is a far cry from the original Mallet myth but may be viewed as its metamorphosis.

For Müller, it is as if Galileo's fault now became that of being a good theologian (so to speak) despite his lack of theological training and position:

> Although he had never undertaken theological studies and indeed had frequently declared his incompetence in this subject, one must agree with Grisar[67] that (except for a few inaccuracies) he advances principles which all theologians would accept today, and which even then (at least in theory) were shared by the more open-minded theologians. But . . . for Galileo to present himself with theological weapons was extremely presumptuous given the situation at the time. There has been much discussion about who brought the question into the theological arena. . . . In 1879 Reusch said that "this was not done by Galilei but by his enemies. In his published writings Galilei did not touch at all on the theological aspect of the question, and if he did it in the letters to Castelli and to Christina of Lorraine, he did it only because he was forced by his enemies."[68]

Müller went on to criticize Reusch's thesis by arguing that Galileo's enemies must be subdivided into at least two groups (philosophers and clergymen), and that although the philosophers Francesco Sizzi and Ludovico delle Colombe preceded Galileo in discussing the role of Scripture in astronomy,[69] he preceded the clergymen Niccolò Lorini and Tommaso Caccini, and in any case he outdid everyone else in exacerbating the problem.

Müller's portrayal of Galileo as a theological intruder (together with the general image) was refined fifty years later by Arthur Koestler in an ingenious, more substantive, but quite sophistical way. Koestler tried to show that the chief aim of the Letter to Christina was to illegitimately shift the burden of proof in the Copernican controversy: "It is no longer Galileo's task to prove the Copernican system, but the theologians' task to disprove it. If they don't, their case will go by default, and Scripture must be reinterpreted."[70]

Just as the time was ripe, after the turn of the century, for an apologetic compendium or synthetic apologia such as Müller's, the time was also ripe for attempts to solve relatively old problems that had never been seriously tackled or systematically dealt with. One of these problems centered on the concept of heresy: what was the notion of heresy according to which the

Inquisition's 1633 sentence had found Galileo "vehemently suspected of heresy"? Was it a legitimate notion? And was this notion the same concept as was presupposed by those authors who subsequently claimed that Galileo had been condemned for "disobedience, not heresy"? In 1912, the Frenchman Léon Garzend published in Paris a well-documented work that addressed this cluster of issues and whose full title conveyed a good idea of its content: *The Inquisition and Heresy: Distinction between Theological Heresy and Inquisitorial Heresy, with regard to the Galileo Affair.*

Garzend's main thesis was that there were two concepts of heresy, a theological and an inquisitorial one. The theological concept was strict and narrow and defined a heresy as a denial of a proposition (1a) explicitly (1b) revealed by God and (1c) officially proclaimed by the Church in a declaration addressed to (1d) all who have been (1e) baptized; this was the concept prevalent among modern theologians as well as those in the seventeenth century. On the other hand, there was an inquisitorial concept which was looser and broader and reflected both the practice and the manuals of the Inquisition; it broadened the concept of heresy to include the denial of propositions that (2a) could be clearly deduced from divine revelations, and/or (2b) embodied common Church teachings, and/or (2c) were clearly contained in Scripture but had not been officially proclaimed by the Church, and/or (2d) were declared articles of faith by lesser Church organs (inquisitors, bishops, popes when not speaking ex cathedra, etc.), and/or (2e) were applicable only to a particular person or group, and/or (2f) were articles of faith but were being denied by unbaptized persons.

One consequence of this thesis was that Garzend could explain a tension between two aspects of the Galilean trial documents. The first aspect referred to the fact that (3a) the Index Decree of 1616 and the last part of the 1633 sentence described the earth's motion as "false and contrary to the Bible" and not as "heretical"; (3b) the 1633 sentence convicted Galileo not of "formal heresy" but of "vehement suspicion of heresy"; (3c) after his abjuration, Galileo was *not* deemed to need absolution for the excommunication automatically incurred by heretics; (3d) the sentence and abjuration took place in private (at the *convent* of Santa Maria sopra Minerva) rather than in public;[71] (3e) during the trial Galileo was never interrogated regarding possible accomplices, as required by canon law in cases of heresy;[72] and (3f) after the condemnation, persons who pleaded in favor of Galileo, as Peiresc did, were not punished for violating the law against pleas for convicted heretics.[73] The second aspect of the trial refers to the fact that (4a) the Inquisition minutes of 16 June 1633 reported a papal decision outlining the conclusion of the trial, including an injunction to never again discuss the topic on pain of being treated as a *relapsed* heretic; (4b) the last part of the sentence specified what Galileo's heresy was (namely, geokineticism and denial of the scientific authority of Scripture); (4c) the same passage stated that Galileo must abjure "the above-mentioned errors and heresies"; and (4d) Galileo was required to recite an abjuration. Garzend's explanation was that the first aspect reflects the theological concept of heresy, the second aspect the inquisitorial concept.

Another consequence of Garzend's thesis was to provide a *novel* answer to the anti-infallibility objection: this was the argument that in the trial of Galileo the Church made various errors, and therefore the doctrine of infallibility is itself erroneous. The usual answer to this objection pointed out that this doctrine remained unrefuted because the infallibility applied only to papal pronouncements ex cathedra, publicly addressed to all the faithful; it did not apply to pronouncements of the Congregation of the Index (such as the anti-Copernican decree of 1616), to declarations of the Inquisition (such as the 1633 sentence against Galileo signed by the cardinal-inquisitors), or to decisions made by the pope when acting as chairman of the Inquisition (such as the papal orders at the Inquisition meeting of 16 June 1633). This usual answer amounted to saying that insofar as the Church committed errors, these were committed by elements of the Church that were supposed to be fallible; so this answer assumed that the Church did commit significant errors. In particular, this answer left the residual difficulty that it was very strange that the Inquisition in general and the pope in particular (as an individual person) should have arrived at a condemnation of Galileo in 1633 which appeared to involve a theological error in interpreting the 1616 decisions and applying them to his subsequent behavior; for indeed, from a strict theological point of view, neither Copernicanism nor Galileo was heretical, nor was Copernicanism declared heretical in 1616, nor was Galileo guilty of formal heresy in 1633.

Garzend's answer to this *residual difficulty* and the novel answer to the anti-infallibility objection was to point out that all this (the non-heretical character of Copernicanism and Galileo) was indeed correct from the point of view of the theological concept of heresy, namely from a strict theological point of view. But in Galileo's trial the pope and the inquisitors were taking the inquisitorial point of view, and Copernicanism *was* an *inquisitorial* heresy and Galileo *was* an *inquisitorial* heretic. Thus, there was no error committed, and the question of infallibility does not arise.

While the scholarly documentation provided by Garzend and his ingenuity are beyond question, I believe he also unwittingly showed something very far from his own explicitly apologetic intention; namely, that the Inquisition practices had no theological justification or were theologically untenable.

Moreover, other unintended consequences follow from Garzend's argument that insubordination was an essential part of strict theological heresy because this heresy reduced to an intellectual error accompanied by a persistent defiance of the Church's injunctions on what to believe. It follows that inquisitorial heresy should not be labeled "disciplinary" (as Garzend himself occasionally designated it), to distinguish it from "formal" heresy, because all heresy is a violation of prescribed discipline; and this conclusion in turn undermines the apologetic line that Galileo was condemned for disobedience and not for heresy, because (even formal) heresy is ultimately disobedience, failing to believe what the Church commands one to believe.

Finally, Garzend also made clear that the concept of heresy subsisted in a cluster of other notions such as erroneous, scandalous, temerarious, near-heretical, dangerous, formally heretical, and suspect (mildly, vehemently, and strongly). Therefore, this apologetic line could also be criticized by saying that the concept of heresy is not univocal, and that there is one special case which is equivalent to the alleged disobedience for which Galileo was convicted.

Chapter 14

A Catholic Hero

Tricentennial Rehabilitation (1941–1947)

In the early 1940s, the tricentennial of Galileo's death occasioned a series of reassessments of Galileo's trial that may be regarded as a semi-official rehabilitation. The rehabilitation was not formal or official because it was not proclaimed either by the pope or by the Congregation for the Doctrine of the Faith (the new name of the Inquisition). On the other hand, it did involve authoritative persons and institutions: the Franciscan friar Agostino Gemelli (O.F.M.), president of the Pontifical Academy of Sciences and of the "Sacred Heart" Catholic University of Milan; the priest and Church historian Pio Paschini, president of the Lateran University (namely, the Roman Seminary); and the Jesuit Filippo Soccorsi, the director of Vatican Radio, whose work was first presented as a lecture to the Royal Academy of Italy, then appeared as articles in the authoritative Jesuit journal *La Civiltà Cattolica,* then in book form with its press, and eventually (in 1964) as a reprint in a volume edited by the Pontifical Academy of Sciences to mark the quadricentennial of Galileo's birth. Although these reassessments were not publicized at the time as a rehabilitation of Galileo, they naturally appear as such to anyone who considers them collectively. Moreover, the fact that there was no rhetoric of rehabilitation makes this development more genuine and important. Indeed it may be regarded both as the first significant development since those discussed in chapter 13 and a preview or anticipation of Pope John Paul II's more explicit "rehabilitation" between 1979 and 1992, to be discussed in a later chapter.

14.1 "Harmony of Science and Religion": Gemelli Reverses Traditional View (1942)

On 30 November 1941, Pope Pius XII attended the inaugural meeting of the Pontifical Academy of Sciences for the academic year 1941–1942.[1] The

agenda included such routine items as a speech by the pope, the presentation of new members, the commemoration of recently deceased members,
and a speech by the president of the academy. The president—Agostino
Gemelli—was a physician, experimental psychologist, and biologist;[2] he
gave a speech that was perfunctory until near the end. The ending of the
speech was about Galileo:

> *A centennial celebration.* This year there will be the Galilean centennial. Our
> academy has a special reason for celebrating this occasion.
>
> Allow me to recall some factual data, although the audience may very well
> remember them. It is known to scholars that at the beginning of the seven-
> teenth century the Roman Federico Cesi, as a consequence of the great inter-
> est felt everywhere for the natural sciences, and together with other equally
> energetic young men (such as G. Heck of Holland, Franco Stelluti of Fabri-
> ano, and Anastasius de Filiis of Terni), founded here in Rome the Lincean
> Academy with the aim of promoting research in the natural sciences, and they
> placed it under the protection of St. John the Evangelist. . . . Our Pontifical
> Academy of Sciences is the direct heir and legitimate continuation of that
> ancient academy, as the late Pius XI recalled in his *motu proprio* of 28 October
> 1936, with which he revived it under the present title of Pontifical Academy of
> Sciences. . . .
>
> Those first academicians who preceded us never thought that to be a
> believer and a Catholic was an obstacle to science; they also prescribed for a
> "Lincean" the rule that was written in the "Linceografo," namely that "the aim
> of the academy is not only to acquire knowledge and understanding of things,
> while living righteously and piously, but also to reveal them to others orally
> and in writing, without harming anyone"; they knew that only with the sup-
> port of high religious officials and of the pope himself had the abbé Coperni-
> cus been able to publish his new worldview; and they knew that insofar as
> Galileo was a scientist and a discoverer of new celestial phenomena, he was
> never persecuted by the Church but instead received great assistance. An
> objective indication of this is the assistance our academy provided to him; nor
> was it the only friend of Galileo, for the great scientist had the support of the
> highly competent physicists and mathematicians who at that time taught at
> the Roman College (among whom was the great Clavius), as well as of author-
> itative and devout friends.
>
> *A work on Galileo Galilei.* It is therefore appropriate that in the present cen-
> tennial anniversary we who are the direct and rightful heirs of the first
> Linceans celebrate the memory of Galileo. A commission of academicians
> was appointed for the explicit purpose of considering men who are especially
> competent to write a work of broad scope about Galileo; it has selected a stu-
> dent of the historical disciplines who is dear to those who practice these disci-
> plines for the elegance with which he is accustomed to present the results of
> his scholarly research: the Most Rev. Msgr. Pio Paschini, distinguished presi-
> dent of the Lateran University. He will not give us a mere biography but
> rather a new portrayal of the figure of Galileo, placing his work in the histori-
> cal context of his time and re-placing the figure of the great astronomer in its
> true light by a documented narration of his life and a rigorous critical exposi-

tion of his discoveries. This contribution to the celebrations will be something like the continuation of that enlightened act of government by which Leo XIII opened the Vatican Archives to scholars studying Galileo's trial. The projected volume will thus be an effective demonstration that the Church did not persecute Galileo but helped him considerably in his studies. However, it will not be a work of apologetics (because this is not the task of scientists) but of scientific and historical documentation. We must add that we will be able to accomplish such a work thanks to the perspicacity of the late Pius XI, who decided to add as special members of our academy scholars of the historical disciplines, that is, the distinguished men who (working in silence and with profit in the Vatican Archives and the Vatican Library) render a service to history, namely to truth, namely to the Church.[3]

Clearly the president of the Pontifical Academy was trying to associate it and Galileo by way of the Lincean Academy. Of course, one could question the historical connection, for Gemelli was giving a somewhat slanted historical sketch. Similarly, he was either begging the question or equivocating when he asserted "that insofar as Galileo was a scientist and a discoverer of new celestial phenomena, he was never persecuted by the Church." But, historical connection aside, there was a principle that they did indeed share, namely the principle of harmony between science and religion. In announcing the new work on Galileo by Paschini, Gemelli did not explicitly say that the work would elaborate the account he himself had just sketched, but he implied it when he said that "the projected volume will thus be an effective demonstration that the Church did not persecute Galileo, but helped him considerably in his studies." On the other hand, Gemelli tried to deny this implication by stating that "it will not be a work of apologetics." Thus, he was falling into an incoherence that seems difficult to explain away.

At any rate, Gemelli had the occasion to elaborate his own view in his lecture at the tricentennial commemoration held at the Catholic University of Milan, whose proceedings were published as a book in 1942.[4] As president of this university, he gave the first of a series of lectures and titled it "Science and Faith in the Person of Galilei." Gemelli's opening remarks were refreshing and contained no trace of incoherence, equivocation, or evasiveness:

[1] Ours is homage not only to the scientist but also to truth and to the believer.

It is homage to the truth because, now that the ideologies that flourished in the nineteenth century have gone out of fashion, and now that the passions that were felt in connection with his name have faded away, Catholics are not afraid to sincerely recognize that the trial against him was an error. This was an error that does not refute either the infallibility of the pope or the authority of the Church; nor was it [2] contrary to the norms of charity observed by the Church in legal proceedings. It was an error of theologians, which "has become a constant warning," as Pastor stated.[5]

Ours is also homage to the believer, as well as to the scientist. Since any prohibition against his works ceased more than a century ago, and since the anticlerical sectarianism that turned his name into a battle cry against the Church has vanished, nowadays we can serenely honor him not only as a great Italian whose genius exalted his motherland in an age of political humiliation, but also as a believer who never doubted the harmony between religion and science, between faith and reason; this was so even though, because of the exceptional circumstances of his time, he more than anyone else could have been tempted not only to doubt it, but also to despair of it.[6]

So, Gemelli had no hesitation in admitting that the condemnation of Galileo was a theological error. This error must have included at least the claim that Copernicanism was contrary to Scripture, which had long been admitted to be erroneous. But Gemelli probably also included the error of holding that Scripture was a scientific authority, for such an admission was relatively easy to make fifty years after Leo XIII's *Providentissimus Deus* (see chapter 13.2); moreover, Gemelli refers to Pastor, and Pastor spoke of theological errors in the plural and says that whereas the 1616 error was the contrariety of Copernicanism and Scripture, the error suggested by the 1633 condemnation was to uphold the astronomical authority of Scripture.[7]

However, Gemelli was also claiming that Galileo's tragedy embodied a great positive lesson: that faith and religion are harmonious with reason and science.

He went on to argue that although Galileo did not provide a decisive demonstration of Copernicanism, neither did Newton, Bradley, or Foucault; on the other hand, Galileo did provide "the convergence of probabilities that were increasingly more and more numerous in favor of the Copernican system";[8] and, in any case, the Ptolemaic arguments were weaker.[9] This interpretation of Galileo's view of the status of Copernicanism was an increasingly widespread realization in the history of the Galileo affair, but Gemelli's reiteration was important and appropriate.[10]

Gemelli had a unified explanation of the condemnation of Copernican books in 1616 and of Galileo in 1633. He attributed it to Galileo's exclusivist epistemology to the effect that "the new science was not only a truth, but the *whole* truth,"[11] and to his enemies' diametrically opposite view, that is, their exclusive and excessive reliance on the speculative method of metaphysics. This was an interesting and relatively original thesis, and Gemelli had elaborated his first version twenty years earlier.[12] Unfortunately, I do not think such an attribution to Galileo is tenable.

With regard to Galileo's faith and religiousness, "his worship of the Author of the universe had a childlike simplicity."[13] Moreover, "he worshipped with a self-conscious humility; he considered himself to be simultaneously privileged and ordered to discharge a great mission: to observe new works of God and reveal them to men";[14] but Gemelli was quick to add that such Galilean religiousness should not be interpreted as pantheism, deism, or indifference. Finally, Gemelli argued that although Galileo was wrong to get involved in theological and scriptural questions and imprudent in the way he discussed them, this interference and this imprudence were "a proof of the sincerity of his Faith."[15]

Gemelli denied that the trial disturbed Galileo's spirit or undermined his faith: "He saw in the conflict the passions of men, not incompatibility of ideas; that is, he saw what we see nowadays; time has shown he was right."[16] Here Gemelli was not only advocating a circumstantialist interpretation, which had gained prominence about a century earlier with Biot and Chasles, [17] but he was attributing it to Galileo himself.

Now, it is certainly true that even after his condemnation, Galileo never interpreted his tragedy as a sign of the incompatibility between religion and natural philosophy. However, it is equally true that he continued to see the incompatibility between the experimental, mathematical, and Copernican-oriented philosophy he practiced and both the naively empirical, qualitative, and geocentric philosophy of the Peripatetic majority on the one hand and the literalist and traditionalist interpretation of Scripture of the conservative theological majority on the other. Thus, to carry circumstantialism too far may be questionable. However, Gemelli was trying to stress that even after the condemnation, Galileo continued to believe in the harmony between science and religion and continued to hold a circumstantialist view with regard to this potential conflict. And Gemelli wanted to stress this apparently in order to claim that Galileo had been right in yet other respects (besides astronomy, physics, methodology, and theology), namely the interpretation of the trial and the relationship between science and religion.

No wonder that Gemelli added that "Galileo seemed to come out of the inevitable conflict as the victim, but in reality he was the victor, and such he felt to be in the bottom of his unperturbed conscience."[18] This was an important point to stress, and when we examined Galileo's reaction to his condemnation we saw (chapter 3.4) that there is evidence supporting this interpretation.

Gemelli concluded his reevaluation as follows:

Thus, whereas for about two centuries Galileo has been a symbol of the revolt of reason and science against dogmatic intransigence, nowadays with a more serene vision we can admit that he is a man who in the innermost part of his soul achieved the harmony between reason and faith, between science and religion, a harmony that takes effect beyond the incomprehension of men and the times. Galileo was profoundly certain of this harmony, and from the point of view of his time this certainty was prophetic; in such certainty lies the secret of his serenity. His admirable faith in the objectivity and unity of truth constitutes for Catholic scientists the most significant teaching and the highest legacy of Galilean thought. His faith in the harmony between Catholicism and scientific progress is a warning which Italians of the age of the Lateran Treaty can understand better than the Italians of the Risorgimento and can follow to use as a mental lever for future progress; for by itself thought is a prodigious force, but combined with Christian faith it can do anything (one may say without fear of exaggerating); it can, with Columbus and Galileo, discover new lands and new heavens.[19]

This account provided a favorable reevaluation of Galileo. The "suspected heretic" had become the embodiment of the harmony between science and religion, and this Galilean lesson was all the more instructive insofar as Galileo had not only preached such harmony, but also practiced it; and what is more, he had continued to uphold it even after his condemnation, when an ordinary believer would have found reason to despair. But in his religious faith Galileo was no ordinary mortal. Here were the seeds of a new appreciation of Galileo. Gemelli was rehabilitating the "suspected heretic."

Another aspect of this account is worth noting. Gemelli was doing more than deny the traditional interpretation of the trial as epitomizing the conflict between science and religion. He was *reversing* that traditional view; the history of the trial not only did not support the conflictual interpretation but proved the opposite. This was also a novel and completely unprecedented account.

14.2 A Model of Religious Faith: Paschini's Preview (1943)

While the person who was president of both the Pontifical Academy of Sciences and the "Sacred Heart" Catholic University of Milan (Gemelli) was publicizing his novel religious appreciation of Galileo, the person who was both the Pontifical Academy's appointee to write a book-length reevaluation and the president of the Lateran University, Paschini, was hard at work on his project. However, that project turned out to be more time-consuming and challenging than anticipated, and it would be published only posthumously in 1964, as we shall see later (chapter 16). In the meantime, spurred by the tricentennial celebrations and probably inspired by Gemelli's appreciation,[20] Paschini found the opportunity to show that he was not idle by accepting an invitation to write a short article on Galileo. It was published in 1943 in the Roman journal *Studium* and was aimed at a general educated audience. It is important both as a preview of Paschini's later, full account and as a sign of the new ecclesiastical appreciation inaugurated by Gemelli.

The article was titled "The Teaching of Galileo: Do Not Be Afraid of the Truth."[21] Although this title was apparently chosen by the journal vice editor,[22] there is no evidence that Paschini objected; furthermore, it reflected the content of the article. Paschini's text was preceded by a short editorial introduction whose most striking point was that "the life and work of Galileo are like a triumphant hymn to truth, a severe and convincing warning never to be afraid of the truth, regardless of how it is revealed to the human mind, as long as one searches for it with humility and dedication and with the constant effort to free oneself of any trace of human pride and of any shady pursuit of selfish aims."[23]

Together with this editorial introduction, the title could be taken to be a double entendre meant to convey that Galileo was not afraid of the truth, even when it seemed to undermine traditional beliefs, including traditional interpretations of Scripture; that this is an instructive and positive lesson which we today can learn from Galileo; and that in particular we should not be afraid of the truth about the Galileo affair, even when the truth is that his condemnation was an error.[24] Despite the lack of concrete detail, the appreciative and positive attitude toward Galileo was obvious.

In Paschini's four-page essay, the account of the Galileo affair took up the second half and read as follows:

[96] One day he was able to transform the telescope, which for a short while had been a toy for curious persons, into an instrument of scientific research; not satisfied with looking at known distant objects as if they were near, one night he pointed it to the sky. That was like the egg of Columbus. No one had thought of it, and it must also be admitted that, without the instrument he built, no one would have seen anything. Galileo's astonished eye immediately saw a myriad stars that had never been perceived by human eyes; then the moon with its mountains and valleys; then Venus sickle-shaped with its phases, Saturn with its spots, and Jupiter with its satellites. The learned world found it hard to believe these things; many obstinately refused to believe any of them. Kepler himself was at first amazed and skeptical, but he soon surrendered to the evidence of the new observations. The Ptolemaic system was hit to the fullest; the Copernican one became definitely more likely, although not yet fully proved. Galileo started searching for decisive proofs, but did not find them. In such a situation, the two systems were, at worst, on an equal footing: if the Copernican system was not fully proved, the Ptolemaic one was much less so; thus, one could discuss them. The Ptolemaics admitted it too, but they had in reserve another argument in their favor. Debatable from a scholarly point of view, the Copernican system was however contrary to the literal expressions of Sacred Scripture, and consequently it was untenable from a theological point of view; thus, it was contrary to the tradition of Church Fathers and theologians. Copernicus too had thought of this difficulty, but he easily thought of the answer; and Paul III, to whom his work had been dedicated, found nothing to object to. But now the situation had changed, and in this regard the change was not for the better. The opposition of the theologians, which did not displease Ptolemaic philosophers, soon became more determined; Galileo's opponents raised the issue before the authorities in Rome. He also went there personally, but despite his efforts, he did not succeed in preventing the condemnation of the heliocentric system as heretical in theology and absurd in philosophy. The Holy Office, from which the con-

demnation originated, did not directly condemn him personally but issued him an injunction not to defend that system. Discussion was not prohibited, as long of course as it was conducted in such a way that the truth of the geocentric system emerged.

Such a result cannot but surprise us nowadays. But at that time, at least in Italy, it surprised and displeased only the few followers of Galileo.

He did not despair of a vindication, obtained by legitimate means (I must add); for he never wavered in his full and unconditional acceptance of Catholic teaching, and he had not been forbidden to continue his investigations. He hoped for an improvement when Maffeo Barberini (originally from Florence, and his friend) became pope with the name of Urban VIII. However, like Cardinal Federico Borromeo and other cardinals and distinguished persons, the new pope was well disposed toward Galileo with regard to his discoveries, but not with regard to his astronomical system. There remained the theological concerns which since 1615 Galileo had in vain tried to dispel. Now he came back into action with the *Dialogue on the Two Chief World Systems,* attempting to proceed with cleverness; in fact, it was clear that his aim was to show the inanity of the arguments in favor of the geocentric system, but it was not his fault if the arguments for the heliocentric system turned out to be more convincing. He also succeeded in snatching the imprimatur from the Master of the Sacred Palace. But as soon as the *Dialogue* was published, Galileo's aim became clear. The Holy Office tried immediately to stop sales and summoned the author to Rome to explain his manner of proceeding. It was easy to see that he had violated the injunction of 1616; as much as he tried to claim his obedience and to declare that he was far from holding the Copernican doctrine, it was evident that he had accepted it with full conviction. Thus, he was obliged to abjure it, and as penalty for the disobedience, he was given a religious penance and detention, first in Siena at the residence of his friend the archbishop and then (at Galileo's request) at his own villa "Il Gioiello" in Arcetri. This occurred in 1633. His detention, which according to the rules of the Holy Office was equated to imprisonment, lasted until his death, despite all his efforts to obtain freedom. The Holy Office was probably afraid that he would resume his propaganda in favor of his ideas and that a pardon could be taken to mean that it had changed its mind in their regard.

Galileo spent the last years of his old age under surveillance by ecclesiastical authorities, who always distrusted him; he endured the disgrace of a condemnation that dishonored him in the eyes of the public; and he was the object of scorn by the masses of the learned world. Afflicted by illnesses and then [97] also by complete blindness, which prevented him from continuing with his studies, still he managed to include in his *Discourse on the Two New Sciences* the results of his speculations in mechanics; also to guide in their studies Evangelista Torricelli and Vincenzio Viviani, who were with him the last few months of his life; and to maintain an active correspondence with some of the most illustrious scientists. The sadness produced in him by so many misfortunes did not cloud even for a moment the lucidity of his always alert mind, nor did it instill in his heart the regret of having followed a course that (although it had earned him much glory) had also been the origin of bitter oppositions. His adversaries may have thought they had cut him down by taking advantage of the weaknesses of his character, and that they had guaranteed forever the victory of their prejudices by having him condemned by the supreme tribunal. Instead they rendered nobler and more sacred his reverence toward the authority of the Church to which he had always belonged; he never cursed against it; he did not resent that authority; he did not lose faith; he remained firm in the sincerity of his feelings and sure of a vindication by Providence.[25]

Paschini's account was full of simplifications, as was appropriate for the audience of educated nonspecialists to which it was addressed. However, these simplifications usually avoided being oversimplifications and often contained important insights, which could be elaborated and documented and which were occasionally relatively novel.

One of these insightful simplifications was to say that Galileo transformed the telescope from a playful curiosity into a scientific instrument to learn new truths about nature. Another was Paschini's claim that the telescopic discoveries implied a reassessment of the relative merits of the Ptolemaic and Copernican systems, with the consequence that heliocentrism was then better supported than geocentrism, while still not conclusively demonstrated; this judgment embodied the insight that the assessment of scientific theories is a comparative or contextual question, as well as a matter of degree rather than an all-or-nothing affair. And this insight formulated the issues in the only way that is really fruitful, by contrast to the many previous discussions which claimed either that Copernicanism was philosophically "false and absurd,"[26] or not yet "demonstrated,"[27] or weaker than geocentrism.[28]

It was surprising to see a churchman of Paschini's stature affirm that between the times of Copernicus and Galileo the situation had changed for the worse with regard to religious interference into philosophical and scientific matters. It was also surprising, although for a different reason, that Paschini should commit the error of saying that heliocentrism was condemned as theologically heretical by the Inquisition; for despite the constant recurrence and wide prevalence of this error, we have seen that Copernicanism was judged heretical by the consultants in 1616, but merely contrary to Scripture by the Inquisition.

Paschini's interpretation of the 1616 warning or injunction to Galileo was revealing. Paschini did speak of a formal injunction rather than a mere warning, but formulated its content merely as the prohibition to defend Copernicanism, whereas "discussion was not prohibited" (96). Apparently he had been convinced by the arguments against the existence and/or propriety of a *special* injunction.

Paschini admitted that in an important sense Galileo did not acquiesce at the results of the 1616 proceedings and that he was looking for an opportunity to resume the fight. But rather than charge Galileo with bad faith, as such apologists as Tiraboschi, Cooper, and Marini had done,[29] Paschini clarified that Galileo's plan was to do it "by legitimate means" (96). And Paschini even described such legitimate means used by Galileo when the election of Urban VIII led him to write the *Dialogue:* Galileo discussed the arguments on both sides, showing that the pro-Copernican arguments were stronger than the pro-Ptolemaic ones; "but it was not his fault if the arguments for the heliocentric system turned out to be more convincing" (96). This was an important and relatively original thesis and could form the basis for defending Galileo even from the charge of violating the warning not to defend Copernicanism.

However, Paschini did not go that far, and all he seemed to want to claim was that this was a legitimate defense of Galileo's behavior. For Paschini also seemed to admit that the Inquisition had a legitimate reason for claiming that Galileo had violated the warning. And it is in light of the clash of such legitimate reasons that Paschini can then give credit to Galileo for having yielded. In such a context, his confession and abjuration then "rendered nobler and more sacred his reverence toward the authority of the Church" (97), that is, they became acts of superior piety and religiousness. Thus Paschini was transforming Galileo into a model of religious faith and in that sense rehabilitating him.

14.3 A Noble Intellectual Sacrifice: Soccorsi Justifies Galileo's Retraction (1947)

Although Paschini sketched an account of the affair portraying Galileo as a model of religious piety, he did not stress the ecclesiastical errors as Gemelli had done. Soon thereafter, the Jesuit Filippo Soccorsi elaborated and documented an account that combined both points of view.[30] Soccorsi's account also originated as a lecture, in a series delivered in Rome in 1942 and sponsored by the Royal Academy of Italy to celebrate the Galilean tricentennial.[31] At the time Soccorsi was the director of Vatican Radio.[32] The account was first published in 1946 in the authoritative and more or less official Italian Jesuit journal *La Civiltà Cattolica.* The following year an expanded version was published in book form by the press of the same organization. Then it reprinted the book in 1963. Finally, in 1964, the book was one of three works published by the Pontifical Academy of Sciences under the collective title of *Miscellanea Galileiana;* the set included Paschini's *Life and Works of Galileo Galilei* (published then for the first time) and was meant to signal the Church's open-minded attitude, especially in the wake of the second Vatican Council (1963–1965). This presentation and publication history suggest that Soccorsi's account found wide acceptance in ecclesiastical circles and came to be viewed as a semi-official statement.

Soccorsi began by admitting that in 1616 a fateful error was committed by various ecclesiastical institutions and persons (although not by the official Church or the pope speaking ex cathedra); the error was to believe and declare that Copernicanism was contrary to Scripture. He used very strong language to describe the error, calling it "a collective illusion" (13).[33] In fact, the phrase he used (abbaglio collettivo) could also be taken to mean "collective hallucination" (but also, more neutrally, "collective error"). And Soccorsi was clear that in the face of such an error the proper task was to try to explain why it happened, rather than to try to justify, excuse, or rationalize it: "Let us observe that, once the first error was committed of judging the Copernican thesis contrary to Scripture, the logical consequences can be explained; but the first error still demands an explanation, not in order to justify it, but to understand it as a psychological act; it was not an error of a single person but a collective error, which necessarily requires a cause that exerted an influence on the mind of many, although not of all. What was this cause?" (16).

Then Soccorsi stated and criticized a number of common explanations. The first explanation alleges that in 1616 Copernicanism was condemned because Galileo did not have a rigorous demonstration. Soccorsi pointed out that this was true but irrelevant; for it is easy to see how the lack of demonstration could have led to doubt; but the Inquisition did not merely express doubt; instead it condemned the geokinetic thesis (16–18).

A second explanation claims that ecclesiastic institutions were reacting against novel and personal interpretations of Scripture, reminiscent of those of the Protestants. Soccorsi argued that this cause was true but insufficient to produce the effect since it should have led merely to a disciplinary decree prohibiting public discussion of novel interpretations of the Bible; but the Index's anti-Copernican decree did much more than that (18–19).

Third, some have blamed the 1616 condemnation on human passions and emotions, such as the envy and revenge by Peripatetic losers and theological fanatics. Soccorsi admitted that there was an element of passion, namely "superficial absolutism, or the emotion that easily warms up the heart when one maintains opinions that are dear and are being contrasted" (22). However, if this is so, then "we are dealing essentially not with emotions but with prejudices" (23); and in this context a prejudice is an *intellectual* error, and the problem is to define its content.

Others have explained the condemnation as resulting from blind faith in the official science of the time. But, Soccorsi objected, theologians could not have been ignorant of the fact that official science was in crisis; they must have been more impressed by the fact that the earth's motion seemed to them "harmful to the veracity of the senses and even more to the veracity of God, who had allowed such language to be used in his name in Scripture" (28).

Then Soccorsi went on to give his own explanation, basing it on the documentation of Bellarmine's lectures at Louvain University in 1571 and his letter to Foscarini in 1615. At Louvain, Bellarmine discussed the question of whether the heavens were solid (so that the fixed stars were attached to the celestial sphere, whose rotation carried them in unison around the central and motionless earth) or fluid (so that each star individually revolved around the earth); he opted for the fluidity of the heavens mostly on biblical grounds. However, he added the following qualification: "But if in the future it will be conclusively proved that the stars move by the motion of the whole sky and not on their own, then one would have to see how to understand the Scriptures so that they do not conflict with an established truth. In fact, it is certain that the true meaning of Scripture cannot conflict with any other philosophical or astronomical truth."[34] In 1615, after Foscarini sent Bellarmine his book that tried to reconcile Copernicanism with Scripture, the cardinal replied with a letter meant for Galileo as well as Foscarini. Soccorsi called attention to a similar qualification contained in that letter: "If there were a true demonstration that the sun is at the center of the world and the earth in the third heaven, and that the sun does not circle the earth but the earth circles the sun, then one would have to proceed with great care in explaining the Scriptures that appear contrary, and say rather that we do not understand them than that what is demonstrated is false."[35]

We have seen (in chapter 5.2) that Bellarmine's hypothetical claim is susceptible of a progressive interpretation to the effect that if the earth's motion is conclusively proved, then the geostatic passages of Scripture will be reinterpreted nonliterally; that this hypothetical gives priority to demonstrated physical truth over scriptural statements; and that this priority implies that Scripture is not a scientific authority and one may disregard it in physical investigation. Auzout in 1665 had hinted at such an interpretation of Bellarmine's claim. However, Soccorsi stressed the fact that Bellarmine was apparently uncomfortable and puzzled about how to interpret Scripture in case of those conclusive demonstrations, and stated that such discomfort reflected "a point of view different from ours" (32). I believe the point of view to which Soccorsi was referring is the view that Scripture is a scientific authority and its statements about the physical world are literally true. So what he seemed to be suggesting was that the condemnation of Copernicanism in 1616 was to be explained as caused by the acceptance of biblical traditionalism; the error that Copernicanism was contrary to Scripture was the effect of the error that Scripture was an authority in natural philosophy.

Of course, after Leo XIII's encyclical *Providentissimus Deus* there was no problem with such an explanation, with blaming the condemnation of 1616 on a principle that had been explicitly and formally discarded. In a sense, Soccorsi's explanation amounted to making explicit one consequence of that encyclical. Nevertheless, it also stressed a connection that had historical reality and had played a role in bringing about a real historical occurrence.

After this explanation (but *not* justification) of the erroneous condemnation of Copernicanism in 1616, with regard to the 1633 proceedings Soccorsi elaborated what may be called an explanation (*and* justification) of Galileo's retraction. This is an aspect of the 1633 trial which many have been ready to understand, forgive, and excuse in an intuitive and commonsensical way; which some, such as Brewster (1835) and Chasles (1862), have been arrogant enough to blame and criticize; but which hardly anyone has ever tried to justify in a serious way that might be linked to a documented explanation of how and why it happened. I believe Soccorsi offered us such a serious justifying explanation. It is the following:

[50] Let us suppose there is a controversy about an issue that is still not mature and not sufficiently clarified. Whether rightly or wrongly, one fears a conflict with the faith. Church authorities believe so, or at least they see a danger. After examining the question, [51] feeling they have the duty to guide believers along safe paths with regard to the faith, they issue a decree that prohibits the suspected doctrine, although it does so in a revocable manner.

Let us also suppose that in fact the decree is erroneous; and let us now consider the attitude which four different categories of persons adopt toward it.

The first category agrees fully with the reasons that persuaded the judges. It therefore accepts the official decree with personal conviction.

The second category consists of persons who are not competent on the matter. Although these persons do not believe the prohibited doctrine is infallibly false, nevertheless they presume it is false and simply reject it without further reflection. If instead they should want to hold it, they would be providing evidence that they do not much love the faith; this would be like the case of someone who proceeds forward recklessly along a road that displays a sign indicating danger ahead, and justifies himself by saying that whoever put up the sign is not infallible.

After these first two categories of persons who are persuaded and who are not competent about the decree, there is the category of persons who are competent on the matter, but give weight to reasons insufficiently [52] appreciated by the judges; they can however sincerely obey the authorities and follow the official decree since it is understood that such assent is revocable and conditional. By this it is also understood that although they reject the prohibited doctrine on the authority of the judges, they nevertheless reserve the right of thinking it may be true; the Church explicitly recognizes they have this right. Among these persons who are competent, it is possible to have an infinite gradation of cases since the weight they give to the arguments favoring the condemned doctrine can vary from a very small to a very great value. Despite these different gradations, the nature of their assent is the same: on the authority of the judges and on the value of the reasons that persuaded the judges themselves, these persons reject the prohibited doctrine, but at the same time they admit that that doctrine may be true or even that it is more probable.

But what should one say about the extreme case when our competent person sees with ineluctable evidence that the prohibited doctrine is certainly true? By going to the limit (a point readily understood by mathematicians), [53] that competent person would not be obliged to be convinced of the truth or possibility of the official decree, but would still be obliged not to impugn it publicly. The passage to the limit invoked here is not an abusive application of a principle of mathematical analysis to moral questions; the situation is properly conceived this way. The assent to the truth of the decree is a conditional one; it is understood (and the Church agrees) that no assent would be required if truly conclusive reasons should demonstrate its falsity; thus, if this condition is already satisfied for some particular person, the assent is not required. The Church knows and approves this. In such a case, she does not demand more than respectful silence. Note that this attitude is not a mask devoid of meaning; it is rather an expression of the sincere recognition of and respect for legitimate authority; it is an admission that certain disciplined behaviors are necessary for social reasons; it is a truly religious assent (as it is labeled) that bows before the Sacred Hierarchy. Its value and necessity can be understood by an obvious comparison with the discipline that is required in any social group, whether civilian or military: imagine the harm if a subject could reject or disparage with impunity the directives of superiors simply by declaring that he had [54] personal reasons for objecting to them! It could be that his evaluation was the correct one; but his protest would still be harmful. What group would be able to resist anarchic breakdown? In the same way, the Church, a group founded on faith, would be subject to fatal dissolution if an authentic guide (just because it was not infallible) could not authoritatively hold the reins. A particular error is less important; but one should not uproot the principle of unity and discipline without which the common interest would be inexorably sacrificed for the benefit of particular interests.

Let us now apply the criteria just discussed to the case of Galileo; he knew that if the authority of the Holy Office required one's obedience, its fallibility did not require one's conviction. To which one of the four categories can we assign our scientist?

No one thinks or has ever thought of assigning him to the category of persons who were not competent on the matter. Perhaps the judges thought that he was in the category of persons who were persuaded of the correctness of the decree, if not since 1616 or when he was writing the *Dialogue,* at least after the 1633 proceedings, [55] for that would justify the declarations he made at the trial.

However, we place Galileo in the category of persons who were competent and appreciated very well reasons contrary to those valued by the judges. There remains to consider what was the degree of his appreciation: did he regard them as clearly apodictic, or did he admit the possibility (however remote) that they might not be on the right track?

Many authors claim that Galileo could not have the full certainty that would inexorably exclude any reasonable doubt. Here it must be noted that the issue is not to estimate the value which the arguments objectively have in themselves, but the weight they had in Galileo's mind, especially at that unpleasant moment when he was facing the Church's Supreme Tribunal, which furthermore was telling him that there were reasons of another kind beyond his competence to exclude the Copernican thesis. Those who want to place Galileo in the category of persons who are competent but not completely free of reasonable doubt claim that the argument which Galileo deemed strongest (namely the one from the tides) [56] was mistaken; and they wonder how an error could give conviction of the truth to the point of certainty. Moreover, the extrinsic authority of real and serious experts of that time, such as the supreme authority of Kepler, indicated that the problem was not completely solved.[36]

Despite such arguments, however, many prefer to place Galileo in the fourth exceptional category of persons who are competent and fully enlightened and convinced. Galileo's mind was formed entirely on novel views and must have regularly experienced incompatibility with the ancient ideas, much as we experience it today. Then, the flaws of the individual arguments must have been counterbalanced by the intuitive genius with which Galileo was ahead of his time and penetrated (with a new intelligence and perceptiveness) the value of the combination of the various arguments.[37] If this is so, we should perhaps conclude that the legitimate meaning of the sentence, regardless of what the judges with their erroneous presumptions thought, was such that he was exempted from an internal conviction and it limited itself to demanding only a religious assent consisting of a sincere respect for the authority of the Church.

We must admit that there may be a difficulty with this interpretation. [57] For Galileo did not appear before judges who asked him his opinion, and so sincerity would require him to explain his internal opinion and his reasons for his different opinion.[38] But could not Galileo have regarded himself legitimately exempted from such an act? He could have said, "For twenty years I have been explaining my reasons, and these judges have not yet understood them. On the other hand, the real experts immediately understood their great strength and probability. Why then waste any more time, and instead run the risk of conveying a false impression, namely the impression of a lack of submission to sacred authority and divine revelation? In such a situation," Galileo could legitimately conclude, "the Church cannot and does not intend to ask more of me."

It would be interesting to know to what category of competent persons Galileo would have been assigned by the priests and friars (such as Msgr. Dini, Fra Maculano, Benedetto Castelli, and others) who knew Galileo's thought very closely and shared his idea that the question [58] should not be resolved with Scripture. We cannot know their judgment; but we must admit that they did not fail to enlighten Galileo's conscience on this subject, indicating to him courses of action in difficult situations for someone who is both a scientist and a Christian.

These considerations confirm the hypothesis that Galileo in one way or another, perhaps by means of a subjective evaluation that was more or less correct but in good faith, acted with sincerity at the crucial moment of the abjuration. Nor can one deny that in subsequent years there followed the effects of that resolute and frank act of will. A purely external and insincere adaptation to a superior coercion would have turned the illustrious scientist into someone who constantly betrayed himself with his contrived and feigned attitude; and we would not have had Galileo be faithfully submissive despite his suffering, or seek consolation in conversation with his daughter Sister Maria Celeste, or (in his confidential statements to Fabri) speak of the comfort of his good conscience in the eyes of God.

[59] In his act of obedience Galileo did not contradict his intellect but made an extremely serious intellectual sacrifice. In 1615 he had written to Dini, "My upbringing and inclination are such that, rather than contradict my superiors, 'I should pluck my eye out so that it would not cause me to sin.' "[39] Later in his life he had to endure blindness, and the light went out for those pupils that had so passionately scrutinized the heavens. Here he was being asked to make a more intimate sacrifice: it certainly was not the light of his mind that went out, but it could not shine with joy in the satisfaction of enlightening the minds of others. The renunciation must have been painful like a deep cut into raw nerves, which had palpitated so much and with so much ambition and feverish impatience for glory. Considered by itself, the renunciation could only cause repugnance in Galileo's soul; but he was able to view it from a broader perspective. He considered the unity [60] of the Church, per the last will and testament of Christ; he considered the practical requirements of the faith, not for himself, but for the masses that needed the guide of an authority; and he considered the necessity of respecting that authority, which was indeed committing its first error in the sixteen centuries of its life, but without which it would be impossible to avoid a daily erosion and dissolution of the framework of doctrines pertaining to the faith, harming interests that were higher and more general than astronomy and worldly glory. It is noble to sacrifice oneself for greater purposes, acting as a good soldier in a disciplined organization.

One cannot deny that such feelings suggest that we are dealing not only with a mind that shines, but also with a heart that beats, with a faith that believes, and with a will that decides. We are dealing with a whole man who deserves admiration both for his high intellect and for the magnanimous strength of his virtue, capable of heroism. Such was Galileo.[40]

So for Soccorsi, Galileo's retraction was a sincere, indeed admirable, act of religious heroism. It was sincere because it did not require an internal adoption of the geostatic belief, let alone an internal reasoned conviction of it, but rather only an intention "not to impugn it publicly" (53) or to observe "respectful silence" (53) on the controversy; and this intention had to be based on appropriate reasons, such as "on the authority of the judges and on the value of the reasons that persuaded the judges themselves" (52), and more generally on "an admission that certain disciplined behaviors are necessary for social reasons" (53). However, such an intention and such a motivation were quite possible.

On the other hand, the stronger the internal reasons supporting one's internal conviction and opposing the judges' belief, the harder it becomes to maintain such a position. And this is where intellectual sacrifice is required, so that the renunciation can be seen to be admirable and even heroic. Now, Galileo definitely belonged to Soccorsi's third or fourth category of respondents to the official judgment. As regards the choice between the two, Soccorsi did not really commit himself. However, he stated (55–56) that there were reasons for thinking that Galileo was strongly inclined to believe heliocentrism but could not exclude all reasonable doubt (which would place him in the third category); and Soccorsi also stated (56) that there were reasons for thinking that Galileo was certain and believed to be in possession of conclusive arguments (which would place him in the fourth category).[41] At any rate, in light of Soccorsi's own treatment of the fourth category as the upper limit of the third one, it is relatively unimportant to determine to which category Galileo belongs. The important thing is to know, understand, analyze, and evaluate properly the Galilean arguments in the *Dialogue*. Two of Soccorsi's analyses, in particular, are worth summarizing here.

One of Soccorsi's points involved Galileo's argument for the earth's motion based on the annual pattern of sunspot motion across the solar disk.[42] Soccorsi gave a sophisticated and plausible reconstruction of this argument, which renders it immune from the many criticisms that are often advanced against it.[43] The key point is that in deciding whether to explain the annual motion of sunspots by attributing motion to the earth or to the sun, one must not disregard the diurnal motion; then can one see the superiority of the geokinetic over the geostatic explanation.[44]

Although Galileo's individual arguments for the earth's motion were stronger than his critics imagined, it remains true that none of them provided a rigorous or conclusive demonstration. This gave Soccorsi the opportunity to make the very shrewd observation that in the situation Galileo was dealing with, the entire case for Copernicanism was stronger than one might think if one combined the evidence by means of any simple additive rule; rather it was a situation where the whole was greater than the sum of the parts, so to speak: "If no one of the arguments constituted a rigorous demonstration, still their combination . . . was not devoid of persuasive force, especially for a mind that could comprehend the synthesis and penetrate it with the new point of view of the new mechanics. . . . This observation allows us to explain how it was possible that there was a profound misunderstanding between Galileo and those old-fashioned minds . . . which . . . at most were concerned with asking the experts whether any one of the proofs was conclusive: in this manner they missed the persuasive force of Galileo's arguments" (80–81).

However, there is a difficulty in this explanation and justification of Galileo's retraction. Soccorsi's account is quite plausible with regard to Galileo's confession (at the second deposition on April 30) that in his *Dialogue* he had unintentionally done something wrong, that he was sorry, and that he was ready to make amends. A similar explanation and justification would apply to Galileo's original acquiescence when informally warned by Bellarmine (in 1616) or his original promise that he would not (publicly) hold or defend Copernicanism. But the abjuration amounted to more than renouncing public advocacy and keeping silent: it involved saying, "With a sincere heart and unfeigned faith I abjure, curse, and detest the above-mentioned errors and heresies."[45] While these words could make Galileo's act insincere, immoral, and sinful, the Inquisition would perhaps fare no better; for as some critics have argued, its imposition of the penalty of abjuration meant tempting Galileo to perjure himself, which is to say leading him into sin; and such temptation is itself sinful, indeed a graver sin. Soccorsi tried to resolve this difficulty as follows:

[100] With regard to a given doctrine, first there can be a speculative judgment, which considers directly the intrinsic truth or falsity of the doctrine. But there can also be a practical judgment, which [101] considers the doctrine in relation to a norm to follow; in our case, such a norm pertains to the faith, and the judgment does not evaluate directly the truth or falsity of the doctrine, but its safety or danger as regards the faith. A doctrine is called "safe" when it is certain that it does not contradict the faith, or when it agrees positively with the faith at least with sufficient probability so that it can be prudently allowed; and a doctrine is dangerous when it does not enjoy such probability.

This is a clear distinction between the two aspects of truth and falsity on the one hand and of safety and danger on the other. In fact, speculatively speaking, a given doctrine cannot change from being true to being false; but a true doctrine may be rightfully judged not safe for lack of sufficient clarity, and then it can become safe after clearer arguments have shed light on the difficulties; this is like a road, which can become safe after daylight arrives, although it was dangerous at night. Even a false doctrine (such as the Ptolemaic one) can be safe, whenever there is no conflict with the faith; in this case the error of those who follow the doctrine in no way involves the faith.

Many decrees of the Holy Office indicate clearly that they intend to judge the safety or danger of a doctrine by explicitly qualifying it as "safe" or "not safe." Other times the decree simply rejects a doctrine by formulating other kinds of censures. And there are theologians, some of great and highest authority (for example, Franzelin and Billot), who claim that all doctrinal decrees of the Roman Congregations have a practical character that properly judges the safety or danger of the doctrine, regardless of what the evaluations and censures are, for these say only to what extent the doctrine agrees or does not agree with the rules of the faith.[46]

This nature of the doctrinal judgments of Roman Congregations [102] is presumably determined by the very nature of these Congregations and of their function, which is not that of defining questions but that of ensuring the well-being and safety of the doctrine of the faith; this is in accordance with article 274,1 of Canon Law (on the Holy Office "of safeguarding the doctrine of the faith and morals").

This is not the place to discuss to what extent the account of Franzelin and Billot can be supported and accepted in its generality, but rather to consider to what extent it can be applied to the case of Galileo, not only from an abstract and theoretical point of view but also from a practical and historical one.

From a theoretical point of view, it is certain that their account gives to the formulations in Galileo's abjuration a less rigid meaning, according to which to reject the Copernican doctrine does not mean to judge it false and erroneous (even with a conditional and revocable judgment), but rather only to consider it to be a dangerous doctrine and for this reason to no longer hold it.[47] Thus, "I abjure" presumably means "I renounce a doctrine which I believe to be not safe as regards the faith," and I do this not by reason of divine revelation, but by reason of the sacred authority of the ecclesiastical institution that clearly has the task of safeguarding the well-being and safety of the doctrine of the faith. Such a formulation does not create any difficulty for the sincerity of the abjuration by someone who may be convinced of the truth[48] of the doctrine. In fact, although Galileo held the Copernican doctrine to be true, he could simultaneously consider it to be not safe as regards the faith, not for him personally, but for the community of believers; given the unclear status of the question and the prestige of the ecclesiastical judges, these believers may not have understood the compatibility of the Copernican doctrine with Scripture, and from the assertion of the doctrine they might have been led (although incorrectly) to doubt Scripture. In other words, Galileo could in good faith recognize that, although the new doctrine was true, it was nevertheless dangerous for the faith to advance that doctrine at that time in those particular circumstances, given that prejudices and panic had placed it in apparent conflict with Scripture.

[103] So far we have examined the theoretical aspect of the question. But the aspect that interests us more is the historical one. The question is to see whether in Galileo's case the sentence and the abjuration really conformed to the criteria described.

It seems that this claim cannot be maintained as regards the judges. Although they did not think they issued an infallible dogmatic definition (for which they were not the appropriate body), nevertheless they must have been definitely persuaded that they had read their opinion in the Sacred Scripture; thus, they proposed it as a truth that had to be accepted without question.

However, despite the erroneous presumption of the judges, Galileo was in a position to view the abjuration in its legitimate meaning and may have interpreted it in the sense of the account expounded above. To corroborate this hypothesis one can adduce the fact, mentioned by Billot, that already in Galileo's time the theologian Johannes Caramuel Lobkowitz distinguished "speculative" authority (which he attributed only to the Roman pontiff speaking ex cathedra) and the "practical" authority of the Congregation. Galileo may have been informed of such a view by his theologian friends. The hypothesis is certainly legitimate, but it belongs more to the realm of possibilities than probabilities.

It is more plausible to suppose that, independently of excessively technical disquisitions on the status of the doctrinal decrees of the Holy Office, Galileo considered the authority and the fallibility of the judges and did not take the verb *abjure* (which in itself has various meanings) in the sense of a speculative acceptance of a decree he regarded as erroneous, but in the practical sense of a renunciation of a true doctrine whose acceptance rendered him suspected of heresy. This supposition too provides a radical solution to the difficulty of the sincerity of the abjuration.[49]

Although Soccorsi's argument is not completely convincing, he was right to stress that the sincerity of Galileo's abjuration depends on the meaning of the word *abjuration,* and that its meaning is not obvious. Moreover, one cannot deny that Soccorsi's solution is ingenious and technically grounded and has some plausibility. He was claiming that when Galileo said, "I abjure the above-mentioned heresies and errors," he did not mean that he was really rejecting the propositions of Copernicus as heretical or erroneous theologically or speculatively (philosophically), but rather that he was abandoning their advocacy, giving deference to the authorities' judgment that they were practically unsafe or dangerous, namely harmful to religion (because those propositions were widely, albeit incorrectly, perceived to be at least contrary to Scripture). Finally, there is no question that Soccorsi's solution to this difficulty was another obvious indication that his account of Galileo's trial was meant to rehabilitate him religiously and morally.

Chapter 15

Secular Indictments

Brecht's Atomic Bomb and Koestler's Two Cultures
(1947–1959)

At about the same time that Galileo was being implicitly rehabilitated by various Catholic persons and institutions as a result of the tricentennial of 1942, he became the subject of unprecedented criticism by various representatives of secular culture. It was almost as if a reversal of roles was occurring, with his erstwhile enemies turning into friends, and his former friends becoming enemies. Several other circumstances add interest and significance to such a development. These critics elaborated what might be called social and cultural criticism of Galileo. They were mostly writers with backgrounds and sympathies subsumable under the left wing of the political spectrum. And the most outstanding and original examples of such criticism were a German playwright, Bertolt Brecht, whose account is found in a drama, and a Hungarian-born writer, Arthur Koestler, who was best known as a novelist but decided to write what was purportedly a history book. In short, we now must examine how Galileo's trial was viewed by secular, socially conscious, left-leaning literary intellectuals in the middle part of the twentieth century.

Although Brecht's drama is the most famous and important work of creative writing on Galileo, it was neither the first nor the last. François Ponsard's *Galilée* was performed for about a month in Paris in March 1867,[1] and in 1875 Eleuterio Llofriu y Sagrera's *Galileo: Episodio Dramático en un Acto y en Verso* was performed with greater theatrical success in Madrid.[2] Theatrical works on Galileo had already been written in the other three major European languages: Italian in 1820; English in 1850; and German in 1861.[3] And since Brecht, there have been at least four other plays in English.[4] I have also come across at least four novels about Galileo.[5]

Similarly, although Brecht and Koestler are the most distinguished, well-known, and accessible secular critics to focus specifically on Galileo's trial, they were perhaps part of a tradition of sorts. They were preceded (in the 1930s) and may have been influenced by Edwin Burtt and Edmund Husserl,[6] who, however, focused on abstract philosophical questions and on Galileo's scientific work in general; and they were followed by Paul Feyerabend, who in the 1980s charged Galileo with "the tyranny of truth."[7]

15.1 Galileo's Social Betrayal: Brecht's Historical Fiction (1947/1955)

Perhaps more people have been led to reflect on Galileo's trial by Brecht's play *Galileo* than by any other single cause. However, the drama had a very modest, indeed obscure, beginning.[8] In November 1938, Brecht wrote a play titled *The Earth Moves,* which was slightly revised within several months and retitled *Life of Galileo;* it was performed for the first time in Zurich on 9 September 1943.[9] At the time of its writing, Brecht had been living in exile in Denmark for several years, having had to leave his native Germany when the Nazis came to power. He was then deeply concerned with the question of whether it was right to escape Germany and thus seek safety and fight Nazism from the outside, or whether it might have been better to remain there and continue the anti-Nazi struggle in a covert manner from within. This concern was reflected in the play's stress on Galileo's external abjuration of Copernicanism in 1633 in order to covertly pursue his work in mechanics and publish the *Two New Sciences* in 1638.

Brecht moved to the United States in 1941,[10] and between 1944 and 1947 (while living in Southern California) he collaborated with the actor Charles Laughton to complete a revision and English translation of the play that amounted to a second version of it. At that time Brecht was deeply affected by the construction and dropping of the first atomic bomb, and the consequent problem of the social responsibility of scientists. Titled simply *Galileo,* the play was performed for the first time at the Coronet Theatre, Los Angeles, on 30 July 1947 and then opened at Maxine Elliott's Theatre, New York, on December 7. Both productions were directed by Joseph Losey, with Laughton playing the leading part. The play received bad reviews, and each run lasted only three weeks. This "American" version of the play was published for the first time in 1953.[11]

In the late 1940s Brecht left the United States for Switzerland,[12] and eventually he moved to East Berlin, in Communist East Germany. There, between 1953 and 1956, he revised the play into a third and last version; and although this time the substantive revisions were not major, the stylistic and linguistic ones were significant because the second version existed only in the Laughton English-language version, and so this final revision involved a translation into German of Laughton's text. Besides his previous social and political concerns, Brecht now worried about such problems as the Cold War, the analogy between Soviet-style Communism and the Catholic Church of Galileo's time, and the building of the hydrogen bomb, including the Robert Oppenheimer affair. Retitled *Life of Galileo,* the revised play was first performed in Cologne on 16 April 1955.[13] The playwright died the following year, before the play opened in Berlin on 15 January 1957 in a production by the Berliner Ensemble, the group of which he had been a member.[14]

From then on, Brecht's *Galileo* started gaining popularity and critical acclaim until it became the classic it is today. For example, on 16 June 1960 it opened at the Mermaid Theatre, London;[15] in December 1962 at the Actor's Workshop, San Francisco;[16] and on 22 April 1963 at the Piccolo Teatro, Milan, directed by Giorgio Strehler.[17] By 30 October 1966, it had undergone at least eighty-nine productions in cities worldwide.[18] And on 13 August 1980, the play opened at the National Theatre, London; this production had been conceived in 1973 and the performance continued for several months; it was popular with the public but received mixed reviews from the critics.[19]

The final version of the play has fifteen scenes. It begins with a depiction of Galileo at home in Padua, explaining some details of the geokinetic theory to Andrea Sarti (the son of his housekeeper) and giving a long speech about the dawn of a new age; this is the longest speech in the play and stresses that "a vast desire has sprung up to know the reasons for everything"[20] and that "our own lifetime will see astronomy being discussed in the marketplaces" (8). Then a young nobleman named Ludovico Marsili arrives to arrange for private lessons from Galileo; before leaving, the young man tells Galileo he has just returned from Holland and reports the invention of a device made with two lenses that magnifies distant objects. Then a financial officer from the University of Padua comes, bringing news that Galileo's request for an increase in salary has been denied. The conjunction of the two visits starts Galileo thinking about constructing a telescope and offering it to his employer.

In the next scene, Galileo demonstrates the telescope to some officials of the Venetian Republic. Here Brecht portrays Galileo as claiming deceptively and cynically that the instrument is a new invention, original to him. In return, the professor receives a huge increase in salary and lifetime tenure.

The third scene takes place in 1610. By means of the telescope, Galileo has just observed many phenomena in the heavens that confirm Copernicanism. He is confident he will be able to convince others; this confidence is portrayed partly as an expression of naïve empiricism and partly as faith in the ability of common people to see the light once it is pointed out to them. Accordingly, Galileo declares: "Today is 10 January 1610. Today mankind can write in its diary: Got rid of Heaven" (24). However, his friend Sagredo warns him of possible difficulties.

In the fourth scene, Galileo has moved to Florence and is trying to explain his discoveries to the ducal court. A mathematician and a philosopher refuse to look through the telescope to see Jupiter's satellites on the grounds that such heavenly bodies are impossible and unnecessary according to Aristotle's philosophy.

Next, the plague is raging in Florence. On its account the grand duke has left the city and sent a carriage to Galileo's house so that he too can leave. But he refuses because he is in the midst of some research, which later leads him to the discovery of the phases of Venus.

The sixth scene takes place at the Jesuit Roman College in 1616. Galileo and several churchmen and scholars are waiting for Professor Christopher Clavius to report his evaluation of Galileo's discoveries. While they wait, various churchmen advance arguments against Galileo: the earth's motion contradicts the Joshua passage in Scripture; the new discoveries "degrade humanity's dwelling place" (52); "heaven and earth are no longer distinct" (52). Moreover, Galileo's removing mankind from the center of the universe "makes him an enemy of the human race" (53); contradicts the fact that "mankind is . . . God's highest and dearest creature" (53); undermines the doctrine that God "would . . . send His Son to such a place" (53); and suggests that "you are fouling your own nest" (53). Then Clavius enters and gives his verdict: "He's right" (54).

In the seventh scene, there is a party at the house of Cardinal Bellarmine in Rome, attended by (among others) Cardinal Maffeo Barberini (the future Pope Urban VIII), Galileo, his daughter, and Ludovico Marsili (to whom she is now engaged). In a conversation with Galileo, Barberini expresses his favorite objection based on divine omnipotence: since God is all-powerful, he could have created a universe with a motionless earth; hence, no matter how much evidence there is in favor of the earth's motion, we can never assert with certainty that the earth *must* move, because do so would be presuming to limit God's power to do otherwise. In another exchange, Bellarmine expresses an instrumentalist interpretation of Copernicanism. Finally, in the name of the Inquisition, Bellarmine warns Galileo to abandon his Copernican views; after some further exchanges, including Bellarmine's clarification that "you are also at liberty to treat the doctrine in question mathematically" (60), Galileo acquiesces.

The eighth scene is a conversation at the Tuscan embassy in Rome between Galileo and a little monk who also appears in the Roman College scene. This monk has decided to give up astronomy as a result of the Index's decree prohibiting Copernican books. He argues that common people (such as his parents) could not tolerate their hard lives if they knew that they were not at the center of God's creation and that God's word was erroneous; in any case, he adds, truth will prevail, so if Copernicanism is really true, eventually it will prevail even without his efforts. To this Galileo replies, "No, no, no. The only truth that gets through will be what we force

through: the victory of reason will be the victory of people who are prepared to reason" (68).

In the next scene, which takes place eight years later, Brecht portrays
Galileo investigating floating bodies in order to stay away from the subject
of the earth's motion, on which he has been silent since 1616. Ludovico
informs Galileo that the pope is dying and that Barberini will probably be
the new pope. As a result, Galileo decides to resume research on the dangerous topic by investigating sunspots and solar rotation. This leads
Ludovico to break off his engagement to Galileo's daughter Virginia.
Despite the inaccuracy of this chronology and the invention of this incident, Brecht does put in Galileo's mouth the insightful remark that "my
object is not to establish that I was right but to find out if I am" (80–81).

The tenth is a street scene during the carnival of 1632. Over the previous
several years Galileo's ideas have become popular with the common people. This is reflected in the words recited by one of the ballad singers.
Galileo's doctrine is allegedly undermining the rule that "around the
greater went the smaller" (82); the result will be that everyone will "say and
do as he pleases" (83); then servants will defy masters, altar boys will defy
priests, tenants will defy landlords, and wives will defy husbands; but "people must keep their place, some down and some on top" (84). Moreover,
the authority of Scripture will be rejected; in fact Galileo is labeled "the
bible-buster" (85).

The next scene takes place at the ducal palace in Florence in 1633, after
Galileo has been summoned to stand trial in Rome. He brings a copy of his
Dialogue to the grand duke and is told that the duke cannot prevent the
trial. But a powerful and wealthy iron founder named Vanni tells Galileo
that all businessmen in Northern Italy are behind him and offers to help
him escape.

In the twelfth scene, an Inquisition official presents to the pope the
charges against Galileo. First, he observes, "a terrible restlessness has
descended on the world. . . . Are we to base human society on doubt and no
longer on faith?" (91). Moreover, "your Holiness's Spanish policy has been
misinterpreted by short-sighted critics" (91). Third, "what would be the
effect if they were to believe in nothing but their own reason, which this
maniac has set up as the sole tribunal?" (92). Next, "they would start wondering if the sun stood still over Gibeon" (92). Fifth, "God is no longer necessary to them" (92). And there is "the abolition of top and bottom" (92).
Seventh, "this evil man . . . writes . . . not in Latin but in the idiom of fish-
wives" (92). Finally, "his book shows a stupid man, representing the view of
Aristotle" (93). The pope decides that "at the very most he can be shown
the instruments" of torture (94).

The thirteenth scene takes place on 22 June 1633, the day of Galileo's
sentencing. A group of his disciples, including Andrea, are waiting for the
conclusion of the trial, confident that he will not recant. They are devas-

tated when they learn that he has abjured. Andrea cries, "Unhappy the land that has no heroes" (98), and he inveighs against Galileo, "Wine-pump! Snail-eater! Did you save your precious skin?" (98). Galileo's only words are, "No. Unhappy the land where heroes are needed" (98).

In the climactic penultimate scene (number fourteen), set a few years after the abjuration, we are at Arcetri, where Galileo is under house arrest. Andrea comes to see Galileo and say good-bye, as he is about to leave for Holland. To his pleasant surprise, Andrea learns that Galileo has completed the manuscript of the *Two New Sciences* and wants him to smuggle a copy out of Italy. This redeems the master in the disciple's eyes. For Andrea, "this alters everything. Everything. . . . You were hiding the truth. From the enemy. Even in matters of ethics you were centuries ahead of us. . . . So in '33 when you chose to recant a popular point in your doctrine I ought to have known that you were simply backing out of a hopeless political wrangle in order to get on with the real business of science . . . which is . . . studying the properties of motion, mother of those machines which alone are going to make the earth so good to live on that heaven can be cleared away" (106).

However, Galileo does not agree with Andrea's assessment. First, Galileo confesses, "I recanted because I was afraid of physical pain. . . . They showed me the instruments" (107). Then he expresses his attitude by telling his disciple, "Welcome to the gutter, brother in science and cousin in betrayal" (107). The point is that Galileo feels he has betrayed the true cause of science, so much so that "I no longer count myself a member" (107). The reason is that "as a scientist, I had a unique opportunity. In my day astronomy emerged in the market place. Given this unique situation, if one man had put up a fight it might have had tremendous repercussions. Had I stood firm the scientists could have developed something like the doctors' Hippocratic oath, a vow to use their knowledge exclusively for mankind's benefit. As things are, the best that can be hoped for is a race of inventive dwarfs who can be hired for any purpose. . . . I handed my knowledge to those in power for them to use, fail to use, misuse—whatever best suited their objectives" (109).

The play ends with a scene in which Andrea crosses the Italian frontier, thus taking the manuscript of the *Two New Sciences* to safety.

The first thing that needs clarification is the issue of truth or accuracy from a historical or factual point of view versus plausibility or effectiveness from a dramatic, theatrical, or fictional point of view. This is necessary because otherwise one will misunderstand the play and raise irrelevant objections against it. It would seem largely misconceived to criticize the play because, for example, contrary to what it states or implies, Galileo did not investigate floating bodies in 1624, after the anti-Copernican decree, but rather mostly in 1611–1612, before that prohibition; similarly, he studied sunspots mostly in 1612–1613, before that decree, and not in 1624, after

Barberini became Pope Urban VIII; Galileo's daughter Virginia was never engaged to be married but rather became a nun in 1616, the year Brecht portrays her as Marsili's fiancée; a fortiori, Galileo's research cannot be blamed for the breakup of her engagement; and the manuscript of the *Two New Sciences* was not smuggled out of Italy by one of Galileo's disciples but rather first (in 1635) by Prince Mattia de' Medici to Germany,[21] and later (in 1636) by the publisher Louis Elzevier to Holland.[22]

But perhaps such criticism (exposing inaccuracies and falsehoods) is largely irrelevant not because it raises factual objections to the play but because the inaccuracies in question are relatively insignificant. Here the inaccuracies just exposed can be usefully contrasted with major ones. For example, in Brecht's play a crucial issue in the controversy is the centrality of the earth: not only whether it is *true* that the sun rather than the earth is at the center, but also whether it is *proper* that this should be so; that is, for Brecht the replacement of geocentrism by heliocentrism was the source of most opposition to Galileo's ideas because abandoning geocentrism was seen as undermining an anthropocentrism that was felt essential for human life to have meaning, at least among common people. However, as a matter of fact this was not a key issue in the Galileo affair; to be sure, it was one of the many minor issues, but not one of the crucial ones. Instead, the most important issue was the question of the scientific (or philosophical) authority of Scripture, which Brecht does mention, but whose significance he does not perceive. Thus, Brecht was clearly wrong on this topic.[23]

Another comparable major error is Brecht's portrayal of Galileo as a naïve empiricist, someone who thinks that sense experience is primary and all-important in scientific investigation, and consequently that simple observation could show that his astronomical views were right and those of his opponents wrong.[24] It is true that Galileo was not the apriorist portrayed by scholars such as Alexandre Koyré, for such apriorist interpretation is the opposite extreme of naïve empiricism and equally wrong.[25] However, the observations to which Galileo was appealing had to be interpreted, and their interpretation required argumentation; this was especially true since his observations involved an artificial instrument that could be questioned for its methodological admissibility and practical reliability and the optical-theoretical details of how it worked.

The question now is whether these major misinterpretations are also irrelevant, as the minor ones admittedly were. Part of this question hinges on the writer's intention: did Brecht claim to be giving a historically accurate portrayal, or did he claim to be writing a work of fiction whose aim was different? Eric Bentley, who has discussed this problem with masterful insight, reported talking to Brecht about it but receiving ambivalent answers.[26] In any case, the more important question is whether, independently of the author's intention, the actual work is trying to do something different from conveying historical information and so is operating by different rules. It seems obvious that Brecht's play does not operate in the domain of history but in that of theater. The hard thing is to define what theater is and how it differs from history.

Such a definition can hardly be elaborated here, but it is instructive to refer to an essay by Bentley intriguingly titled "The Science Fiction of Bertolt Brecht." Paraphrasing the views of Aristotle, Gotthold Lessing, and Luigi Pirandello, Bentley claims "that drama has a different logic than that of fact. History can be (or appear to be) chaotic and meaningless; drama cannot. Truth may be stranger than fiction; but it is not as orderly. . . . The truth doesn't have to be plausible but fiction does."[27] For Bentley, works such as George Bernard Shaw's *Saint Joan* and Brecht's *Galileo* embody the following paradox: "The historical truth, rejected for its implausibility, has the air of an artifact, whereas the actual artifact, the play, has an air of truth. The villains of history seemed too melodramatic to both authors. The truth offended their sense of truth, and out of the less dramatic they made the more dramatic."[28] In short, to paraphrase Pascal's aphorism about the reasons of the heart, drama or theater aims at its own truth that (historical) truth does not know.

Having clarified (I do not mean solved) this problem of historical accuracy versus dramatic plausibility, let us examine Brecht's play on its own terms. In this regard, an invaluable source is a series of comments made by the playwright himself on various occasions. As long as we understand that his saying something about the play does not automatically make it so, his remarks are very instructive.

Is the play a tragedy (i.e., a pessimistic play) or a comedy (i.e., an optimistic play)? Brecht himself once raised this question[29] when he suggested that his play may not be a tragedy because, although its ending (scene 14) is negative, its beginning (scene 1) stresses the dawn of a new age; and although the traditional criterion attaches greater importance to the ending than to the beginning, he did not write the play with such rules in mind; indeed, "the play shows the dawn of a new age and tries to correct some of the prejudices about the dawn of a new age."[30] I would add that even if one emphasizes the ending, the climactic penultimate scene embodies two messages: Andrea's positive one that Galileo has succeeded in finishing his *Two New Sciences*, which contains an epoch-making contribution to the science of motion and machines, and so will not only advance human knowledge but also human well-being; and Galileo's own pessimistic view that he betrayed science because he started a struggle, gave up during the fight by submitting to authority, and thus started the trend of scientists' abdicating their social responsibility and letting authorities decide how to use the discoveries they make. Moreover, the existence of these two themes, in the context of our own theme of retrying Galileo, suggests that Brecht's play may not be as anti-Galilean as many[31] have alleged.

On another occasion, Brecht made the extremely important clarification that his play was not a depiction of the struggle between science and religion but rather of the conflict between science and authority in general. He explicitly declared that it would be an impoverishment of the play to see it as advancing an anticlerical or anti-Catholic message, for in it "the Church functions, even as it is opposed to free investigation, simply as authority . . . the play shows the temporary victory of authority, not the victory of the priesthood."[32] Furthermore, the excessively religious interpretation of the play would be more than an aesthetic and theatrical error, inasmuch as it would be a social and political error: "It would be highly dangerous, particularly nowadays, to treat a matter like Galileo's fight for freedom of research as a religious one; for thereby attention would be most unhappily deflected from present-day reactionary authorities of a totally unecclesiastical kind."[33] Brecht wrote these reflections in 1939 while in exile, and so he had in mind, at least, the extrapolation of the Galilean fight to a struggle against the Nazi government of his native Germany. However, it is unclear how seriously he meant these words then or whether he continued to believe them later when he moved to Communist East Germany; for although such a move undoubtedly had a complex motivation, one reason may have been an insensitivity or a blindness to the striking parallels between the Church of Galileo's time and Soviet-style Communism. As Brecht's otherwise sympathetic translator and editor put it, "The parallels are too clear: the Catholic Church is the Communist Party, Aristotle is Marxism-Leninism with its incontrovertible scriptures, the late 'reactionary' pope is Joseph Stalin, the Inquisition the KGB. Obviously Brecht did not write it to mean this, and if he had seen how the local context prompted this interpretation he might have been less keen for the production to go on. But as things turned out it has proved to be among the most successful of all his plays in the Communist world."[34] In conclusion, on the one hand, Brecht's invitation that we think of authority simpliciter rather than of religious authority is an insightful (although not totally original) interpretation of Galileo's trial and has great proclerical potential; on the other hand, Brecht's failure to follow his own advice outside the theater in his own practical life suggests that either he may not have put into practice what he was preaching or that what he was preaching was something different from what he said it was.

We now come to what is commonly regarded as the play's central message, a message that is both socially laden and anti-Galilean, and one that is not only found in the play but also in Brecht's reflective pronouncements: "Galileo's crime can be regarded as the 'original sin' of modern natural sciences. . . . The atom bomb is, both as technical and as a social phenomenon, the classical end-product of his contribution to science and his failure to contribute to society."[35] And he is to be blamed not just for the bomb, but also for the undesirable consequences of "what is called the Industrial Revolution. In a sense Galileo was responsible both for its technical creation and for its social betrayal."[36] To be sure, Brecht tried to qualify such a judgment, claiming that "one can scarcely wish only to praise or only to condemn Galileo";[37] and such a qualification seemed to reflect the twofold character of the play's climax. But Brecht also qualified this qualification when he also said, "In the Californian version . . . Galileo interrupts his pupil's hymns of praise to prove to him that his recantation had been a crime, and was not to be compensated by this work, important as it might be. *In case anybody is interested, this is also the opinion of the playwright.*"[38]

One question about the issue of Galileo's crime involves historical truth versus dramatic plausibility. The betrayal which Brecht attributed to Galileo is also one of which Galileo accuses himself at the end of the play. Historically speaking, it is not at all likely that Galileo would have accused himself in this manner. On the contrary, his fight against religious and biblical interferences in the search for truth suggests that he would have opposed social and political interference, whether advanced by authorities and institutions or by individuals or persons; that is, he would have wanted to distinguish the question of whether a proposition is factually or scientifically true from the question of whether its truth or acceptance is socially harmful or beneficial. The same conclusion is suggested by his frequent criticism of teleological and anthropocentric ways of thinking, which reduce to arguing that something is true because it is useful and false because it is useless.[39] It is more likely that the historical Galileo would justify himself in the manner in which the play's Andrea justifies the master's submission to authority.

However, let us now consider dramatic plausibility. Is it dramatically plausible for Galileo to accuse himself in his long speech at the climax of the play? Conversely, would it have been implausible for Galileo to justify himself with the argument with which Andrea justifies him? In any case, why? I fail to see why such self-justification would have been implausible. I do not see anything in the play to preclude it. Nor do I find any good reason in Brecht's reflections on the play. On the contrary, Galileo's self-accusation in the climactic scene is out of character with his words and actions in the rest of the play, and so it strikes me as dramatically flawed.

This criticism has been eloquently made by Bentley, and his argument deserves discussion. He compares the first (1938–1943) and the second (1943–1947) versions of the play, and for the purpose of this discussion we may take the third version (1953–1956) as essentially equivalent to the second. For Bentley, "*Galileo I* is a 'liberal' defense of freedom against tyranny, while *Galileo II* is a Marxist defense of a social conception of science against the 'liberal' view that truth is an end in itself."[40] In terms of dramatic action, "the sense of the earlier text is: I should not have let my fear of death make me overlook the fact that I had something more to defend than a theory in pure astronomy. The sense of the later text is: To

be a coward in those circumstances entailed something worse than cowardice itself, namely treachery."[41]

Here it should be stressed that Brecht's second version was advancing a
novel and original assessment: the accusation was not merely, or primarily,
cowardice, which of course went back at least to Brewster (1835); the crime
was betrayal or treachery insofar as Galileo started a fight but fled in the
midst of it. Brecht's own words on this point are worth quoting:

> The issue in Galileo's case is not that a man must stand up for his opinion as
> long as he holds it to be true. . . . The man who started it all, Copernicus, did
> not stand up for his opinion. . . . But unlike Copernicus who avoided a battle,
> Galileo fought it and betrayed it. . . . A new class, the bourgeoisie with its new
> industries, had assertively entered the scene; no longer was it only scientific
> achievements that were at stake, but battles for their large-scale general
> exploitation. . . . The new class, clearly, could exploit a victory in any field
> including that of astronomy. . . . Galileo became anti-social when he led his
> science into this battle and then abandoned the fight.[42]

Is this charge of treachery dramatically plausible? In particular, was the
second version a dramatic improvement over the first? Bentley did not
think so:

> As far as this penultimate scene is concerned, it is not clear that, in making it
> more ambitious, Brecht also improved it. To show the foulness of Galileo's
> crime, he has to try to plumb deeper depths. The question is whether this
> befouled, denatured Galileo can be believed to be the same man we have seen
> up to then. The impression is, rather, of someone Brecht arbitrarily declared
> bad at this stage in order to make a point. Which would be of a piece with
> Communist treatment of the Betrayal theme generally. One moment, a Tito is
> a Jesus, and the next a Judas. There is, perhaps, an intrusion of unfelt Com-
> munist clichés about traitors and renegades in the later *Galileo*. One cannot
> find, within the boundaries of the play itself, a full justification for the viru-
> lence of the final condemnation.[43]

One final point deserves discussion. Although Brecht's accusation is nei-
ther historically accurate nor dramatically plausible as a self-accusation by
Galileo, it may have some other kind of validity, perhaps philosophical.
That is, if we think of it merely as an accusation by Brecht himself against
Galileo, then it may be regarded as a social-philosophical criticism of
Galileo, and it may not be devoid of all validity. But that is not to say that it
is valid, for the criticism would have to withstand (among others) the Galilean arguments (which we have inherited) against the theological-pastoral,
teleological, and anthropocentric ways of thinking that were the traditional
and historical precursors of Brecht's social philosophy. On the other hand,
such a historical context suggests the possibility that some other historical
agent or dramatic character might have formulated (some appropriate version of) Brecht's socialist accusation against Galileo. Perhaps, if the play's

Andrea had advanced that complaint against his master, its dramatic plausibility might have been enhanced. And if that accusation could be attributed
to some appropriate historical agent, even the historical accuracy of such a
Brechtian account might be vindicated.

15.2 Galileo's Blame for "Science versus Religion": Koestler's Fictional History (1959)

In 1959, the novelist Arthur Koestler (1905–1983) caused a sensation with
a book titled *The Sleepwalkers: A History of Man's Changing Vision of the Universe.*[44] In a sense, the work was part of the increasing concern with, and discussion of, the problem of cultural fragmentation that prompted C. P.
Snow to deliver, in the same year, at Cambridge University, a lecture titled
"The Two Cultures and the Scientific Revolution."[45] One scholar has
described that discussion as the "Great Debate of Our Age"[46] and has
claimed that it elicited abusive criticism the like of which is seldom found in
literature.[47] The two cultures to which Snow was specifically referring were
the attitudes, thought, and behavior patterns of natural scientists on the
one hand and those of what he called literary intellectuals on the other.
Koestler too was concerned with the question of the sciences versus
the humanities, but he was even more concerned with another cultural
division.

Koestler's *Sleepwalkers* discussed two main themes. One of them was the
interaction of science and religion from antiquity to the seventeenth century, in which regard he advanced one of his chief theses: that is, in reality
science and religion share a deep commonality, and yet in the modern age
they experience a separation that is at best unfriendly and often worse.[48]
The second main theme was the history and nature of intellectual or scientific discovery; in this regard, Koestler claimed that "the history of cosmic
theories, in particular, may without exaggeration be called a history of collective obsessions and controlled schizophrenias; and the manner in which
some of the most important individual discoveries were arrived at reminds
one more of a sleepwalker's performance than an electronic brain's" (15).
Besides being important in its own right, the sleepwalker thesis helped
Koestler justify his other thesis about the deep-structural unity of science
and religion. The sleepwalker thesis also had an important corollary: the
debunking of science; for Koestler, this criticism was unintentional, but the
consequence was inescapable (15).

This general scheme worked as described for Koestler's treatment of
Kepler. The account of Kepler not only took up half of his book (and so was
about three times longer than the account of Galileo) but it was very sympathetic and sensitive, so much so that it is clear that Koestler had no intention of debunking Kepler. And in his analysis of the workings of Kepler's

mind, Koestler made a plausible and insightful case for the mystical origin and sleepwalking character of Kepler's discoveries.

However, Koestler's treatment of Galileo was very different. He even admitted and explained his anti-Galilean "bias" in a passage (425–26) which some may take as a candid confession, but others will regard as a rhetorical ploy that equivocates on the word *bias* (which can mean an unjustified or a justified conviction). His anti-Galilean animus did not derive from proclerical bias, for he was equally critical of the Church on account of the Inquisition's violence against heretics. Instead, Koestler confessed, "I find the personality of Galileo equally unattractive, mainly on the grounds of his behaviour towards Kepler" (425). Indeed, Galileo did not answer most of Kepler's letters; did not keep Kepler properly informed of his telescopic discoveries; did not properly appreciate Kepler's laws of planetary motion; and ridiculed Kepler's attempt to explain the tides as due to the moon's influence.

Koestler also had other, more objective and less personal reasons for disliking Galileo. One was that Galileo did *not* exemplify "the unitary source of the mystical and scientific modes of experience" (426). For Koestler, Galileo "was utterly devoid of any mystical, contemplative leanings, in which the bitter passions could from time to time be resolved; he was unable to transcend himself and find refuge, as Kepler did in his darkest hours, in the cosmic mystery. He did not stand astride the watershed; Galileo is wholly and frighteningly modern" (363). I believe Koestler is basically correct about Galileo's "modernity" and insensitivity to mystical experience and that these traits provide an "objective" factor that helps to explain his coldness toward Kepler. However, the important point here is that Koestler was frightened of such modernity. Moreover, apparently it did not matter to Koestler that Galileo was also deeply convinced of the harmony between science and religion. What mattered more was that Galileo's reasons were different: for Galileo science and religion were harmonious because they both derived from God;[49] whereas for Koestler, their harmony stemmed from mystical experience.[50]

There was a third, even more concrete, reason for Koestler's dislike of Galileo. It was connected with the "resentment" (425) Koestler felt that the clash between Galileo and the Church had occurred at all when it could have been avoided. In a way, this thesis was Koestler's peculiar way of expressing what I have called (chapter 11.4) the circumstantialist approach to Galileo's trial. However, Koestler's circumstantialism was such that Galileo received most of the blame. In fact, Koestler ended his account of the trial by saying that after Galileo's death in 1642, "when his friends wanted to erect a monument over his grave, Urban told the Tuscan Ambassador that this would be a bad example for the world, since the dead man 'had altogether given rise to the greatest scandal throughout Christendom.' That was the end of the 'perilous adulation,' and the end of one of the most disastrous episodes in the history of ideas; for it was Galileo's ill-conceived crusade which had discredited the heliocentric system and precipitated the divorce of science from faith" (495).

In summary, Koestler's account of Galileo's trial is embedded in the following scheme. There is in modern culture a separation between science and religion. The cultural, social, and human costs of such a separation are disastrous. This disaster is all the more acute inasmuch as the separation perverts the real nature of science and religion, which is such that they share a deep underlying unity; this unity lies in a certain type of mystical experience. The Galileo affair "is one of the historic causes of that division" (425); and Galileo himself, his behavior and his personality, is the chief cause of the affair.

It is perhaps no exaggeration to say that Koestler's criticism is the most serious indictment of Galileo since the original trial. It is certainly comparable with Brecht's holding Galileo responsible for the atomic bomb and the harmful consequences of the Industrial Revolution. Koestler's indictment is more serious insofar as it is advanced in a work that appears to be a history book; that has all the trappings of scholarship (references, footnotes, etc.); and that is well written, highly readable, and widely accessible. Moreover, Koestler's book is full of arguments that have the appearance of strength; I do not say that his account is well argued, for in this case the appearance does not correspond to reality. In short, Koestler was a sophisticated sophist.

But let us look at the details. His account of the trial has two main parts, corresponding to the two phases that climaxed with the prohibition of Copernicanism in 1616 and the condemnation of Galileo in 1633. As regards the earlier phase, it is well known that the Inquisition proceedings began in earnest when, on 7 February 1615, the Dominican friar Lorini filed a written complaint with the Inquisition, attaching Galileo's Letter to Castelli as incriminating evidence. Then on March 20 another Dominican friar, Caccini, made a personal appearance before the Roman Inquisition; in his deposition he charged Galileo with suspicion of heresy, based not only on the content of the Letter to Castelli, but also on hearsay evidence of a general sort and of a more specific type, involving two other individuals, named Ferdinando Ximenes and Giannozzo Attavanti. Koestler stressed the fact that the Inquisition consultant who examined the Letter to Castelli wrote that in its essence it did not deviate from Catholic doctrine; and that the cross-examination (in November) of the two witnesses, Ximenes and Attavanti, exonerated Galileo since the hearsay evidence of his utterance of heresies was found to be baseless. On the basis of these facts, Koestler concluded that "Lorini's denunciation had fallen through . . . Caccini's charges of heresy and subversion were a fabrication, and the case against Galileo was again dropped . . . Galileo himself had been cleared of all charges against his person" (442). This interpretation led Koestler to the question of why three months later Copernicanism was banned and Galileo warned to abandon it. The short answer is that it was all due to what Galileo said and did in those three months.[51]

To elaborate, first Koestler attributed to Galileo a motivation that amounted to a kind of hubris: "He had said that Copernicus was right, and whoever said otherwise was belittling his authority as the foremost scholar of his time. That this was the central motivation of Galileo's fight will become increasingly evident" (438). Then there was his refusal to compromise: by 1615 Galileo's attitude "was to refuse any compromise on the Copernican system. Copernicus did not mean it to be understood merely as a hypothesis. It was to be accepted or rejected absolutely" (446). To support this allegation, Koestler referred to Galileo's Letter to Dini of 23 March 1615, where indeed there is a sentence to this effect. Koestler also cited other evidence suggesting that a compromise was being proposed by such churchmen as Cardinal Bellarmine, Cardinal Maffeo Barberini, and the Jesuit Christopher Grienberger (professor of mathematics at the Roman College). "But Galileo was by now beyond listening to reason. For, by accepting the compromise, he would . . . be 'laughed out of court'. Therefore he must reject it. . . . He must insist that the Church endorse it, or reject it, absolutely."[52] One reason why Galileo allegedly could afford to be uncompromising was that he felt he had a "secret weapon," or a conclusive physical proof of Copernicanism based on explaining the tides as caused by the earth's motion.[53] Thus, even though the charges against him had already been dismissed, in December 1615 Galileo went to Rome to bring his fight to the center of power.

However, Galileo's effort misfired and turned out to be counterproductive, yielding instead a condemnation of the doctrine and a restrictive warning to himself. There were two main reasons for Galileo's failure, according to Koestler.[54] One was that the secret weapon turned out to be ineffective; the tidal argument was actually invalid and apparently unconvincing. That argument tried to show that the only way to explain why the tides occur was in terms of the earth's motion; the ebb and flow of sea water was the result of the combination of the earth's diurnal axial rotation and annual heliocentric revolution; for terrestrial points facing the sun (and thus experiencing daylight) the two motions are in opposite directions and thus yield a resultant equal to the annual minus the diurnal speed; for terrestrial points on the opposite side from the sun (experiencing nighttime) the two motions are in the same direction and so yield a resultant equal to the annual plus the diurnal speed. Koestler claimed that Galileo's theory improperly dismissed the fact that "in the *Astronomia Nova,* Kepler had published his correct explanation of the tides as an effect of the moon's attraction. . . . It contradicted Galileo's own researches into motion . . . and postulated that there ought to be *only one high tide,* precisely at noon—whereas everybody knew that there were two, and that they were shifting around the clock. The whole idea was in such glaring contradiction to fact, and so absurd as a mechanical theory—the field of Galileo's own immortal achievements—that its conception can only be explained in psychological terms" (453–54). And Koestler elaborated his psychic analysis of Galileo: "He had improvised this secret weapon in a moment of despair;[55] one would have expected that once he reverted to a normal frame of mind, he would have realized its fallacy and shelved it. Instead it became an *idée fixe,* like Kepler's perfect solids. But Kepler's was a creative obsession: a mystic chimera whose pursuit bore a rich and unexpected harvest; Galileo's mania was of the sterile kind. The tides, as I shall presently try to show, were an indirect substitute for the stellar parallax which he had failed to find" (464). The connection is this: "The fallacy in Galileo's reasoning is *that he refers the motion of the water to the earth's axis, but the motion of the land to the fixed stars.* In other words, he unconsciously smuggles in the absent parallax through the back door. No effect of the earth's motion relative to the fixed stars could be found. Galileo finds it in the tides, by bringing the fixed stars in where they do not belong. The tides became an *Ersatz* for parallax" (465–66).

The other reason Koestler advanced for the failure of Galileo's effort in the winter of 1615–1616 was his flawed rhetoric. This allegedly consisted of various tactics that were outwardly brilliant, momentarily devastating, and entertainingly appealing, but unfortunately excessively aggressive, offensive, and shortsighted. Koestler traced Galileo's rhetoric all the way back to his first controversy over the proportional compass with Baldessar Capra in 1607. Then he proceeded to claim that "in his later polemical writings, Galileo's style progressed from coarse invective to satire, which was sometimes cheap, often subtle, always effective. He changed from the cudgel to the rapier. . . . But behind the polished façade the same passions were at work which had exploded in the affair of the proportional compass: vanity, jealousy and self-righteousness combined into a demoniac force, which drove him to the brink of self-destruction" (363). As regards that fateful winter, Koestler quoted Galileo's own assertion that he was journeying to Rome so that "I could use my tongue instead of my pen."[56] And Koestler told us that "his method was to make a laughing stock of his opponent—in which he invariably succeeded, whether he happened to be in the right or in the wrong. . . . It was an excellent method to score a moment's triumph, and make a lifelong enemy. It did not establish his own point, but it destroyed his opponent's" (452).

This account of the first phase of the trial is highly ingenious, relatively original, more or less coherent, and apparently plausible. In particular, Koestler's stress on the fact that the pre-1616 proceedings had almost exonerated Galileo is important. In fact, I believe it is still an unsolved problem that the trial proceedings include a consultant report on the Letter to Castelli that found Galileo's hermeneutics unobjectionable.[57] But

Koestler's account is mostly a fabrication that can give the appearance of being historically accurate by constantly perpetrating the straw-man fallacy; that is, attributing to the individual under scrutiny silly and untenable views by interpreting relevant texts in the most uncharitable manner and most unfavorable light.

The straw-man fallacy is most obvious in Koestler's account of the tidal argument. What he presented was an incoherent piece of reasoning that caricatures the original; but the presentation proves only the superficiality of Koestler's reading, not the fallaciousness of Galileo's reasoning. There is no question, of course, that Galileo held that the tides are caused by the earth's motion, and that we now know this is not true. But the falsity of a claim cannot be equated with the fallaciousness of the reasoning from which the claim is inferred, or of the reasoning used to draw other consequences from that claim. Koestler's evaluation was also anachronistic and wise after the event.

Moreover, Koestler constantly repeated that Galileo regarded his tidal argument as conclusive and decisive; and yet the exposition he wrote for Cardinal Orsini in January 1616 is full of expressions of tentativeness and only claims probability at most. For example, Galileo says, "I would be greatly inclined to agree that the cause of tides *could* reside in some motion of the basins containing seawater; thus, attributing some motion to the terrestrial globe, the movements of the sea *might* originate from it. If this did not account for all particular things we sensibly see in the tides, it would thus be giving a sign of not being an adequate cause of the effect; similarly, if it does account for everything, it *may* give us an indication of being its proper cause, or at least of being *more probable* than any other one advanced till now."[58]

Koestler seemed to interpret the italicized terms expressing possibility and probability as meaning certainty and necessity. In so doing he was repeating a practice Duhem had started.[59] It is no more admissible in Koestler's case than it was in Duhem's.

However, the straw-man fallacy is found elsewhere in Koestler's account. For example, to support his claim that in oral argument Galileo used the shortsighted and ineffective rhetorical tactic of ridiculing an opponent and making an enemy of him, Koestler quoted from an often-cited letter (dated Rome, 20 January 1616) by one Antonio Querengo to Cardinal Alessandro D'Este: "Although the novelty of his opinion leaves people unpersuaded, yet he convicts of vanity the greater part of the arguments with which his opponents try to overthrow him. . . . What I liked most was that, before answering the opposing reasons, he amplified them and fortified them himself with new grounds which appeared invincible, so that, in demolishing them subsequently, he made his opponents look all the more ridiculous."[60] The technique that Querengo liked best is actually a very sound, wise, and proper one; it really amounts to being concerned to *avoid* the straw-man fallacy; that is, before criticizing an opponent, it is a sign of a serious critic to first strengthen the opposing argument as much as possible and interpret it in the most charitable manner; by so doing, one's criticism will really undermine the argument, rather than destroying one's own caricature invented to make one's own task easy. Despite what Querengo says, in such a situation the opponent is not made to "look all the more ridiculous" but rather is portrayed as someone who holds plausible arguments and good reasons, which are nevertheless invalid and incorrect. However, Koestler completely overlooked this aspect of Galileo's tactics and instead focused on other aspects that were indeed less effective and more shortsighted. But again, what that shows is Koestler's superficiality and penchant for straw-man reconstructions rather than a real flaw in Galileo's behavior.

Finally, we have to question Koestler's claim that Galileo refused any compromise.[61] Here, the main difficulty is that Koestler failed to realize that there were several distinct compromises, and so he equivocated; his argument reduces to saying that because Galileo rejected one particular compromise, he was rejecting any compromise. In the process, Koestler failed to see that there were some compromises which Galileo himself was proposing and which were rejected by the Church. One compromise which Galileo clearly rejected was the instrumentalist one, according to which one could regard the earth's motion as a mere instrument of astronomical prediction and mathematical calculation, without any pretensions to describe physical reality; this was the compromise endorsed by Bellarmine and corresponded to one meaning of the word *hypothesis*. But this does not imply that Galileo was rejecting the probabilist or fallibilist compromise, according to which (given the admitted lack of a conclusive demonstration) the earth's motion should be treated as a description of reality that *might be* true (or false), but which was *not yet* known to be true (or known to be false), and which therefore required further research to determine which was the case; this compromise corresponded to the other meaning of the word *hypothesis* and was endorsed by Galileo in the essay entitled "Considerations on the Copernican Opinion," which contains his answer to Bellarmine.[62] Galileo also proposed the compromise that Copernicanism should not be condemned (nor endorsed) by the Church; he did this in his replies to the scriptural objection, which he criticized as invalid and abusive, and hence as not worthy of the Church's endorsement by means of a prohibition.[63] A related compromise was the suggestion that either Galileo's biblical critics should be prevented from using scriptural passages against his physical theories, or he should be allowed to answer them; but that it was quite improper to allow them free rein while silencing him and his supporters. This suggestion was contained in almost everything Galileo wrote, said, and did during that period.[64] In short, it was not Galileo who refused any compromises but rather Koestler who refused to see them.

In summary, although Koestler's account of the first phase of the affair does fit into his scheme of holding Galileo responsible for the modern cultural division between science and religion, it does not fit the facts of the case. Let us now see whether Koestler's view of the second phase fares any better.

Since the condemnation of 1633 was occasioned by Galileo's publication of the *Dialogue* in 1632, Koestler's assessment of this book is an important part of his assessment of the trial. Koestler claimed that the *Dialogue* actually discredited the author and the Copernican cause. He began finding problems with the work in the very process of its composition, which took about six years, with frequent interruptions; although there were causes beyond Galileo's control, "one may also suspect that he was hampered by a recurrent psychological blockage, a repressed doubt in the soundness of his 'conclusive proof'" (473). Koestler also faulted Galileo for being unable to reject circular orbits and adopt Kepler's ellipses (476). Then Koestler blamed Galileo for a presentation of the Copernican doctrine that ignored the many epicycles that it still needed, "and here Galileo is downright dishonest . . . his account . . . was not a simplification but a distortion of the facts, not popular science, but misleading propaganda" (476–77). The comparison of the two chief world systems, Ptolemaic and Copernican, was one-sided and ineffective because "he keeps silent about the fact that the Tychonic system fits the phenomena equally well" (477) vis-à-vis the Copernican system. In discussing the pro-Copernican argument based on the motion of sunspots, Koestler charged Galileo with believing that it was impossible for the sun to maintain its axis of rotation parallel to itself, which was required in the geostatic explanation; but here "one simply gapes . . . in a later section Galileo discusses at great length the reasons *why* the earth moves thus, and explains that the preservation of the fixed tilt of its axis 'is far from having any repugnance or difficulty in it.' . . . There can also be little doubt that the sunspot argument was a deliberate attempt to confuse and mislead."[65] Finally, a passage in the *Dialogue* expressing puzzlement about the details of Mars's orbit prompted Koestler to say: "This was written some twenty years after Kepler's determination of the Martian orbit laid a new foundation for planetary theory. The truth is that after his sensational discoveries in 1610, Galileo neglected both observational research and astronomic theory in favour of his propaganda crusade. By the time he wrote the *Dialogue* he had lost touch with new developments in that field, and forgotten even what Copernicus had said" (479). If such a reading of the *Dialogue* were accurate and fair, then one might begin to take seriously Koestler's conclusion that "Galileo's ill-conceived crusade . . . discredited the heliocentric system and precipitated the divorce of science from faith" (495).

Moreover, for Koestler the Galilean crusade was ill-conceived for other reasons besides the astronomical, logical, and intellectual flaws of the *Dia-*

logue; he was guilty of serious social and psychological failures. For he managed to alienate and antagonize almost everyone he dealt with: the Jesuits, the Roman censors of his book, the pope, and the Inquisition officials and judges at the trial. As regards the Jesuits, it was his controversy with Scheiner over sunspots and with Orazio Grassi over comets that poisoned the atmosphere: "The attitude of the *Collegium Romanum* and of the Jesuits in general changed from friendliness to hostility, not because of the Copernican views held by Galileo, but because of his personal attacks on leading authorities of the Order" (470). Furthermore, the lengthy negotiations to get permission to publish the *Dialogue* were described by Koestler in such a way as to suggest that Galileo "had obtained the *imprimatur* by methods resembling sharp practice" (483). Koestler also adopted the Urban personal-insult thesis: "It did not require much Jesuit cunning to turn Urban's perilous adulation into the fury of the betrayed lover" (483).

As regards his behavior at the trial, one of the things Koestler stressed in Galileo's first deposition (12 April 1633) was his answer to the question of why, when seeking the book's imprimatur, he had not mentioned that in 1616 he had received a warning not to hold or defend the earth's motion. Galileo had answered, "I did not judge it necessary to tell it to him, having no scruples since with the said book I had neither held nor defended the opinion of the earth's motion and sun's stability; on the contrary, in the said book I show the contrary of Copernicus's opinion and show that Copernicus's reasons are invalid and inconclusive."[66] Koestler commented that this answer "was suicidal folly. Yet Galileo had had several months' respite in which to prepare his defence. The explanation can only be sought in the quasi-pathological contempt which Galileo felt for his contemporaries. The pretence that the *Dialogue* was written in refutation of Copernicus was so patently dishonest that his case would have been lost in any court" (485).

In the second deposition (30 April), Galileo changed his mind and confessed his guilt for the crime of having defended Copernicanism (while claiming that this transgression was unintentional). Koestler did refrain from explicitly criticizing the defendant at this point, but he was very much interested in giving an explanation: "His panic was due to psychological causes: it was the unavoidable reaction of one who thought himself capable of outwitting all and making a fool of the Pope himself, on suddenly discovering that he had been 'found out.' His belief in himself as a superman was shattered, his self-esteem punctured and deflated" (489). Then, in a postscript to his confession, Galileo had volunteered adding a "day" to the *Dialogue* to make it explicit that he did not hold the geokinetic opinion and to give more explicit criticism of the Copernican arguments. Since the Inquisition ignored this offer, this aspect of the trial enabled Koestler to advance a novel and ingenious contextualization of Galileo's abjuration of his Copernican conviction: "Up to the age of fifty Galileo had been hiding that conviction, and at his trial he had twice offered to add a chapter to the *Dia-logue* refuting Copernicus. To recant in the Basilica of the Convent of Min-erva, when everybody understood that this was an enforced ceremony, was certainly much less dishonourable for a scholar than to publish a scientific work contrary to his convictions. One of the paradoxes of this perverse story is that the Inquisition had thus in fact saved Galileo's honour in the eyes of posterity—no doubt unintentionally" (494).

Koestler's anti-Galilean compendium echoed, expanded, and rivaled the previous lists compiled by such synthesizers as Marini (1850) and Müller (1909).[67] However, Koestler's purpose was not primarily to debunk Galileo but rather to explain the existence of the modern cultural gap between science and religion as originating largely with him. And the connection is that his condemnation could and would have been avoided if he had not insulted the intelligence of his contemporaries with the fallacies, sophistries, confusions, and errors of the *Dialogue*, and had not offended their sensibilities with his insults, lies, deception, dishonesty, hypocrisy, false promises, and so on.

However, the fact that Koestler's debunking of Galileo subsists in this wider context means that it is liable to an additional potential criticism, as we shall see presently, after evaluating the other two parts of his account of the 1633 trial. Let us begin with Koestler's interpretation of the *Dialogue*. His reading of this work is essentially a series of illustrations of the straw-man fallacy, whereby he caricatures inaccurately, unfairly, and unchari-tably the texts in question. There is no point in examining each instance,[68] but one in particular deserves attention. In his discussion of the pro-Copernican argument from the motion of sunspots, Koestler made it sound as if Galileo found "very hard and almost impossible to believe"[69] that in the geostatic explanation the sun would keep its rotational axis always parallel to itself as it revolved around the earth. However, it was not the sun's axial parallelism per se which Galileo found hard to believe, for he himself explained that this could happen if the sun had a fourth motion; it was such a fourth motion which he found hard to believe because it would be ad hoc and violate the principle of economy. Moreover, as Soccorsi had pointed out in his superb analysis,[70] we must also take into account the diurnal motion, which implies that the fourth solar motion would in actuality have to have a daily period to compensate for the sun's diurnal motion. Of course, such considerations would considerably complicate Koestler's dis-cussion, and might conflict with his aim of addressing a general audience. On the other hand, other parts of Koestler's work (especially those on Kepler) contain even more subtle and technical complications; and at another level, the context is one where Koestler was charging that the *Dia-logue* "was not a simplification, but a distortion of the facts, not popular science, but misleading propaganda" (476–77). In such a context, the introduction of those complications was absolutely essential. I leave it to the reader to infer further conclusions besides the one drawn here, that Koestler was unwilling or unable to read the *Dialogue* as one ought.

As regards the other series of critiques in Koestler's account of the 1633 trial, they amounted to what could be called a character assassination. Like Marini and Müller before him, Koestler engaged in the practice of viewing every aspect of Galileo's activities in the worst possible light. Some of his techniques deserve exposure, and then I can leave further evaluation to the reader's own sensibilities. Consider, for example, Koestler's account of Galileo's change of mind between the first and the second deposition. After quoting Galileo's April 30 confession, Koestler remarked, "I have criticized Galileo freely, but I do not feel at liberty to criticize the change in his behaviour before the Inquisition. He was seventy, and he was afraid. That his fears were exaggerated, and that his self-immolatory offer (which the Inquisitors discreetly allowed to drop as if it had never been made) was quite unnecessary, is beside the point. His panic was due to psychological causes" (489). And then Koestler gave the psychic analysis[71] of Galileo quoted above, attributing to him the feeling of being a "superman" who was "capable of outwitting all" but who had just been "found out" (489). Other authors have openly, if harshly and unfairly, criticized Galileo's change of mind. Now, it seems that while pretending to refrain from criticizing the scared septuagenarian, Koestler was merely refraining from doing so openly, but in the process he was covertly advancing his own deeper and more damaging criticism; for Koestler was advancing an *explanation* that implied as unfavorable a *criticism* of Galileo's behavior as any that have been advanced in a direct manner.

Next, a question must be raised about the internal coherence of Koestler's account of Galileo's condemnation, specifically the connection between such secondary debunking and the primary thesis that this condemnation was one of the historic causes of the divorce between science and religion. The difficulty stems from the fact that Koestler's assessment of the *Dialogue* (and more generally of Galileo's other contributions, except for the dynamics of the *Two New Sciences,* which Koestler judged to be scientifically significant but did not discuss) was so negative that, if accurate, the work and the author would hardly deserve to be described as scientific but would more properly be viewed as instances of sophistry and charlatanry; in that case, what the condemnation would contribute to cause would be the separation of religion from sophistry or charlatanry. The difficulty is that the more Koestler debunked Galileo as a scientist, the more he undermined his own case that the Church's condemnation had anything to do with bringing about the much-dreaded divided house of faith and reason.

In conclusion, Koestler's account of the Galileo affair displays considerable ingenuity, apparent plausibility, literary readability, interdisciplinary understanding, and a deep and wide-ranging cultural sensitivity. But this appreciation cannot be extended to include historical accuracy, sound interpretation of texts, or fair evaluation of issues and persons. However, it would be a mistake to conclude that it tells us nothing of value about the Galileo affair. Indeed, perhaps Koestler's work tells us more about the author than about Galileo's trial; but it is important that Galileo can elicit so much dislike and that his condemnation can motivate a Koestler to construct around it an edifice like *The Sleepwalkers*. This fact is a sign of the unique status which the Galileo affair possesses and which I am trying to document in this book.

Chapter 16

History on Trial

The Paschini Affair (1941–1979)

We have seen that in 1941, to mark the tricentennial of Galileo's death, the Pontifical Academy of Sciences commissioned Pio Paschini to write a book on Galileo's life and work and their historical background and significance. We have also seen that although in 1943 Paschini managed to make a small contribution to the silent rehabilitation of Galileo occasioned by that tricentennial, this book was not published until 1964.[1] Now it is time to discuss the reasons for the delay, the reasons for the posthumous publication, and the controversy generated by such a publication.[2]

16.1 Silencing a Historian:
Paschini's Letters (1941–1946)

Paschini was born in 1878 near Udine in the northeastern region of Italy. In 1900 he was ordained a priest,[3] and in 1906 he started teaching Church history at the Udine Seminary.[4] His scholarly focus was local Church history. After Pius X condemned sixty-five "modernist" propositions with the encyclical *Lamentabili Sane Exitu* in 1907,[5] Paschini came under suspicion as a modernist sympathizer but did not get into serious trouble.[6] In 1913, he was appointed professor of ecclesiastical history at the Roman Seminary (also known as the Lateran University), and he moved to Rome. Paschini earned the appointment over another candidate named Angelo Roncalli, who was under even greater suspicion of modernism;[7] Roncalli later became Pope John XXIII.

Paschini's Roman appointment coincided with a reorganization of several Roman seminaries by Pius X. Among other things, the Roman Seminary was supposed to become a national institution. The pope personally chose Paschini, whom he admired for his integrity, scholarship, teaching ability, and orthodoxy; with regard to the latter, by that time Paschini had been able to convince his superiors that he was no modernist.[8]

When he moved to Rome, Paschini stopped using Latin and started lecturing in Italian in his classes, without disciplinary repercussions. He stayed away from Vatican politics and from careerism. He expanded his scholarly interests from local Udine topics to more general questions of sixteenth-century Roman ecclesiastical history. In his historical approach, he stayed away from apologetics and fideism. He found little occasion for pastoral work, focusing instead on scholarship and study. He became a good and typical member of what one scholar has called the "holy Roman republic,"[9] which included Protestant scholars such as Ludwig von Pastor.

In 1919, Paschini took a leading role when the Roman Seminary began sponsoring and publishing a series of Lateran Lectures and the journal *Lateranum*. This practice, which was soon emulated by many Vatican schools, led to several advancements in his career. In 1932, he was appointed president of the Roman Seminary. During the academic year 1932–1933 he was appointed *Privatdozent* in modern history at the Faculty of Letters and Philosophy of the (state) University of Rome, at a time when one member of the faculty council was the leading philosopher and Fascist sympathizer Giovanni Gentile.[10] In 1937, Paschini became the Vatican representative to the International Commission on the Historical Sciences. And in 1941, he was invited to give a lecture inaugurating the new academic year and celebrating the tenth anniversary of the papal constitution *Deus Scientiarum Dominus;* the lecture, which was published in *L'Osservatore Romano,* was relatively progressive and raised some eyebrows.[11]

When Paschini was approached in November 1941 by the Pontifical Academy about writing a book on Galileo's life and works, at first he refused since he felt the topic was outside his field. To be sure, the previous year Paschini had published *Rome in the Renaissance,* which was thoroughly documented and incisively argued and had been well received despite the fact that it was often critical of the Church of that time;[12] and this work of ecclesiastical history overlapped chronologically and thematically with the projected Galilean work; but Paschini knew he had no scientific background. A revealing glimpse into his attitude is given by a letter he wrote to his friend Msgr. Giuseppe Vale on 4 December 1941, four days after his appointment to the project had been announced at the Academy meeting of 30 November:

> [175] You may have read in *L'Osservatore* of this past Monday about the assignment I have been given as regards the life of Galileo. It had been mentioned to me about a few weeks ago by Msgr. Mercati in the name of the Pontifical Academy of Sciences, but I had declined on the grounds of being already overworked with my regular duties and of the fact that the project involves an age and a topic that have never been part of my research.

[176] I thought that was the end of the matter, but then a little more than a week ago Fr. Gemelli came to my house begging me with great enthusiasm not to repeat my negative answer; for the Academy had confidence in my work and was ready to grant me the necessary financial support, as well as other assistance that might facilitate my work. I was unable to refuse and asked him to find a way of relieving me of some of my tasks.

He told me he would talk to the pope about the matter. But I do not see how this could come about, and so I do not have much hope. Similarly I do not have much hope of getting some research assistants. I only expect to get help in secondary matters, such as bibliographical assistance.

They want a Galileo in the context of his time, with regard to both religious and scientific life. This is certainly not easy and does not allow me to benefit from a research assistant at my side. In any case, I have begun to do research and to reread Galileo's letters, which can be read with pleasure. But I am apprehensive about dealing with the two chief world systems, with the physical cosmology of the time; they are abstruse and boring. But I must be patient. I have no deadline, but in 1942 there is the tricentennial of his death, and so it would not be appropriate to take too long.[13]

Msgr. Angelo Mercati was a friend of Paschini and a member of a small committee charged by the Academy to select an appropriate scholar for the Galilean project. Mercati held the position of director of the Vatican Secret Archives and was the brother of Cardinal Giovanni Mercati. At the time Angelo Mercati was working on a related topic, for the previous year (1940) he had discovered the Inquisition's summary of Giordano Bruno's trial and was in the process of publishing it.[14]

Despite Paschini's pessimism, some assistance did come. An assistant, Michele Maccarrone, was appointed to teach his course at the seminary.[15] As we will see later, Maccarrone is also the person to whom Paschini bequeathed the manuscript of his book and who was instrumental in having it published in 1964.

Paschini had too much scientific knowledge to learn, too much historical material to go through, and too much integrity to finish within the tricentennial year and thus produce a work that would have been too superficial, although (as we saw in chapter 14.2) he did write (in 1942) and publish (in 1943) a short popular article, and although this article was full of insights and sketched a relatively original rehabilitation of Galileo. Finally, on 23 January 1945, Paschini completed his book manuscript and submitted it to the Church authorities for approval.[16]

The manuscript was given for a review of its scientific parts to Giuseppe Armellini, director of the Vatican Astronomical Observatory and a member of the Pontifical Academy of Sciences; his scientific evaluation was generally positive. However, other reviewers, who evaluated the historical aspects, objected to many of Paschini's judgments. For example, allegedly he had been too hard on and unfair toward the Jesuits.[17] But this was just the tip of an iceberg that remained mostly submerged. We will try to ascertain these objections presently. However, they were such that Paschini was unwilling to change the substance of his judgments. So the Academy refused to publish the manuscript and forwarded the case to the Vatican Secretariat of State. There the case was handled by Giovambattista Montini, who was a deputy secretary and a consultant to the Holy Office and would later become Pope Paul VI. Then Pope Pius XII, who originally had been very interested in and favorable to the project in general and Paschini in particular, decided to forward the case to the Holy Office. They received it in July 1945.[18]

For the next twelve months Paschini inquired about the precise nature and substance of the objections against his manuscript. He was in touch with all three of the institutions involved: the Academy of Sciences, the Secretariat of State, and the Holy Office; and he met and exchanged correspondence with an official from each: Gemelli, Montini, and the assessor Msgr. Alfredo Ottaviani, respectively.[19]

Apparently Paschini was never given any written evaluations of his manuscript, but he was able to talk to the officials involved. He related the substance of these discussions in various letters he wrote. From these letters, we can get a glimpse of the objections raised against his manuscript. These letters are also valuable because they contain his answers to those objections. The first relevant letter is one he addressed to Deputy Secretary Montini after a meeting they had had and is dated 12 May 1946:

[202] I do not conceal from you that the oral communication you conveyed to me a few days ago, regarding my work on Galileo, has filled me with disappointment [203] and bitterness. As you know, I would have never thought of undertaking such a work on my own initiative. In fact, I knew very well that there had not yet been an end to the polemical echoes of those distant events and to the passionate conflicts that have accompanied them until our own day. Thus, when Msgr. A. Mercati wanted to entrust me with the assignment in the name of the Pontifical Academy of Sciences, I declined without any further ado. However, I was unable to resist a second and more pressing invitation by Fr. Gemelli, in the hope of accomplishing a work that would be useful and advantageous to the Holy Church. At that time Fr. Gemelli told me that what was wanted was a contribution to the clarification of Galileo's life and scientific activities that was far from the preconceptions of any party, namely impartial toward all sectarian or ideological interests. They thought I was the right person for this. I had no reason to turn myself into a champion and apologist of Galileo, about whom I knew very little, just enough to discharge my teaching duties. . . .[20] In all my publications I have aimed to proceed with absolute impartiality, and so I was extremely surprised and disgusted that I should have been accused of having produced nothing but an apology for Galileo. In fact, this accusation profoundly attacks my scientific integrity as a scholar and teacher: in the entire course of my publications and teaching, I can say I have always felt it was my duty to let truth speak out and be free of any obstacle created by ignorance or partisanship; and in my study of Galileo too, I think I have not failed in my aim, which is the aim of any honest person.

... I worked on the topic for more than three years, overcoming considerable difficulties, reading the works and correspondence of Galileo and works about him with the firmest aim at objectivity; and I think I can assert, without boasting, that I know Galileo better than whoever criticizes me, perhaps without having read the whole manuscript; for, I realize, it is not pleasant to read a manuscript as closely argued as mine. . . . They oppose me with the already superseded difficulty that Galileo had not advanced conclusive proofs for his heliocentric system. I knew that very well; but the traditional theory then prevalent in the schools did not have proofs in its favor either. . . . In the document you read to me, some sentences from my book were mentioned against me. I would not have refused to tone down or modify some expressions which they wanted corrected, and indeed I would have gladly done it; but regarding the ones that were mentioned, I had written them precisely in order to distinguish clearly the responsibilities of the 1633 consultants from those of the Church and of the Inquisition itself; it is not my fault if those consultants appear to lose face today, as everybody knows. As regards the other comments you read to me, you know I did not understand very well the tenor of some of them, and at this moment I do not recall some others;[21] on the other hand, I cannot take more of your precious time for a personal matter of mine. Still I am sorry I wasted so much of my time without profit for a work which I had been burdened with solely in order to please others. Please sympathize with these simple words of mine, and keep up your precious benevolence toward me, and be sure that I am always . . .[22]

It is easy enough to sympathize with the victim of this episode of the four-centuries-old Galileo affair. However, Montini's predicament was not qualitatively different from that of any editor who has had a manuscript refereed and has received a negative review; no responsible editor could ignore such a review. Of course, the editor can do many things in response, such as seek additional reviews, give the author the opportunity to reply to criticism, and so on. Thus, although one wishes that Paschini had been provided with a written copy of the negative review and had been given the time to reflect on it and perhaps answer in writing, his meeting with Montini may be viewed as the official's attempt to inform the author of the criticism and to give him the opportunity to reply.

A general and crucial objection was that Paschini's book was an apology for Galileo. Paschini defended himself cogently from this charge by elaborating his open-mindedness and objectivity. We may add that his situation was arrestingly analogous to that of Galileo and the charge that his *Dialogue* was a defense of Copernicanism. Paschini himself had suggested in his 1943 article that it was not Galileo's fault if the geokinetic arguments were stronger than the geostatic ones. All that could be expected of someone who was discussing the question was that he present all the arguments for both the geokinetic and geostatic theses, that he make an honest effort to understand them, and that he analyze them adequately and fairly; unless one is prevented from evaluating the respective arguments, a conclusion will follow to the effect that one side or the other is stronger or that they are about equally strong. There is nothing objective or impartial in saying that they are equally strong if in fact they are not. Honesty required Galileo to point out that geokineticism was stronger than geostaticism, just as honesty required Paschini to say what he believed—that Galileo's condemnation had been erroneous and that he was in fact a model of religiousness.

Another criticism of Paschini's book stemmed from the claim that Galileo did not provide a conclusive demonstration of Copernicanism. This was a standard proclerical argument that had germinated as far back as Bellarmine's letter to Foscarini (1615) and could be traced through such authors as Lazzari's consultant report (1757) and Olivieri's summary of the Settele affair (1820c). In replying that the lack of conclusive demonstration applied with even greater force to geocentrism, Paschini was making a point that was becoming increasingly well established; it had been made in Galileo's own reply to Bellarmine[23] and could be traced through such authors as Auzout (1665), Gemelli's lecture (1942b), Paschini's own article (1943), and Soccorsi's account (1947).

The only other specific objection that can be inferred from this letter is that presumably Paschini had been too harsh in criticizing the 1633 consultants. If it was meant literally, such an objection would refer to the three consultants who in April 1633 wrote evaluations stating that in the *Dialogue* Galileo clearly defended the earth's motion and came very close to holding it.[24] But perhaps the point was meant to also include the 1616 consultants who assessed Copernicanism as contrary to Scripture, heretical, and philosophically false and absurd. In any case, Paschini's reply pointed out that by being tough on the consultants, he was lessening the blame on the Church as a whole and on the Inquisition as an institution. He did not explicitly say so, but the proclerical implications of such a shift of the blame were obvious. Although such a line of defense was not original to Paschini, it certainly was a well-established part of traditional defenses; so it should have been appreciated by Montini and the other officials.

A few days after writing the letter to Montini, Paschini wrote about the same meeting and the same problem to his friend Vale. This letter is dated 15 May 1946:

[72] Let me tell you what has been happening with my *Galileo*. After requesting a resolution from the higher authorities (since I thought they were deliberately drawing out the matter for too long), Msgr. Montini (who is very much on my side) sent for me and read to me a kind of decision of the Holy Office; as you know he had wanted to see the manuscript under the pretext of examining the appropriateness and manner of publication. It said that my work was an apology for [73] Galileo; it made some comments on a few of my sentences; it objected that Galileo had not given the proofs of his system (the usual sophism); and it concluded that publication was not appropriate. The manuscript was returned to me. There was no condemnation or censure

against me, aside from what I have just said. From all indications, I have become firmly convinced that in reality the Holy Office from the very beginning did not want at all such a publication. The Pontifical Academy of Sciences did want it; the Holy Father approved it; but the Holy Office did not; the latter was only too happy to find a pretext for letting the matter fall by the wayside. I was sorry that Fr. Gemelli had so little foresight and did not act at all like a gentleman. Imagine if I, who have never been an apologist for anybody, could have been one for Galileo! However, to expect that I should speak ill of him just to please them is something else; I will never do that. If centuries ago they made a big mistake (and that is not the only one), we today should commit dishonesty! And to think that they had requested me again and again to be objective, impartial, etc.! It is as clear as daylight that the Jesuits and the Dominicans lose face in this affair, and they do not like to be told this. However, maybe they were not directly involved (in fact, I do not know who read my manuscript); maybe the cause is the prejudice that superiors are always right, especially when they are wrong, as someone said. So I have worked for three years to arrive at this beautiful result, and further to earn the blame of being unable to do what I was supposed to do. I replied to Msgr. Montini by letter (as he had suggested), so that there would remain some record of my thoughts on the matter; and I did it with some diplomacy but with firmness. I am owed an adequate compensation, and we shall see what will be decided in that regard. Enough for now.[25]

This letter generally confirms the previous account and the previous charges, and it adds two new and important details. The first is that Paschini was apparently perceived as having been too critical of the Jesuits and the Dominicans. The complaint about the criticism of the consultants mentioned in the letter to Montini overlaps with the present one. Here it should be stressed that Paschini was an ordained priest of the regular clergy and did not belong to any particular religious order. The language of this letter reflected the attitude which many regular priests have traditionally displayed toward the Jesuits and the friars of the special orders. In his book, Paschini had had no compunction about criticizing Jesuits and Dominicans, and now he was not about to change his statements to please them, their sympathizers, and their supporters. This issue thus reflects one of the long-standing divisions within the Catholic Church that prevents it from being the monolithic institution envisaged by some.

The other important detail in the letter reflects the same issue. This letter makes clear that there were three institutions that became involved in Paschini's book project: the Pontifical Academy of Sciences, the Vatican Secretariat of State, and the Holy Office (or Inquisition). Paschini was confessing to his friend that although he had the support of the first two, the third opposed him. Presumably even Pope Pius XII was personally in favor, but as head of the Church he could not simply or arbitrarily overrule the opposing elements of the bureaucracy. The bureaucracy had its own procedures that had to be respected. Such limitations on the power of a pope are analogous to those that were revealed in the Settele affair in 1820, even though in the end that affair had a progressive (albeit double-edged) outcome.

Sometime after his meeting with Montini, Paschini was able to meet with the Inquisition's assessor, Ottaviani. On 4–5 July 1946 he wrote about it to his friend Vale:

[77] I can tell you that I was called by Msgr. Ottaviani regarding my *Galileo;* I had been notified about it by my superior,[26] Cardinal Marchetti Selvaggini, to whom I had expressed my complaints on the matter and who, of course, had shown little enthusiasm on my behalf. Ottaviani proposed buying the manuscript from me to compensate for the harm I would suffer from the lack of publication. I immediately said, No; I had no intention of allowing the fruit of three years of intense work to be lost, for once I sold the manuscript, it would have ended up in the bowels of a warehouse and nobody would have heard of it again. Moreover, although I did not say this, I knew they wanted to reduce everything to silence with just a few thousand lire. I said frankly that I would arrange for the publication of my work when and in the manner I deemed appropriate. I was told immediately that they wanted to be at least notified. I answered that I could do that, but that I remained profoundly disillusioned that I should be treated in such a manner, after I undertook a work which I would have never thought of doing and which I accepted only after Fr. Gemelli, with the approval of the pope, reiterated the invitation from the Academy of Sciences. I added I knew very well that I was getting into a big imbroglio. I had the fault of yielding not to my ambition, but to the appeals made to me based on the seriousness of my scholarship; at that time I was told they were looking for a scholar who was level-headed and balanced and who would clarify the personality of Galileo as a scholar and a Christian. I did my best to act in accordance with this trust, and now I was told that I had sided too much with Galileo, which is absolutely false. Certainly I could not, in order to please those who had a vested interest, falsify the results of my inquiry. That is, I could not repeat what had been done by Fr. Müller with a shameful book like his;[27] for one must have the courage to tell the truth even when it turns out to be bitter.[28] On the other hand, I added, if it was deemed appropriate to introduce in the manuscript some clarifications and some softening of the language, I would never have refused to do it, as long as one remained within the limits of historical honesty; for I did not claim to be infallible. It was objected that I had based myself too much on Italian sources and studies, and that abroad there might be some who had [78] reservations; I answered that foreigners had always had many prejudices against our things, not just today but also at the time of Galileo and even earlier; moreover, the sources are mostly Italian, and it was not my fault that this was so. Finally, I was told one was afraid that some assertion I made could be twisted into a proven accusation against the Holy Office; I answered that this fear gave me too much importance, and I pointed out that it reflected the outlook of someone who must have been living on the moon and that there was no danger that this might happen because of my work; on the other hand, there were many unjust accusations against the celebrated tribunal, and my investigation at

least had the merit of trying to set the record straight, namely in its true light. This, in summary, is what he told me. At the end Msgr. Ottaviani asked me whether there was someone I trusted for a new examination of the issue; I proposed Fr. Cordovan, who was not a professional historian, but was an intelligent and level-headed man, and had written for *L'Osservatore* an article that had not displeased me. He answered he would pass on the information; more than a week has elapsed, and I have heard nothing; but one must not be in a hurry; nor have I heard anything from Fr. Gemelli, although one can understand that he wants to wash his hands after getting me into the imbroglio. Naturally, for now do not say anything to anybody about this; I wrote it to you so that I would not forget about it myself.

P.S. Last night at a meeting I saw Msgr. Angelo Mercati, who told me that regarding my affair his brother the cardinal does not intend to yield victory "to those people" . . .[29]

This meeting with Ottaviani suggested that the situation was still fluid. The Inquisition seemed to display some flexibility, as shown by its willingness to consider a financial settlement and a reevaluation of the manuscript by additional readers. And Paschini seemed flexible when he expressed his willingness to tone down some of his assertions, although he refused to compromise on substantive matters. This letter also revealed some additional criticism of Paschini's manuscript, which he again cogently answered.

Despite the apparent flexibility of the Inquisition's assessor (Ottaviani), the apparent favor of the deputy secretary of state of the Vatican (Montini), the original sponsorship by the Academy of Sciences, and the initial support of the pope himself, the book was not published at that time. On 8 August 1946, Montini informed Paschini by letter that the Holy Office had shelved the question; that the Secretariat of State was reconfirming the nonpublication and closing the case; and that Paschini would be paid the sum of 20,000 lire in compensation for his expenses, his work, and his troubles (besides some unknown amount previously given).[30]

Paschini's manuscript remained unpublished for the rest of his life, and he remained silent about the matter. However, he did not receive any formal reprimand, and his career continued to advance at its previous modest pace. For example, he served as editor-in-chief of the *Enciclopedia cattolica*. This position even enabled him to revisit the topic, for he authored the Galileo entry in volume 5, which appeared in 1950; but it was a self-censored essay, purely descriptive and unpolemical.[31] Paschini died in December 1962.

16.2 "Rehabilitating" a Historian: The Pontifical Academy's Edition of Paschini's *Galileo* (1964)

In his last will and testament, Paschini named his former student Maccarrone as the legal heir of his manuscript on Galileo. In 1963 Maccarrone undertook an effort to have it published by approaching all the offices and institutions with which his mentor had dealt two decades earlier. He began with the Vatican deputy secretary of state, from whom he received encouragement. Then he sought an opinion from the dean of Church history at the Jesuit Gregorian University in Rome, whose judgment was also favorable. Maccarrone then approached the Pontifical Academy of Sciences, which appeared interested in publishing the book on the occasion of the four-hundredth anniversary of Galileo's birth (in 1564); it commissioned the Belgian Jesuit Edmond Lamalle to make appropriate revisions and updates and to edit the book. Maccarrone also contacted the Holy Office, which advanced no objections to its publication but had some reservations about its lack of novelty and the timeliness of its appearance. On 31 July 1963, he had an audience with Pope Paul VI (the former deputy secretary of state, Montini), informing him of the steps he had taken. In the autumn of 1963, the pope reconvened the Second Vatican Council, which had been started by John XXIII; some of its discussions focused on the relationship of the church and the world, the freedom of scientific research, and the condemnation of Galileo; and it would soon emerge that the publication of Paschini's book was felt to be appropriate and useful in the context of these discussions.[32]

On 15 February 1964, *L'Osservatore Romano*, dedicated a full page to commemorate the four-hundredth anniversary of Galileo's birth.[33] On March 4, the Holy Office gave its unconditional approval for the publication of Paschini's book, leaving to the Academy the task of its editing and publication. In the summer, Lamalle made his revisions to Paschini's manuscript. On 2 October 1964, the book was published, in the Academy's series *Scripta Varia*, as volumes 1 and 2 of three volumes of *Miscellanea Galileiana*, with an "Introductory Note" by Lamalle. On November 13, there was a presentation of the book at the Roman bookshop "Paesi Nuovi," with talks by Luigi Firpo, Paolo Brezzi, and Daniel O'Connell.[34]

Lamalle's introduction did not tell the story of Paschini's book, but it did state that the Academy was trying to make up in the quadricentennial of 1964 for what had been impossible, on account of World War II, at the tricentennial of 1942. However, "since in the meantime the author had deceased, at the considerable age of 85 (on 14 December 1962), without having revised his manuscript, the president of the Academy asked the author of these lines to see what Msgr. Paschini's posthumous work needed so that it could be published."[35]

After this introduction to his introduction, Lamalle devoted the main part to discussing the weaknesses of Paschini's work.[36] First, the book was written two decades earlier, and the history of science had made considerable progress since then. Second, Paschini was writing outside his specialty, which was Church history of the sixteenth century. Third, he relied exclusively on the National Edition of Galileo's complete works edited by Favaro

(1890–1909), thus neglecting other important sources. Fourth, Paschini's method was that of textual analysis and quotation. Fifth, he relied on Favaro not only for the documentation, but also "for his erudition, his points of view, and for the kind of problem he would formulate."[37] Finally, the result of all this was "to see the facts, especially regarding controversial questions, only through the eyes of Galileo and his disciples."[38]

Then Lamalle went on to briefly discuss his own editorial practices. First, it would not have been appropriate to modify the book's perspective. Nor would it have been proper to write a second book in copious and extended footnotes. He claimed that "our changes, both in the text and in the notes, have been deliberately very discreet; they were limited to corrections that seemed to us to be indispensable and to a minimum of bibliographical updating."[39]

Lamalle's opening historical remarks are strikingly misleading, both by the omission of the real story and by his blaming of the nonpublication on the war. But perhaps one should not be too harsh on Lamalle, who had no access to the documents recording the story. Moreover, the publication did not originate with him but with the Academy and other officials, and so he was merely their appointee. The bulk of Lamalle's introduction, discussing the weaknesses of Paschini's work, could be interpreted as a subtextual justification of the original nonpublication and of the later editorial revisions. Nevertheless, in the concluding section Lamalle indicated that it would have been impossible and improper for him to completely rewrite the book, and so his revisions had been minor and minimal. At that time there was no reason to doubt his word.

In the meantime, during the Second Vatican Council, there was a movement for the rehabilitation of Galileo. In March 1964, the French Dominican Dominique Dubarle forwarded to the pope a request, endorsed by many scientists and academics, for a "solemn rehabilitation of Galileo";[40] the pope forwarded the request to the Inquisition, which decided (May 15) that they had already acted on the matter by approving Paschini's manuscript for publication.[41] However, the Galileo affair was also on the formal agenda of the Council.[42]

The topic was discussed in connection with the question of the relationship between the Church and culture, and the autonomy of science in particular. Several delegates proposed that the Church officially recognize the condemnation of Galileo as an error and that Galileo be rehabilitated. One of these Galilean supporters was Fr. André Charrue, bishop of Namur (Belgium).[43] Another was Cardinal Suenens, archbishop of Mechlin, which includes the University of Louvain, Belgium; he was quoted as saying, "I beg you, Fathers, let us not have a new trial of Galileo. One is enough for the Church."[44] At the session of 4 November 1964, Arthur Elchinger (assistant bishop of Strasbourg) discussed what he called the "tragic incomprehension" of scientists for religion, stemming from many cases for which the

Galileo affair had become the symbol; correspondingly, the rehabilitation of Galileo would effectively symbolize the Church's coming to terms with that problem: "The rehabilitation of Galileo on the part of the Church would be an eloquent act, if accomplished humbly but correctly. Such a decision, if enacted by the supreme Authority of the Church, could not fail to redound to the Church's own credit, since with such an action it would reclaim the trust of the contemporary world and would perform a great service to the cause of human culture."[45]

As a result of such discussions, on 11 February 1965 a reference to the condemnation of Galileo was inserted in the draft of a committee report being prepared for approval by the Council. It said: "May we be permitted to deplore certain mental attitudes which are alien to healthy scientific research and which in centuries past showed themselves visible perhaps internal to the Church itself. Giving birth as they did to disputes and controversies, these mental attitudes were the cause whereby many ended by opposing science to faith with most grave damage to both. On the other hand, these errors are easily understood, given those times, and they were not exclusive to Catholics, since similar attitudes were present in other religions. Still, it is necessary that we do our best, insofar as human frailty permits, that such errors, as for example the condemnation of Galileo, are never repeated."[46]

This text was discussed at a more general meeting on April 1. But there was opposition, with some delegates recommending that Galileo should not be mentioned at all and others suggesting that scholarly experts be consulted. Finally, a compromise was worked out: the explicit mention of Galileo in the text would be dropped, but a footnote reference to Paschini's book would be added. The minutes of that meeting contain the following abbreviated notes that reveal the rationale underlying the compromise: "Galilei.—Inopportune to speak of this in the document.—Let us not force the Church to say: I made a mistake. The matter should be judged in the context of the time. In Paschini's work everything is said in the true light."[47] The originator and moving force of the compromise was Msgr. Pietro Parente, who was the cochairman of that committee; but he was also the assessor of the Holy Office who, although not opposed to the publication of Paschini's manuscript, had expressed reservations in 1963 when Maccarrone approached him.[48] The reasons for his change of mind are unclear but would merit further reflection after the discovery that the 1964 Vatican edition of Paschini's manuscript was adulterated, of which more presently.

After further discussions and endorsements at various sessions, such a compromise found its way into the final official statement of the Second Vatican Council, the constitution *Gaudium et Spes* approved on 7 December 1965. The wording of the relevant passage was: "One can, therefore, legitimately regret attitudes to be found sometimes even among Christians, through an insufficient appreciation of the rightful autonomy of science, which have led many people to conclude from the disagreements and controversies which such attitudes have aroused, that there is an opposition between faith and science."[49] And there was a footnote to this passage that said, "Cf. Pio Paschini, *Vita e opere di Galileo Galilei*, 2 vols., Vatican City: Pontifical Academy of Sciences, 1964."[50]

Paschini's literary heir, Maccarrone, interpreted this citation as a posthumous vindication and triumph for Paschini, pointing out that "the rare privilege of a specific mention in the Council's final constitution, which was unique among twentieth-century authors other than popes, has conferred on him a crown of everlasting glory."[51] Maccarrone also argued that this reference to Paschini was a better response by the Church to the Galilean problem than "to redo the trial of Galileo, as some proposed and as others requested after the Council."[52] His reason was that "this would have been anachronistic and useless,"[53] whereas Paschini's work represented "the superseding of the apologetic position that had prevailed in so much Catholic historiography, especially in the delicate Galileo question."[54]

There is no question that Paschini received a personal vindication, both through the indefatigable efforts of his faithful disciple Maccarrone and from the convergence of the world-historical circumstances of the Second Vatican Council. However, regarding the more general historiographical implications of the Paschini affair, Maccarrone's assessment turned out to be more ironical than prophetic. How ironic this assessment was will be apparent after I examine what the 1964 Vatican edition of Paschini's *Galileo* did to the original manuscript.

16.3 Adulterating Historiography: Bertolla's Recovery of the Genuine *Galileo* (1978)

In 1978, a conference was held in Udine to commemorate the centennial of Paschini's birth, and the proceedings were published the following year.[55] As one would expect, several contributions (though not all) focused on the history of Paschini's work on Galileo.[56] For example, Maccarrone contributed the account which I have already referred to and which concluded with the vindication claim just quoted. Moreover, the occasion afforded a perfect opportunity for scholars to study the archival materials that were held in or near Udine. One set of such materials was the correspondence between Paschini and his friend Giuseppe Vale, who lived in Udine while Paschini lived in Rome. They exchanged a total of 877 letters,[57] and I have already utilized the two dated 15 May and 4–5 July 1946, containing crucial information about the original nonpublication of Paschini's work.

The Library of the Udine Seminary held the original manuscript of Paschini's book. Maccarrone, who had inherited it, donated it to the library after the book was published. One of the conference participants, Pietro Bertolla, decided to examine Paschini's original manuscript and compare it with the published book. The results were surprising.

It was already known, from Lamalle's introduction to the Academy edition of Paschini's book, that the editor had made some changes. The number of changes was not large, about one hundred in a two-volume work of more than seven hundred pages;[58] many of the changes were minor and could be easily classified as what is normally understood as editorial. However, many of Lamalle's emendations partially or wholly changed Paschini's judgments. These changes usually involved four topics. Regarding Aristotelianism and the Jesuits, the changes toned down Paschini's negative remarks and added favorable judgments. As regards Galileo's precursors and rivals, the changes had the effect of diminishing the novelty, originality, and importance of Galileo's work. With regard to Galileo's interaction with the Inquisition, the changes made it appear in a better light and Galileo in a worse one. Finally, in a few cases, the changes not only completely altered Paschini's judgments, but also reversed them by turning them into the opposite of what Paschini had said. Let us examine the more significant of the changes.

One of these occurred in Paschini's discussion of the earlier proceedings against Galileo (1613–1616), where Paschini commented on the views of the relationship between scriptural interpretation and scientific investigation expressed in Galileo's letters to Castelli and Christina. In his manuscript, Paschini had written the following: "In this regard, it is undeniable that he [Galileo] expounded the right principles, whereas the theologians appeared preoccupied with preserving the rules that for them had become traditional through Scholasticism. Thus, his opponents took his words seriously and provoked a decision by those same theologians that, given the context, could only be the one we shall describe presently. Therefore, it was not Galileo who moved the scientific debate into the scriptural field, but it was the Peripatetics in alliance with the theologians (as it often happened) who did so."[59]

As we have seen (chapters 8.3 and 13.4), the question of who first injected scriptural hermeneutics into the astronomical discussion was a crucial issue, for it has consequences about which of the two sides subscribed to bad theology (using Scripture as an astronomical authority) and bore the responsibility for the 1616 disaster. The corresponding passage in the printed book reads:

> [317] In this regard, it is undeniable that on the whole Galileo expounds the right principles, that is, those that later prevailed among Catholic exegetes. But it would be completely unhistorical to label the attitude of the theologians who rejected his conclusions simply intellectual myopia or stubborn attachment to tradition. They were indeed mistaken, but they were not ignorant; they had received the best education then available and had consulted,

when needed, the best-known specialists; for example, as we have seen, this was done by a man of such a high mind as Bellarmine, who was free and open. Moreover, the preoccupation of defending the veracity of Sacred Scripture in all its statements (and not only in the statements on faith and morals) was in itself justified and praiseworthy, as was praiseworthy the defense of the robust tradition that relied on the Church Fathers and on the great Scholastics. To allow an interpretation of a scriptural passage (nowadays one would speak of a "literary genre") different from the literal meaning, which always holds the presumption, the reason must be conclusively binding. Since one was dealing with the physical system of the world, such conclusiveness could only derive from sure experimental discoveries. [318] Similarly, to abandon the Peripatetic deductions in the interpretation of the physical world, which were favored by the fact that they corresponded to immediate sense experience, philosophers required an experimental proof that such experience was inadequate. Bellarmine had said explicitly that if one advanced true proofs of the immobility of the sun and rotation of the earth, one would have "to proceed with great care" in interpreting Sacred Scripture.[60] Now, as regards these conclusive experimental proofs which the theologians were demanding, Copernicus had not given them, and Galileo had promised them but was unable to deliver them. Now we know that his intuitions were right (disregarding particular theses) and anticipated the possibility of demonstration. In this situation, once the debate was brought into the scriptural field, the decision could only be the one we shall describe presently. Galileo defended himself with remarkable ability and clarity, but these qualities were not enough to save him.[61]

The published text edited by Lamalle did admit that Galileo was right, and his theological opponents wrong, about the theological principles of scriptural interpretation; and this admission corresponded to Paschini's text. But Lamalle had added a qualification intended to justify the theologians' error: they were well educated, well informed, and well intentioned, and had good reasons; these reasons stemmed from the fact that Galileo lacked conclusive proofs for the earth's motion, and the burden of proof was on the proponents of a nonliteral interpretation of the geostatic passages of Scripture. Correspondingly, Galileo did not have good reasons for his correct hermeneutical principles.

Aesthetically and logically, one can perhaps admire the elegance of Lamalle's qualification. It did not formally contradict Paschini's claim but added another dimension to the discussion, a dimension that is relevant and even important. That is, in a controversial situation like the Galileo affair, it is important to distinguish two senses of being right or wrong. First, one can be right (or wrong) with regard to the substantive content of the claim one makes or belief one holds; in this sense Galileo was unquestionably right both about the physical claim (heliocentrism) and about the methodological principle (denial of scientific authority of Scripture). Second one can be right (or wrong) about the reasons one has for making the claim one makes or holding the belief one holds; and clearly it is logically possible, and indeed empirically common, that people hold right beliefs for wrong reasons, or wrong beliefs supported by good reasons. What Lamalle was trying to do was to place Galileo in the former category (right beliefs for wrong reasons) and his theological opponents in the latter (wrong beliefs supported by good reasons).

After such an aesthetic and logical appreciation, I would nevertheless want to argue, first, that Lamalle's interpretation is substantively untenable. But this is not the place to elaborate this argument.[62] The more important question here is whether it was ethically proper for Lamalle, as editor, to add a qualification pointing in a direction opposite to the aim of the original author. And this question must be judged in the light of two factors. The first is the previous history of the manuscript; that is, the book was previously refused publication because Paschini was unwilling to change pro-Galilean theses such as the one in question here; thus we may safely conclude that Paschini would not have approved of this qualification. The second factor was Lamalle's complete silence about such changes, except for the remarks he makes in the introduction, which do not mention specifics; that is, in the published book, Lamalle gave no indication of how, and how much, he was qualifying and adding to Paschini's original judgment.

Another example of an inadmissible change involved Paschini's discussion of the February 1616 consultants' assessment of Copernicanism as theologically heretical or erroneous and philosophically false and absurd. After quoting from the consultants' report, Paschini had asked what reason might have led them to this philosophical assessment. To answer this question, he thought he could do no better that to quote from a classic defender and supporter of that assessment, Riccioli: "We do not know the reasons that motivated the Sacred Congregation to condemn the opinion of the earth's motion and sun's rest as philosophically false and absurd. But if we may be allowed to guess them, perhaps they had to do with this: philosophizing physically, and not whimsically as if one were dealing with a mathematical or metaphysical possibility, one must base the conclusions about the natural motion and rest of bodies on the evidence of sensation; now, from universal and constant sense experience the whole human species is entitled to affirm that the sun moves and the earth stands still."[63]

In his manuscript, Paschini had commented: "It would be really difficult to imagine a more childish reason, but it was one of those that were current at that time."[64] This comment did not appear in the published book; instead there was a footnote to the Riccioli passage that read:

> This argument may seem childish to someone who takes it outside its historical context. But it touches the crux of the conflict, in which there was a confrontation of two kinds of reasoning and two types of experience. For ancient Aristotelian physics, experience was the rudimentary one given by the everyday contact of the senses with objects and formulated in ordinary language; it

was approximately the experience of the children of today, before it is set straight by the warnings of adults or the study of physics. As R. Lenoble says, "At the time of Galileo, it was adults who conceived the world as children do today." That experience disappeared, to be replaced by modern scientific experience (incorrectly labeled Baconian), which substitutes spontaneous perceptions with experiments; these are produced and reproduced, quantitative and no longer qualitative, and so they are subject to the rigor of mathematical controls and to the utilization of instruments that become increasingly more delicate in order to make up for the inadequacy of the senses. This new experience was invading the scene, but still elicited many objections. Cf. R. Lenoble, "Histoire de la Pensée Scientifique," in *Histoire de la Science*, op. cit., pp. 467–71.[65] Lenoble shows very well how, for the Peripatetics of that time, the propositions cited were "philosophically absurd."[66]

The effect of the deletion of Paschini's judgment and the insertion of Lamalle's footnote was to replace a negative evaluation of a traditional argument with a positive appreciation of the same argument. Admittedly, Paschini's evaluation was a cursory dismissal without a supporting justification, whereas Lamalle's reverse evaluation was supported with a nonnegligible argument. However again, one's aesthetic sensibility and logical judgment rejoice at such a clarification and appreciation of the anti-Copernican argument from sense experience, but one's ethical sense cringes at the deception of portraying the original author as saying what the editor himself wants to say (in a situation where the two things are opposite to one another). In other words, if Lamalle were writing his own book, or presenting commentary on the book being edited, such additions and substitutions would be unobjectionable. But Lamalle and the Pontifical Academy were pretending to be presenting and publishing the dead author's own work. Here it must be stressed that it would have been quite feasible and proper for them to publish the original manuscript intact (except for "merely" editorial corrections of typographical errors and the like), and then have a second set of notes (besides Paschini's own) in which Lamalle made all the "corrections" he wanted. Why was this not done?

Rather than try to answer this question, let us go on to examine the most important other discrepancies. At the end of his discussion of the first phase of the original Galileo affair, Paschini commented: "Thus ended what was improperly called the first trial of Galileo. I say improperly because the proceedings, which had begun with a denunciation against him and his writings, left out his person and his writings in order to be directed against the Copernican doctrine and to arrive at a condemnation in a decree pronounced with a levity that was wholly unusual on the part of the austere tribunal. What is worse is that one never revisited that decree with a weightier examination. The Peripatetics had won and did not want to let go so soon of the victory. As regards Galileo, he was silenced by means of an injunction, as one says in legal terminology."[67]

In the corresponding paragraph of the published book, instead of the second and third sentences of this passage, one found the following: "I say improperly because the proceedings, which had began with a denunciation against him and his writings, left out his person and his writings in order to be directed against the Copernican doctrine and to arrive at a condemnation in an unfortunate decree; this decree appears surprising today considering that it came from such a balanced and austere tribunal, but it should not be surprising if we consider it in the context of the doctrines and the scientific knowledge of that time."[68]

In the original manuscript, Paschini had said that the anti-Copernican decree of 1616 was careless, which was bad enough; but it was never carefully reexamined, which was even worse. In the published book, Lamalle made Paschini say that although the decree was unfortunate, it was uncharacteristic of the Inquisition and understandable in the historical context. While the change in this case is not from an assertion to its opposite, it is from a mostly unfavorable assessment to a mostly favorable one.

From the examples given so far, one would expect to find additional questionable emendations in the discussion of the 1633 trial. Indeed, in his manuscript, Paschini had advanced this final judgment on the condemnation of Galileo: "Thus came to a conclusion what was the true trial of Galileo. Regarding the responsibility, one can frankly say that 'the persons who are most to blame in the eyes of history are the defenders of an outdated school who saw the scepter of science slipping from their hands and could not bear that the oracles coming out of their lips should no longer be religiously listened to, and so they used all means and all intrigues to regain for their teaching the credit it was losing. One of the chief means used were the Congregations and their authority, and the latter's fault was to have allowed themselves to be so used.' "[69] And Paschini indicated that his quotation came from the *Revue d'Histoire Ecclésiastique*, volume 7, 1906, page 358.[70]

In the corresponding passage of the published book, one finds this assessment instead:

[548] Thus came to a conclusion what was the true trial of Galileo. In order not to have a completely inaccurate idea of it, one must be careful not to include in one's account certainties and points of view that have triumphed only in the following centuries. Because of not having heeded such a warning, in the 1700s and 1800s one easily believed that Galileo had advanced dazzling proofs of his theories and that his judges had closed their eyes so as not to see them; thus everything reduced to a struggle between genius and ignorance or fanaticism.[71] "However, one was dealing with a great struggle, for it involved a drama of the mind. Scientific reason took a bold step, although without advancing decisive proofs; and such a giant step necessitated a recombination of the familiar images that were connected to the representation of the universe, in the mind of the scientist as well as in that of the man on the

street. [549] If we admire the greatness of the learned man who risked everything for the success of his intuitions, one must also understand that men with
a different background could not risk the adventure (and that was their great
responsibility)." What was wrong was to later become rigid in the mistaken
position.[72]

A footnote indicated that the quotation was taken from a 1957 work by
Robert Lenoble, which had been quoted before.[73] Then it stated that "the
old schemes have been dragged into our century,"[74] and as an example it
quoted the passage Paschini had quoted in his manuscript from the 1906
article of the *Revue d'Histoire Ecclésiastique*.[75] The footnote concluded that
"no serious historian could still endorse simplifications of this kind."[76]

Here Lamalle's editing accomplished the following. Paschini's manuscript concluded his account of Galileo's trial by endorsing a judgment
published by Delannoy in a French journal in 1906: the 1633 condemnation was blamed on the traditional professors of natural philosophy (who
were defending their turf and monopoly) and on the congregations of the
Index and Inquisition (which allowed themselves to be exploited by those
professors). The published book quoted this same judgment in order to dismiss it as outdated, oversimplified, and unhistorical; instead it endorsed
Lenoble's 1957 judgment that Galileo's condemnation was the result of a
struggle between a familiar and established and a new and bold worldview,
at a time when disagreements were natural, unavoidable, and legitimate;
and the text seemed to suggest that the real mistake came later, with the
subsequent attempts to uphold the mistaken worldview.

Although the interpretation endorsed by Lamalle contains an important
and insightful contribution to a full understanding of the Galileo affair,
one could question whether the older view endorsed by Paschini is as
inadequate and useless as Lamalle made it sound. However, the important
point to stress is that even if Lamalle's view were right and Paschini's view
wrong, one would still have to question the editor's right to make the
change and to do so silently. In fact, paradoxically enough, it seems that the
more nearly correct Lamalle was on the substance and merits of the issue
(namely the interpretation, explanation, and assessment of Galileo's condemnation), the more ethically incorrect he was on the editorial issue, for
the silent replacement of the incorrect interpretation by the correct one
introduced an element of misrepresentation of the author being edited
and published.

Despite what Lamalle said in his introduction about the impropriety of
rewriting Paschini's book, he did rewrite it; or at least he rewrote Paschini's
account of the Galileo affair. That is, Lamalle rewrote Paschini's book not
so much from the quantitative point of view of bulk, but from the qualitative point of view of the key points and crucial theses. His not taking the
credit for the changes may be viewed as a sign of modesty, but if so it was
false and misplaced modesty. Given the previous suppression of the manu-

script, the only proper thing to do would have been to leave the allegedly incorrect interpretations and assessments intact and to add new editorial notes exposing the alleged limitations of Paschini's views.

As mentioned above, the adulteration of Paschini's manuscript in the published book was first exposed in 1978 at the centennial conference and was published the following year in the volume of proceedings. Since then most scholars have agreed that Lamalle's emendations were improper; indeed they have condemned such a practice.[77] A few have tried to explicitly defend their legitimacy.[78] Still others aggravate the original adulteration with silence in the context of referring to Paschini's published text; that is, they quote Paschini as an authority to support their own claims, without mentioning that what they are quoting are not really Paschini's own judgments but Lamalle's emendations.[79]

We cannot pursue any further this ongoing controversy about a recent controversy (the Paschini affair, 1941–1979) that is part of the modern controversy (the Galileo affair, 1633–1992) about the original controversy (Galileo's trial, 1613–1633). But I will give the last word to one of the defenders of Lamalle and the Vatican edition of Paschini's book, Maccarrone. As we have seen, Maccarrone was the disciple of Paschini who substituted for him in his teaching while he was researching his book from 1942 to 1944; he inherited the manuscript and succeeded in having it published in the official and celebratory way we have seen; he became Paschini's biographer;[80] and he made Lamalle revise the original draft of his introduction to make it less critical. Thus, Maccarrone's devotion to Paschini cannot be questioned, and so one might ask (rhetorically, at least), if Maccarrone did not complain about Lamalle's revisions of Paschini's book, why should anyone else?

Referring to Bertolla's exposure of Lamalle's changes, Maccarrone claimed that "such documentation does not constitute at all an act of accusation against the editor, but rather proves his vast knowledge of the subject and of the history of science at the time of Galileo."[81] Regarding the last emendation discussed above, Maccarrone disapproved of Lamalle's remark that "no serious historian could still endorse simplifications of this kind."[82] However, Maccarrone countered that "it is not legitimate to shout that it is a falsification and to disturb Cicero and Leo XIII! 'Serious historians,' to whom Fr. Lamalle is referring, will be able to judge for themselves."[83] Maccarrone also explicitly rejected Pietro Nonis's charge that Lamalle's revision was "a filter equivalent to a real manipulation, scholarly harmful, and morally illicit";[84] for Maccarrone, "the comparison of the original text and that of the published book allows anyone who so desires to 'establish the truth' on that point."[85]

I am unimpressed by Maccarrone's defense, but there it is, for serious historians to judge for themselves.

Chapter 17

More "Rehabilitation"

Pope John Paul II (1979–1992)

In 1978, for the first time in the two-thousand-year history of the Catholic Church, a compatriot of Copernicus was elected pope. He was also the first non-Italian to occupy the post since the condemnation of Galileo. Karol Wojtyla took the name of John Paul II, partly to commemorate his short-lived predecessor (John Paul I) and partly to join the late pope in signaling a departure from tradition. Wojtyla was a Polish nationalist whose background made him appreciate freedom from foreign domination; a staunch anticommunist whose experience enabled him to appreciate individual freedom—of religion and of conscience; and something of a philosopher who could sense the importance of freedom of thought. On the basis of such considerations, one could have guessed that sooner or later he would revisit Galileo's trial.

This happened sooner than expected, perhaps due to the unfolding of geopolitical events such as the Iranian Revolution in 1979. Since this revolution was inspired and led by fundamentalist religious leaders of the Shiite sect of Islam, and since it brought to new heights the extent of atrocities committed in the name of revolution, some people were beginning to acquire new reasons for associating religion with violence, intolerance, and fanaticism. The new Polish, philosophic, and freedom-loving pontiff must have been waiting for an opportunity to make an appropriate statement or take some appropriate action. The one-hundredth anniversary of Albert Einstein's birth provided the opportunity.

17.1 Admitting Wrongs versus Admitting Mistakes: The Einstein Centennial Speech (1979)

On Saturday, 10 November 1979, the Pontifical Academy of Sciences held a meeting to commemorate the centennial of Einstein's birth. In attendance were such luminaries of physics as Paul Dirac and Victor Weisskopf. The pope delivered a speech for the occasion that was later published by *L'Osservatore Romano* with the title "Deep Harmony Which Unites the Truths of Science with the Truths of Faith."[1] Besides expressing a tribute to Einstein and discussing this general topic, the pope brought up the Galileo affair. Of course, by now we know that this Academy had been directly involved with Galileo at least twice before (in 1942 and in 1964),[2] even if not in 1610 when Galileo was inducted into the Lincean Academy, a precursor of the Pontifical Academy.

John Paul began by saying that "the Apostolic See wishes to pay to Albert Einstein the tribute due to him for the eminent contribution he made to the progress of science, that is, to knowledge of the truth present in the mystery of the universe."[3] This definition of science quickly led to the reflection that "the search for truth is the task of basic science. . . . Basic research must be free with regard to the political and economic authorities, which must cooperate in its development, without hampering it in its creativity or harnessing it to serve their own purposes. Like any other truth, scientific truth is, in fact, answerable only to itself and to the supreme Truth, God, the creator of man and of all things."[4] Here the pope was advancing a theological justification for freedom of thought and of inquiry.

The independence of science from politics and economics, and the theological justification of this independence, quickly led to the question of the relationship between science and religion. On this topic, the pope said, "The Church willingly recognizes, moreover, that she has benefited from science. . . . The collaboration between religion and modern science is to the advantage of both, without violating their respective autonomy in any way. Just as religion demands religious freedom, so science rightly claims freedom of research."[5] After reminding the audience that such freedom had been affirmed by the First Vatican Council (1870) and reaffirmed by the Second Vatican Council (1963–1965), the pontiff went on to reaffirm the reaffirmation: "On the occasion of this solemn commemoration of Einstein, I would like to confirm again the declarations of the Council on the autonomy of science in its function of research on the truth inscribed in creation by the finger of God. The Church, filled with admiration for the genius of the great scientist in whom the imprint of the creative Spirit is revealed, without intervening in any way with a judgment which it does not fall upon her to pass on the doctrine concerning the great systems of the universe, proposes the latter, however, to the reflection of theologians to discover the harmony existing between scientific truth and revealed truth."[6]

If nothing else, talk of the chief world systems was bound to bring Galileo to mind. And indeed, next John Paul discussed the Galileo affair. After that discussion, the pope went on briefly to announce the award of the Pius XI Medal for scientific contributions to one Dr. Antonio Paes de Carvalho.

And he concluded by reiterating the Church's support for science, quoting a previous president of the Academy and expanding on what he had stated. The account of the Galileo affair, however, was clearly the dominant theme of the Einstein centennial speech. Let us focus on that account.[7]

At one point John Paul stated that Galileo "had to suffer a great deal at the hands of men and organisms of the Church."[8] The pope was admitting that Galileo had been treated unjustly and that an injustice had been committed. To be sure, the pope was making the usual and important distinction between the Church as such on the one hand and ecclesiastical persons and institutions on the other; and of course, he was attributing the injustice not to the former but to the latter. However, the pope's statement was more than an admission of error, and seemed to be an admission of wrongdoing. Even an admission of error would have been significant since it was completely unprecedented for a pope to make such a statement,[9] although error had been admitted by many churchmen before; but the admission of wrongdoing signaled a new open-mindedness and sensitivity.

To speak of Galileo's "suffering," as the pope did, implies that his treatment was undeserved or illegitimate. Moreover, the pope implicitly called Galileo's treatment an instance of unwarranted interference.[10] And John Paul was implicitly "deploring" Galileo's treatment by recalling that the Second Vatican Council had "deplored" such interferences.[11] Indeed such expressions (*suffering, unwarranted,* and *deploring*) suggested that the pope was not merely admitting some unpalatable fact but also condemning it. In short, the condemnation of Galileo was itself being condemned!

The reference to the Second Vatican Council was in part an appeal to authority to help John Paul justify what he was saying and doing about Galileo. On the other hand, for this appeal to have the desired probative function, the pope had also to *interpret* the previous action of that council in the desired manner. He did so by mentioning the footnote reference to Paschini's *Galileo* in the text of the constitution of the Council, *Gaudium et Spes.* This mention amounted to applying the general principle of noninterference to Galileo's case, and thus implied an admission that Galileo's trial was an instance of unwarranted interference of churchmen and church institutions into the affairs of science. Certainly this interpretation of the Paschini reference was plausible, but it was not the only one. For in that episode (see chapter 16.2), the footnote was the result of a compromise that removed Galileo's name from the text of *Gaudium et Spes;* so it was a way of hiding the injustice done to Galileo, a way of avoiding having to admit what John Paul seemed to have no difficulty admitting.

In fact, the pope was explicit that he wanted "to go beyond this stand taken by the Council."[12] And so he issued a call for further studies of the Galileo affair. Such studies, he said, should be guided by three things: bipartisan collaboration between the Galilean scientific side and the ecclesiastical religious side; open-mindedness on the part of each to its own wrongs and the merits of the other side; and validation of the harmony between science and religion. The pope promised his full support for such studies and later followed up on his promise.

Of these three guidelines, the first two were relatively unobjectionable. But the third one seemed to amount to prior commitment to a particular substantive thesis on the subject. Of course, one could hardly expect that a pope should not subscribe to the thesis of harmony. But should one not be open-minded enough to allow the possibility that science and religion might be discordant? Or at least to acknowledge that, although there is evidence that they are harmonious, there is also evidence that they are not? And even if, generally speaking, science and religion are harmonious (in the sense that most of the evidence and arguments point in the direction of harmony), might it not be the case that the Galileo affair is one of the exceptions to this rule, namely that it is one of the few cases that exemplify conflict?

John Paul apparently wanted to foreclose all those possibilities. In fact, besides his call for deeper studies, there was another way in which he wanted to go beyond the Vatican Council. He was ready and willing to advance some important particular theses about the Galileo affair. That is, not only was Galileo's condemnation an error; not only was it an injustice; not only was it deplorable; the real tragedy was that when seen in its true light the Galileo affair really showed the harmony between science and religion, the opposite of what it has traditionally been taken to prove. In other words, the traditional interpretation is not only to be denied but also to be reversed.

We have seen that Gemelli, the president of the Pontifical Academy of Sciences at the time of the tricentennial in 1942, had elaborated this view in his lecture to the Catholic University of Milan.[13] Thus, John Paul was echoing Gemelli's account; knowingly or not, with or without a direct influence, as the case may be. However, the pope's endorsement of the reversal of the traditional interpretation was extremely significant, given his position and authority and the context of his speech.

John Paul also gave a justification of the harmony interpretation similar to that of Gemelli. The Galileo affair, the pope said, supported the harmony between science and religion for three reasons. First, Galileo himself believed that science and religion are harmonious; he gave a cogent justification of this view based on the divine origin of faith and Scripture on the one hand and human reason and physical nature on the other; and what is more, this justification corresponds to the one advanced by the Vatican Council. I suppose the connection here is that, by studying the Galileo affair carefully, one can hardly avoid learning the insightful and essentially correct reasons Galileo had for holding that science and religion are harmonious. Second, Galileo conducted his scientific research in the spirit of religious service and worship, for in his scientific research he felt he was trying to read, understand, and praise the work of God. This second reason again echoed Gemelli's account, which contained nearly identical language; and it added another dimension to the thesis of Galileo as a model of religiousness, stressed in Paschini's 1943 article and in Soccorsi's 1947 book; they had focused on the piety of Galileo's submission to Church authority. Third, Galileo's ecclesiastical troubles gave him the opportunity to elaborate important epistemological principles about scriptural interpretation; and these principles correspond to the correct ones later clarified and formulated by the Church. John Paul did not explicitly mention Leo XIII's *Providentissimus Deus* (1893) but rather Pius XII's *Divino Afflante Spiritu* (1943); the latter encyclical was an elaboration, continuation, and fifty-year commemoration of the earlier one.[14]

John Paul was aware that his account did not solve all the problems of the Galileo affair, but he felt it held the key for properly understanding the rest. The pope had revived what I have called the tricentennial rehabilitation of Galileo and updated it and placed it in a more contemporary and more authoritative context. It is not surprising that the speech was widely reported at the time,[15] and continued to be commonly interpreted later,[16] as a "rehabilitation" of Galileo.

With regard to such a rehabilitation thesis, a clarification is in order. It involves what might be called the reverse issue of a point made in the context of the objection that the 1616 condemnation of Copernicanism was endorsed by Pope Paul V and the 1633 condemnation of Galileo was endorsed (indeed ordered) by Pope Urban VIII, and so the doctrine of papal infallibility is untenable. The initial reply to this is that those popes made those decisions in the course of conducting ordinary Church business; so those decisions were not papal pronouncements ex cathedra; and infallibility applies only to the latter. Similarly, John Paul's Einstein centennial speech was not a papal declaration ex cathedra, but rather a statement of John Paul's personal opinion; now, the personal judgment of a pope is certainly authoritative, but it does not ipso facto have the force of law.

Another way of looking at the rehabilitation issue is this. Galileo was condemned at a formal trial held by the relevant ecclesiastical institution (the Inquisition). Although such a formal act does not have the status of infallibility because the doctrine of infallibility does not apply to the Inquisition, the condemnation was an official act and could be undone only by some analogous or corresponding official act. A papal speech at the Pontifical Academy of Sciences does not have the requisite status. Since the notion of "rehabilitation" carries with it some connotation of formal or official action, John Paul's speech did not really rehabilitate Galileo.

On the other hand, it was an important and revealing action. One might even speak of an "informal" rehabilitation, if this phrase is not a contradiction in terms.

A second criticism is in order, this one directed at the idea or project of vindicating the harmony between science and religion based on the Galileo affair, that is to say, at the Gemelli-Wojtyla thesis reversing the traditional interpretation. To begin with, science and religion are not and should not be treated as self-subsisting entities that exist in some relationship to one another in some Platonic heaven of abstractions, but rather as concrete historical entities that interact dynamically and reciprocally in various ways.[17] Taking such a realistic or dynamic approach, in the Galilean controversy we can plausibly take the Copernican theory of the earth's motion to represent science and the Bible to represent religion. It is indeed true that Galileo claimed there was no real incompatibility between the two (his key reason being that the Bible does not aim to convey scientific information); but it is equally true that the Inquisition was claiming that the apparent conflict between Copernicus and the Bible was real. It follows that there was an irreducible conflictual element in the Galileo affair between those like Galileo, who believed that there was *no* conflict between Scripture and science, and those like his inquisitors, who believed that there *was* a conflict.

17.2 Rethinking versus Retrying Galileo: The Vatican Study Commission (1981–1992)

John Paul's 1979 speech did not end the centuries-old controversy but in fact started a new subepisode in the Galileo affair. The pope seized every opportunity to advocate and publicize the views he had expressed in that speech. Moreover, he followed up on his promise to encourage and support new studies.

Thus, in October 1980, news media reported a statement made at the synod of bishops in Rome by Paul Poupard, then vice president of the Vatican Secretariat for Nonbelievers: "According to the wishes of the Pope, research has begun on the case of Galileo to consider this fact with complete objectivity."[18] Newspapers embellished the story with such headlines as: "Vatican Reviewing Galileo's Conviction for Heresy" (*New York Times,* October 23); "Vatican Opens Study on Clearing Galileo" (*Los Angeles Times,* October 24); and "World Takes Turn in Favor of Galileo: Vatican Agrees to Reopen Heresy Trial of 17th Century Astronomer" (*Washington Post,* October 24). It was not clear what had happened or what further action had really been taken, other than that a commission was going to be appointed to study the matter.

On 3 July 1981, a clearer and more concrete action was taken, although the public did not know about it at that time. On orders from John Paul, the Vatican secretary of state (Cardinal Agostino Casaroli) wrote Poupard a letter appointing a Vatican commission to study the Galileo affair.

The structure and membership of the commission reflected the Church's apparent seriousness about the matter. It was headed by a cardinal (Gabriel Garrone), with secretarial assistance provided by the administrative office of the Pontifical Academy of Sciences. It was subdivided into four subcommittees: exegetical, cultural, scientific-epistemological, and historical-juridical. The exegetical subcommittee was chaired by a Jesuit who was bishop of Milan (Carlo M. Martini). The cultural subcommittee was presided by a French bishop who was President of the Secretariat for Nonbelievers (Poupard). The scientific and epistemological subcommittee had two cochairmen: Carlos Chagas was president of the Pontifical Academy of Sciences, and the Jesuit George Coyne was director of the Vatican Astronomical Observatory. The historical and juridical subcommittee also had two cochairmen: Michele Maccarrone and Edmond Lamalle, both of whom had been instrumental in the rehabilitation of Paschini and publication of his *Galileo*. Lamalle soon had to be replaced by Mario D'Addio, professor of the history of political doctrines at the University of Rome.[19] Despite the commission's high profile, however, it did not include any experts in Galilean scholarship, any non-Catholics, or many laymen (only two). This reflected the Church's traditional approach to such questions. In fact, on the occasion of the tricentennial commemoration, the task of writing a reexamination of Galileo had been assigned to Msgr. Paschini, who was not a Galileo specialist.

Besides the appointment of the commission and its subdivision into four subcommittees, the most important part of Casaroli's letter was their charge. The aim was neither the review or revision of the original trial nor the rehabilitation of Galileo, but rather the "rethinking" of the Galileo affair: "The aim of the various groups should be to rethink the whole Galileo question, with complete fidelity to historically documented facts and in conformity to the doctrine and culture of the time, and to recognize honestly, in the spirit of the Second Vatican Council and of the quoted speech of John Paul II, rights and wrongs from whatever side they come. This is not to be the review of a trial or a rehabilitation,[20] but a serene and objectively founded reflection, in the context of today's historical-cultural epoch."[21] After all the media talk of rehabilitation that had followed the pope's original speech of 1979, and all the journalistic hype about a retrial of Galileo that had followed the October 1980 announcement, such a clarification was essential. Of course, one wishes that at the time the public had been informed of the July 1981 appointments and of the programmatic clarification, but this disclosure and this clarification were not given until 1983, when Poupard edited an anthology on the cultural aspect of the Galileo affair;[22] it had a preface by commission's president, Garrone, containing the information and the programmatic clarification.[23] However, in such a context, the clarification was easy to miss; thus the talk, perception, and expectation of a rehabilitation or retrial continued[24] until the official closing of the case in 1992, and they have not completely subsided yet. Casaroli's letter was published for the first time five years later by Michael Segre in the official journal of the History of Science Society *(Isis)*.[25]

The "rethinking" of the affair was to be free and objective, unprejudiced and open-minded. However, it was also to be guided by the idea and project outlined by John Paul in his 1979 speech, namely that the Galileo affair illustrates the harmony between science and religion. As I discussed earlier, there is some tension between these two requirements, and Casaroli's letter did nothing to resolve that tension. And again, this ambiguity is reminiscent of the one embodied in the task assigned to Paschini for the tricentennial commemoration.

At any rate, as a result of the July 1981 action, most of the persons appointed became active studying and publishing on the Galileo affair and promoting its study by the organization of conferences and the edition and sponsorship of relevant works. In 1983, D'Addio started publishing his own *Considerations on Galileo's Trial*;[26] its most crucial and revealing conclusion was probably its endorsement of Gebler's thesis that the special injunction transcript, although authentic, was legally worthless and inadmissible at the trial.[27] Poupard commissioned an international group of scholars to write essays on the cultural debate of the past 350 years and edited them into a collection entitled *Galileo Galilei: Toward a Resolution of 350 Years of Debate—1633–1983;* it was issued first in French in 1983, then in Italian in 1984, and in English in 1987.[28] Although some of the contributors were good scholars who specialized in relevant fields, the whole was less than the sum of the parts and quite disappointing.[29] In 1984, Coyne organized an international scholarly conference in the pope's hometown of Cracow, Poland, on the pope's own thesis of *The Galileo Affair: A Meeting of Faith and Science*, as the subsequent volume of proceedings announced in its title.[30] Most of the contributors elaborated this papal theme, but Paul K. Feyerabend raised a dissenting voice, although an iconoclastically anticlerical and anti-Galilean one; he advanced a social criticism of Galileo and an appreciation of Bellarmine from the same social point of view, thus suggesting a criticism of John Paul and the present Church that now sided with Galileo on such issues.[31]

Furthermore, the Vatican Observatory, under Coyne's direction, started a monograph series of Galilean studies. Five such monographs had been published by 1992, when John Paul decided to close the case; but the series continued, issuing at least four other monographs after that date. The first was a booklet on Galileo and the Council of Trent, more useful for its summary of criticism of various traditional accounts than for presenting its own.[32]

The second monograph was edited by Ugo Baldini and Coyne and was descriptively titled *The Louvain Lectures (Lectiones Lovanienses) of Bellarmine and the Autograph Copy of His 1616 Declaration to Galileo*.[33] It provided two new documents, the first of which suggested that Bellarmine was more opposed to Galileo than previously thought, whereas the second suggested the reverse. In fact, Bellarmine's Louvain lectures in 1571 defended the anti-Aristotelian thesis of the fluidity of the heavens on scriptural grounds, thus exhibiting a commitment to the philosophical (scientific) authority of Scripture, which goes counter to the progressive interpretation of Bellarmine which authors such as Duhem have advanced.[34] On the other hand, the handwritten draft of Bellarmine's certificate to Galileo, dated May 1616, revealed that the cardinal had been gracious enough to revise its wording to make Galileo appear in a better light.[35]

The third of the Vatican Observatory monographs was the proceedings of the Cracow conference.[36] The fourth was a booklet on the idea of unification in Galileo's epistemology, which had no apparent connection with the Galileo affair.[37] The fifth was a collection of essays on Galileo's trial by a distinguished (and non-Catholic) historian of science, Richard S. Westfall; the essays made up in cogency and general interest whatever they have lacked in thematic unity.[38] Westfall argued that Bellarmine was primarily a biblical literalist and traditionalist in scientific methodology, and so neither an epistemological instrumentalist nor a hypothetico-deductivist, and consequently Duhem's interpretation was a one-sided oversimplification. Further, the Jesuits' shift from support of Galileo (in 1610–1616) to opposition (in 1630–1633) was due less to intellectual differences than to intellectual rivalry, less to competition over material support from patrons than to competition over hegemony and leadership, and less to impersonal social forces than to Galileo's personality flaws of egotism and insufferability. The publication of the *Dialogue* was due more to the logic of the situation in the institution of patronage than to cognitive intellectual factors. And Galileo's atomism could not have been the root cause of the trial of 1633, partly because his interest in it (while real) was minor, partly because there was an insufficient perception of its incompatibility with the doctrine of the Eucharist, and partly because the historical record shows that the principal agents were worried about other issues instead.[39]

The four post-1992 monographs of the Vatican Observatory series are all versions of a major historical and critical reexamination of the Galileo affair authored by an Italian ex-Jesuit named Annibale Fantoli.[40] The English-language version appeared under the title *Galileo: For Copernicanism and for the Church*, and the book is available also in its original Italian, in Russian, and in French, with revised editions in Italian and English. The book articulates an account that is generally pro-Galilean and critical of the Church; but the work is well documented, well argued, and well balanced; it would deserve extended discussion here were it not for the fact that it really falls beyond the scope of my investigation in this chapter.[41] In fact, while Fantoli's work does have an editorial connection with John Paul's reexamination of the Galileo affair, even its first edition appeared after the papal closure of the case; and so it is obviously no part of the developments leading up to that closure. On the contrary, and more important, Fantoli is critical of the content and form of that closure,[42] and so its discussion belongs to the post-1992 period of ongoing discussion and controversy.

Besides the scholarly efforts by members of the papal commission, and besides the publication of the Vatican Observatory series, John Paul's action in 1981 encouraged other research and publications sponsored and published by various ecclesiastical institutions, chiefly the Pontifical Academy of Sciences. One of the most important undertakings was a new search of Church archives in Rome for documents relating to Galileo's trial, resulting in a new edition, by Sergio M. Pagano with the assistance of Antonio G. Luciani, published by the Academy in 1984.[43] No significant discoveries were made, but a few minor documents were found, generally confirming what was known. Moreover, while the volume observed the highest scholarly standards,[44] it also provided a much more accessible and readable collection of the documents than the 1908 edition included in volume 19 of Antonio Favaro's National Edition of Galileo's complete works. The document search and the accessible edition were very much in the spirit of John Paul's project.

In 1986, the Academy published a short monograph by Rinaldo Fabris on the exegetical question.[45] This question had been recognized as a crucial aspect of the Galileo affair both in John Paul's 1979 speech and in Casaroli's 1981 memorandum. Fabris was a competent and reputable biblical scholar;[46] the work was full of useful historical information; and it contained a preface by Carlos Chagas (president of the Academy) stressing John Paul's explicit exegetical vindication of Galileo's hermeneutics. However, the monograph left unresolved several tensions, which therefore took on the character of incoherences: it pointed out that Galileo's views on scriptural interpretation were both rooted in tradition and novel; that contemporary theologians were split about whether Scripture was a philosophical authority; and that Galileo's writings on the topic both argue against the astronomical authority of Scripture and assume it in order to develop Copernican interpretations of problematic passages.

One of the most massive works produced in the present context dealt with the Settele affair in 1820 (see chapter 10). The work contained both the voluminous original documents and a historical interpretation, with the interpretive part due mostly to the German Jesuit Walter Brandmüller and the documentary part mostly to the German scholar Egon Johannes Greipl.[47] Although printed and formally published in 1992 by Leo S. Olschki of Florence, the book was explicitly sponsored (and copyrighted) by the Pontifical Academy.[48] And although extremely valuable for the documentation and historical information provided,[49] it is less so for its interpretative thesis that the Settele affair of 1820 ended the Galileo controversy. This thesis was also misleadingly advertised in the title, which translates as *Copernicus, Galilei, and the Church: The End of the Controversy (1820): The Proceedings of the Holy Office.*

Thus, there is no question that the Vatican Commission generated a considerable body of work in Galilean studies, nor that many of these works contained valuable and useful contributions. However, one may question whether such work as a whole amounted to a rethinking of the Galileo affair, let alone a retrial and rehabilitation (which it was not even supposed to do). For by and large, it amounted to a reaffirmation, repetition, and reinforcement of the thesis that science and religion can be in harmony and Galileo's work and even Galileo's trial can help us see such harmony. This had been John Paul's thesis in the Einstein centennial speech, appropriated from various sources (chiefly Gemelli) generated during the tricentennial rehabilitation of 1942. On the other hand, the fact that the pope had publicly elaborated such a view in 1979, and that for about a dozen years afterwards there was a Vatican commission doing the same, represented an important sociocultural fact.

Actually, during the same period there was also a relatively novel and apparently influential reinterpretation, to which I now turn.

17.3 The "Right to Make Mistakes": Brandmüller's New Apology (1982/1992)

Besides collaborating with Greipl to produce their massive work on the Settele episode, Brandmüller also worked on a more general and interpretive work on the Galileo affair whose key theme was the "right to make mistakes."[50] The book was first published in German in 1982, then translated into Spanish in 1987, and then revised, expanded, and translated into Italian in 1992. Because of the novelty of its key thesis, and because of the influence of the work in general, Brandmüller's work deserves extended discussion.

One of Brandmüller's main theses[51] was that the Galileo affair ended in 1820 when, occasioned by the Settele episode, the Church withdrew its earlier condemnation of the Copernican doctrine. In fact, there was a close relationship between the reason why in 1613–1633 the Inquisition was right in condemning Galileo for treating the earth's motion as a fact (thesis) rather than as a hypothesis, and why it was right in 1820 to withdraw its condemnation of Copernicanism. The reason is that Galileo did not provide a valid scientific proof of the earth's motion, but this demonstration was available in 1820 after a number of other discoveries: Newton's universal gravitation (1687), Bradley's stellar aberration (1729), Guglielmini's eastward deflection of falling bodies (1789–1792), and Calandrelli's annual stellar parallax (1806).

This thesis may be regarded as an uncritical acceptance and historical extrapolation of Olivieri's account of the Settele affair. The thesis is not literally true since, as we saw in chapter 10.1, the date in question cannot be 1820 but at best must be 1822 or 1835. For although Settele's own astronomy textbook was approved in 1820, the controversy reemerged in April 1822 in connection with a journal publication of an extract from that book; and the general issue about the status of the Copernican doctrine was not resolved until September 1822, when the Inquisition ruled and the pope approved allowing the doctrine as a thesis. Moreover, and more important, although the resolution of the doctrinal issue ended the issue stemming from the 1616 prohibition of Copernicanism, it did not in the least resolve the issue of the 1633 condemnation of Galileo the person. Instead it proved beyond any doubt that the two are and were distinct (although of course related) issues.

Brandmüller also advanced what may be regarded as a novel apologia for the Inquisition. He did admit that the Inquisition was wrong insofar as it condemned Galileo on the grounds (among others) that the earth's motion contradicts Scripture, but he argued that it would be improper to blame the Inquisition since to do so would be a denial of its right to commit its own errors.[52]

To elaborate, there exists a right to make mistakes because the future development of knowledge cannot be predicted, and so one cannot be blamed for not knowing at any given time what is discovered afterward. In the matter of Galileo's trial, the Church cannot be denied this right any more than Galileo can. With regard to Galileo, "although Galileo's propaganda for the theories of Copernicus was corroborated in subsequent years by the studies of Newton, Bradley, and others, nevertheless he was mistaken precisely in regard to the evidentiary value of the arguments which he advanced in favor of Copernicus."[53] On the other hand, the Inquisition erred in saying that heliocentrism was contrary to Scripture. This error was not so much that the inquisitors "decided to respond with theological means to a problem in astrophysics,"[54] for this criticism would be anachronistic. Rather, "what must be criticized is that Galileo's theological opponents were unable to interpret adequately the literal meaning of Sacred Scripture by appropriating the principles of interpretation already employed (for example, by Caetani), or by understanding (as other contemporaries did) the fact that in the cited passages Sacred Scripture was using ordinary language. Here if anywhere, one can detect an error of the Inquisition. The error of the Inquisition was precisely this."[55]

Certainly one has the right to make mistakes. And certainly this right cannot be denied to either Galileo or his opponents. However, Brandmüller seemed to be giving a "liberal" justification of the Inquisition, that is a justification based on the ethical, political, and philosophical liberalism of such authors as John Milton, Voltaire, John Stuart Mill, and Benedetto Croce. But it is unclear that such liberal principles are a part of Catholic doctrine. Insofar as they are not, it would follow that a Catholic could not consistently justify the Inquisition in this manner, although a liberal could.

Another possible objection would be that, in granting Galileo the right to make mistakes, Brandmüller attributed mistakes to him too cavalierly, too superficially, and too anachronistically. The main mistake Brandmüller had in mind is one of scientific method: Galileo allegedly believed that his pro-Copernican arguments were conclusive demonstrations, whereas in fact they were not. Of course, this is a controversial question, but Brandmüller fell far short of the sophistication of Soccorsi's analysis, and in general I would argue it is Brandmüller and not Galileo who misjudges the logical and evidentiary strength of the Galilean arguments by erroneously magnifying their weaknesses and underestimating their strength.[56]

A third difficulty with Brandmüller's "right to make mistakes" involves the question of how this right relates to the duty to admit one's mistakes. The right to make mistakes is often equated by Brandmüller and his followers with the right to deny one's mistakes. For example, Brandmüller implicitly denies that the Inquisition, the Index, and their consultants erred in claiming the earth's motion to be philosophically false and absurd; and this claim is a mistake because (as Paschini saw in his 1943 essay)[57] the telescopic discoveries changed the epistemological status of Copernicanism to equally probable or more probable as compared to geocentrism. Instead Brandmüller likes to focus on the opposite error, the claim that in 1616 the earth's motion was demonstrably true, and he likes to attribute this error to Galileo; but such an attribution is untenable (as the probabilistic language of Galileo's "Discourse on the Tides" shows).[58]

On the basis of his account, Brandmüller liked to belabor "the paradox of a Galileo who makes mistakes in the field of science and of a curia that makes mistakes in the field of theology. Vice versa, the curia was right in the scientific field, and Galileo was right in the interpretation of the Bible."[59] This thesis is reminiscent of Duhem's claim that although Galileo and Kepler were right in physical and astronomical theorizing, Bellarmine and Urban were right in logic and epistemology. It may be said to reflect the desire to avoid one-sidedness and to partition rights and wrongs regardless of the source, and to that extent it embodies a laudable aim. But of course, in so doing one's apportionment must be accurate, and Brandmüller's is not.

17.4 Undoing a Rehabilitation: Poupard's Commission Report (1992)

On Saturday, 31 October 1992, there was a meeting of the Pontifical Academy of Sciences at which Pope John Paul heard and accepted the Vatican

Commission's report on the Galileo affair. The report was given in a speech by Cardinal Poupard,[60] president of the Pontifical Council for Culture, who had been a member of that commission since its inception and had later succeeded Cardinal Garrone as chairman of the commission.

Poupard began by recalling that he was reporting on an undertaking that went back thirteen years, starting with the pope's Einstein centennial speech in 1979 and being practically organized with the appointment of the commission two years later.[61] In summarizing the task of the commission, Poupard did not point out that it embodied two unresolved tensions or ambiguities. First, the commission was to undertake a calm, objective, and bipartisan investigation of the Galileo affair; but it was also supposed to rethink the topic in light of John Paul's thesis that far from epitomizing the conflict between science and religion, the Galileo affair illustrated their harmony. Second, juridical questions were to be studied by one of the four subcommittees, and yet a retrial or formal rehabilitation was being explicitly excluded from the commission's task.

The cardinal went on to describe very succinctly the commission's investigations and the publications they yielded.[62] The footnotes certainly provided a useful and comprehensive list. Two substantive results were elaborated at greater length, the first pertaining to the original controversy and Bellarmine's role, the second involving the historical aftermath.

According to Poupard, Bellarmine deserves credit for appreciating the importance of asking whether Copernicanism was demonstrably true, and whether it was compatible with scriptural statements. Presumably, he also understood the proper relationship between these two aspects of the problem: that is, if there were a demonstration that heliocentrism is true, then geostatic statements in Scripture should be interpreted nonliterally; but as long as there is no proof of heliocentrism, those scriptural statements can and should be interpreted literally. And presumably Bellarmine realized that Galileo had not provided a conclusive proof of the earth's motion. By contrast, Galileo did not realize that he lacked such proofs, and in particular that the tidal argument which he regarded as conclusive was not so.

By portraying Bellarmine as a shrewd methodologist in this manner, Poupard was apparently relying on and following Brandmüller's account. In his book on the right to make mistakes, Brandmüller had given a similar assessment of Bellarmine, quoting the same passage.[63] And of course, we know that such a view of Bellarmine goes back much further and can be traced to Duhem, among others (see chapter 13.3). However, by endorsing Brandmüller, Poupard was also ignoring, and indeed contradicting, two other relevant studies stemming from the commission. Both the work by Baldini and Coyne on Bellarmine's Louvain lectures of 1571 and one of Westfall's *Essays on Galileo's Trial* indicated that Bellarmine was a biblical traditionalist for whom Scripture was an astronomical authority, so that in his view the geostatic statements in Scripture guaranteed both the truth of geocentrism and the fact that no demonstration of heliocentrism would ever be discovered. Whether one praises Poupard for relying on Brandmüller or blames him for discarding Westfall's interpretation and the Baldini-Coyne documents, it is perhaps more important to point out that Poupard was in the eternal predicament of nonexperts who rely on specialists: often the specialists disagree, and then one can pick and choose among them to justify whatever conclusion one wants.

For his account of Galileo's theory and practice of demonstration, Poupard also relied largely on Brandmüller's work.[64] Such an interpretation represents a regress of several steps compared to the analysis made by Soccorsi in 1947 and traceable as far back as Auzout in 1665. Among other things, Poupard was also ignoring Soccorsi's point that Galileo's lack of conclusive proof was true but irrelevant, since this lack could not (and should not) have motivated a condemnation or prohibition; and he was also ignoring Paschini's point that if heliocentrism lacked demonstration, so too did geocentrism, and hence the question reduced to which side had the better and stronger arguments. But with respect to Galileo's view of his demonstrations, the situation is different from Bellarmine's view of the logical situation, because among the studies produced by the Vatican commission there were none that followed the Auzout-Soccorsi-Paschini line, and so perhaps the nonexpert Poupard may be excused for having had no choice. On the other hand, the failure to commission, sponsor, or encourage any study along these lines could be taken as a sign that the commission had not been as objective and bipartisan in its investigations as it was supposed to be. Moreover, there was one contributor to Poupard's own anthology who had mentioned such an alternative interpretation; in his essay "Galileo and the Professors of the Collegio Romano at the End of the Sixteenth Century," William Wallace had argued that Galileo was well aware that his Copernican arguments in the *Dialogue*, while strong, were not demonstrative and conclusive.[65] Admittedly, this claim was not stressed by Wallace and was relatively minor in the context of his work, but the existence of Wallace's essay brings home the point that in compiling his report as he did, Poupard was choosing sides, and not always the best documented or best argued.

Regarding the historical aftermath, Poupard gave an account[66] claiming that the 1633 condemnation of Galileo had been "reformed" several times: in 1741–1744 with the Church's imprimatur for the publication of the *Dialogue;* in 1757–1758, with the abolition of the general prohibition of Copernican books; in 1820–1822, with the explicit permission of books advocating the earth's motion as a thesis; and in 1835 with the removal from the *Index* of Copernicus's and Galileo's books. Poupard's account was largely an abstract of Brandmüller's interpretation,[67] augmented by imprecisely reported dates and carelessly described events.[68] Poupard was also gratuitously extrapolating from considerations that affected the 1616 condemnation of the Copernican doctrine to those that affected the 1633 condemnation of the person Galileo. The point is that whereas the developments mentioned by Poupard can be taken to show that the anti-Copernican decree had been reformed, retracted, repealed, or undone, they do not even begin to show that the condemnation of Galileo had been similarly revised. If anything, the effect was to generate a new problem crying out for resolution: How can one reform the 1616 decree that provided the basis for the 1633 condemnation of Galileo without doing something about rectifying the latter?

In his conclusion,[69] Poupard stressed the theological error committed by Galileo's opponents and judges: they failed to grasp that Scripture is not a scientific authority. He had an interesting explanation of this failure: a unified conception of the world that allowed (indeed encouraged) the transposition and confusion of the domains of scientific observation and religious faith. This explanation was, again, a thesis that had been advanced by Brandmüller, who had used the same language of a "unitarian" worldview but had also used such labels as *totalitarianism* and *globalism*.[70] Finally, for Poupard, that failure in turn explained why Galileo's judges had committed the subjective error of judgment that Copernicanism was contrary to Scripture and Catholic tradition.

The report was thus flawed in several ways, which reduced to an uncritical acceptance of Brandmüller's account. While acknowledging the theological errors of Galileo's judges (which the pope had done earlier in no uncertain terms), Poupard had shown his pseudo-Solomonic bipartisanship by explaining that Galileo too had committed significant errors, which Bellarmine had pointed out to him. Under orders from the pope not to stage a formal retrial to rehabilitate Galileo, Poupard had in effect subjected him to an informal retrial that upheld the original conviction. Moreover, since Galileo's errors turned out to be scientific-methodological and those of the inquisitors were theological-scriptural, the result was again the Brandmüller paradox. But let us see how the pope responded to Poupard's report and what he said and did to close the case.

17.5 Closing a "Case": The Pope's Complexity Conference Speech (1992)

Poupard's report was followed by a speech from the pope.[71] As with John Paul's 1979 speech, although the Galileo affair was the main topic of a plenary session of the Academy, it was not the only one. This time the other, more current and routine topic was the nature of complexity as studied in mathematics, physics, chemistry, and biology: "The emergence of the subject of complexity probably marks in the history of the natural sciences a stage as important as the stage that bears relation to the name of Galileo, when a univocal model of order seemed to be obvious. Complexity indicates precisely that, in order to account for the rich variety of reality, we must have recourse to a number of different models."[72] However, this leads to the fragmentation of knowledge and to the philosophical problem of keeping such fragmentation under control: "Contemporary culture demands a constant effort to synthesize knowledge and to integrate learning. Of course, the successes that we see are due to the specialization of research. But unless this is balanced by a reflection concerned with articulating the various branches of knowledge, there is a great risk that we shall have a 'shattered culture,' which would in fact be the negation of true culture. A true culture cannot be conceived of without humanism and wisdom."[73]

These reflections allowed the pope to move on easily to the topic of Galileo, although we shall see presently that there was a more substantive connection than might appear at first. His speech continued for several more paragraphs, in which John Paul anticipated that the next great problem in the relationship between science and religion would be likely to emerge in the area of biology and genetics. But he reaffirmed his belief that scientific research leads one to God. Quoting Einstein's aphorism that "what is eternally incomprehensible in the world is that it is comprehensible,"[74] John Paul concluded that "this intelligibility, attested to by the marvelous discoveries of science and technology, leads us, in the last analysis to that transcendent and primordial Thought imprinted on all things."[75]

Let us now examine the central part of the speech.[76] In the opening paragraph, the pope expressed the proper thanks and appreciation to Poupard, but he did not simply endorse his report. John Paul was expressing gratitude to all members of the commission and to all experts who had participated in its projects. He explicitly mentioned the publications produced by the commission and its conclusions in general, but not Poupard's report as such. To be sure, later in his speech the pontiff mentioned and endorsed some specific theses from the report, but he was not rubber-stamping the whole document. Thus, the pope was acknowledging that the commission had finished its work, but he was drawing his own conclusions.

In the rest of the opening section (no. 4), John Paul went on to reiterate the theme of the science-religion interaction, which he had stressed in his original 1979 speech, which had been studied by the commission, but which had hardly been touched upon in Poupard's report. And the pope added two important reflections. First, it may be that "one day we shall find ourselves in a similar situation,"[77] and so the lessons of the Galileo affair may be useful, relevant, and applicable in the future; and, as mentioned, at the end of his speech the pope explicitly mentioned the worrisome future area (biology and genetics). Second, when he cryptically said that "the approach provided by the theme of complexity could provide an illustration of this,"[78] he was probably referring to the analogy he had suggested at the very beginning of his speech between Galileo's breakthrough and the contemporary study of complexity. And the analogy is that on the one hand, "complexity indicates precisely that, in order to account for the rich variety of reality, we must have recourse to a number of different models";[79] that is, one must abandon a unified approach and adopt a multifaceted one; on the other hand, as the central part of the speech argued,[80] one of Galileo's merits over his theological opponents was to introduce a separation between the scriptural message and physical investigation against the unified world view of his opponents.

John Paul went on to comment on the epistemological dimension of the Galileo affair. The key problem was the role of Scripture in physical science, the nature of scriptural interpretation, and its relationship to scientific investigation. The pope's memorable judgment was that "Galileo, a sincere believer, showed himself to be more perceptive in this regard than the theologians who opposed him."[81] However, the lesson from this aspect of the episode had nothing to do with a conflict between science and religion but rather involved the epistemology of interdisciplinary interaction: "The birth of a new way of approaching the study of natural phenomena demands a clarification on the part of all disciplines of knowledge."[82]

This epistemological conclusion was an insightful observation. It was an original insight vis-à-vis Poupard's report (although of course many scholars have discussed this aspect of the Galileo affair).[83] The favorable assessment of Galileo's hermeneutics was even more important, on account of its explicitness and authoritativeness: although such an assessment had been common ever since Leo XIII's *Providentissimus Deus,* although it was implicit in John Paul's 1979 speech, and although it was also implicit (but only implicit) in Poupard's report, such explicitness on the part of a supreme pontiff ought to carry supreme weight. To be sure, it was not a papal declaration ex cathedra, carrying the trait of infallibility; but it was a judgment on a subject which, however controversial and however capable of being examined with evidence and argument, a pope should know something about.

Equally authoritative but much more novel was John Paul's analysis of the pastoral dimension of the affair.[84] Catholic authors have usually argued that although Galileo may have been right in astronomy and biblical hermeneutics, he was definitely wrong from the pastoral point of view; this requires that the mass of believers not be scandalized or misled by new discoveries, and so the dissemination of truth (if not its pursuit) must be careful not to upset popular beliefs too suddenly and must be mindful of the social and practical consequences of truth. Even such a shrewd and pro-Galilean writer as Soccorsi had been sensitive to such pastoral considerations, although he had used them not to criticize Galileo but rather to justify his abjuration (by attributing them to Galileo's own deliberations).[85] Instead of siding with Galileo's opponents, John's Paul's solution to the pastoral issue was "that the pastor ought to show a genuine boldness, avoiding the double trap of a hesitant attitude and of hasty judgment, both of which can cause considerable harm."[86] He was not reversing the traditional anti-Galilean solution, but rather he was denying it and pointing out that the correct pastoral position is one of arriving at a judicious mean between the two extremes of too much conservation and too much innovation. Thus, while he was not really siding with Galileo on the historical substantive issue, his rejection of the opposite side was contextually a pro-Galilean position.

The pope went on to accept some of Poupard's specific conclusions. One was the thesis about the unity of culture in Galileo's age, together with the explanation that "this unitary character of culture, which in itself is positive and desirable even in our own day, was one of the reasons for Galileo's condemnation."[87] And the connection between cause and effect here was that cultural unitarianism led to a failure to distinguish scriptural interpretation from scientific investigation, and so to an illegitimate transposition from one domain into the other.

John Paul also seemed to endorse Poupard's reference to Bellarmine. But the pope traced Bellarmine's key point to Saint Augustine. Thus, the endorsement was partial and apparently diluted.

Similar remarks apply to the Poupard-Brandmüller thesis that the 1633 sentence was "reformed" in subsequent history and that "the debate . . . was closed in 1820."[88] This could be taken as an instance of uncritical acceptance by the pope of an untenable and misleading thesis. But he was so cursory about it that one gets the impression that he mostly wanted to use it to add further support to his own historical cultural thesis: that the Enlightenment fabricated the myth that Galileo's trial illustrates the conflict between scientific progress and the Catholic Church, but that this conflict is a thing of the past.[89] However, we have seen (in the course of this book) that although such an interpretation was endorsed and embellished by the Enlightenment,[90] it was not created then but goes back to the Strasbourg edition of Galileo's *Dialogue* (1635) and *Nov-antiqua* (1636) and to Milton's *Areopagitica* (1644).[91] Moreover, although it is undeniable that this interpretation has extreme and distorted versions, it has an irreducible kernel of truth centering on the conflict between Scripture and Copernican astronomy; the fact that this conflict was a mental illusion does not render illusory the historical conflict between people who believed in it and people who did not.

John Paul ended the central part of his speech on the more positive note of trying to derive some more lessons from the Galileo affair. The last and most important lesson was an idea that may be called methodological pluralism, "that the different branches of knowledge call for different methods."[92] And this not only connected with the theme of complexity mentioned at the beginning but was also illustrated by Baronio's principle.[93]

Given that this principle eloquently summarized Galileo's position, the pope was thus ending on a note that was doubly Galilean, in style as well as in substance.

Thus, in this speech the pope was acknowledging the completion of the commission's work, as reported by Poupard. He was reiterating his own earlier view that a key lesson of the Galileo affair is the harmony between science and religion. He was clearly and explicitly praising Galileo's biblical hermeneutics, thus finalizing what might be called the *theological* rehabilitation of Galileo. John Paul was placing such a theological rehabilitation in the context of a broader philosophical appreciation, one along the lines of the epistemology of interdisciplinary relations, the other in line with methodological pluralism. And he was giving an unprecedented pastoral interpretation of the affair which, while not implying that Galileo was right on the pastoral issue, did suggest that he was no more wrong than his ecclesiastical opponents.

John Paul did not, however, explicitly endorse Poupard's report. Although he accepted some particular conclusions, in the context of the papal speech those theses lost the anti-Galilean flavor and implications they possessed in Poupard's speech. If this interpretation of John Paul's speech is correct, and if it is correct to say that the Vatican commission studies had been acquiring an increasingly anti-Galilean tone and apologetic flavor, then perhaps one may conjecture that the pope was closing the Galileo case because he wanted to close the retrial of Galileo at the hands of people such as Poupard and Brandmüller.

Epilogue

Unfinished Business

Many people were disappointed or dissatisfied with the process or the ending of Pope John Paul II's rehabilitation of Galileo in 1979–1992.[1] Some have gone so far as to claim that the primary effect of the whole episode has been to generate a new myth about the Galileo affair, the myth that the Church has rehabilitated Galileo.[2] One need not agree with either the milder expression of disappointment or the stronger mythologization charge. Instead, one could adopt the view, articulated in the last chapter, that there was a rehabilitation, but that it was informal, partial, incomplete, not unopposed, and not unprecedented; indeed it was the sort of rehabilitation one would have expected in light of the previous four centuries of the Galileo affair. However, it should be obvious that this affair did not end in 1992, as claimed in the title given to Poupard's commission report by *L'Osservatore Romano*, any more than the anti-Copernican ban ended in 1820, as claimed in the title of Brandmüller and Greipl's 1992 book on the Settele affair.[3] The case closed by Pope John Paul in 1992 was the process he had himself started in 1979, which is merely a subepisode of the cause célèbre studied in this book.

Although history did not end in 1992, this book will have to do so. This end date is amply justified by the closure of John Paul II's rehabilitation process and the historical and cultural importance of his effort. This chronological limit may be regarded as a reason for my not having included extended discussions of later accounts of the affair by such scholars as Antonio Beltrán Marí, Francesco Beretta, Mario Biagioli, Richard Blackwell, Annibale Fantoli, Rivka Feldhay, John Heilbron, and William Shea and Mariano Artigas. But this is not the only reason.

Another reason stems from the fact that the key theme of this book is "retrying" Galileo. In fact, as befits works of scholarship, these recent works are characterized by a balanced and objective analysis of the evidence that avoids the kind of partisanship endemic to the idea and the practice of retrying Galileo. Thus, my thematic delimitation is also a reason that has steered me away from trying to integrate into my story some recent (but pre-1992) scholarly accounts. To just mention the main contributions during the previous half a century, this reason applies to the works of Stillman Drake, Ludovico Geymonat, Jerome Langford, Howard Margolis, Guido Morpurgo-Tagliabue, and Giorgio de Santillana.

It should be noted, however, that what I have not done, because of the chronological and thematic boundaries of this book, is to examine such works as *primary sources* that might define some subepisode(s) of the subsequent Galileo affair. But my references to them make it clear that I have used such works as *secondary sources* in the normal course of my investigation.

If this is right, that is, if the exclusion (from the status of primary sources) of such scholarly works of the last half a century was suggested by their character of objective scholarship, then another consequence follows. That is, the rise of professional disciplines such as the history of science and the fact that the Galileo affair has become a topic of scholarly research by a body of professionals are bound to have a consequence for this cause célèbre. In short, it would be useful to explore the effect that the professionalization of the history of science has had or is having on the Galileo affair. But this topic will have to wait for some future investigation.[4]

Other topics will have to wait, too, which brings me to a third delimitation of this book, namely its character of "survey." As I stated in the introduction, this book aims to be an introductory survey of the sources, facts, and issues of the Galileo affair from his condemnation in 1633 to Pope John Paul's rehabilitation in 1992. By a survey, especially an introductory one, I mean an examination of the material that aims to be general and comprehensive and to lay the groundwork for deeper probing later. Although there is no necessary conflict between breadth and depth, there is a practical limitation involving length, bulk, and space: this introductory survey already makes up a long enough book. There is also a methodological limitation. That is, especially when the material has never been surveyed as a whole, after the first survey has been completed, a pause is needed before one undertakes further steps leading to the assimilation, analysis, interpretation, and critical evaluation of that material. Although such deeper probings are beyond the scope of the present work, I look forward to undertaking them in the future.

For example, in the introduction I formulated a thesis about the implications of the original Galileo affair (1613–1633) for the interaction between science and religion and for the dialectic of conservation and innovation, and a thesis about the implications of the subsequent affair (1633–1992) for the same interaction between science and religion and that between cultural myths and historical facts. I also formulated a thesis about the formative role played by Mallet du Pan's (1784) myth in the subsequent

Galileo affair, involving the subdued character of the controversy in the preceding century and one-half, and the heated debates thereafter. But such theses were advanced as tentative suggestions rather than firm generalizations established by my survey; only future deeper probing of the surveyed material and further reflection on the particular low-level conclusions can substantiate them and reveal others.

Or consider the main textual sources that have been reproduced here, from Cardinal Antonio Barberini's memorandum to nuncios and inquisitors (1633) to Soccorsi's justification of Galileo's retraction (1947). I have been concerned primarily to translate them into English, place them in their historical contexts, and point out the main elements of their content that connected them to the trial and the retrying of Galileo. Despite inclinations to the contrary, I have mostly refrained from sustained analysis or criticism. Accordingly, their deeper scrutiny is among the tasks for the future.

Or consider these facts: that in 1633 the Church undertook unique attempt to publicize the condemnation of Galileo; that a polarization of sides took place with the Strasbourg editions of 1635–1636 and Riccioli's apology in 1651; that during the papacy of Benedict XIV, there occurred a partial unbanning of Galileo's *Dialogue* in 1744 and of Copernicanism in general in 1758; that the publication of the Vatican file in 1867–1878 yielded a consensus that there had been a miscarriage of justice in 1633 but no subsequent tampering with the evidence in the file; that on the occasion of Galileo's tricentennial of 1942 there was an implicit but real clerical rehabilitation of Galileo; that in the middle part of the twentieth century important elements of lay culture advanced and popularized new indictments against Galileo; and so on. Although I regard such facts to have been relatively well established, I have not tried to explain why they happened in terms of their historical causes, nor have I tried to ascertain their deeper significance in terms of overarching principles or general patterns; not because I believe that causal explanation and interpretive generalization have no place in historical inquiry, but because lack of time and space forces me to postpone such investigations.

Or consider, aside from the issue of the interaction between science and religion, what can we learn from the Galileo affair about the relationship between truth and rationality, that is, between beliefs and evidence or reasons for beliefs? Can we say that although Galileo turned out to be right and the Inquisition wrong in their beliefs about the earth's motion, he was wrong and it was right in their evidence or reasons for their beliefs? Can we say at least that the Inquisition's reasons were better than his? Similarly, can we say that although Galileo turned out to be right and the Inquisition wrong on the theological or hermeneutical principle that Scripture is not a scientific authority, he was wrong and it was right in their supporting reasons? Can we say at least that it was right and he was wrong with regard to the practical pastoral harmfulness of that principle at that time? And even if Galileo was right on the question of scientific fact, on the question of hermeneutical principle, and on the question of grounding rationale, can we perhaps say that the Inquisition was right and he wrong in their epistemology or methodology?

Or again, besides the interaction of science with religion and truth with rationality, should the lessons of the affair be sought in other areas, such as the interaction of individual freedom with institutional authority, science with political power and expediency, and science with social responsibility?

And besides the original affair (1613–1633), how is the subsequent affair (1633–1992) to be viewed? What are the latter's implications for the science versus religion question? What are its implications for the evolution of cultural myths and their interaction with historical facts? Why have so many myths arisen regarding the trial? Have both sides of the controversy been equally prone to myths? Has the number of myths decreased with time, or is it mostly their content that has changed? What exactly is a cultural myth? Are such myths inescapable in human thinking? Do myths disappear by being confronted with facts or by being displaced by other myths?

My survey provides the basis for working out the answers to such questions, but such working out is a further undertaking that will take time and effort.

Thus, the business of the Galileo affair is indeed unfinished in more ways than one. That is, the Church has more work to do for the rehabilitation of Galileo, which is to say for its own rehabilitation through the admission of its errors or injustices toward him. And scholars have work to do to develop a historical understanding and a critical evaluation of the affair, both the original trial from 1613 to 1633 and the subsequent and continuing cause célèbre.

Finally, I should mention two recent developments which require me to qualify some of the clarifications expressed above, but which also reinforce the modest conclusion just reached. The first concerns a new account of Galileo's trial advanced by the Italian historian Pietro Redondi in a work entitled *Galileo Heretic*, first published in 1983. Redondi claimed that the charge triggering the 1633 trial was that in *The Assayer* (1623), Galileo held and defended the doctrine of atomism, which Church officials regarded as incompatible with a key doctrine of the faith; that is, the dogma of transubstantiation and the Eucharist, according to which during Mass, after the bread and wine are consecrated, their substance changes into the body and blood of Christ. Thus the trial and conviction for defending and discussing Copernicanism were a cover-up to shield Galileo from the more deadly charge of undermining the doctrine of the Eucharist. This account did cause a sensation for a few years. Redondi's book was quickly translated into English and French; there were intense and heated discussions among scholars; and educated laypersons read the book and followed the controversy from the sidelines. But a scholarly consensus soon emerged that Redondi's account was untenable, being contradicted by almost all the evidence that had accumulated in the past four centuries; the account was judged to be an ingenious fabrication based on a new genuine document discovered by Redondi in the Inquisition archives in Rome; this is an unsigned and undated complaint against *The Assayer* on the grounds that its atomism undermines transubstantiation and the Eucharist.[5]

The Redondi controversy had the proper credentials for being included in my story. His account was unprecedented and staggeringly original. It constituted a perfect example of a "retrial." It had the potential for historiographical or metahistorical lessons because it stemmed from the discovery of a new document; because it raised the whole issue of scholarly access to Church archives; and because it raised the issue of scholars' reaction to novel ideas. It had apologetic relevance since the 1633 trial was being interpreted as a step by Church officials to defend Galileo from what would have been a clear and undisputed heresy, and so there was the suggestion that this was one more instance of Church kindness to Galileo and support for science. And there was even drama, in the engrossing manner in which Redondi told his story. Finally, in the context of my investigation, the untenability of Redondi's account is not a good reason for exclusion since it should be obvious that my survey is meant to include the cultural effects of Galileo's trial (independently of their intrinsic truth value), and thus has included such things as lies, forgeries, myths, fiction, and outrageous constructions.

On the other hand, the Redondi controversy was too short-lived. It is too recent, and so one does not yet have sufficient perspective. And I faced the practical and methodological requirement of being selective in order to produce a work of manageable proportions. So Redondi had to be added to the long list of authors and episodes that (while interesting) are secondary in importance: Auguste Comte, John Henry Newman, Raffaele Caverni, Edmund Husserl, the ultramontanism debate between William Ward and William Roberts, the vicissitudes of the mechanical and astronomical proofs of the earth's motion from James Bradley and Giambattista Guglielmini to Friedrich Bessel and Léon Foucault, and so on.

The second development worth mentioning involves Dava Sobel's book *Galileo's Daughter.* This work was first published in 1999 and so falls in the period that is outside the scope of my investigation, but it has become so popular that it raises questions that are central to my investigation. Thus my chronological boundary could have been extended, were it not for the practical requirement of selectivity and manageable length.

The book is subtitled *A Historical Memoir of Science, Faith, and Love.* The author is a science journalist who, several years earlier, published *Longitude,* which told the story of the solution of the problem of determining longitude at sea and became an international best-seller. Although *Longitude* was telling a relatively unknown story, whereas the new book focuses on the relatively well known story of Galileo's life, Sobel's talent for storytelling and popularization comes through in the later book as it did in the earlier one. And indeed *Galileo's Daughter* also became an international best-seller.

The title notwithstanding, the book is a biography not of Galileo's daughter but rather of Galileo, stressing his scientific work, his trial, and his relationship to his elder daughter. It is a popular rather than scholarly work, but it is based on research and study undertaken by the author herself, and it shows acquaintance with some primary sources and with relevant scholarly work. It is thus surprisingly accurate as compared with the typical popularization. It does not provide a novel view of Galileo's scientific work or trial, but the focus on Galileo's relationship to his daughter does provide an engrossing human-interest story and a refreshing portrayal. Moreover, because, as we saw earlier (chapter 3.2), Galileo's daughter became a nun at the age of sixteen and lived in a convent for the rest of her life, the focus on their relationship and on the daughter's devotion and love does have implications for the way one views Galileo's attitude toward religion.

An indication of the book's cultural significance is the fact that it was made into a movie, as indeed Sobel's earlier work on longitude had been. The movie version of *Galileo's Daughter,* a two-hour docudrama titled "Galileo's Battle for the Heavens," was first broadcast nationally in the United States by PBS television stations on 29 October 2002. It was billed as a NOVA production for WGBH/Boston; written and produced by David Axelrod, of Green Umbrella, Ltd., in Los Angeles. The production credits state that it is based on and adapted from Sobel's book. And generally speaking this is correct, although it is obvious that NOVA and the producer consulted many other scientists, scholars, and institutions. By and large, the movie does follow the book with regard to the balance of topics, such as Galileo's life, scientific work, trial, and his daughter and her letters to him. However, whereas the book is unusually accurate and judicious by the standard of popular historical books, the movie falls short; it contains several factual errors and misleading interpretations, although I would judge their number and frequency to be about average for the genre.

Another indication of the book's status as a cultural sensation is the extent of popular response to it. In November 2002 the Amazon.com website listed no fewer than 170 customer reviews. By comparison, for Thomas Kuhn's *Structure of Scientific Revolutions,* which is probably the best-selling and most frequently cited scholarly book of all time, the site listed only sixty customer reviews. Similarly, the Amazon.com sales rank for *Galileo's Daughter* at that time was 179, as compared to 1,829 for Kuhn's book.

Despite its popular success, many readers criticize Sobel's book for being more about Galileo than about his daughter. One type of criticism has been to object that the author missed a good opportunity; for example, one dis-appointed reader objected that "Sobel's book perpetuates the stereotype of the woman who can only be viewed through the lens of the great men around her."[6] At the opposite extreme there have been insinuations that the key to the book's success was a consequence of the ideology of affirma-tive action applied to books, their subject matter, and their authors.[7] My main point here is simply that, intentionally or otherwise, Sobel's book has injected the Galileo affair into feminist issues and vice versa; that this was to be expected, namely that in a society and culture significantly concerned with feminist issues, sooner or later such a discussion would happen; and that by now it is a familiar phenomenon for the Galileo affair to be inter-preted or exploited in light of the ideas or concerns of the historical period or context.

Thus, although an extended discussion of Redondi's *Galileo Heretic* and Sobel's *Galileo's Daughter* as primary sources could not be included in my survey for the (chronological and methodological) reasons mentioned, they reinforce my modest conclusion about the unfinished business of the Galileo affair. For we can add a third item to the future agenda, besides the further Church rehabilitation and the deeper scholarly assimilation men-tioned earlier: that is, historical development is not yet finished with the Galileo affair; the cunning of history is likely to utilize it again in new and unpredictable ways.

NOTES

INTRODUCTION

1. Among the most important scholarly contributions, see Riccioli 1651 (in chapters 2.5 and 4.3); Viviani 1654 (in chapter 5.1); Auzout 1665 (chapter 5.2); Arnauld 1691 (chapter 6.1); Tiraboschi 1793 (chapter 8.4); Venturi 1820b (chapter 9.4); Marini 1850 (chapter 11.3); Wohlwill 1870, Gherardi 1870 and Gebler 1879a (chapter 12); Duhem 1908 and Garzend 1912 (chapter 13); Soccorsi 1947 (chapter 14.3); and Paschini 1964a,b (chapter 16). Besides the post-trial letters to and from Galileo himself, Descartes's 1633–1634 letters to Mersenne, and Leibniz's correspondence, significant correspondence includes Viviani 1690 (cf. chapter 5.1); Barbier 1811 (cf. chapter 9.1); Marini 1817a (cf. chapter 9.2); Venturi 1820a and Delambre 1820 (cf. chapter 9.3); and Paschini 1941, 1946a–c (cf. chapter 16.1). Among the ecclesiastical documents, the most important are Giovasco 1742 (cf. chapter 7.1); Lazzari 1757 (cf. chapter 7.2); the numerous 1820 writings of Anfossi, Olivieri, and Grandi, especially Olivieri 1820c (cf. chapter 10); and Casaroli 1981 (cf. chapter 17.2).

2. See Brandmüller 1992b, 127–203; D'Addio 1993, 206–29; Fantoli 1996, 487–532; Feldhay 1995, 13–25; Gebler 1879a, 299–344; Langford 1971, 159–88; and Santillana 1955a, 322–30.

3. Absolutely indispensable works, especially for their documents, are Baldini 2000c; Bertolla 1979; Brandmüller and Greipl 1992; Favaro 1887a,c, 1891c; Maccarrone 1979a,b; Maffei 1987; Mayaud 1997; Monchamp 1893; Motta 2000; Pagano 1994; Pesce 1987, 1991a; Segre 1997; and Simoncelli 1992. For other valuable works, see Baldini 1996b, 2000a; Baldini and Spruit 2001; Beltrán Marí 1998, 2001a,b; Benítez 1999b; Beretta 1999a,b, 2001; Bertoloni Meli 1988, 1992; Blackwell 1998a,b; Borgato 1996; Borgato and Fiocca 1994; Brooke and Cantor 1998; Bucciantini 1994b, 1997, 1998, 2001; Carroll 1995, 1997, 1999, 2001; Casanovas 1999; Coyne forthcoming; Crombie 1956a,b; Del Prete 2001; Doncel 2001; Fantoli 2001; Favaro 1885a,b, 1887–1888, 1891d; Feldhay 2000; Galluzzi 1977, 1993a,b, 1998, 2000; Garcia 2000, 2001; Garin

1984; Gebler 1877 (xx–xxxii), 1879a (334–40); Hall 1979, 1980; Heilbron 1999 (176–218); Howell 1996a,b, 2002; Lerner 1998a,b, 1999, 2001a,c,d, 2002b; Lindberg and Numbers 1986, 1987; Maffei 1975; McMullin 1980, 1998; Mercati 1926–1927; Monchamp 1892; Motta 1993, 1996, 1997b, 2001; Navarro Brotons 1995, 2001; Numbers 1985; Pagano 1984 (10–26); Pantin 1999, 2000, 2001; Pepe 1996a,b; Pesce 1991b, 1992a,b, 1995a, 1998, 2000, 2001; Poupard 1983, 1984, 1987; Redondi 1994; Segre 1989, 1991a, 1998, 1999; Shea 1991 (317–39); Simoncelli 1988, 1993; Stevart 1871, 1890; Stoffel 2001; Tabarroni 1983; Wallace 1981a, 1982, 1983b, 1987, 1995, 1996 (392–96), 1999; and Westman 1984, 1986.

4. See, for example, Beretta 1999a, 446–54; Garzend 1912; L'Epinois 1878, 263–68; Mivart 1885; Roberts 1870, 1885; and chapter 13.4.

5. For some classic sources, see D'Alembert 1751c, xxiii–xxiv, and Comte 1835; for some vulgarizations, see Draper 1875 and White 1896, 1: 130–52; interesting twists can also be found in such modern contemporary icons as Russell (1935, 31–43), Einstein (1953), and Popper (1956); some recent and sophisticated views are advanced in Blackwell 1998a, Feyerabend 1985, and Pera 1998; for an appreciative and critical analysis of these, see Finocchiaro 2001b.

6. See, for example, Coyne, Heller, and Zycínski 1985; Gemelli 1942b; and John Paul II 1979a, 1992a.

7. The nonmonolithic character of the Catholic Church has been stressed in various ways by other authors, such as Segre (1991b, 30) and Feldhay (1995); the latter emphasizes the disputes between Jesuits and Dominicans, in regard to which I would want to point out that these two orders were not themselves monolithic either.

8. One author who has recognized the importance of the dialectic of conservation and innovation in the history of science is Kuhn (1977).

9. Gemelli (1942b) and John Paul II (1979a; 1992a).

10. For some accounts that stress the mythological dimension of the Galileo affair, see Benítez 1999, 85–110; Carroll 1995; Finocchiaro 2002b; and Lessl 1999; see also chapters 6.2, 8.3, 11.4, and 13.1.

11. See, for example, Finocchiaro 1986a, 1989, 27–33.

12. See, for example, Müller 1911, 139–40, and Koestler 1959, 437.

13. See, for example, Drake 1976, 1980, 1999 (1: 153–56); but this view goes back much further, for example, to L'Epinois 1867 (143–45), 1877, 1878.

14. The more important and extensive of these translated excerpts and their location here are as follows (chronologically arranged): A. Barberini 1633a-c, chapter 2.1; Buonamici 1633, chapter 2.4; Guiducci 1633b, chapter 2.2; Carafa 1633, chapter 2.3; Descartes 1633, 1634a,b, chapter 3.1; Renaudot 1633, chapter 2.5; G. Galilei 1634a,b, 1635a, chapter 3.4; Peiresc 1634, 1635, chapter 3.3; Pieroni 1637, chapter 4.1; Micanzio 1639, chapter 4.1; Auzout 1665 (58–66), chapter 5.2; Leibniz 1679–1686, 1688, chapter 5.3; Viviani 1690, chapter 5.1; Giovasco 1742, chapter 7.1; Calmet 1744, chapter 7.1; D'Alembert 1754, chapter 6.4; Lazzari 1757, chapter 7.2; Tiraboschi 1793, chapter 8.3; Barbier 1811, chapter 9.2; Marini 1817a, chapter 9.3; Venturi 1820a, chapter 9.3; Venturi 1820b, chapter 9.4; Olivieri 1820c, chapter 10.2; Gemelli 1941, chapter 14.1; Paschini 1941, 1946a–c, chapter 16.1; Paschini 1943, chapter 14.2; and Soccorsi 1947 (50–60, 100–103), chapter 14.3.

CHAPTER 1

1. Favaro 19: 402–6; Finocchiaro 1989, 287–91. As Lerner (1998b; 2002b) has stressed, there exists as yet no really critical edition of the Inquisition's sentence, although considerable work was done by such authors as Sandonnini (1886) and Favaro (1887c; 1887–1888).
2. Genovesi 1966, 46–51.
3. Favaro 19: 403, lines 15–27.
4. Favaro 19: 277–98, 307–11; Finocchiaro 1989, 134–35, 136–41.
5. Favaro 11: 605–6; Favaro 5: 281–88; Finocchiaro 1989, 47–54.
6. Favaro 19: 403, lines 28–38.
7. See, for example, Galilei 1997, Finocchiaro 1989, 15–25.
8. Favaro 19: 403–4, lines 39–51.
9. Favaro 19: 404, lines 52–56.
10. Favaro 19: 404, lines 57–68.
11. Favaro 19: 404, lines 69–83.
12. Favaro 19: 404–5, lines 84–97
13. Favaro 19: 348; Finocchiaro 1989, 153.
14. Favaro 19: 405, lines 97–101.
15. Favaro 19: 405, lines 102–5.
16. Finocchiaro 1989, 290; cf. Favaro 19: 405, line 103.
17. Favaro 19: 405, lines 106–16.
18. Favaro 19: 405, lines 117–26.
19. Cf. Masini 1621; Limborch 1692, 1731; Garzend 1912; Genovesi 1966.
20. Favaro 19: 405–6, lines 126–37.
21. Favaro 19: 406, lines 138–48; for some discussions of this issue, see Cantor 1864; Pieralisi 1875, 218–24; Santillana 1955a, 310–11; Langford 1971, 153; Redondi 1983, 328; Beretta 2001, 568 n. 98.
22. Finocchiaro 1989, 291; cf. Favaro 19: 405, lines 118–19.
23. Finocchiaro 1989, 289; cf. Favaro 19: 404, lines 53–56.
24. Favaro 19: 406–7; Finocchiaro 1989, 292–93.
25. Favaro 19: 406, lines 160–63.
26. Favaro 19: 406, lines 155–60.
27. Favaro 19: 406, lines 163–65.
28. Favaro 19: 407, line 167.
29. Favaro 19: 407, lines 167–68.
30. Favaro 19: 407, lines 170–79.
31. For example, Genovesi 1966, 268.
32. The text of this document may be found in Favaro 19: 322–23; Finocchiaro 1989, 148–50.
33. I owe such information about these Roman congregations to Mayaud 1997, 37–56.
34. For the full story behind my simplifications about Copernicus's *Revolutions*, see Westman 1975a, 1987, 1990, and forthcoming.
35. For information about Zúñiga and about several other matters in this section I am indebted to Mayaud 1997, especially pp. 44–45, although our interpretations tend to differ.
36. Finocchiaro 1989, 149; cf. Favaro 19: 323, lines 45–46.

37. Finocchiaro 1989, 149; cf. Favaro 19: 323, lines 46–47.
38. Bucciantini 1995, 88; Brandmüller and Greipl 1992, 443; Mayaud 1997, 59, 64–69.
39. Mayaud 1997, 52, 88.
40. Favaro 19: 400–401; Finocchiaro 1989, 200–202.
41. Copernicus 1992a, 5, lines 36–47. Cf. Copernicus 1976, 26–27 for a more literal translation, which for example retains the word *mathematics* (or its variants) in its four occurrences in this passage, instead of translating it into *astronomy* as Edward Rosen does in the quoted text. Cf. Favaro 19: 400, lines 14–15, and Finocchiaro 1989, 200–201.
42. In the Index document from which I am quoting, this reference reads "book 1, chapter 1, page 6," which is incorrect, as Favaro (19: 400) points out.
43. Cf. Copernicus 1992a, 11, lines 40–44; Copernicus 1976, 40; Favaro 19: 400, lines 16–18; Finocchiaro 1989, 201.
44. Cf. Copernicus 1992a, 16, lines 10–15; Copernicus 1976, 44; Favaro 19: 400, lines 23–26; Finocchiaro 1989, 201. Here Copernicus was quoting Virgil, *Aeneid,* III, 72.
45. Cf. Copernicus 1992a, 17, lines 30–32; Copernicus 1976, 46; Favaro 19: 400–401, lines 27–28; Finocchiaro 1989, 201.
46. Cf. Copernicus 1992a, 17, lines 38–41; Copernicus 1976, 46; Favaro 19: 401, lines 29–30; Finocchiaro 1989, 201.
47. Cf. Copernicus 1992a, 17, lines 44–46; Copernicus 1976, 46; Finocchiaro 1989, 201; Favaro 19: 401, lines 31–33.
48. Cf. Copernicus 1992a, 20, lines 35–38; Copernicus 1976, 49; Favaro 19: 401, lines 34–35; Finocchiaro 1989, 201.
49. Cf. Copernicus 1992a, 20, lines 38–39; Copernicus 1976, 49; Favaro 19: 401, lines 35–36; Finocchiaro 1989, 201.
50. Cf. Copernicus 1992a, 22, lines 36–37; Copernicus 1976, 51; Favaro 19: 401, lines 37–38; Finocchiaro 1989, 201.
51. Copernicus 1992a, 22; cf. Copernicus 1976, 51.
52. Cf. Favaro 19: 401, lines 39–40, and Finocchiaro 1989, 201.
53. Cf. Copernicus 1992a, 208; Copernicus 1976, 217; Favaro 19: 401, lines 41–42; Finocchiaro 1989, 201–2.
54. Cf. Finocchiaro 1989, 22, and Favaro 19: 401, lines 41–42.
55. Quoted from Finocchiaro 1989, 200; cf. Favaro 19: 400.
56. Finocchiaro 1989, 289; cf. Favaro 19: 404, lines 54–55.
57. Finocchiaro 1989, 149; cf. Favaro 19: 323.
58. Finocchiaro 1989, 149; cf. Favaro 19: 323.
59. Finocchiaro 1989, 289; cf. Favaro 19: 404.
60. Finocchiaro 1989, 290; cf. Favaro 19: 405.
61. Finocchiaro 1989, 291; cf. Favaro 19: 405.

CHAPTER 2

1. It is puzzling that Motta (1993) should claim otherwise.
2. For this and related issues, see Sandonnini 1886; Favaro 1887b, 122–26; Favaro 1887–1888; Cioni 1908; Favaro 1908c; and Lerner 1998b, 2001.

3. Cf. Firpo 1993, 102–3; Schoppe 1600.

4. The last clause of Galileo's abjuration (just before his signature) speaks of the "convent" rather than the church of Santa Maria sopra Minerva (Favaro 19: 407, lines 182–83); and in any case this private venue was in accordance with inquisitorial practice for the particular crime of which Galileo was being convicted (Garzend 1912, 54 n. 91 bis).

5. There was actually a third relative who was made cardinal by Pope Urban VIII: Antonio Barberini, another nephew, who was Francesco's younger brother. When an explicit distinction had to be made between the two Antonios, they were called Antonio Barberini Senior and Antonio Barberini Junior. Such papal "nepotism" had been widely criticized and managed to be injected into the Galileo affair: the *Dialogue* showed at the bottom of the title page the publisher's trademark, which happened to be three dolphins arranged in a circular pattern with the mouth of one near the tail of the next; in 1632, one of the slanders against Galileo circulated by his enemies and other gossipmongers was that this picture satirized the pope's having carried nepotism to new heights; this allegation was soon cleared up by demonstrating that the three dolphins had for a long time been the publisher's trademark and had nothing to do with the Barberinis or with Galileo's satire; but it is a good example of the poisoned atmosphere. Cf. Magalotti 1632; Pieralisi 1875, 360–61; Pastor 1938b, 38–58.

6. I have translated this memorandum from the Italian text provided by Favaro (15: 169), who printed the manuscript copy found in the State Archives of Modena; but for my translation I have also examined the Latin text provided by Riccioli (1651, 2: 497), who printed the copy sent to the Inquisitor of Venice, and I have examined the Italian text given by Pagano (1984, 244–45), who printed the manuscript copy sent to the Inquisitor of Siena, now available in the Inquisition Archives in Rome. The variations are insignificant, except as noted below.

7. *Maintains* is my translation of the Italian word *sostenta,* found in the Modena copy in Favaro 15: 169, line 3; this corresponds to the Venice copy in Riccioli 1651, 2: 497, which has *sustineatur;* but the Siena copy in Pagano 1984, 244, has *tratta* ("treats of"), which conveys a significantly stronger prohibition. However, such variations were relatively common at the time, when scribes made multiple copies.

8. Strangely as it may be, the original letter gives the title of the book simply as *Galileo Galilei Linceo.*

9. Here, both the Modena-Favaro and the Siena-Pagano copies have the harsher Italian *carcerato,* whereas the Venice-Riccioli copy has the milder Latin *inclusus.* My rendition *detained* has been influenced by the latter.

10. Favaro 15: 169.

11. Favaro 19: 282–83, 360–61; Pagano 1984, 154, 229.

12. Favaro 19: 284, 363; Pagano 1984, 156, 231–32.

13. We also have evidence that the Florentine nuncio helped to promulgate the condemnation by sending copies to other nuncios, although it is unclear whether he was acting on his own initiative or on orders from Rome; cf. Favaro 15: 260–61; Nelli 1793, 2: 555.

14. Favaro 19: 363–90; Pagano 1984, 158–203.
15. Favaro 19: 369, 375, 376–77, 383–84; Pagano 1984, 168–69, 178–79, 180–82, 192–93; these negligent inquisitors were those of Florence, Ferrara, Faenza, Como, and Pavia.
16. Favaro 19: 383–85; Pagano 1984, 192–94.
17. As Westman (1984) points out, most copies of the *Dialogue* had gone out to readers outside the universities; but apparently the Church felt most vulnerable at the universities because of their teaching function.
18. Cf. Guiducci to Galileo, 20 August 1633, in Favaro 15: 230–31.
19. Filippo Pandolfini (1575–1655) held several offices in the Tuscan government and translated into Latin Galileo's *Discourse on Bodies in Water, Sunspot Letters,* and *Assayer;* cf. Favaro 20: 502. Niccolò Aggiunti (1600–1635) was tutor to various Medici princes and, from 1626, professor of mathematics at Pisa; cf. Favaro 20: 363–64. Francesco Rinuccini (1603–1678) received a degree in law from Pisa and from 1637 to 1642 served as grand ducal ambassador to the Republic of Venice; cf. Favaro 20: 521–22. Dino Peri (1604–1640) became professor of mathematics at Pisa in 1636, after the chair had become vacant upon Aggiunti's death.
20. Apparently more than fifty professors had been summoned and were in attendance, as reported by the Florentine inquisitor to Rome on 27 August; cf. Favaro 19: 369 and Pagano 1984, 168–69.
21. Here I omit seven lines from the letter that are not directly relevant.
22. Ascanio Piccolomini.
23. Favaro 15: 240–42.
24. Buonamici to Galileo, 3 September 1633, in Favaro 15: 245–46.
25. Lagonissa 1633; Kellison 1633; Libri 1841a, 34; Monchamp 1892, 115–22; Favaro 19: 380–81, 392–93; Pantin 2001, 634; Stimson 1917, 75–76; Pagano 1984, 188, 205–6.
26. Favaro 19: 373–74; Pagano 1984, 176.
27. Favaro 19: 390; Pagano 1984, 202.
28. Monchamp 1892, 122–26.
29. Monchamp 1893, 14–17; cf. Favaro 19: 412–13.
30. For further details, see Monchamp 1893, 20–26, to which I am indebted, although I do not share all its claims.
31. Favaro 19: 412, lines 11–12; cf. Monchamp 1893, 15.
32. Favaro 19: line 20; cf. Monchamp 1893, 15.
33. Favaro 19: 404–6, lines 50 and 159; cf. Finocchiaro 1989, 289, 292.
34. Favaro 15: 169, lines 4–7.
35. Descartes reached the same conclusion; see chapter 3.1.
36. Favaro 19: 415–17.
37. Favaro 19: 384–85; Pagano 1984, 194.
38. Favaro 19: 411–12.
39. Favaro 19: 390; Pagano 1984, 202.
40. Belli 1633; Dibner 1967, 169; cf. Lerner 1998b, 622.
41. Cf. Favaro 1902a; also important are the reports of the Tuscan ambassador in Rome to the Tuscan secretary of state; see, for example, the letters of 26 June and 3 July 1633 in Favaro 15: 165, 170–71.
42. Buonamici to Galileo, 3 September 1633, in Favaro 14: 245–46.

43. Favaro 19: 407–11. A flawed version of Buonamici's account was first published by Nelli (1793, 2: 544–50); abridged versions of Nelli's copy were later published by Venturi (1818–1821, 2: 177–79) and Albèri (1842–1856, 9: 449–52); T. Martin (1868) criticized that version as inauthentic or apocryphal; Guasti (1873) proved that Buonamici had indeed written some such account, but did not locate the original manuscript; Favaro found the original, as well as the manuscript used by Nelli and another manuscript containing corrections in Buonamici's own hand, and published the genuine version in 1902 (Favaro 1902a, 696–99).

44. Here and in the rest of this text, my *spyglass* translates Buonamici's *occhiale,* and my *telescope* his *telescopio.*

45. Here Buonamici must have been referring to the *Sunspot Letters,* which was published in 1613, and not to the *Dialogue,* which appeared in 1632, although the content description corresponds more to the latter than to the former.

46. The controversy *de auxiliis* was a dispute between Dominicans and Jesuits which raged for decades in the late sixteenth and early seventeenth centuries. It involved subtle theological points about the nature of grace, predestination, free will, and personal merit and eternal salvation. It threatened to split the Catholic Church until 1607, when Pope Paul V put an end to the controversy by decreeing that for the time being it did not have to be authoritatively resolved and that in the future neither side must act as if the other side was heretical; then in 1611 the Inquisition prohibited all writings on the subject unless they received its own special approval. For more details, see Pastor 1891–1953, 25: 248–51; Feldhay 1995.

47. Christopher Scheiner (1573–1650), an Austrian Jesuit who discovered sunspots at about the same time as Galileo but tried to interpret them in an Aristotelian, Ptolemaic framework; cf. Gorman 1996.

48. "Mostro" was the nickname of Dominican friar Niccolò Riccardi, who held the title of Master of the Sacred Palace, a position whose main task was the censorship of books published in the city of Rome.

49. Giovanni Ciampoli (1589 or 1590–1643), a clergyman confidant of Maffeo Barberini when he was cardinal, and his correspondence secretary when he became Pope Urban VIII.

50. Not exactly right; born in 1564, Galileo was sixty-nine in 1633.

51. *Free custody* is my literal translation of Buonamici's oxymoronic *libera custodia;* but his meaning is that at one point during the 1633 trial Galileo was detained at the Inquisition palace but allowed to lodge in the apartment of the prosecuting attorney; Galileo was not under surveillance by guards, but he was ordered not to leave. This condition actually lasted for eighteen days, from 12 April to 30 April, rather than eleven, as Buonamici went on to state.

52. Favaro 19: 407–11.

53. Favaro 5: 315; Finocchiaro 1989, 92.

54. Recall that it was Buonamici who provided Galileo with a copy of the sentence and abjuration, sending them together with his own account on 3 September 1633; cf. Favaro 15: 245–46.

55. Favaro 19: 413–15; cf. Favaro 16: 18–19; Lerner 1998b (614 n. 27), 2001 (519–23); Motta 1993, 24; Nellen 1994, 56 n. 22; Renaudot 1634, 531ff.; and Ross 1646, 9. Lerner 2001a, 522–23, indicates that Cardinal Armand Jean

Richelieu, chief minister of France, controlled the *Gazette,* and so its publication of the Galilean sentence had political connotations.

56. Here I have changed the original French singular *Maistre;* cf. Favaro 19: 414, line 33.

57. Favaro 19: 413–15.

58. Favaro 19: 414, lines 22–23.

59. See P. Lansbergen 1630; Froidmont 1631; J. Lansbergen 1633; Froidmont 1634; and cf. Cavalieri 1633; Favaro 1887b, 111 n. 2; Hooykaas 1976, 40; Lerner 2001a; McColley 1938; Montucla 1799–1802, 2: 298; Pantin 2001, 634; Venturi 1818–1821, 2: 133–35.

60. Mersenne 1634, 214–18; cf. Pessel 1985, 385–93, 425.

61. Favaro 19: 403, lines 33–38; Finocchiaro 1989, 288.

62. Mersenne 1634, 217; cf. Pessel 1985, 386–87.

63. Favaro 19: 406, lines 157–59; Finocchiaro 1989, 292.

64. Mersenne 1634, 226; cf. Pessel 1985, 391.

65. For more details, see De Waard and Beaulieu 1932–1988, 4: 74–76, 156–57, 270–71; Favaro 1887b, 109; Favaro 16: 119; Lenoble 1943, xx, 399–401; Lerner 1998b (614–16), 2001a (524–26); Mersenne 1634; Mersenne 1932–1988, 4: 267–70; Pessel 1985, 403–25; J. Russell 1989, 371.

66. Pessel 1985, 341–46.

67. Pessel 1985, 403–8.

68. Pessel 1985, 353–55.

69. Pessel 1985, 411–14.

70. Pessel 1985, 377–81.

71. Pessel 1985, 417–22.

72. Pessel 1985, 383–85.

73. Pessel 1985, 423–25.

74. I have consulted three copies of Mersenne's book, and they are all semi-expurgated versions. The original and the fully expurgated versions are given in Pessel 1985, 211–425.

75. Lerner 2001a, 522–23; as Lerner points out, the *Mercure,* like the *Gazette,* was under Richelieu's control.

76. For more details, see Boulliau 1639; Morin 1642; T. Martin 1868, 387; Boyer 1970, 348; Nellen 1994; Lerner 2001a.

77. Polacco 1644; Montucla 1799–1802, 2:300; Venturi 1818–1821, 2: 133.

78. Riccioli 1651, 2: 496–500.

CHAPTER 3

1. Descartes leaves this quotation in Latin *(nonumque prematur in annum),* without giving any reference. It is from *De Arte Poetica,* 388. I have used the translation by Christopher Smart in Horace (1863); cf. www.perseus.tufts.edu. I thank my colleague Thomas Osborne for locating this reference.

2. Favaro 15: 340–41; cf. Descartes 1897–1913, 1: 270–73.

3. Favaro 16: 56; cf. Descartes 1897–1913, 1: 280–84.

4. Descartes to Mersenne, 14 August 1634, in Favaro (16: 124–25) and in Descartes (1897–1913, 1: 303–6).

5. Favaro 7: 372–83.

6. Favaro 7: 117–24, 244–73, 383–99.

7. Here Descartes quoted the Latin *Bene vixit, bene qui latuit* (Ovid, *Tristia*, III, 4, verse 25).

8. Favaro (16: 89, n. 1) suggests that this churchman may be G. Wendelin; Lerner (2001a, 533 n. 77) states that it was Ismaël Boulliau; but it seems to me that Descartes could be referring in a very indirect way to Mersenne himself.

9. Descartes here quoted the Latin clause "*quamvis hypothetice a se illam proponi simularet*"; cf. Favaro 19: 413, lines 28–29; Monchamp 1893, 16; and chapter 2.3.

10. Favaro 16: 88–89; cf. Descartes 1897–1913, 1: 284–91.

11. Favaro 19: 413, lines 28–29; Monchamp 1893, 16; chapter 2.3.

12. Favaro 16: 88, lines 7–8.

13. Descartes to Mersenne, 14 August 1634, in Favaro 16: 125, lines 30–39 (cf. Descartes 1897–1913, 1: 303–6); Favaro 19: 412–13, lines 26–36; Monchamp 1893, 16.

14. In Favaro 15: 125.

15. Descartes to an unknown correspondent, autumn 1635, in Favaro 20: 608; and in Descartes 1897–1913, 1: 321–24.

16. Descartes, *Discourse on Method*, VI, in Haldane and Ross 1955, 1: 119.

17. Descartes, *Method*, VI, in Haldane and Ross 1955, 1: 119.

18. Descartes, *Method*, VI, in Haldane and Ross 1955, 1: 119–22.

19. Descartes, *Method*, VI, in Haldane and Ross 1955, 1: 119.

20. Descartes, *Method*, VI, in Haldane and Ross 1955, 1: 122.

21. Descartes, *Method*, VI, in Haldane and Ross 1955, 1: 124.

22. Descartes, *Method*, VI, in Haldane and Ross 1955, 1: 126.

23. Descartes, *Method*, VI, in Haldane and Ross 1955, 1: 127–28.

24. Descartes, *Method*, VI, in Haldane and Ross 1955, 1: 128.

25. Descartes, *Method*, VI, in Haldane and Ross 1955, 1: 128–29.

26. Descartes, *Principles of Philosophy*, III, 28, in Miller and Miller 1991, 94–95.

27. In Miller and Miller's English translation of the *Principles of Philosophy*, phrases within braces are additions taken by them from the 1647 French translation of the original 1644 Latin edition; cf. Miller and Miller 1991, xi.

28. Descartes, *Principles*, III, 26, in Miller and Miller 1991, 94.

29. Gaukroger 1995, 12.

30. Descartes, *Principles*, III, 43, in Miller and Miller 1991, 104.

31. Descartes, *Principles*, III, 44, in Miller and Miller 1991, 105.

32. Descartes, *Principles*, III, 44, in Miller and Miller 1991, 105.

33. Favaro 7: 29–30, 133, 298, 382–83, 487–88; cf. Finocchiaro 1980, 12–18.

34. In Heilbron's inimitable words, this was "the convenient fiction that it [the earth's motion] was a convenient fiction" (1999, 22).

35. Favaro 1916b, 8; Motta 1993, 171.

36. See Gaukroger 1995, especially 11–12, 292.

37. I am referring to Sobel 1999, whose best-seller status bears witness to the persistent power of the Galileo affair over people's minds; the much-advertised reevaluation of the affair suggested by this book perhaps lies in the direction of stressing that if Galileo could create, nurture, and inspire such a human being as Sister Maria Celeste, then his many detractors (in his own time as well as in ours) must be missing something. For other accounts of this relationship,

see Arduini 1864; Allan-Olney 1870; M. C. Galilei 1883; Favaro 1891; Saverio and Rossi 1984; M. C. Galilei 1992.

38. Favaro 13: 116–17.

39. Favaro 15: 352–53.

40. Galileo to Bocchineri, 27 April 1634, in Favaro 16: 84–85.

41. Maria Celeste Galilei to Galileo, in Favaro 15: 292–93; and in Sobel 1999, 312–14; 2001.

42. For more details, see Gassendi 1641, 1657; Favaro 20: 504, 1917b; Rizza 1961, 1965; Brown 1974; P. Miller 2000.

43. Favaro 1917b, 630–36.

44. Peiresc to Dupuy, 15 January 1634, in Favaro 16: 18–19; cf. Favaro 1917b, 614–19.

45. Peiresc to Galileo, 26 January 1634, in Favaro 16: 27–28.

46. The entire letter, of which Favaro gives only the second half, is quoted in Pieralisi 1875, 304–10.

47. Favaro (16: 170 n.) identifies these men as Girolamo Aleandro and Lorenzo Pignoria.

48. *Philosophical play* is my rendition of *scherzo problematico;* see the discussion in n. 53 below.

49. Peiresc 1634.

50. F. Barberini to Peiresc, 2 January 1635, in Favaro 16: 187.

51. Peiresc 1635.

52. But two and a half centuries afterward, Peiresc's plea was criticized at length by Pieralisi (1875, 301–40), a clergyman who was director of the Barberini library in Rome.

53. Favaro 16: 170, line 43; Peiresc repeated the philosophical-play interpretation in his letter to Galileo of 1 April 1635 (in Favaro 16: 145–48, at p. 247, line 68). Peiresc was not the only one to suggest such an interpretation of the *Dialogue.* Tommaso Campanella (1568–1639), author of the first explicit *Apology for Galileo* (1622), advanced a similar suggestion in a letter to Galileo written immediately after reading the book in August 1632. Making an explicit comparison between Galileo's and Plato's literary style, as well as between Salviati's and Socrates' maieutic method of discussion, Campanella (Letter to Galileo, 5 August 1632, in Favaro 14: 366) also called the book a "philosophical comedy," using a phrase *(comedia filosofica)* that is more explicit but less tragic than Peiresc's *scherzo problematico.* I thank Robert Westman for having called to my attention Campanella's description and having alerted me that my translation of Peiresc's phrase is not a literal one; cf. Westman (1984, 334).

54. Favaro 16: 171, 17: 26–27; cf. Wolynski 1877.

55. See Wolynski 1872–1873, 17: 3–22; cf. Ladislaus to Galileo, 19 April 1636, in Favaro 16: 420–21; Galileo to Ladislaus, July–August 1636, in Favaro 16: 458–59; and chapter 4.1.

56. For more details on Diodati, see Favaro 1888; Gardair 1984; Garcia 2001; and chapter 4.2.

57. Francesco Niccolini (1584–1650).

58. This was not the same as the Tuscan ambassador's residence, which was Palazzo Firenze; so for this short period (about two weeks) there was a change

from the situation before the sentence. See Shea and Artigas 2003, 30, 179–80, 195.

59. Ascanio Piccolomini.

60. Libert Froidmont; cf. Froidmont 1631, 1634.

61. Cf. Rocco 1633.

62. Melchior Inchofer; cf. his *Tractatus Syllepticus* (1633).

63. G. Galilei 1634a.

64. Favaro 19: 393; cf. Pieralisi 1875, 254–61.

65. Favaro 15: 344, 363.

66. G. Galilei 1634b, 115–17, lines 1–71.

67. Roberto Galilei to Galileo Galilei, 7 February 1635, in Favaro 16: 206–7.

68. G. Galilei 1635a.

69. Micanzio to Galilei, 10 February 1635, in Favaro 16: 208–10; cf. Favaro 16: 229–30, 236–37.

70. Cf., for example, Favaro 16: 300–302.

71. Favaro 16: 500–501, 507 n. 3, 512–13; Favaro 20: 610.

72. Favaro 18: 14–15, 23–24, 26.

73. Favaro 18: 227, 237.

74. Castelli to Galileo, 22 December 1635, in Favaro 16: 363–64.

75. Cf. Favaro 7: 488–89; Finocchiaro 1997, 33–34, 306–8.

76. Pieralisi 1875, 365–66.

77. Magalotti to Guiducci, 7 August 1632, in Favaro 14: 382–83.

78. Favaro 16: 461, 454–55; cf. Wolynski 1877.

79. Galileo to Micanzio, 28 June 1636, in Favaro 16: 444–45, lines 20–22.

80. Favaro 17: 237–38.

81. Favaro 17: 254–55, 272; cf. Favaro 17: 237–38.

82. Favaro 19: 290.

83. Favaro 17: 290, 310–11, 312, 312–13, 320, 321, 324.

84. Favaro 16: 463–68.

85. Favaro 17: 66, 119; cf. Frisi 1775, 87.

86. Favaro 17: 356.

87. Favaro 17: 357, 366, 369–72, 372–73; cf. Favaro 18: 140–41 and Favaro 19: 397–98.

88. Favaro 17: 374, 375–76, 381, 382–83, 386, 393, 398, 406, 410; Favaro 19: 395.

89. T. Martin (1868, 231) claims plausibly that both factors played a role.

90. Favaro 16: 201–2.

91. Favaro 16: 317–18.

92. Favaro 16: 292–93, 329–30.

93. Favaro 16: 340–44; cf. Favaro 16: 327–28, 335–37, 350–54.

94. Favaro 16: 354–55.

95. Favaro 16: 436, 450–53; Favaro 1916c, 487.

96. Favaro 16: 444–45.

97. Favaro 16: 454–55, 456.

98. Favaro 16: 520.

99. Favaro 16: 500–501, 507 n. 3.

100. Favaro 17: 75–76.

101. Favaro 17: 176.

102. Favaro 17: 183–84; cf. Favaro 16: 191–92.

103. Cf. Madden 1863, 35.
104. Favaro 8: 472, 19: 622, 20: 555.
105. Favaro 18: 303, 364, 364–65.
106. Favaro 18: 360, 19: 567, 626.
107. Favaro 18: 372; Motta 1993, 82–85.
108. Galileo to Dal Pozzo, 20 January 1641, in Favaro 18: 290–91.
109. Favaro 16: 116, lines 41–42.

CHAPTER 4

1. My account here paraphrases the text found in Poisson 1670a, 235–37, quoted in Lerner 2001a, 546–47; cf. also Montucla 1799–1802, 2: 297; Venturi 1818–1821, 2: 196; Drinkwater Bethune 1832, 191; Favaro 1916d, 8–9; Crombie 1961, 2: 223; Lerner 2001a, 540–41.
2. Favaro 1916d, 9.
3. Lerner 2001a, 546 n. 1. Lerner's opinion seems to be based in part on a 1666 letter by Launoy noting that thirty years earlier he had defended a colleague from the charge of Copernicanism by advancing an argument similar to the one reported by Poisson; cf. Lerner 2001a, 529–30.
4. It is unclear what this professor was referring to; he might have been referring to the condemnation of 1277, but that was decreed by the University of Paris and not by an ecumenical Church Council.
5. My account here relies on Pagano 1984, 39–40; Pardo Tomás 1991, especially pp. 23–29, 75–81, and 183–90; Navarro Brotons 1995; and Navarro Brotons 2001, especially pp. 816–18.
6. Favaro 19: 390; Pagano 1984, 202.
7. Favaro 19: 416, lines 41–44; Pardo Tomás 1991, 183–90.
8. Cf. Pieroni to Galileo, 11 August 1635, in Favaro 16: 300–2.
9. Favaro (17: 131 n. 1) identified him as Francis Dietrichstein.
10. See Favaro 19: 550 for the text of this endorsement, signed by Gio. Tomaso Manca de Prado, O.P., and dated 18 November 1636.
11. Pieroni enclosed copies of the imprimaturs in his letter.
12. See Favaro 19: 550 for the text of this endorsement by Gio. Ernesto Platais, dated 20 November 1636; the bishop's point here stems from the fact that Manca de Prado had not included the book's title in his endorsement.
13. The identity of this town is unclear; the Italian Nissa translates into Nis, but Nis is a town in Serbia and not in Silesia.
14. See Favaro 19: 551 for the text of this endorsement by Gualterus Paullus, S.J., dated 29 April 1637.
15. Leon. Mylgiesser, doctor of medicine; cf. Favaro 19: 551.
16. Favaro (17: 131 n. 7) identifies this person as Ernest Adalbert D'Harrach.
17. Pieroni 1637, lines 1–46.
18. See Favaro 19: 551.
19. Wolynski 1872–1873, especially 16: 231–71, 17: 3–22.
20. Pieroni to Galileo, 11 August 1635, in Favaro 16: 301–2, lines 46–51.
21. Favaro 16: 301, lines 65–68.
22. Ladislaus IV to Galileo, 19 April 1636, in Favaro 16: 420–21.
23. Galileo to Ladislaus IV, July–August 1636, in Favaro 16: 458–59.

24. See Wolynski 1873–1873, 17: 3–22.

25. Cf. Favaro 16: 73–74, 19: 286, 393–94 and chapter 3.4.

26. Micanzio to Galileo, 29 April 1634, in Favaro 16: 86–87.

27. Cf. Favaro 16: 208–10, 229–30, 236–37.

28. Micanzio to Galileo, 24 March 1635, in Favaro 16: 139.

29. Cf. Favaro 17: 356, 357, 366, 369–72, 372–73, 18: 140–41, 19: 397–98, and chapter 3.4.

30. On 23 July, 17 September, and 8 October 1639; cf. Favaro 18: 74–75, 104–5, 112–13.

31. Venice and the king of France, respectively.

32. Micanzio to Galileo, 17 September 1639, in Favaro 18: 104–5, lines 9–26.

33. Sarpi 1616.

34. From Finocchiaro 1989, 225; cf. Galileo to Diodati, 15 January 1633, in Favaro 15: 25–26.

35. Finocchiaro 1989, 225; cf. Favaro 15: 25–26.

36. Favaro 1916b, 38–46; Favaro 15: 218.

37. For this and other details about Bernegger's life, I rely on Favaro 1916b; other details relevant to this story are in Favaro 1916c.

38. Nonnoi 2000, 202; cf. Sackenreiter-Zeyssolff 1984.

39. Favaro 16: 158, 238, 258; cf. Favaro 1916b, 38–46; Garcia 2000, 308–20; Nonnoi 2000, 187–207.

40. See G. Galilei 1635b.

41. For example, the *Systema Cosmicum* incorporated a correction of some numbers in the 1632 edition, and this correction was not in the published list of errata but was sent by Galileo to Castelli to ensure it would be included in the copies circulating in Rome; for this and other details, see Garcia 2000.

42. Garcia 2000, 320 n. 44; cf. G. Galilei 1635b, title page. As Pantin (1999) points out, this quotation had been adopted earlier by Kepler.

43. Garcia 2000, 320 n. 44; cf. G. Galilei 1635b, title page.

44. G. Galilei 1635b, 459–64 and 465–95, respectively.

45. G. Galilei 1635b, 459.

46. This is claimed by Favaro 1916b, 47.

47. G. Galilei 1635b, title page; cf., for example, Cinti 1957, 197.

48. Bernegger, "Benevole Lector," in G. Galilei 1635b, preface, p. 3 (unnumbered); here translated and quoted from Garcia 2000, 317.

49. See Howell 2002, 109–35, for a fuller account of Kepler's views on the relationship between astronomy and Scripture.

50. Kepler 1661, 461.

51. Kepler 1661, 466.

52. The following account is adapted from Westman 1984, 338–39.

53. Motta 2000, 74; Pantin 1999, 261.

54. See G. Galilei 1636; these may also be consulted in Favaro 16: 194–96 and 389–90.

55. G. Galilei 1636, 17; cf. Motta 2000, 101; Favaro 5: 319; Finocchiaro 1989, 96.

56. The literature on this topic is enormous; good places to start would be Biagioli 2003; Carroll 2001; Finocchiaro 1986a; McMullin 1998; Motta 2000; Pesce 2000; Rossi 1978; Stabile 1994. Almost all these studies make no use of the text in the original edition of the *Nov-antiqua* and rely on the slightly different

text published by Favaro (5: 309–48), except for Pesce and Motta, who have argued convincingly that the original text is at least as valuable. I agree and offer the following example of how the original text can help us avoid some unnecessary difficulties.

One of the most problematic passages in Galileo's Letter to Christina is the following: "In regard to those propositions which are not articles of faith, the authority of the same Holy Writ should have priority over the authority of any human writings containing pure narration or even probable reasons, but no demonstrative proofs; this principle should be considered appropriate and necessary inasmuch as divine wisdom surpasses all human judgment and speculation" (Finocchiaro 1989, 94; cf. Favaro 5: 317); this passage suggests that biblical authority has priority over propositions supported by testimony or probable arguments; but this thesis seems inconsistent with other parts of the Letter, where Galileo denies the philosophical (scientific) authority of Scripture; one of these parts is Baronio's aphorism (and the supporting argument) that "the intention of the Holy Spirit is to teach us how one goes to heaven and not how heaven goes" (Finocchiaro 1989, 96; cf. Favaro 5: 319); this inconsistency is regarded by some authors as evidence for the "incoherence" of Galileo's position. However, the version in the *Nov-antiqua* says: "In regard to those propositions which are not articles of faith, the authority of the same Holy Writ should have priority over the authority of any human *sciences* that are written not with a demonstrative method but with pure narration or even with probable reasons" (G. Galilei 1636, 14; Motta 2000, 97–98; italics mine); this wording suggests that biblical authority has priority over propositions treated of in the nondemonstrative sciences; this claim is not inconsistent with Baronio's principle because biblical authority would still *not* have priority over *not-yet-proved* propositions in the *demonstrative* sciences; and, of course, the geokinetic proposition fits the latter case. Galileo's notion of a nondemonstrative science here corresponds to the one he advanced in a famous passage of the *Dialogue,* where he drew a distinction between the natural sciences, in which demonstration is essential, and disciplines such as "law or other human studies" (Favaro 7: 78; Finocchiaro 1997, 101), in which rhetoric is important.

57. I do not want to convey the impression that this division of labor was neater than it really was. In fact, in the *Nov-antiqua*, Galileo also argued that Copernicanism was not really contrary to Scripture; for example, that the Joshua passage (when literally interpreted) contradicted the Ptolemaic system but was in accordance with Copernicanism (when plausibly developed); cf. Galilei 1636, 52–60; Favaro 5: 343–48; and Finocchiaro 1989, 114–18. Similarly, neither Foscarini nor Kepler argued merely for the conclusion that Copernicanism was compatible with Scripture, but rather also (although less centrally) for the principle denying the astronomical authority of Scripture; for example, Kepler came close to Baronio's aphorism in the Introduction to *Astronomia Nova,* where he said that "I answer in one word, that in theology the weight of authority, but in philosophy the weight of reason is to be considered" (Kepler 1661, 467).

58. Finocchiaro 1989, 291; Favaro 19: 405.

59. Here my summary relies on Garcia 2000, 328–32.

60. Pantin 1999, 261.
61. I am using this term in a loose sense that stresses the importance of individual freedom; I do not mean to equate *liberal* with *libertine* in the seventeenth-century sense; for a criticism of the "libertine" interpretation of Galileo, see, for example, Pantin 2000.
62. Sirluck 1959, 162.
63. Milton 1644, 1959; for a good interpretation and evaluation of these arguments, as well as an account of the historical context, see Sirluck 1959.
64. Milton 1644 (24), 1959 (537–38).
65. Milton 1644 (24), 1959 (537–38); I have modernized the spelling.
66. Cf. Favaro 19: 9 n.3; Harris 1985, 5–6; Chaney 1991, 141.
67. I owe much of this account to Montucla 1799–1802, 2: 296–97; T. Martin 1868, 241; Pesce 1991a, especially pp. 77–79; Motta 1993, 89–90; Pesce 2000, 63–64.
68. See Morin 1631, 1634; Gassendi 1642; Morin 1643; Gassendi 1649; Morin 1650.
69. Gassendi 1649, appendix, pp. 1–60, 61–63, 65–78, and 79–95, respectively; I take this information from Pesce 1991a, 78 n. 45.
70. In Favaro 5: 281–88 and 309–48, respectively; cf. Finocchiaro 1989, 49–54 and 87–118.
71. Dini to Galileo, 7 March 1615, in Favaro 12: 151–52; Finocchiaro 1989, 58–59.
72. Psalms 19: 1–6, in the King James Version: "The heavens declare the glory of God; and the firmament sheweth his handywork. . . . In them hath he set a tabernacle for the sun, which *is* as a bridegroom coming out of his chamber, *and* rejoiceth as a strong man to run a race. His going forth *is* from the end of the heaven, and his circuit unto the ends of it."
73. Salusbury 1661–1665, 1: 1–424, 425–60, 461–67, 468–70, 471–503, respectively.
74. Milton 1644 (24), 1959 (537).
75. Salusbury 1661a, which is printed on two unnumbered pages just before the *Dialogue* in Salusbury 1661b; in the following quotations, I have modernized the spelling.
76. Cf. Pieralisi 1875, 365–66.
77. Most of the published books are listed in the bibliography in Carli and Favaro 1896, but the following are not: J. Lansbergen 1633; Ward 1635; Descartes 1637; Morin 1640; White 1642; Argoli 1644; Descartes 1644; Wendelen 1644; Mousnier 1646; Hevelius 1647; Renieri 1647; Wendelen 1647; Mousnier and Fabri 1648; Le Tenneur 1649; Morin 1650; Varenius 1650.
78. Inchofer 1635; Cavalieri 1642; Hobbes 1642; Le Tenneur 1646.
79. Especially Inchofer 1633, 1635; Accarisius 1637; Campanella 1637; Parasin 1648; cf. Pesce's works, especially Pesce 1987, 1991a.
80. Chiaramonti 1633; Rocco 1633; Barenghi 1638; Wilkins 1638, 1640a; Digby 1644; Ross 1646; Morin 1650.
81. J. Lansbergen 1633; Ross 1634; Linemannus 1635; Ward 1635; Ross 1636; Wilkins 1636; Wilkins 1640a; Hobbes 1642; Deusingius 1643; Ross 1646; Hevelius 1647; Parasin 1648; Varenius 1650.

82. Ross 1646, 9; a trace of this type of attempt to delegitimize Galileo is found even in Riccioli, as suggested by his stress on the abjuration in his chronology and biographical glossary (Riccioli 1651, 2: xxviii, xxxiv).

83. Galluzzi 2000, 509.

84. Galluzzi 2000, 539.

85. Here I rely mostly on Galluzzi 1993a, 2000; but cf. Baliani 1638, 1646; Cabeo 1646; Caramuel Lobkowitz 1644; Gassendi 1642, 1646; Gassendi 1649; Huygens 1646; Le Cazre 1645a,b; Le Tenneur 1646, 1649; Mersenne 1647; Morin 1643; Mousnier 1646; Mousnier and Fabri 1648; Palmerino 1999; Torricelli 1644.

86. Favaro 18: 378–79, 19:558.

87. Favaro 18: 378–79, 19:558, 596, 1891c (380); Galluzzi 1993b (146), 1998 (418).

88. Nelli 1793, 2: 850–51.

89. Favaro 19: 559–62; cf. Nelli 1793, 2: 852; Venturi 1818–1821, 2: 324; Favaro 1891c, 377–78.

90. On 12 January 1642; cf. Favaro 18: 378.

91. Favaro 18: 378–79, 379–80.

92. 29 January 1642; cf. Favaro 18: 380, 381–82.

93. Favaro 18: 383 [#4204]; 19: 535–37; cf. Drinkwater Bethune 1832, 299; Brewster 1841, 113; Venturi 1818–1821, 2: 324; Favaro 1891c, 378–80, 384–88.

94. Erythraeus 1643, 279; cf. Brucker 1766–1767, tome 4, part 2 (= vol. 5), 634 n. e; Nelli 1793, 1: 25; Carli and Favaro 1896, 41, 91, 112; Motta 1993, 102.

95. Viviani 1654, 1717.

96. Brucker 1742–1744.

97. Brucker 1766–1767, tome 4, part 2 (= vol. 5), 634 n. e; and tome 5 (= vol. 6), appendix, 916.

98. Diderot and D'Alembert 1751–1780, 1: 790.

99. Viviani's biography of Galileo was written in 1654, first published in 1717, and then reprinted in the 1718 edition of Galileo's works; cf. Viviani 1654, 1717; G. Galilei 1718.

100. Nelli 1793, 1: 25–26.

101. Nelli 1793, 1: 26 n. 1.

102. Riccioli 1651, 2: 193–535.

103. Riccioli 1651, 1: xxvi–xxviii.

104. Riccioli 1651, 1: xxviii.

105. Riccioli 1651, 1: xxxiv.

106. Riccioli 1651, 2: 280–89; cf. Stimson 1917, 79–84.

107. Stimson 1917, 79–84. For the critique of the pro-Copernican arguments, see Riccioli 1651, 2: 311–407; for the anti-Copernican arguments, see Riccioli 1651, 2: 408–78.

108. Riccioli 1651, 2: 478.

109. Here I rely on Pesce 1987, 266–68, and Delambre 1821, 672–81; cf. Riccioli 1651, 2: 479–95; Dinis 1989, 239–55; Baldini 1996a.

110. One is tempted to label Riccioli's position "fundamentalism." This label might be regarded as anachronistic, since "fundamentalism" was a Protestant movement that emerged in the early twentieth century. However, one of the fundamental tenets of Christianity, according to this movement, is the infallibility or

inerrancy of the Bible in scientific and historical matters. Thus, Riccioli was a fundamentalist in this sense. Nevertheless, I relegate such a suggestion only to the title of this chapter.

111. Riccioli 1651, 2: 290; here quoted and translated from the French text in Delambre 1821, 1: 672.

112. Here quoted from Stimson 1917, 79–80; cf. Riccioli 1651, 2: 500; it should be noted that Stimson gives p. 496 as the reference of her quotation, but the passage in question is found on p. 500.

CHAPTER 5

1. Favaro 8: 472, 19: 622, 20: 555; for this and other details about Viviani, see also Favaro 1912.

2. In Salvini 1717, 397–431; cf. Favaro 19: 599–632; Viviani 1992.

3. For some of the controversies surrounding Viviani's biography, see Favaro 1915b; Segre 1989.

4. Viviani 1654, 617.

5. Favaro 1887b (113, 126–27), 1912 (111); Besomi and Helbing 1998, 938.

6. Southwell to Viviani, February 1662, quoted in Motta 1993, 87, from Favaro 1885b, 35.

7. The details of this story are given in the account and the documents found in Favaro 1887c.

8. Baldigiani to Viviani, 26 May 1678, in Favaro 1887c, 128–29.

9. Viviani to Baldigiani, 14 June 1678, in Favaro 1887c, 135–37.

10. In Favaro 1887c, 135.

11. In Favaro 1887c, 135.

12. Cf. Guicciardini to Cosimo, 4 March 1616, in Favaro 12: 241–43; Fabroni 1773–1775, 1: 53–57; chapter 8.2.

13. In Favaro 1887c, 143–44.

14. In Favaro 1887c, 144.

15. In Favaro 1887c, 144.

16. Bertoloni Meli 1988, 33; Robinet 1988, 231–33.

17. Bertoloni Meli 1988, 29–36, especially p. 35

18. Favaro 1887c, 152 n. 1.

19. Viviani's term here is indeed *scienziati*.

20. This was a point Galileo had made in the Preface to the *Dialogue* to explain his motivation, and his having done so is important even though it failed to impress the authorities; cf. Favaro 7: 29–31, 19: 352; Finocchiaro 1989 (266), 1997 (77–82).

21. Viviani 1690; the available manuscript breaks off here in midsentence.

22. Bertoloni Meli 1988, 29–36; Robinet 1988, 114–18.

23. Besomi and Helbing 1998, 930–40.

24. Baldigiani 1693, 156; Berti 1876a, 152–53; Favaro 1887b, 121, 155–56.

25. Fahie 1903, 404; Galluzzi 1993b, 1998; Nelli 1793, 2: 854–67; J. Russell 1989, 382; Viviani 1701, appendix, p. 122.

26. Fabri 1661 (= Divini 1661).

27. Fabri 1660 (= Divini 1660); Huygens 1660.

28. "A Further Account" 1665, 74–75; I have made a few stylistic emendations for

the sake of clarity. Cf. Divini 1661 (= Fabri 1661), 49. I first learned about this passage from Mayaud 1997, 261–62, 327–28. Fabri quotes Virgil, *Aeneid*, III, 72; this was the same verse quoted by Copernicus in *Revolutions*, book 1, chapter 8 (see chapter 1.4, this volume).

29. Quoted from Finocchiaro 1989, 68; cf. Favaro 5: 172.

30. Guiducci to Galileo, 6 September 1624, in Favaro 13: 203; cf. Garzend 1912, 462–63; Pieralisi 1875, 152–53; W. Ward 1865, 406.

31. Berti 1876a, 121–25.

32. See, for example, Amort 1734; Auzout 1665; Bailly 1785, 2: 131–32; Bouix 1869, 2: 465; Lalande 1771, 1: 539–41; Lazzari 1757, folio 491 v; Leibniz 1679–1686 (31–2), 1688 (200–202), 1704c (514–15, 517–21); Mayaud 1997, 327–34; Montucla 1758a, 1: 541–42; Pieralisi 1875, 130; Roberts 1885, 35; Ward 1871b, 162–63; Wolff 1735, 3: 607; and "A Further Account" 1665, 74–75.

33. Auzout 1665.

34. Here I omit a paragraph in which Auzout quoted Fabri's statement.

35. Here I omit eighteen lines in which Auzout discusses Fabri's change of mind with regard to the rings of Saturn.

36. This phrase is quoted from the Inquisition's sentence of 1633; cf. Favaro 19: 403, line 34; Finocchiaro 1989, 288.

37. This was not really a separate treatise but rather section 4 of book 9 in volume 2 of the *Almagestum Novum;* cf. Riccioli 1651, 2: 290–500 and chapter 4.3.

38. Here I omit thirty-three lines in which Auzout criticizes Riccioli's unanswered objections to the earth's motion.

39. Auzout 1665, 58–66; I am indebted to Mayaud 1997, 261–61, 327–26, for my first exposure to this important text.

40. See, respectively, Lazzari 1757 and chapter 7.2; and Olivieri 1820c, 1840, and chapter 10.

41. In Favaro 5: 285–88, 343–48; Finocchiaro 1989, 52–54, 114–18.

42. "A Further Account" 1665, 75; I have retained the original spelling and punctuation.

43. Baruzi 1907.

44. For more details, see Daville 1909; Robinet 1988.

45. Robinet 1988, 206.

46. Chapter 4.7 in Robinet 1988.

47. For more details, see Bertoloni Meli 1988; Robinet 1988, 96–118; Mayaud 1997, 329–34.

48. Grua 1948, 1: 30–34.

49. Translated from the French text transcribed and published by Mayaud (Rome: Editrice Pontificia Università Gregoriana, 1997), 329–30; cf. Grua 1948, 31–32.

50. Fabri 1661, 49; chapter 5.2.

51. Grua (1948, 32 n. 86) refers to Mersenne 1623, without giving anything more specific; Bertoloni Meli (1988, 22 n. 10) cites Mersenne 1636, 76, but quotes no passage; Mayaud (1997, 329 n. 7) quotes a sentence from Mersenne (1623, column 901) to substantiate the first part of Leibniz's attribution, but admits that he has not found the second part anywhere in Mersenne.

52. Auzout 1665, 60; cf. my translation in chapter 5.2 above.

53. Leibniz 1684, 47–48.

54. Ernst von Essen-Rheinfels to Leibniz, 1/11 November 1684, in Rommel 1847, 2: 49–51.

55. Rommel (1847, 1: 200 n.) identifies this person as Claudius Franciscus Milliet de Challes (d. 1678), professor of mathematics at Turin. Bertoloni Meli (1988, 22), who refers to him by the variant name Claude Dechales, adds that he was the author of *Cursus seu Mundus Mathematicus* (Lyons, 1674) and that Leibniz is almost quoting a passage from pp. 289–90. Mayaud (1997, 330 n. 8) actually quotes the passage.

56. In Rommel 1847, 2: 200–202.

57. For more information on both Kepler and the Letter to Christina, see chapter 4.2.

58. Here my account relies on Robinet 1988.

59. Robinet 1988, 4–5.

60. Bertoloni Meli 1988, 36–40.

61. For example, an essay in dialogue form titled *Phoranomus, seu de Potentia et Legibus Naturae* (cf. Bertoloni Meli 1988, 25 n. 30; Robinet 1988, 85 n. 26) and one titled *Dynamica de Potentia et Legibus Naturae Corporeae* (cf. Bertoloni Meli 1988, 32 n. 63; Robinet 1988, 101, 259–67).

62. Leibniz 1689a,b (= 1689c); cf. Robinet 1988, 96–118.

63. Bertoloni Meli (1988, 29–36) and Robinet (1988, 114–16) concur about this addressee.

64. This corresponds to Leibniz 1689b,c; Couturat 1903, 590–93; Robinet 1988, 111–14; Ariew and Garber 1989, 90–94.

65. Ariew and Garber 1989, 91; cf. Leibniz 1689b, 111.

66. Ariew and Garber 1989, 93; cf. Leibniz 1689b, 113.

67. Bertoloni Meli (1988, 27) calls Leibniz's effort "equilibrist"; Robinet (1988, 116) speaks of "diplomatic equilibrium."

68. This corresponds to Leibniz 1689a; Gerhardt 1849–1863, 6: 145–57; Robinet 1988, 107–11. But there are two variants of the manuscript (cf. Robinet 1988, 102).

69. For example, it was no longer true, as it had been once, that the earth's motion was philosophically absurd, as even Riccioli stated or implied (Leibniz 1689a, 107; Bertoloni Meli 1988, 29–30); and it was inappropriate to search for metaphysical certainty on this issue (Leibniz 1689a, 108; Bertoloni Meli 1988, 31–32).

70. Quoted in Bertoloni Meli 1988, 31; cf. Leibniz 1689a, 108; I have made some minor emendations to the quoted translation.

71. Bertoloni Meli 1988, 33 n. 69.

72. Leibniz to Magliabecchi, 20/30 October 1699, in Dutens 1768, 5: 127–29, at p. 128; cf. Mayaud 1997, 332.

73. I am quoting the paraphrase given by Baruzi (1907, 283).

74. Quoted in Baruzi 1907, 283 n. 3.

75. Bertoloni Meli 1988, 41.

76. Remnant and Bennet 1997, 509–21; cf. Leibniz 1704a.

77. Remnant and Bennet 1997, 512–13; cf. Leibniz 1704a, 512–13.

CHAPTER 6

1. The translation in Pascal 1744 (2: 314) reads "have proved," whereas Pascal 1962 (53) says, "Si l'on avait des observations constants qui prouvassent que c'est elle qui tourne"; hence my emendation.
2. Emended from "it's" in Pascal 1744, 2: 314.
3. Comma in the 1744 translation.
4. Colon in the 1744 translation.
5. Comma deleted from the translation in Pascal 1744, 2: 314.
6. Pascal 1744, 2: 314–15; cf. Pascal 1962 (53), 1967 (295–96).
7. Cf. Copernicus 1976, 1992a; and chapter 1.4.
8. G. Galilei 1636, 8; Favaro 5: 314; Finocchiaro 1989, 91; Motta 2000, 93.
9. Descartes 1634b; cf. chapter 3.1.
10. Leibniz 1679–1686, 1688.
11. Cf. Pascal 1744, 286–87.
12. Pascal 1744, 314.
13. Arnauld 1691, in Arnauld 1775–1783, 9: 307–14.
14. Here I rely on the historical and critical introduction in Arnauld 1775–1783, 8: xiii–xxvi.
15. In Arnauld 1775–1783, 8: 467–767, 9: 1–428.
16. Arnauld 1775–1785, 9: 284; cf. 9: 284–317.
17. Arnauld 1775–1783, 9: 307; cf. 9: 307–14.
18. Arnauld 1691, 307.
19. Arnauld 1691, 308.
20. Arnauld 1691, 308.
21. Arnauld 1691, 309.
22. Arnauld 1691, 311.
23. Arnauld 1691, 312.
24. Arnauld (1691, 312) gives as reference "the third chapter of the third book of Kings," which appears to be incorrect. The passage can be found in 3 Kings 7:23 (Douay), or 1 Kings 7:23 (King James). The passage also recurs in 2 Paralipomenon 4:2 (Douay), or 2 Chronicles 4:2 (King James). I thank Aaron Abbey for this correction and these references.
25. In 1691, Martin E. van Velden, professor of mathematics at Louvain, was disciplined by the university senate for advocating Copernicanism; he appealed the decision to the civil authorities and then to Brussels' papal nuncio, but without success; while this incident indicates that freedom of thought did not flourish in Belgium at that time, it is also important to note that the censorship was being applied by an autonomous academic institution, and it was the accused who requested the intervention of church and state. Cf. Monchamp 1892, 182–346; Motta 1993, 67; Stevart 1871, 1890; Stimson 1917, 76–77.
26. Bernini 1709, 615; cf. Carli and Favaro 1896, 99; Cooper 1838, 72; Müller 1911, 455.
27. As late as Haeckel 1878–1879, 33; cf. Reusch 1879, 266; Müller 1911, 455 n. 2.
28. Nelli 1793, 2: 537–38; cf. Fabroni 1773–1775, vol. 2 (1775).
29. Favaro 1912, 121–22; Nelli 1793, 2: 761–62, 874–75.

30. For these and other details of this story, see Galluzzi 1993b and 1998, to which I am much indebted in this account.
31. Favaro 19: 291–92; Pagano 1984, 14–15.
32. Favaro 19: 399; Pagano 1984, 216.
33. Gebler 1879a, 311; Fahie 1903, 405; Galluzzi 1993b (174), 1998.
34. In Favaro 19: 399; Pagano 1984, 215–16.
35. Nelli 1793, 2: 537–38; cf. Fabroni 1773–1775, vol. 2 (1775).
36. Estève 1755, 1: 289f.; already Andres (1776, 24) dismissed it as false; cf. also Redondi 1994, 97.
37. See *Encyclopedia Britannica*, 11th ed.
38. Baretti 1757, 49–56.
39. Baretti 1757, 52–53.
40. Baretti 1757, 52.
41. For example, on 16 March 1847, the French artist Joseph-Nicolas Robert-Fleury exhibited in Paris a painting which depicted a scene at Galileo's trial and generated wide discussion in France. The catalogue accompanying the show described the scene as being that of Galileo's abjuration. It stated that after reciting the abjuration on his knees, upon getting up he felt some remorse for having perjured himself, and so he tapped the floor with his foot and uttered the words *e pur si muove*. Cf. Redondi 1994, 75–83.
42. Barni (1862a) expresses the point by saying that the myth is historically false but philosophically true.
43. Favaro 1911a,b; Fahie 1929, 72–75, plate 16; Drake 1978, 356–58.
44. For more details, see Baldini 1994.
45. Fabroni 1773–1775, 1: 47 n. 1.
46. The pun can be better appreciated in Latin: "*Viri Galileai, qui statis aspicientes in coelum?*"
47. Favaro 12: 123, 19: 307.
48. See, for example, Frisi 1775, 59; Targioni Tozzetti 1780, 1: 58; Nelli 1793, 1: 395 n. 3; Gebler 1879a, 51; Koestler 1959, 439–40.
49. Cf. *Lettres Philosophiques*, in Voltaire 1877–1883, 22: 75–188.
50. Voltaire 1877–1883, 22: 127 n. 4.
51. Gillespie 1974.
52. Voltaire 1734a,b.
53. Voltaire 1734a, 169.
54. Voltaire 1734a, 169.
55. Here the translator says "demonstrated by irrefragable proofs" (Voltaire 1734a, 167), whereas the original French says simply *pour avoir demonstré le mouvement de la terre* (Voltaire 1734b, 129); hence my emendation.
56. Voltaire 1734a, 167; cf. Voltaire 1734b, 129.
57. For example, this view reemerged in Draper 1875. Voltaire's remark was also echoed (although not repeated verbatim) in 1887, when a marble column commemorating Galileo was inaugurated in Rome, near the Medici Palace; cf. "Epigrafi ed offese"; Favaro 1915a, 104–5; Berggren and Sjöstedt 1996, 145–47; and chapter 13.1.
58. Voltaire 1951, 1–5.
59. Voltaire 1901, 23: 279; cf. Voltaire 1877–1883, 14: 535.

60. Voltaire 1877–1883 (14: 534–39), 1901 (277–86), 1951 (352–56).
61. Emended from "Louis XIV" in Voltaire 1901, 23: 277; cf. Voltaire 1877–1883, 14: 534.
62. Emended from "Canon Thorn" in Voltaire 1901, 23: 277; cf. Voltaire 1877–1883, 14: 534. Obviously this is a reference to Nicolaus Copernicus.
63. Voltaire 1901, 23: 277; cf. Voltaire 1877–1883 (14: 534), 1951 (352).
64. This is the *Essai sur les Moeurs et l'Esprit des Nations et sur les Principaux Faits de l'Histoire depuis Charlemagne jusqu'à Louis XIII,* in Voltaire 1877–1883, vols. 11 and 12; for the dating of this work, see vol. 11, ix. In the various English translations of Voltaire's works, this *Essai sur les Moeurs* does not seem to have been translated in its entirety or under a corresponding title; rather its chapters overlap with those of *The General History and State of Europe* (Voltaire 1754–1757) and with *Ancient and Modern History* (Voltaire 1901, vol. 26).
65. Chapter 121 in Voltaire 1877–1883, 12: 241–50. The similarly titled Chapter 100 of *Ancient and Modern History* (Voltaire 1901, 26: 310–20) does not contain the Galileo passage. In the *General History and State of Europe* (Voltaire 1754–1757), the original is split into two chapters, chapter 30 of part 3 (pp. 207–15) and chapter 1 of part 4 (pp. 1–6); the Galileo passage occurs in the latter, but in a very faulty translation that will not be used here.
66. Voltaire 1877–1883, 12: 249.
67. Cf. Peiresc 1635 and chapter 3.3.
68. Voltaire 1877–1883 (20: 120–22), 1824 (5: 111–14); for the dating of this fragment, see Voltaire 1877–1883, 20: 120 n. 2.
69. Emended from "Leibnitz" in Voltaire 1824, 5: 113
70. Emended from a lowercase initial ("dominican") in Voltaire 1824, 5: 113.
71. Although Voltaire perhaps meant to be primarily satirical, he happened also to be perceptive and prophetic, for in 1769 an English priest named Dr. Turberville Needham, while lecturing at the University of Lisbon by invitation of the King of Portugal, was confronted by an Inquisition officer about whether he held the Newtonian system as a thesis or as a hypothesis; Needham answered, "As a hypothesis," but left Portugal as soon as he could. Cf. Blackburne 1770, lxxxiii–lxxxiv n.; Sharratt 1974, 46.
72. Emended from "hardihood" in Voltaire 1824, 5: 114, in accordance with the original French *effronterie* in Voltaire 1877–1883, 20: 122.
73. Emended from "apprehend" in Voltaire 1824, 5: 114, in accordance with the original *craindre* in Voltaire 1877–1883, 20: 122.
74. Semicolon substituted for the period in Voltaire 1824, 5: 114, corresponding to the colon in Voltaire 1877–1883, 20: 114.
75. "Anitus" in Voltaire 1824, 5: 114. Anytus was an Athenian politician who secretly maneuvered to have Socrates indicted, tried, and executed in 399 B.C.
76. Voltaire 1824, 5: 113–14; cf. Voltaire 1877–1883, 20: 122.
77. Voltaire 1877–1883 (19: 501–2),1824 (1: 365–66); for the dating, see Voltaire 1877–1883, 19: 501 n. 1.
78. Voltaire 1877–1883, 19: 501–2.
79. Diderot and D'Alembert 1751–1780.
80. Frisi 1777; cf. Frisi 1756, 1775; Casini 1985, 1987.
81. Frisi 1766; cf. Frisi 1777, 176.
82. D'Alembert 1751c, 1963; cf. Schwab 1963.

83. Schwab 1963, xi.
84. Schwab 1963, xi, xiii.
85. For more details, see also Schwab's introduction in Schwab 1963, ix–l.
86. D'Alembert 1751c, xxiii; Schwab 1963, 70–71.
87. D'Alembert 1751c, xxiii–xxiv; Schwab 1963, 71–74.
88. Schwab 1963, 71; cf. D'Alembert 1751c, xxiii.
89. Schwab 1963, 72; cf. D'Alembert 1751c, xxiii.
90. Schwab 1963, 72; cf. D'Alembert 1751c, xxiii.
91. Schwab 1963, 73; cf. D'Alembert 1751c, xxiv.
92. Schwab 1963, 74; cf. D'Alembert 1751c, xxiv.
93. Schwab 1963, 72; cf. D'Alembert 1751c, xxiii.
94. Schwab 1963, 74; cf. D'Alembert 1751c, xxiv.
95. D'Alembert 1751c, xxiv; Schwab 1963, 73–74.
96. In Schwab 1963, 73; D'Alembert (1751c, xxiv) was paraphrasing Ecclesiastes 3:11, a verse also quoted by Calmet (1744, 1); cf. chapter 7.1.
97. D'Alembert 1751b, 790.
98. D'Alembert 1751a; Schwab (1963, 73 n. 18) states that the article included an attack on a defense of pope Zachary's condemnation published in 1708 in the *Journal de Trévoux*.
99. "Encyclopédie, ou Dictionnaire Raisonné des Sciences, des Arts et des Métiers," *Journal de Trévoux*, October 1751, article 111, pp. 2250–95. The *Encyclopedia* had been previewed there earlier that year in "Nouvelles Littéraires," *Journal de Trévoux*, January 1751, article 12, pp. 176–91, at pp. 188–89.
100. *Journal de Trévoux*, October 1751, pp. 2278–82; cf. Thorndike 1924, 369; Schwab 1963, 73–74 n. 18.
101. Mercati 1942; but cf. the criticism in Finocchiaro 2002c. I say "rediscovered" because the summary of Bruno's trial was first discovered in the Vatican Secret Archives in 1886–1887, but Pope Leo XIII ordered that the discovery not be divulged and the document not be given to anyone; so the document had to be rediscovered in 1940 by Angelo Mercati, the prefect of the Vatican Secret Archives; see Blumenberg 1987c, 371.
102. Assuming it was written by D'Alembert.
103. D'Alembert 1764. Similarly, in the article "Florence" in the sixth volume (whose author I cannot identify), Galileo was described as "immortal for his astronomical discoveries and persecuted by the Inquisition" (Diderot and D'Alembert 1751–1780, 6: 877).
104. This date suggests that D'Alembert was acquainted with the text of the Inquisition's sentence.
105. Cf. Voltaire 1824 (5: 113–14), 1877–1883 (20: 122); and section 3 of this chapter.
106. This point had become very widely acknowledged; but it is also an indication of D'Alembert's conciliatory tone.
107. This is D'Alembert's statement of the principle of accommodation.
108. Cf. Pascal 1744 (2: 314–15), 1962 (53), 1967 (95–96); and section 1 in this chapter.
109. This paragraph indicates that the Jesuit conspiracy theory was continuing to spread.
110. Joshua 10:12.

111. D'Alembert 1754. The article continued for three more short paragraphs in which D'Alembert summarized very cryptically some of the arguments supporting the Copernican system and some of its modifications by Kepler and Newton.
112. Quoted in Drinkwater Bethune 1832, 194, from Newton 1739–1742.
113. Garzend 1912, 259–60.
114. Mayaud 1997, 177. I owe to this important book my first acquaintance with D'Alembert's article.

CHAPTER 7

1. Haynes 1970, 152–57, 226–27.
2. Haynes 1970, 180.
3. Haynes 1970, 184.
4. Benedict XIV 1748; and *Observations sur le Bref de N.S.P. le Pape Benoit XIV au Grand-Inquisiteur d'Espagne* 1749.
5. Baldini 2000b, 283–95.
6. Mayaud 1997, 203.
7. Ambrogi to the Roman Inquisition, 29 September 1741, in Mayaud 1997, 130–31. Mayaud has published the relevant documents for the first time and has provided an extremely valuable historical account and critical analysis; although I disagree with some of his interpretations, my account here would have been impossible without his book.
8. Mayaud 1997, 131–32.
9. Ambrogi to Inquisition, 10 February 1742, in Mayaud 1997, 135–37.
10. Translated from the anonymous Italian text transcribed and published by Mayaud (Rome: Editrice Pontificia Università Gregoriana, 1997), 136–37.
11. Viviani 1654 (617), 1717; cf. chapter 5.1.
12. Mayaud 1997, 139–40, 143.
13. The roman numerals in this document are my additions and are meant to divide it into two parts.
14. The original text reads "November," which is obviously erroneous; cf. Mayaud 1997, 137.
15. Edme Pourchot (or Purchotius), *Institutiones Philosophiae,* 5 vols. (Padua: Typographia Seminarii, apud Joannem Manfré, 1738); this was also the publisher and printer of G. Galilei 1744a; see Mayaud 1997, 125–29, especially 127–28.
16. Mayaud 1997, 142.
17. The declaration in question was the previously quoted apologetic editorial preface. The omissions in question were really mentioned not there but in another part of the project description sent by the Paduan inquisitor; this description explained that this edition of the *Dialogue* was being modeled in part on the 1710 semiclandestine edition, but was planning to add only Galileo's Letter to Christina out of all the additional essays (by Foscarini, Kepler, and Zúñiga) published there. See Mayaud 1997, 137.
18. Translated from the anonymous Italian text transcribed and published by Mayaud (Rome: Editrice Pontificia Università Gregoriana, 1997), 137–39.
19. Mayaud (1997, 142, 143) to whom we owe the transcription and publication and French translation of these documents, is much more critical of this mem-

orandum, branding it as incompetent, ridiculous, and derisory. He is equally critical of the bureaucratic incompetence throughout the episode.

20. Mayaud 1997, 144–46.

21. T. Martin 1868, 256.

22. Mayaud 1997, 146.

23. Mayaud 1997, 149–50.

24. Mayaud (1997, 149 n. 47) notes that he is uncertain about the spelling of this surname, which is difficult to read in the manuscript.

25. Translated from the Italian text of Giovasco (1742), as transcribed and published by Mayaud (Rome: Editrice Pontificia Università Gregoriana, 1997), 146–49.

26. For the details, see Bucciantini 1995.

27. For relevant details on Toaldo, see Toaldo 1744a–c; Fabroni 1802; Lorenzoni 1913; Restiglian 1982; Fantoli 1996, 495–96, 549; Heilbron 1999, 313, 319, 323–24, 326.

28. G. Galilei 1744a, vol. 1, unnumbered page following p. 601; G. Galilei 1744a, vol. 4, unnumbered last page of book; Mayaud 1997, 120.

29. Besomi and Helbing 1998, 955–59.

30. In G. Galilei 1744a, vol. 4; English translation in Fantoli 1996, 495–96.

31. G. Galilei 1744a, vol. 4; Fantoli 1996, 495–96; Favaro 7: 29; Finocchiaro 1997, 78.

32. G. Galilei 1744a, vol. 4; Fantoli 1996, 495–96; G. Galilei 1632a; Finocchiaro 1997, 359.

33. Mayaud 1997, 145.

34. To be more precise, each of the first three volumes carried the title *Works / of / Galileo Galilei / Divided into Four Volumes / in this New Edition Augmented / by Many Unpublished Things;* then there was a line indicating the volume number; but the fourth volume had an additional line: *Containing the Dialogue.* See G. Galilei 1744a; Cinti 1957, 335–39; Mayaud 1997, 119.

35. However, Carli and Favaro (1896, 122, entry #479) list a separate edition of just the fourth volume of the 1744 *Works,* published the same year with the same publisher but with the full title on the title page (G. Galilei 1744b).

36. Calmet 1744; cf. Calmet 1720, 1734; T. Martin 1868, 256; Fahie 1903, 427; Mayaud 1997, 122–23.

37. Here Calmet (1744, 1) quoted the Latin verse "Mundum tradidit disputationi eorum," for which he gave the reference Ecclesiastes 3:11. I have translated his Vulgate version literally, to convey his point better than is done by the Douay version, which reads: "He has made everything appropriate to its time, and has put the timeless into their hearts, without men's ever discovering, from beginning to end, the work which God has done." In this case, the King James Version offers a slightly better translation than the Douay Version, although still far from the Latin: "He hath made every thing beautiful in his time: also he hath set the world in their heart, so that no man can find out the work that God maketh from the beginning to the end."

38. Here Calmet (1744, 2) quoted Ecclesiastes 1:13 in Latin; I have used the Douay Version; the King James Version reads: "This sore travail hath God given to the sons of man to be exercised therewith."

39. Here Calmet (1744, 2) quoted "Multa abscondita sunt majora his; pauca vidimus operus eius," citing Ecclesiasticus 43:36. The Douay version has only thirty-five verses in the book of Ecclesiasticus (or Sirach), which is of course distinct from Ecclesiastes; but Calmet's verse seems to correspond to Ecclesiasticus 43:34, which is what I have quoted. The King James Version does not include this book.

40. Calmet (1744, 2) gave the title as *Cartesius Mosaisans.*

41. Calmet 1744, 1–3; this was the prologue of the essay, before Calmet undertook a detailed exegesis.

42. Here Calmet (1744, 20) quoted Ecclesiastes 12:13 in Latin; I have used the Douay Version; the King James Version reads: "Let us hear the conclusion of the whole matter: Fear God, and keep his commandments: for this is the whole duty of man."

43. Here Calmet (1744, 20 n. 2) referred to Augustine, *De Genesi ad Litteram,* book 2, chapter 9.

44. Calmet 1744, 19–20; here Calmet's essay ended, and the last two paragraphs are what I have called its epilogue.

45. G. Galilei 1636, 15; Favaro 5: 318; Finocchiaro 1989, 95; Motta 2000, 99.

46. Leo XIII 1893, paragraph 18, p. 334; cf. chapter 13.2.

47. Contrast pp. 145 and 137 in Mayaud 1997.

48. Mayaud 1997, 145.

49. Mayaud 1997, 149; see my quotation above.

50. Mayaud 1997, 150.

51. Mayaud 1997, 142, 159, 161.

52. Mayaud 1997, 149 n. 48, 161.

53. For all these facts, see Baldini 2000b, 304–5.

54. Beretta 1999a, 465–66; Baldini 2000b, 304–6.

55. Lazzari 1757; Baldini 2000b, 281–82, 307–28.

56. Gebler 1879a, 312–13; Favaro 19: 419; Mayaud 1997, 197; Baldini 2000b, 305.

57. Mayaud 1997, 203.

58. For example, Mayaud 1997, 189.

59. Mayaud 1997, 193–96, 202–12.

60. Mayaud 1997, 197, 205, 212.

61. This was discovered by Baldini and published in Baldini 2000b, 307–28; Mayaud was apparently unaware of it, even though he studied and published most of the other extant documents for the first time, in Mayaud 1997, 190–93, 196–201.

62. Translated from the Italian (and Latin and French) text of Lazzari (1757), as transcribed and published by Baldini (Padua: Cooperativa Editrice Libraria Università di Padova, 2000b, 307–28). Lazzari included many quotations from other works, usually in Latin but also in French and Italian; unless otherwise noted, I have translated these quotations directly from the text as found in Lazzari's memorandum; the original language may be presumed to be Latin, unless noted otherwise. Lazzari usually did not give precise titles of books or specific page numbers for his citations, but Baldini has provided them; in my translation, I have utilized the information he provided. Thus, I am indebted to Baldini in several ways, and my account here would have been impossible

without his contribution. I thank Leen Spruit for first calling to my attention Lazzari's memorandum and Baldini's edition.

63. Although such bracketed numbers usually designate the page numbers of the printed source from which I quote or translate such texts (here, Baldini 2000b, 307–28), in this document they designate the manuscript folio pages of Lazzari's (1757) manuscript, as given by Baldini.

64. As Baldini (2000b, 307 n. 61) says, here Lazzari had a slip of the pen and wrote "May" instead of "March."

65. The number for this first part of Lazzari's memorandum is missing in Baldini's transcription (2000b, 308), but since those for the second and third parts are explicitly given (Baldini 2000b, 313, 323), I add it here.

66. Here Lazzari (1757, 308) quoted a Latin passage from the anti-Copernican decree of the Index of 1616, inserting an ellipsis in place of the titles and description of the works (by Copernicus, Zúñiga, and Foscarini) being prohibited. I quote the English translation from Finocchiaro 1989, 149; cf. Favaro 19: 323; Pagano 1984, 103.

67. Baldini (2000b, 309 n. 63) gives the precise reference: Gassendi 1654, 6.

68. Baldini (2000b, 309 n. 64) gives the precise reference: Kepler 1619b, 185.

69. Baldini (2000b, 309 n. 64) has traced this reference to an earlier edition: *Commentarii Collegii Conimbricensis Societati Iesu in Quatuor Libros de Coelo Aristotelis Stagiritae* (Lugduni, 1594), 389.

70. Cf. Biancani 1620, 75, per Baldini 2000b, 309 n. 66.

71. Cf. Charpentier 1560, ff. 28r–29r, per Baldini 2000b, 309 n. 67.

72. Cf. Tassoni 1620, 146, per Baldini 2000b, 310 n. 70.

73. Cf. Lipsius 1604, 99–100, per Baldini 2000b, 310 n. 71.

74. Psalm 74:4 in the Douay Version of the Bible; cf. Psalm 75:3 in the King James Version: "I bear up the pillars of it." Cf. *Commentarii Collegii Conimbricensis*, p. 300, per Baldini 2000b, 311 n. 72.

75. Ecclesiastes 1:5 (Douay); "The sun also ariseth, and the sun goeth down" in the King James Version.

76. Baldini (2000b, 311 n. 74) gives the precise reference: Valles 1587, 458.

77. Here Lazzari (1757, 311) was quoting a passage in Italian from G. Galilei 1636, 10; I have taken the quotation from Finocchiaro 1989, 92; cf. Favaro 5: 315; Motta 2000, 95.

78. Here Lazzari (1757, 311–12) was quoting, again in Italian, from G. Galilei 1636, 11; I have taken the quotation from Finocchiaro 1989, 92; cf. Favaro 5: 315–16; Motta 2000, 95.

79. Ellipsis in Lazzari 1757, 312; Baldini (2000b, 312 n. 77) gives the precise reference: Varenius 1650, 54.

80. Here Lazzari (1757, 312) was quoting from Riccioli 1651, 2: 494.

81. Varenius 1650, 52, per Baldini 2000b, 312 n. 80.

82. See chapter 4.3.

83. Cf. Milliet de Chales 1674, 3: 297–300, per Baldini 2000b, 313 n. 81.

84. Baldini (2000b, 313 n. 82) gives the more precise reference: Bacon 1624, 202.

85. Huygens 1698, 14, per Baldini 2000b, 314 n. 84.

86. Here Lazzari (1757, 314) was quoting from the French text in D'Alembert (1754, 173–74); I reproduce the English translation from chapter 6.4.

87. Here Lazzari (1757, 514–15) quoted an Italian passage from the Italian translation of this dictionary; Baldini (2000b, 315 n. 87) gives the precise reference: Chambers 1749, 8: 61.

88. The Minim Fathers François Jacquier (professor of experimental physics at the University of Rome from 1746) and Thomas Le Seur (professor of applied mathematics from 1749); cf. Baldini 2000b, 315–16 n. 90. They were the coeditors of the famous edition of and commentary to Newton's *Principia* in 1739–1742.

89. Baldini (2000b, 315–16, n. 90) identifies the last two as Christopher Maire (1697–1767) and Benedetto Stay (1714–1801). The Jesuit Father Ruggiero G. Boscovich (1711–1787) is of course an important figure in his own right and may have had an important role in convincing Pope Benedict XIV to relax the prohibitions, although there is no concrete and conclusive evidence; cf. Casini 1993; Mayaud 1997, 169–75; Baldini 2000c.

90. This last clause is my admittedly free translation of Lazzari's (1757, 316) impossible *e così pure "far la testa o la barba" e "far la casa e la scrittura."*

91. "A Further Account" 1665; see chapter 5.2.

92. Wolff 1735, 3: 607, per Baldini 2000b, 319 n. 94.

93. Baldini (2000b, 320 n. 95) gives the precise reference: Châtelet 1743, 63.

94. Keill 1742, 245, per Baldini 2000b, 320 n. 96.

95. Bradley 1729; cf. Bradley 1748, 1832–1833.

96. Baldini (2000b, 320 n. 98) gives the more precise reference: MacLaurin 1749, 269–89.

97. Sfondrati 1695, 95–96, per Baldini 2000b, 325 n. 104.

98. Here Lazzari (1757, 326) was quoting from the French text in D'Alembert 1754, 174; I reproduce the English translation from chapter 6.4.

99. Baldini (2000b, 326 n. 105) gives the precise reference: Serry 1742, 1: 130.

100. Serry 1742, 1: 130, per Baldini 2000b, 326 n. 105.

101. In Baldini (Padua: Cooperativa Editrice Libraria Università di Padova, 2000b), 307–28.

102. For comments from a different perspective, see Baldini 2000c, 328–32.

103. See Finocchiaro 1986a and chapter 4.2.

104. Lazzari 1757, p. 321, f. 493r.

105. Lazzari 1757, p. 322, f. 493v.

106. Lazzari 1757, p. 322, f. 493v.

107. Lazzari 1757, p. 322, f. 494r.

108. In the "Considerations on the Copernican Opinion" (1615), Galileo answered one of the points made by Bellarmine in his letter to Foscarini by saying: "It is true that it is not the same to show that one can save the appearances with the earth's motion and the sun's mobility, and to demonstrate that these hypotheses are really true in nature. But it is equally true, or even more so, that one cannot account for such appearances with the other commonly accepted system. The latter is undoubtedly false, while it is clear that the former, which can account for them, may be true. Nor can one or should one seek any greater truth in a position than that it corresponds with all particular appearances" (in Finocchiaro 1989, 85; cf. Favaro 5: 369). For an interpretation of the *Dialogue* along these lines, see Soccorsi 1947, especially pp. 80–81; cf. Finocchiaro 1980, 1986a.

109. Mayaud 1997, 189.
110. Olivieri 1820c, section 58; cf. chapter 10.2.

CHAPTER 8

1. Lalande 1771, 1: 536–41, paragraphs 1103–4; cf. Gebler 1879a, 313; Brand-müller 1992b, 162; Maffei 1987, 17; Mayaud 1997, 215–16.
2. Cf. Mayaud 1997, 213–33; Baldini 2000b, 338–39.
3. Motta 1997b, 136, 167 n. 96.
4. Andres 1776; cf. Andres 1782–1799; Navarro Brotons 2001, 825–26.
5. Cifres 1998, 83; Motta 1993, 17–18.
6. Nelli 1793, 1: 129 n. 1.
7. Tiraboschi 1782–1797, 8 (1785): 147–49 n.
8. See, for example, Bérault-Bercastel 1778–1790, 21: 140–46; Feller 1797, 4: 251–53; Gamba 1819, 89–94; Venturi 1818–1821, 2: 179–82; Tiraboschi 1822–1826, 14: 257–60 n.; Feller 1832, 6: 26–28; G. Galilei 1842–1856, 7: 40–43; Paladini 1890, 171–74. Cf. Nelli 1793, 1: 129 n. 1; Albèri 1842–1856, 7: 40 n. 1; Pieralisi 1875, 20; Favaro 1905, 144–45; Mayaud 1997, 224 n. 35.
9. My account relies primarily on Albèri (1842–1856, 7: 40 n. 1) and Favaro (1905, 144–45), and to a lesser extent on Chasles (1862, 115–20); Fiorani (1969, 77–78 n. 158, 157; 1973a,b); Libri (1840–1841, 209 n. 1); T. Martin (1868, 160, 193, 212, 393, 400); Mayaud (1997, 224 n. 35); Monsagrati (2000); and Nelli (1793, 1: 129 n. 1).
10. I retain the spelling of this name used by Albèri, T. Martin, and Nelli, even though the *Dizionario biografico degli italiani* (Fiorani 1973a,b) and Fiorani 1969 prefer an initial C (Caetani) rather than G.
11. Nelli 1793, 1: 129 n. 1.
12. Viviani 1654, 629.
13. See Venturi 1818–1821, 1: 19–20; Favaro 1906, 1907e; Casini 1985, 40 n. 23; and Segre 1991, 123–24.
14. Albèri 1842–1856, 7: 40 n. 1.
15. Albèri 1842–1856, 7: 40.
16. In Finocchiaro 2002b, 758–60.
17. Finocchiaro 2002b, 758; cf. Albèri 1842–1856, 7: 40.
18. Finocchiaro 2002b, 750; cf. Albèri 1842–1856, 7: 43.
19. Finocchiaro 2002b, 759; cf. Albèri 1842–1856, 7: 41.
20. Finocchiaro 2002b, 758; cf. Albèri 1842–1856, 7: 40.
21. Finocchiaro 2002b, 759; cf. Albèri 1842–1856, 7: 40.
22. Finocchiaro 2002b, 759; cf. Albèri 1842–1856, 7: 40.
23. Finocchiaro 2002b, 759; cf. Albèri 1842–1856, 7: 41.
24. Finocchiaro 2002b, 759; cf. Albèri 1842–1856, 7: 41.
25. Finocchiaro 2002b, 759; cf. Albèri 1842–1856, 7: 40.
26. Fantoli 1997, chapter 6, section 5; Fantoli 1997 is the second *Italian* edition; Fantoli's second *English-language* edition (1996, 407) did not yet contain this information but rather assumed incorrectly that Maculano was the commissary in autumn 1632. Beretta (2001, 566) is also explicit about this change.
27. Favaro 15: 40–41; Favaro 20: 393; cf. Beretta 2001, 566.

28. Favaro 14: 401 n.1; cf. Fantoli 1997, chapter 6, section 5.
29. Finocchiaro 2002b, 760; cf. Albèri 1842–1856, 7: 43.
30. *Libertà della campagna,* in Viviani 1654, 617, lines 597–98; see also chapter 5.1.
31. Nelli 1793, 2: 830; Favaro 19: 391.
32. Targioni Tozzetti 1780, 1: 125–37, 142–43.
33. Guicciardini to Cosimo, 4 March 1616, in Fabroni 1773–1775, 1: 53–57; in Favaro 12: 241–43; cf. Santillana 1955a, 116–17, 119; Gebler 1879a, 91–93.
34. Cf. Favaro 5: 377–95; Finocchiaro 1989, 119–33.
35. Galileo's Letter to Castelli had been published in Gassendi 1649, appendix, 65–78. Galileo's Letter to Christina had been published in G. Galilei 1636; in Salusbury 1661–1665, vol. 1, part 1, pp. 425–60; and in G. Galilei 1710. See chapter 4.2.
36. See Mallet du Pan 1784; this article was first brought to my attention by Redondi (1994, 98 n. 81).
37. Mallet du Pan 1784, 122.
38. Mallet du Pan 1784, 122.
39. Mallet du Pan 1784, 127.
40. Cf. *Nouvelle Biographie Générale* (Paris, 1863), 33: 78–81; and *Enciclopedia italiana* (Rome, 1934), 22: 24–25.
41. Ferri 1785, 54.
42. Although this question had not yet become an explicit issue of the subsequent affair and could not be settled then, the Inquisition sentence of 1633 (as noted in chapter 1.1) contained language that suggested torture: "Because we did not think you had said the whole truth about your intention, we deemed it necessary to proceed against you by a rigorous examination. Here you answered in a Catholic manner, though without prejudice to the above-mentioned things confessed by you and deduced against you about your intention" (Finocchiaro 1989, 290; cf. Favaro 19: 405).
43. This particular claim can be found in Estève 1755, 1: 289f; cf. Andres 1776, 24, and chapter 6.2.
44. Cf. Bernini 1709, 615, and chapter 6.2.
45. Ferri 1785, 56.
46. Ferri 1785, 57.
47. Mallet's thesis was held, elaborated, and modified in such works as Bergier 1788–1790a,b, 1823; Bérault-Bercastel 1790; Feller 1797, 1832; Cooper 1838, 1844; Purcell 1844b; Marini 1850, 39 n. 2, 141; Reumont 1853; Madden 1863; "Epigrafi ed offese" 1887; Duhem 1908; Sacchi 1913; Carrara 1914; for more details, see Finocchiaro 2002b.
48. Bérault-Bercastel 1790, 140.
49. Bérault-Bercastel 1790, 146.
50. Bérault-Bercastel 1790, here quoted from the English translation in Madden 1863, 144–45.
51. Bérault-Bercastel 1790, 141.
52. Mallet du Pan 1784, 124; Favaro 12: 241–43.
53. Bergier 1788–1790a, tome 3, column 464, capitals in the original.
54. Bergier 1788–1790a, tome 3, column 464.
55. Bergier 1788–1790b.
56. Bergier 1788–1790b, tome 7, p. 369.

57. Although I have been able to consult only the eighth edition of this encyclopedia, published in 1832, the article in question may be dated to 1797, the
date of the second edition, because this edition mentions Mallet, whereas the
first one appeared before Mallet's article; Feller died in 1802. Cf. Feller 1782,
1797, 1832.

58. The claim had been advanced in Erythraeus 1643; Brucker (1742–1744)
repeated it but later retracted it (Brucker 1766–1767, vol. 4, part 2, p. 634);
cf. Nelli 1793, 1: 25–26. See also chapter 6.2.

59. See Tiraboschi 1792, 1793.

60. See Tiraboschi 1772–1782, 8: 123–44; cf. Tiraboschi 1782–1797, 8: 143–
72.

61. For more information on Tiraboschi and a focus on different aspects of his
activities, see Tiraboschi 1778, 1779, 1782, 1785, 1900; Mamachi 1785; Mauri
1833; Bertoni 1937; Mayaud 1997, 222–26; Motta 1997b; Venturi Barbolini
1997.

62. Tiraboschi 1792, 365–72; cf. Dreyer 1953, 282–95, 317–19; Fantoli 1996,
24–25.

63. Baldigiani 1678; cf. chapter 5.1.

64. "Revelation" in the King James Version.

65. I have been unable to identify this author and retain the spelling given by Tiraboschi.

66. A follower of John Wycliffe (1320?–1384).

67. This was actually the second time, the first having been in 1587; see Drake
1978, 12–13; Shea and Artigas 2003.

68. Belisario Vinta (1542–1613), the Tuscan secretary of state.

69. The Jesuit Christopher Clavius (1538–1612), mathematician and astronomer
at the Roman College.

70. Here Tiraboschi (1793, 375 n. 1) cited Fabroni (1773–1775, 1: 32–33); cf.
Galileo to Vinta, 1 April 1611, in Favaro 11: 79–80.

71. Here Tiraboschi (1793, 376, n. 1) cited "Atti e Memorie dell'Accademia del
Cimento," in Targioni Tozzetti 1780, tome 2, part 1, pp. 19–20; cf. Favaro 11:
87–88, 92–93.

72. Acts 1: 11 (King James Version); "Men of Galilee, why do you stand looking up
to heaven?" in the Douay Version. Cf. Fabroni 1773–1775, 1: 47 n. 1, and
chapter 6.2.

73. Here Tiraboschi (1793, 376 n. 2) cited Fabroni (1773–1775, 1: 35–37); cf.
Galileo to Curzio Picchena, 8 January 1616, in Favaro 12: 222–23.

74. Here Tiraboschi (1793, 376 n. 3) cited the first edition of his own *Storia della
letteratura italiana* 8: 125 (Tiraboschi 1772–1782); see also Tiraboschi 1882–
1897, 8: 146 n. 3; cf. Querengo to Alessandro D'Este, 1 January 1616, in
Favaro 12: 220.

75. Tiraboschi (1793, 377) mistakenly wrote "Ferdinando" instead of Cosimo
II, who ruled Tuscany from 1609 to 1621; Ferdinand I (Cosimo's father)
ruled from 1586 to 1609; Ferdinand II (Cosimo's son) ruled from 1621 to
1670.

76. Here Tiraboschi (1793, 377 n. 1) cited Fabroni (1777–1775, 1: 53–57); cf.
Guicciardini to Cosimo, in Favaro 12: 241–42, lines 1–5.

77. Cf. Favaro 12: 242, lines 20–21.

78. Tiraboschi (1793, 378) cited Targioni Tozzetti 1780, tome 2, part 1, p. 22; Galileo's Letter to Castelli had also been published in 1649, in Gassendi 1649, appendix, 65–78; cf. chapter 4.2.

79. Here Tiraboschi (1793, 379 n. 1) referred to Fabroni (1773–1775, 1: 48–51, 51–53); he mistakenly identified the addressee as Belisario Vinta (who had died in 1613) instead of Curzio Picchena. These letters were dated 6 and 12 March 1616; see Favaro 12: 243–45, 247–49; Finocchiaro 1989, 150–51, 151–53.

80. Here Tiraboschi (1793, 379 n. 2) referred to Fabroni (1773–1775, 2: 294).

81. Here Tiraboschi (1793, 379) mistakenly wrote "1636," which I take to be a slip of the pen or a misprint. In this same sentence, however, he betrayed other lapses, as when he implied that Galileo began writing the book after returning home in 1616, whereas this happened after his next trip to Rome in 1624. However, at that time there was a gap in the documentation and knowledge about Galileo's activities in the period from 1616 to 1630, a gap that was not filled by Fabroni's recent collection (1773–1775).

82. Quoted from Finocchiaro 1997, 77–78; cf. Favaro 7: 29.

83. Quoted from Finocchiaro 1997, 78; cf. Favaro 7: 29.

84. Quoted from Finocchiaro 1997, 78; cf. Favaro 7: 29.

85. Quoted from Finocchiaro 1997, 81; cf. Favaro 7: 30.

86. Giovanni Ciampoli (1589 or 1590–1643), then correspondence secretary of Pope Urban VIII.

87. Here Tiraboschi (1793, 381 n. 1) cited Fabroni (1773–1775, 2: 276, 286, 295).

88. Here Tiraboschi (1793, 382 n. 1) cited Fabroni (1773–1775, 2: 291); cf. Favaro 15: 41; Finocchiaro 1989, 242–43.

89. Here Tiraboschi (1793, 382 n. 2) cited Fabroni (1773–1775, 2: 308).

90. Here Tiraboschi (1793, 382 n. 3) cited Targioni Tozzetti (1780, tome 2, part 1, p. 126).

91. Francesco Niccolini (1584–1650), the Tuscan ambassador to Rome in 1621–1643.

92. Tiraboschi 1793. An anonymous English translation of Tiraboschi's lecture was published in London in 1900 (Tiraboschi 1900); although I have consulted it, I decided against reproducing it here because it would have required such extensive corrections, emendations, and adaptations that it would have been more time-consuming than making my own brand-new translation. The 1900 English translation seems in any case to be extremely rare and very little known; there is a copy in the British Library (# 3940 a 10) but not in the Library of Congress or the libraries of Harvard University or the University of California; and the only reference to it I have ever seen is in Taylor 1938, 207.

93. Viviani 1654, 617; see chapter 5.1.

94. Ferri 1785, 54.

CHAPTER 9

1. The start of the revolution in 1789 also happened to coincide with the acceleration of important developments in one strand of the intellectual aftermath of Galileo's trial, namely the demonstration and experimental confirmation of

the earth's motion—the key physical doctrine that had been condemned. This story includes Guglielmini's attempt to measure the eastward deviation of falling bodies, the invention of Foucault's pendulum, and Hagen's experiments at the Vatican Observatory. Cf. Acloque 1982; Baldini 1996b, 50–51; Benzenburg 1804, 1845; Bertoloni Meli 1992; Borgato 1996; Borgato and Fiocca 1994; Brandmüller and Greipl 1992, 163, 168; Calandrelli 1806a,b; Finocchiaro 2001a; Foucault 1851, 1878; Gauss 1803; Guglielmini 1789, 1792, 1994; Hagen 1911; Hall 1903, 1904, 1910; Laplace 1803, 1805; Tadini 1796a–c, 1815; Wallace 1999, 8.

2. Madden 1863, 165.

3. Madden 1863, 163–67.

4. Favaro 1887a, 183; Pagano 1984, 11; cf. Drinkwater Bethune 1832, 187 n.; Redondi 1994, 70–71.

5. Favaro 1887a, 201–2 (# vii).

6. Favaro 1887a, 183, 197, 200 (# v).

7. Cf., for example, Pagano 1984, 1–4.

8. Lowercase initials in the original: *écriture sainte.*

9. Barbier 1811.

10. Cf. Mallet du Pan 1784 and chapter 8.3.

11. Cf. Tiraboschi 1793 and chapter 8.4.

12. Cf. Voltaire (1824, 5: 113–14) and chapter 6.3.

13. Favaro 1887a, 184, 199–200 (# iv–v).

14. Favaro 1887a (184 n. 4, 226 n. 1), 1902b (764 n. 2); cf. Venturi 1818–1821, 2: 197–99; Delambre 1821, 1: xxiii–xxix; Scartazzini 1877–1878, 10: 420–21; and sections 3 and 5 of this chapter.

15. Cf. Favaro 19: 293–97; Pagano 1984, 63–68; Finocchiaro 1989, 281–86.

16. Cf. Favaro 19: 297–98; Pagano 1984, 69–71; Finocchiaro 1989, 134–35.

17. Cf. Favaro 19: 299–305; Pagano 1984, 71–77; Finocchiaro 1989, 45–54.

18. Cf. Favaro 19: 305; Pagano 1984, 68–69; Finocchiaro 1989, 135–36.

19. Cf. Favaro 19: 306; Pagano 1984, 77–78.

20. Cf. Favaro 19: 306–7; Pagano 1984, 79–80.

21. Cf. Favaro 19: 307–11; Pagano 1984, 80–85; Finocchiaro 1989, 136–41.

22. Cf. Favaro 19: 311; Pagano 1984, 85.

23. Cf. Favaro 19: 311–12; Pagano 1984, 86–87.

24. Favaro 1902b, 764 n. 2.

25. Quoted from Finocchiaro 1989, 134; cf. Favaro 19: 297.

26. Marini 1850, 40–41.

27. Cifres 1998, 75, 82.

28. Cifres 1998, 81.

29. Favaro 1887a, 201 (# vi).

30. Favaro 1887a, 201–2 (# vii).

31. Favaro 1887a, 202 (# viii).

32. Favaro 1887a, 204 (# xii); Gebler 1879a, 320.

33. Favaro 1887a, 206–7 (# xv).

34. Favaro 1887a, 208 (# xviii).

35. Favaro 1887a, 209 (# xxi); Marini 1850, 147.

36. Favaro 1887a, 210 (# xxii).

37. Favaro 1887a, 210 (# xxiii); Marini 1850, 147–48.

38. Favaro 1887a, 211.
39. Favaro 1887a, 212–20; cf. Marini 1850, 148–52; Pagano 1984, 24.
40. Marini 1817a.
41. Favaro 1887a, 228; Gebler 1879a, 342; Gherardi 1870, 23.
42. Mercati 1926–1927; cf. Pagano 1984, 24.
43. Favaro (1887a, 223 n. 1) identifies this person as Carlo Altieri (d. 1837), a Benedictine friar who was at one time director of the Vatican Secret Archives and later theologian to the Austrian court.
44. Kepler 1619a; cf. Bucciantini 1994a.
45. Venturi 1820a.
46. See chapters 2.5 and 4.3; cf. Venturi 1818–1821, 2: 170–76.
47. Gaetani 1785; cf. Venturi 1818–1821, 2: 179–82; chapter 8.1.
48. Buonamici 1633; cf. Venturi 1818–1821, 2: 177–79; chapter 2.4.
49. Cf. Venturi 1818–1821, 1: 203–8, 212–18, 224–52, respectively.
50. G. Galilei 1813–1814; cf. Favaro 6: 507 n. 2.
51. Venturi 1818–1821, 2: 6–45; Favaro 6: 509–61; Finocchiaro 1989, 154–97.
52. Cf. Favaro 19: 402–6; Finocchiaro 1989, 287–91; chapter 1.1.
53. Gaetani 1785; cf. Venturi 1818–1821, 2: 179–82; chapter 8.1.
54. Favaro 19: 336–47; Pagano 1984, 124–37; Finocchiaro 1989, 256–62, 277–81.
55. Favaro (1887a, 224) gives the year as 1830, but this is an obvious misprint.
56. Favaro (1887a, 226 n. 1) did not reprint those French extracts but stated that the archival material he consulted included those documents as attachments to the letter.
57. Delambre 1820.
58. Favaro 1902b, 765 n. 2.
59. Venturi 1820b; Favaro 1887a, 226 n. 3; cf. Venturi 1818–1821, 2: 192–97.
60. Venturi 1820c, 226.
61. Again, Venturi was writing this before Delambre found the French extracts of the file and sent them to him.
62. Venturi's letter to Delambre identifies this person as the Benedictine friar Carlo Altieri.
63. Niccolò Riccardi (1585–1639), a Dominican friar who from 1629 to 1632 was Master of the Sacred Palace, the chief censor for books printed in the city of Rome.
64. Here Venturi (1818–1821, 2: 193) referred to pp. 150 and 159 of his book (volume 2); cf. Niccolini to Cioli, 11 September 1632, in Favaro 14: 388–89; Finocchiaro 1989, 232–34, at p. 233. Subsequent parenthetical references in this text are also Venturi's and also refer to his own book.
65. Francesco Niccolini (1584–1650), Tuscan ambassador to Rome in from 1621 to 1643.
66. For the second of these two quotations, Venturi (1818–1821, 2: 193 n. *a*) cited the "Life of Monsignor Ciampoli," in Targioni Tozzetti 1780, vol. 2.
67. Guido Bentivoglio (1577–1644), cardinal since 1621 and one of the ten judges at the 1633 trial.
68. Here Venturi was referring to Castelli's letters to Galileo of 22 December 1635, 12 July 1636, and 9 August 1636, now in Favaro 16: 363–64, 449–50, 461, respectively; cf. chapter 3.4.

69. Venturi (1818–1821, 2: 194) mistakenly wrote 1638 instead of 1636.
70. Actually, this occurred in spring 1624; see, Favaro 13: 175, 182–85.
71. Cf. Niccolini to Cioli, 11 September 1632, in Favaro 14: 389; in Finocchiaro 1989, 233. This Jesuit was Melchior Inchofer.
72. Cf. Buonamici 1633, in Favaro 19: 407–11 and chapter 2.4.
73. Cf. Luke Holste to Peiresc, 7 March 1633, in Favaro 15: 62.
74. This story was told in the apocryphal letter to Renieri; cf. Gaetani 1785 and chapter 8.1.
75. Although Venturi (1818–1821, 2: 195) had quotes around this sentence, it is not a direct quotation but a paraphrase; cf. Galileo to Leopold, 23 May 1618, in Favaro 12: 389–92, and in Finocchiaro (1989, 198–200). Thus I have translated this sentence directly from Venturi's own paraphrase.
76. Quoted from Finocchiaro 1989, 155; cf. Favaro 6: 510.
77. Quoted from Finocchiaro 1989, 156; cf. Favaro 6: 511.
78. Although Venturi (1818–1821, 2: 195) had quotation marks around this passage, it is not a direct quotation but a paraphrase from several sentences in that preface; cf. Favaro 7: 29–30; Finocchiaro 1997, 77–81. Thus I have translated this passage directly from Venturi's own paraphrase.
79. Cf. Angeli 1667, 1668a,b, 1669.
80. We know today that Galileo was not all that prompt in obeying the summons, which was first given to him on 1 October 1632; he did not leave Florence until 15 January 1633; cf. Favaro 19: 331–32, 15: 27.
81. Niccolini to Cioli, 19 June 1633, in Favaro 15: 160, and in Finocchiaro 1989, 254–55.
82. Andrea Cioli (1573–1641), Tuscan secretary of state from 1627.
83. Cf. Montucla 1799–1802, 2: 297 and chapter 4.1.
84. Venturi 1820b.
85. Pieralisi 1875, 365–66; cf. chapter 3.4.
86. Venturi 1818–1821, 1: 274.
87. Venturi 1820c, 226.
88. Venturi 1818–1821, 2: 197–99.
89. Delambre 1821, 1: 616–72.
90. Delambre 1821, 1: xx–xxxii.
91. Delambre 1821, 1: xxv–xxvii; Venturi 1818–1821, 2: 197–99; cf. Favaro 19: 295–97; Pagano 1984, 66–68; Finocchiaro 1989, 284–86.
92. Venturi 1818–1821, 1: 197.
93. Delambre 1821, 1: xxiv–xxv.
94. Peiresc to F. Barberini, 5 December 1634, in Favaro 16: 169–71, at p. 170, line 43 *(scherzo problematico)*; cf. chapter 3.3.
95. Venturi 1818–1821, 2: 199.
96. Delambre 1821, 1: xxvii.

CHAPTER 10

1. My account relies primarily on the documents published in Maffei 1987 and in Brandmüller and Greipl 1992, without which it would have been impossible. Besides the interpretive parts of these books, other especially useful works are Brandmüller 1992; Pagano 1994; Baldini 1996b; Mayaud 1997.

2. Maffei 1987, 283–421, which I also designate as Settele 1820–1833 and occasionally as Settele, *Diario*.

3. Maffei 1987, 427–50, and Brandmüller and Greipl 1992, 352–80, which I also designate as Olivieri 1820c and occasionally as Olivieri, "Ristretto," or "Summary." Although the edition by Brandmüller and Greipl is useful in many ways, it cannot replace Maffei's invaluable facsimile edition.

4. The facsimile edition is in Maffei 1987, 451–580, the critical edition in Brandmüller and Greipl 1992, 133–484.

5. Respectively: Settele 1820a (March), in Maffei 1987, 463–70, and in Brandmüller and Greipl 1992, 167–77; Settele 1820b (August), in Maffei 1987, 535–38, and in Brandmüller and Greipl 1992, 288–89; Settele 1820c (August), in Maffei 1987, 543–45, and in Brandmüller and Greipl 1992, 303–5.

6. Respectively: Anfossi 1820b (May), in Maffei 1987, 536–37, and in Brandmüller and Greipl 1992, 182–84; Anfossi 1820c (August), in Maffei 1987, 548–54, and in Brandmüller and Greipl 1992, 310–17; Anfossi 1820d (September), in Maffei 1987, 452–63, and in Brandmüller and Greipl 1992, 336–49.

7. Olivieri 1820a (June), in Maffei 1987, 471–533, and in Brandmüller and Greipl 1992, 225–87.

8. Grandi 1820a (August), in Maffei 1987, 539–42, and in Brandmüller and Greipl 1992, 294–98.

9. Anfossi 1822; Grandi 1820b; Olivieri 1821a,b, 1823a,b, 1840, 1841a–c; Purgotti 1823.

10. Maffei 1987, 18; Settele 1818, 1819.

11. Settele 1820–1833, 285–86; Brandmüller and Greipl 1992, 73.

12. Settele 1820–1833 (296, 297–98), 1820a; Brandmüller and Greipl 1992, 75, 167–77.

13. Settele 1820–1833, 301; Brandmüller and Greipl 1992, 75, 177–78.

14. Settele 1820–1833, 305; Brandmüller and Greipl 1992, 77, 178–79.

15. Settele 1820–1833, 302–7; Brandmüller and Greipl 1992, 75–76.

16. Settele 1820–1833, 305, 307.

17. Settele 1820–1833, 310, 313; Brandmüller and Greipl 1992, 77–78.

18. Settele 1820–1833, 313; Brandmüller and Greipl 1992, 77, 79.

19. Anfossi 1820b; Settele 1820–1833, 313–14; Brandmüller and Greipl 1992, 78 n. 28, 182–84.

20. Settele 1820–1833, 317–18.

21. Olivieri 1820a; Settele 1820–1833, 318, 319; Brandmüller and Greipl 1992, 184–224, 225–87.

22. Settele 1820–1833, 323–25.

23. Settele 1820–1833, 326; Brandmüller and Greipl 1992, 287–88.

24. Settele 1820–1833, 330; Bonora 1872, xxv; Brandmüller and Greipl 1992, 80.

25. Olivieri 1820d, 535–38, 577; Settele 1820–1833, 331–32; Brandmüller and Greipl 1992, 80, 288–89.

26. Brandmüller and Greipl 1992, 80.

27. Brandmüller and Greipl 1992, 81–82, 294–98.

28. Settele 1820–1833, 338; Favaro 19: 420; Maffei 1987, 19; Brandmüller and Greipl 1992, 82–83, 299–301; Mayaud 1997, 243–44.

29. Settele 1820–1833, 344; Brandmüller and Greipl 1992, 83–84, 303–5.
30. Anfossi 1820c; Olivieri 1820d, 548–54; Settele 1820–1833, 344, 345; Brandmüller and Greipl 1992, 310–17.
31. Olivieri 1820b; Settele 1820–1833, 345; Brandmüller and Greipl 1992, 317–25.
32. Olivieri 1820d, 545–46; Settele 1820–1833, 345; Brandmüller and Greipl 1992, 306–7.
33. Olivieri 1820d, 546–47; Brandmüller and Greipl 1992, 84–85, 307–10, 310–17.
34. Settele 1820–1833, 348; Brandmüller and Greipl 1992, 85.
35. Settele 1820–1833, 353; Brandmüller and Greipl 1992, 85–86, 327–28, 329.
36. Anfossi 1820d; Olivieri 1820d, 452–63; Brandmüller and Greipl 1992, 86–88, 331, 336–49.
37. Settele 1820–1833, 361; Brandmüller and Greipl 1992, 87, 333–35.
38. Settele 1820–1833, 369; Brandmüller and Greipl 1992, 90, 326, 328–29, 351.
39. Olivieri 1820d, 451; Brandmüller and Greipl 1992, 349–50.
40. Olivieri 1820c,d; Settele 1820–1833, 374; Brandmüller and Greipl 1992, 351–79, 393.
41. Brandmüller and Greipl 1992, 380–84.
42. Brandmüller and Greipl 1992, 90–93, 386–94.
43. Brandmüller and Greipl 1992, 394–95.
44. Brandmüller and Greipl 1992, 93–94, 396.
45. Settele 1820–1833, 361, 384, 385; Maffei 1987, 41; cf. Settele 1819, 303.
46. Settele 1820–1833, 410; Brandmüller 1992, 186–87; Brandmüller and Greipl 1992, 119, 412–26.
47. Settele 1820–1833, 411–12; Anfossi 1822; Brandmüller and Greipl 1992, 87 n. 71.
48. Settele 1820–1833, 411.
49. Olivieri 1822.
50. Settele 1820–1833, 413, 418; Favaro 19: 421; Brandmüller 1992, 187–88; Brandmüller and Greipl 1992, 426–28; Mayaud 1997, 244–45.
51. Favaro 19: 421; Brandmüller and Greipl 1992, 429.
52. Settele 1820–1833, 411–14; Brandmüller 1992, 188; Brandmüller and Greipl 1992, 120.
53. Settele 1820–1833, 414–17; Brandmüller and Greipl 1992, 128 n. 48.
54. Brandmüller and Greipl 1992, 440–62.
55. Brandmüller 1992, 190; Brandmüller and Greipl 1992, 124, 462–80, 480 (# 68).
56. Brandmüller 1992, 190; Brandmüller and Greipl 1992, 481.
57. Mayaud 1997, 271–72.
58. Settele 1820–1833, 420.
59. Stimson 1917, 99; cf. Flammarion 1872, 196–98.
60. Settele 1820–1833, 421.
61. Such abbreviated references are in Olivieri's original compilation; they refer to twenty-three numbered documents covering 119 pages that were attached to a 23-page opinion with which the compilation began. By "Summary" (abbreviated "Sum.") Olivieri means the attachments, although I am calling "Summary" his own 23-page opinion on the events, issues, and documents;

one can also call *Summary* the entire booklet comprising both (plus two lengthy and detailed tables of contents).

62. In the original, "S.A.P." is an abbreviation of "Sacred Apostolic Palace"; it is used so frequently that I have kept the abbreviation in my translation.

63. Here I am calling "Insert" the "Paragrafo d'inserzione" (Settele 1820c) that is usually called "Nota" in Olivieri's original text; unfortunately, he also usually calls "Nota" Anfossi's "Nota aggiunta" (Anfossi 1820b), which to avoid confusion I call "Appendix."

64. As was customary in earlier times, Olivieri usually gives his references by citing only the *beginning* page of a document; I have generally followed his practice, except in cases like the present one, when it is useful to have an idea of the length of the document. This essay is Anfossi's "Motives" (1820c), in Maffei 1987, 548–54, and in Brandmüller and Greipl 1992, 310–17.

65. In their edition, Brandmüller and Greipl (1992, 352) commit a typographical error here when they leave out the words "ossia il detto paragrafo d'inserzione. Non tardò poi, che appena."

66. This is Anfossi's *Reasons* (1820d) in Maffei 1987, 452–63, and in Brandmüller 1992, 336–49.

67. Here and elsewhere, the indefinite references using 'f.' and 'ff.' after the page numbers are from the original.

68. To repeat and summarize these charges, the Master (1) violated the pope's order to observe silence; (2) objected to Settele's "Insert" after it had been approved by the Inquisition; (3) printed a book without imprimatur; and (4) omitted the date and place of publication and the name of the publisher.

69. Here I am calling "Appendix" the "Nota aggiunta" (Anfossi 1820b), which the Master published at the end of his reprint of his *Fisiche rivoluzioni* (Anfossi 1820a).

70. Here I am calling *Reasons* the printed booklet entitled *Ragioni per cui . . .* (Anfossi 1820d), which the Master gave to the pope in September 1820 and whose publication embodied a fourfold criminal violation, according to Commissary's Olivieri's account.

71. Here I am calling "Motives" the unpublished essay entitled "Motivi per cui . . ." (Anfossi 1820c), which the Master wrote in August 1820 before the Inquisition's decision of 16 August; it may be regarded as a shorter and earlier draft of *Reasons*, with whose content it overlaps.

72. Emphasis in the original; the same applies throughout this document, although such emphasis was usually displayed with capital letters because italics (rather than quotation marks) were usually used to designate quoted text.

73. The original said "1615," but this is obviously a typographical error or slip of the pen.

74. Brandmüller and Greipl (1992, 356 n. 529) identify this work as Ch. Coquelines, *Bullarium Magnum* (14 vols., Rome: Mainardi, 1739–1762).

75. The secretary of the Index was an ordinary clergyman rather than a cardinal, whereas the head ("prefect") of the Index and the secretary of the Inquisition were both cardinals.

76. These "Reflections" were Olivieri's first and longest essay on the subject (1820a), elicited by Anfossi's "Appendix" (1820b), which was his first and shortest note; the exchanges continued with Anfossi's "Motives" (1820c),

answered by Olivieri's "Comments" (1820b); with Anfossi's *Reasons* (1820d), rebutted in Olivieri's "Summary" (1820c); the latter was answered by Anfossi's "Brief Reply" (1820e).

77. This appeal (Settele 1820b), as well as the first (Settele 1820a), was included by Olivieri among the attachments to his "Summary."

78. Anfossi's "Motives" (1820c) had nine parts, each called a motive; so in his criticism Olivieri usually indicated which of these numbered motives he was discussing.

79. Here Olivieri's original text said "immobility" (*immobilità*), which is obviously a typographical error or slip of the pen; this slip is not corrected or noted by Brandmüller and Greipl (1992, 362).

80. Anfossi's *Reasons* was divided into seventeen numbered sections, and in his criticism Olivieri designated them with the symbol §; so whenever Olivieri indicated that he was discussing some point made by Anfossi in § such and such, even when *Reasons* was not explicitly mentioned, it should be understood that Olivieri was referring to this booklet.

81. Here Anfossi (as quoted by Olivieri) was quoting the Index's 1620 decree correcting Copernicus's *Revolutions;* cf. Favaro 19: 400 and Finocchiaro 1989, 200.

82. Here Anfossi seemed to be saying that what was condemned was not the *proposition* that the earth moves, but the geokinetic *arguments* of its defenders. It is ironic that in the preface to the *Dialogue,* Galileo claimed, "I have in the discussion taken the Copernican point of view, proceeding in the manner of a pure mathematical hypothesis and striving in every contrived way to present it as superior to the viewpoint of the earth being motionless, though not absolutely but relative to how this is defended by some who claim to be Peripatetics" (Finocchiaro 1997, 78; cf. Favaro 7: 29–30). That is, Galileo claimed to be advocating not the superiority of the geokinetic *proposition,* but the superiority of the geokinetic *arguments;* in other words, he claimed to be criticizing not the geostatic proposition, but the geostatic arguments of the Peripatetics. Galileo was making the distinction in an attempt to qualify his commitment to Copernicanism and criticism of geocentrism, whereas Olivieri was making the same distinction in order to qualify the Inquisition's (reverse) commitment to geocentrism and condemnation of Copernicanism.

83. As Brandmüller and Greipl indicate (1992, 363 n. 561), this biography was actually written by Brenna (1778) but published in a multivolume series edited by Fabroni (1778–1805).

84. Anfossi was saying something very plausible with his conditional claim that if Copernicanism was contrary to Scripture in 1616, it would also be contrary to Scripture in 1820; but such a conditional works both ways. For as we saw Auzout (1665a) argue in chapter 5.2, given such a conditional claim, once it can be shown that Copernicanism is demonstrably true, and so not contrary to Scripture, one can also conclude that it was not contrary to Scripture in 1616, and indeed that it was never contrary to Scripture.

85. Olivieri was exaggerating in calling a syllogism the inference from philosophical falsity and absurdity to theological heresy or error, although later he did try to elaborate such a "syllogism" (Olivieri 1840, 57–65); however, his main point is that such an argument tries to justify theological heresy or error on the basis

of philosophical falsity and absurdity, and so once this philosophical premise no longer holds, one can no longer justify the theological judgment.

86. This sentence does not seem to make sense, but this is what the original says.

87. Douay Version, Psalm 92:1; in the King James Version, Psalm 93:1: "The world also is stablished, that it cannot be moved."

88. On that page, Olivieri referred more precisely to Saint Thomas Aquinas, *Summa Theologica*, part 1, question 70, article 1, reply to objection 3. I have taken the translation from the Encyclopedia Britannica Great Books edition (Chicago, 1952), 19: 364.

89. Natalis Alexander, *Historia Ecclesiastica Veteris Novisque Testamenti* (Paris, 1730), "Dissertatio XIII"; cf. Anfossi 1820b (May), in Maffei 1987, 536; cf. Brandmüller and Greipl 1992, 367 n. 569, 183 n. 141, 101 n. 23.

90. Anfossi had quoted this author in §12 of *Reasons* (1820d, 459); Brandmüller and Greipl (1992, 369 n. 574, 345 n. 490) identify this reference more precisely as Jamin 1872, 245.

91. In his "Reflections" (§36), Olivieri (1820a, 487) had given a more precise reference; this is then further specified by Brandmüller and Greipl (1992, 369 n. 578, 243 n. 267, 107 n. 47) as Montfaucon 1706, iv–vi.

92. I found this second quotation in Andres 1782–1799, 4: 352; cf. Andres 1776.

93. Calmet 1744; cf. chapter 7.1.

94. Here I have corrected Olivieri's reference "§15, Sum. 16."

95. Brandmüller and Greipl (1992, 372 n. 594, 347 n. 492, 100 n. 18) refer more precisely to Nieuwentijd 1727, 390.

96. As previously noted, Fabroni (1778–1805) was merely the editor, whereas the author was Brenna (1778).

97. Ellipsis in Olivieri's text.

98. Here Olivieri was making a plausible distinction, which, as I argued in chapter 7.2, was implicit in the consultant Lazzari's (1757) recommendation to delete from the *Index* the general prohibition of all books that teach the earth's motion and the sun's immobility; this further undermines the thesis (in Mayaud 1997, 189) that the partial retraction of the 1758 *Index* was illogically incomplete; as Olivieri stated in the next paragraph (§59), he was acquainted with Lazzari's recommendation.

99. Here Olivieri's reference was simply "ibid.," which in this case is misleading or ambiguous. Brandmüller and Greipl (1992, 374 n. 601) give the incorrect reference of "Motives," no. 6; this is probably because they redescribe all of Olivieri's references to his own "Summary" in order to make them correspond to their own rearrangement of the material.

100. The Italian language in this sentence in Olivieri (1820c, p. 445, §59) is obscure and ambiguous; its meaning becomes clear in light of a later discussion by Olivieri (1840, 37–38); my translation incorporates such a clarification.

101. See Lazzari 1757 and chapter 7.2.

102. Olivieri's statement of this exclusionary rule consisted of a quotation that had too many ellipses and attempted to bring together phrases from separate passages; I found the result unintelligible and thus have paraphrased what I take to be his main point.

103. Here Olivieri mistakenly wrote "p. 17," which I take to be a typographical error or slip of the pen; the sentence occurs in §17 of Anfossi's *Reasons,* but on p. 12 of the attachments which Olivieri appended to his "Summary."

104. Quoted from Finocchiaro 1989, 200; cf. chapter 1.4.

105. Olivieri's "Reflections" (1820a, in Maffei 1987, 523), Settele's "Insert" (1820c, in Maffei 1987, 545), and Brandmüller and Greipl (1992, 285 n. 339) indicate that this refers to Gerdil 1806, 258.

106. Tiraboschi 1785, in Tiraboschi 1782–1797, 10: 388–440.

107. Here I am correcting Olivieri's reference of "(Sum., p. 17)."

108. Guglielmini 1789, 1792.

109. Calandrelli 1806a,b.

110. See §23 above.

111. See §13 above.

112. This paragraph was added by Olivieri on the page that followed his signature, which was a page containing the table of contents of his "Summary" (exclusive of the attachments, which had a separate table of contents); see Olivieri 1820d, in Maffei 1987, 450.

113. Cf. Pino 1802; Brandmüller and Greipl (1992, 379 n. 622, 264 n. 304, 44 n. 73) erroneously identify this work as Pino 1806, which is a single-volume work.

114. Translated from the Italian text of Olivieri (1820c), edited in facsimile by Maffei (1987, 427–50), published, and permission to reprint kindly granted, by Edizioni dell'Arquata di M. R. Trabalza; I have also consulted the text in Brandmüller and Greipl 1992, 351–80. In my translation, I have not reproduced the usual source page numbers in square brackets, since Olivieri's own paragraph numbers render those superfluous.

115. For example, Olivieri 1840, especially pp. 7–8, 100.

116. See Galileo's letters to Giovanni Battista Baliani of 1614 (Favaro 12: 33–36) and 1630 (Favaro 14: 158), and the discussion in *Two New Sciences* (Favaro 8: 123–24); cf. Roberts 1870, 103–4, and Favaro 1908.

117. See Galileo's discussions in the *Dialogue,* in Finocchiaro 1980 (206–22), 1997 (155–71, 212–20); cf. Govi 1872.

118. As mentioned in a previous note, he did so later, in Olivieri 1840, 57–65.

119. For more details on this issue, see Beretta 1999, 469–86. The evolution of the concept also suggests that it is perhaps anachronistic to defend the doctrine of papal infallibility by arguing that the 1616 condemnation (although erroneous) was not a papal decree ex cathedra and so is not relevant to the doctrine; this suggestion has been explicitly drawn and defended by Beretta (1999, 451 n. 35).

CHAPTER 11

1. *Encyclopedia Britannica,* 11th ed. I thank John Brooke for the information about Brewster's religious affiliation.

2. Guicciardini 1616 and chapter 8.2; Baldigiani 1678 and chapter 5.1; and Tiraboschi 1793 and chapter 8.4.

3. Brewster 1841, 57–59.

4. Brewster 1841, 93–95.

5. Libri 1841a–c; 1842.

6. Libri 1840–1841, 212–19; Peiresc 1634, 1635; G. Galilei 1634b; cf. chapter 3.3 and 3.4.
7. Libri 1840–1841, 208–11.
8. Libri 1841c, 235 n. 1, 261 n. 2.
9. Libri 1841c, 261 n.2.
10. Libri 1841a (34–37), 1841c (259–66).
11. Masini 1621.
12. Frisi 1774–1784, in Fabroni 1778–1805, 20: 214–29.
13. Libri 1841a, 46–47.
14. Drinkwater Bethune 1832; Whewell 1837. Cooper's article led Whewell to revise his original account; see Whewell 1847, 1857a,b.
15. Mallet du Pan 1784; cf. chapter 8.3.
16. Cooper 1838, 72–73.
17. See especially Cooper 1838, 88, 95 n., 108 n.; cf. chapter 8.3.
18. Quoted from Cooper 1838, 104; italics in the original.
19. See Adams 1843; Bemis 1956, 518–20; Cooper 1844; Portolano 2000; Purcell 1844b.
20. I owe this information to Dr. Philip S. Shoemaker (private correspondence); he refers to the *Daily Cincinnati Enquirer,* 13 November 1843.
21. Adams 1843, 54.
22. Adams 1843, 54–55.
23. Purcell indicates that here he is quoting from Hallam 1837–1839, 4: 16.
24. Purcell gives no bibliographical reference for this quotation; his article refers to two different editions of Brewster's account (1835, 1841), which have almost identical text.
25. Here Purcell (1844, 19–20) quoted a long paragraph from the *Edinburgh Review.*
26. Purcell 1844, 19–20.
27. Adams was here confusing the 1633 proceedings with the 1616 ones. It was the 1633 proceedings which concluded with an Inquisition sentence signed by seven (out of ten) cardinal-inquisitors. On the other hand, it was in February 1616 that the physical absurdity and the religious erroneousness of the earth's motion were declared (unanimously) in a committee report of eleven Inquisition consultants, but the cardinal-inquisitors did not endorse that report and so did not themselves issue a formal decree. See Finocchiaro 1989, 146, 291, and chapter 1.
28. Adams 1843, 53–54.
29. This was the expression used by a well-disposed Catholic critic of Marini, Biot (1858c, 36).
30. Marini 1850, 141.
31. See, for example, the criticism by Biot (1858c, 36–44).
32. Marini 1850, 54–57.
33. Marini 1850, 56, 59, 61–62; cf. Favaro 19: 361–62; and Finocchiaro 1989, 286–87.
34. Marini 1850, 57; Venturi 1818–1821, 2: 193; cf. chapter 9.4.
35. Marini 1850, 57–58.
36. There is evidence that this rule was indeed observed, as one can see from the documentation now available for no less a famous trial than that of Giordano

Bruno. Cf. Firpo 1993, 96–98, 327–29; Mendoza 1995, 262–64; and Finocchiaro 2002c.

37. Marini 1850, 58–59.

38. Marini 1850, 64–68.

39. Marini 1850, 105–7.

40. Marini 1850, 98–102, 140–41.

41. For Mallet, see Marini 1850, 39 n. 2, 141; for Bergier, see Marini 1850, 39 n. 2,
94; for Feller, see Marini 1850, 54; and for Cooper, see Marini 1850, 6.

42. Marini 1850, 141.

43. Marini 1850, 94; cf. Guicciardini to Cosimo II, 4 March 1616, in Fabroni
1773–1775, 1: 53–57, and in Favaro 12: 241–43; Mallet du Pan 1784, 124; and
chapter 8.3.

44. See Favaro 19: 283; Pagano 1984, 154; and chapter 12.2; the crucial phrase is
"*et e contra*" (italics added).

45. Marini 1850, 54.

46. Marini 1850, 53.

47. Marini 1850, 5.

48. G. Galilei 1636, 17; cf. Motta 2000, 101; Favaro 5: 319; Finocchiaro 1989, 96;
and chapter 4.2.

49. Marini 1850, 53.

50. Biot 1858b, 3: 36.

51. Biot 1816, 437–44.

52. Biot 1858a–c.

53. Biot 1858b, 3: 44.

54. Biot 1858b, 2: 444.

55. Biot 1858b, 2: 457, 3: 44.

56. Biot 1858b, 2: 457.

57. Biot 1858b, 3: 42–43.

58. Chasles 1862, 135; here he also referred approvingly to Reumont and Rosini
as sharing the same approach; as usual, Chasles gave no precise references, but
I presume he is referring to Reumont 1849, 1853, and Rosini 1841.

59. Chasles 1862, 277.

60. Chasles 1862, 278–80, with omissions of text indicated by ellipses.

61. Barni 1862a, 236.

62. Parchappe 1866, 12.

63. T. Martin 1868.

64. Ponsard 1867; Redondi 1994, 102–8.

65. Madden 1863, 62–66, 157–81; cf. Gherardi 1870, 7–8 n.1, 15–19.

66. Madden 1863, 194.

67. Madden 1863, 148–57; in this section, further references to this passage are
given in parentheses in the text.

68. Olivieri 1841a,b; Olivieri 1840, in Bonora 1872, 1–133.

CHAPTER 12

1. Favaro 1887a, 194; Gebler 1879a, 325; Scartazzini 1877–1878, 10: 419.

2. Cantor 1864, 187; cf. Favaro 1887a, 194; Gebler 1879a, 325; Scartazzini
1877–1878, 10: 419.

3. Wolynski 1878, 12–13; cf. Favaro 1902b, 776 n. 2.

4. L'Epinois 1867, 1877; he also published his own interpretation and evaluation (L'Epinois 1878).

5. Berti 1876b, 1878; cf. Berti 1876a, 1877, 1882.

6. Pieralisi 1876, 1879; cf. Pieralisi 1858, 1875.

7. Gebler 1877; cf. Gebler 1876, 1879a,b.

8. Later Favaro, while editing the National Edition of Galileo's works, was given access to the Galileo file and other archives; Favaro included the documents in volume 19 (293–399) but also issued them separately earlier (Favaro 1902c, 1907c). In 1984 there was an official Vatican edition (Pagano 1984).

9. L'Epinois 1867, 146–71, 68–145, respectively; the most important of the shorter documents were given in footnotes to the interpretive account.

10. As we saw in chapter 9.1, Marini (1850, 40–41) reported that the Italian historian Carlo Denina (who lived in Paris between 1810 and 1814) read the Vatican file of Galileo's trial and told Napoleon that it contained nothing worthy of being republicized.

11. Very little is known about Henri de L'Epinois; he is not listed in the *Index Biographique Français,* the *Nouvelle Biographie Générale* (1859ff.), the *Dictionnaire de Biographie Française,* the *Dictionary of Scientific Biography,* or the *Enciclopedia Italiana.* Gherardi (1870, 10 n. 1) calls him "a clerical author indeed, but one of the good ones."

12. For the biographical details in this paragraph, I am indebted to Schütt 2000; cf. Renn 2000b; Wohlwill 2000.

13. Cf. Wohlwill 1909, 1926, and 1870, 1877b, respectively.

14. Gio. Garzia Millini (or Mellini), who at the time was the cardinal secretary of the Inquisition (Favaro 19: 298, 306, 311–13; Viganò 1969, 120).

15. Here quoted from Finocchiaro 1989, 147; cf. L'Epinois 1867, 98 n. 1; Favaro 19: 321; Pagano 1984, 100–101.

16. "And thereafter, indeed immediately" is my translation of the phrase "et successive, ac incontinenti" (folio 43v), which both Favaro (19: 322 line 7) and Pagano (1984, 101) transcribe without the comma. At this point the Latin text is somewhat ambiguous, and the sequence of events is far from clear; cf., for example, Drake 1965; Gebler 1879a, 76–84; Geymonat 1962, 112–15; Morpurgo-Tagliabue 1981, 19–27; Langford 1971, 92–97; Santillana 1955a (125–31), 1965, 1972.

17. Here quoted from Finocchiaro 1989, 147–48; cf. L'Epinois 1867, 98 n. 2; Favaro 19: 321–22; Pagano 1984, 101–2.

18. Nelli 1793, 1: 413 n. 2; Venturi 1818–1821, 1: 273; Marini 1850, 101–2.

19. Having lingered in Rome for some time after the Index's Decree (March 5), Galileo soon received at least two letters from friends in Pisa and Venice informing him that such rumors were circulating in these cities (Favaro 12: 254, 257–59). He showed these letters to Bellarmine (Favaro 12: 257 n. 2; Baldini and Coyne 1984, 6), and in this first sentence the cardinal was essentially reporting their content. This certificate makes it obvious that the two were still on good personal terms despite their theological, philosophical, and scientific disagreements. The extent of Bellarmine's respect and goodwill is further supported by the autograph draft copy of this certificate, which is found in the Roman Archive of the Society of Jesus and has been published by Baldini and

Coyne (1984, 25); the draft shows that Bellarmine deleted two sentences and inserted another to make the end result more favorable to Galileo. For related studies of Bellarmine, see Baldini 1984, 1990; Blackwell 1991; Coyne and Baldini 1985.

20. Here quoted from Finocchiaro 1989, 153; cf. Favaro 19: 348; Pagano 1984, 138. L'Epinois (1867, 166) did *not* reproduce the text of Bellarmine's certificate but merely referred to Marini (1850, 101).

21. Venturi 1818–1821, 2: 197–99; Delambre 1821, 1: xx–xxxii; cf. chapter 9.5.

22. L'Epinois 1867, 159–64; cf. Favaro 19: 336–42; Pagano 1984, 124–30; Finocchiaro 1989, 256–62.

23. Wohlwill 1870, 1–15.

24. Wohlwill 1870, 16–22.

25. Wohlwill 1870, 23–30.

26. Wohlwill 1870, 31–50.

27. Wohlwill 1870, 51–63.

28. Wohlwill 1870, 64–79.

29. Wohlwill 1870, 80–85.

30. Chapter 11.5; cf. Madden 1863, 62–66, 159; Gherardi 1870, 7–8 n.1, 15–19.

31. Gherardi 1870, 23.

32. Mercati 1942, 15–18.

33. Gherardi 1870, 40–41.

34. Favaro 19: 282–83; cf. Gherardi 1870, 81–82; L'Epinois 1867, 129 n. 4; Pagano 1984, 229.

35. Marini 1850, 54, and chapter 11.3.

36. See, respectively, Favaro 19: 399, Pagano 1984, 215–16, and chapter 6.2; and Mayaud 1997, 137–39, and chapter 7.1.

37. L'Epinois 1867, 129 n. 4; Favaro 19: 360–61; Pagano 1984, 154.

38. Here quoted from Finocchiaro 1989, 148; cf. Gherardi 1870, 29–30; Favaro 19: 278; Pagano 1984, 223–24.

39. Gherardi 1870, 40–43.

40. L'Epinois 1878.

41. Berti 1876b, 1877, 1878.

42. There was, in fact, an inquisitorial rule stipulating that "frail seniors older than sixty are not to be tortured, but may be threatened at the discretion of the inquisitor"; quoted in Müller 1911, 457 n. 2, from Bordoni, *S. Tribunal Indictum in Causis Fidei,* Rome, 1648, 576.

43. Pieralisi 1875, 197–98; cf. Favaro 15: 106–7; Finocchiaro 1989, 276–77. This letter may now be read in conjunction with another one recently discovered and published: Maculano to Barberini, 22 April 1633, in Beretta 2001, 571; see also the reinterpretation by Beltrán Marí (2001b).

44. Scartazzini 1877–1878, 5: 1–15, especially 12–15.

45. Scartazzini 1877–1878.

46. Cf. L'Epinois 1867, 168–69; Favaro 19: 361–62; Pagano 1984, 154–55; Finocchiaro 1989, 286–87.

47. Finocchiaro 1989, 287.

48. Finocchiaro 1989, 287.

49. Finocchiaro 1989, 287.

50. Wohlwill 1877b; a good appreciative discussion of this work is found in Scartazzini 1877–1878, 5: 221–49, 6: 401–23; a more critical discussion appears in Gebler 1879a, 253–63.

51. See, for example, Gebler (1879a, 255–56), who quoted Limborch (1692, 322).

52. Scartazzini 1877–1878, especially 4: 829–61, 5: 221–49, 10: 417–53.

53. These numbers reflect the pagination used at the time of Scartazzini and later printed in Favaro's edition (19: 293–399); thus here I ignore the other three sets of page numbers on the documents of the Vatican file, including the modern pagination added in 1926 and used by Pagano (1984); for more details on the origin, meaning, and correspondence of these four sets of numbers, see Gebler 1879a, 330–33; Pagano 1984, 1–4, 55–60; Beretta 1999a, 454–60, 487–89.

54. Scartazzini 1877–1878, 4: 856–60.

55. Scartazzini 1877–1878, 5: 225–33.

56. Gebler 1877, 1–184; Gebler (1877, xvi–xix) also gave a complete list of which folios were paired in the same folded sheet.

57. A brief biography of Gebler may be found in *Ausburger Allgemeine Zeitung,* 6 December 1878, which is abridged in Gebler 1879a, xi–xvi.

58. Gebler 1879a, 334–40; cf. Gebler 1877, xx–xxxii.

59. Not only is the original signed document not in the file, but there is no copy, either. This is a puzzling lacuna for which there is no satisfactory explanation even today. It may have something to do with the fact that the condemnation was followed by an unprecedented effort by the Church to publicize the sentence, as we saw in chapter 2. On the other hand, such publicity ensured that many copies of the sentence survived and that its text (while subject to minor variations) is on the whole incontrovertibly authentic. For more details on such issues, see Sandonnini 1886; Favaro 1887b, 122–26; Favaro 1887–1888; Beretta 1999a; Lerner 1998b.

60. Gebler 1879a, 338.

61. Gebler 1879a, 337.

62. Gebler 1879a, 336–37.

63. Gebler 1879a, 337.

CHAPTER 13

1. This period also includes the work and activities of Raffaele Caverni (1837–1900), who published a monumental history of the experimental method in Italy, focused on Galileo. My reason for omitting it from my discussion is that neither in that work nor elsewhere does Caverni seem to have published anything significant on the Galileo affair, and in this book I focus on the reception and aftermath of Galileo's condemnation rather than of his scientific work per se. This is not meant to deny that Caverni's account of Galileo's scientific work may have an indirect connection with the Galileo affair insofar as Caverni's account was so negative toward Galileo's scientific accomplishments that it produced a reaction that led to his isolation and marginalization in Galileo scholarship. Add to this the fact that Caverni was a priest; that he tried unsuccessfully to be part of the editorial team publishing the National

Edition of Galileo's (1890–1909) complete works; and that Favaro explicitly acted to exclude him, although he was objective enough to vote to award Caverni's book an academic prize. Then the question of Caverni's connection with the Galileo affair ought not to be regarded as closed, and its deeper study may well reveal significant links. For some relevant details, see Caverni 1891–1900; Favaro et al. 1890; Favaro 1920; *Dizionario biografico degli italiani* 23 (1979): 83–88; Bucciantini 1997; Renn 2000b; Schütt 2000; Castagnetti and Camerota 2000a,b; Castagnetti 2000; and Renn 2001, 323–421.

Similar remarks apply to Husserl's philosophical critique of Galileo's world-view: there is no question of its significance for the philosophical interpretation and evaluation of Galileo's science, nor of its influence among philosophers; but I am not aware of any significant writings by Husserl explicitly on Galileo's trial or of any attempts to establish its indirect connection to the Galileo affair; cf. Husserl 1936, 1962, 1970; Agosti 1964; Gurwitsch 1967; Kvasz 2002.

2. *Catholic Encyclopedia* (www.newadvent.org/cathen/15552c.htm, consulted on 9 January 2003).

3. G. Ward 1871a,b.

4. Roberts 1885.

5. Wegg-Prosser 1889.

6. Broderick 1987, 292; Garzend 1912, 258. This Church council also deserves mention in our story insofar as a small minority opinion was expressed in favor of Galileo when, at the session of January 4, the bishop of Savannah proposed the rehabilitation of Galileo; cf. Berti 1876b, lxi n. 60; Genovesi 1966, 37. But such a step was dramatically premature.

7. Grisar 1878a,b, 1882; Reusch 1875, 1879; cf. S. Harris n.d.

8. For more details on this development, see Berggren and Sjöstedt 1996, 145–47.

9. "Epigrafi ed offese," *L'Osservatore Romano,* 23 April 1887. In the spirit of factual corrections mentioned at the beginning of this article, one could object that during the 1633 trial Galileo was not really hosted at Villa Medici but at the residence of the Tuscan ambassador, Palazzo Firenze. Villa Medici was indeed the location of Galileo's detention: the June 22 sentence first condemned him to indefinite imprisonment but the following day was immediately amended by commuting such imprisonment to house arrest at Villa Medici; and this served as Galileo's prison from June 23 to July 6, when the location was again changed to house arrest at the residence of the archbishop of Siena. On another occasion Villa Medici did serve as Galileo's comfortable and pleasant residence, but that was on his third trip to Rome, from 10 December 1615 to 4 June 1616, during the earlier phase of the Inquisition proceedings. For more details about Galileo's visits to Rome, see Shea and Artigas 2003, 30, 74, 106–7, 134–35, 179–80, 195.

10. See Mallet du Pan 1784 and chapter 8.3; Marini 1850 and chapter 11.3.

11. Bucciantini 1997, 427–28.

12. Bucciantini 1997, 428.

13. Favaro 1902b (757–58), 1902c.

14. Favaro 1907c.

15. Bucciantini 1997, 443 n. 60; Ratzinger 1998, 184; cf. Favaro 1902b, 780 n. 1.

16. Favaro 1902b, 802 n. 3.
17. Cioni 1908, reprinted in Cioni 1966.
18. Lämmel 1927, 1928.
19. Much more can be learned in this regard from the works of Mauro Pesce; see his entries in the bibliography. See also Dubois 1655; Wittich 1659; Muratori 1714, 1779, 1964a–c; Jemolo 1923; Fabris 1986, 1992a–d; Del Prete 2001.
20. S. Harris n.d., 16 and n. 24.
21. Motzkin 1989, 210.
22. Ratzinger 1998, 181, 182.
23. Bucciantini 1997, 443 n. 60; Ratzinger 1998, 184.
24. Blumenberg 1987c, 371.
25. Mercati 1942.
26. Blind 1889, 106–7, 113; Gatti 1997, 58.
27. Schlereth 1978; Zahm 1896.
28. Leo XIII 1893, paragraph 9, p. 329.
29. Leo XIII 1893, paragraph 25, p. 337.
30. Leo XIII 1893, paragraphs 20–21, pp. 335–36.
31. Leo XIII 1893, paragraph 20, p. 335.
32. Leo XIII 1893, paragraph 15, p. 332.
33. The crucial passage consists of paragraphs 18–19, in Leo XIII 1893, 334–35 (= Carlen 1981, 2: 334–35).
34. Leo XIII 1893, paragraph 17, p. 334.
35. Leo XIII 1893, paragraph 18, p. 334.
36. G. Galilei 1636, 17; cf. Motta 2000, 101; Favaro 5: 319; Finocchiaro 1989, 96.
37. Leo XIII 1893, paragraph 18, p. 334.
38. Leo XIII 1893, paragraph 18, p. 334.
39. Leo XIII 1893, paragraph 18, p. 334.
40. Leo XIII 1893, paragraph 18, pp. 334–35.
41. Cf. Finocchiaro 1986a, 1995a for a reconstruction of Galileo's argument.
42. Augustine, *De Genesi ad Litteram,* i, 21, 41; Favaro 5: 327 (or Finocchiaro 1989, 101); Leo XIII 1893, paragraph 18, p. 334.
43. Augustine, *De Genesi ad Litteram,* ii, 9, 20; Favaro 5: 318 (or Finocchiaro 1989, 95); Leo XIII 1893, paragraph 18, p. 334.
44. See, for example, Dubarle 1964, 25; Fantoli 1996, 502–3; Langford 1971, 66; Martini 1972, 444; Pesce 1987, 283–84; Poupard 1984, 13; Viganò 1969, 234.
45. John Paul II 1979a–d, 1992a–c.
46. For more on Duhem, see Agassi 1957; Brenner 1990; Goddu 1990; Maiocchi 1985, 1990; R. Martin 1987, 1991; Stoffel 2001.
47. For a fuller discussion of an apologetic interpretation of Duhem's *To Save the Phenomena,* see Maiocchi 1985, 268–77.
48. See R. Martin 1987, 1991, for more details. Antimodernism was the movement originating from Pope Pius X's 1907 encyclical *Lamentabili Sane Exitu,* which condemned sixty-five "modernist" propositions, listing them in a new "Syllabus of Errors"; these propositions were taken mostly from the works of Alfred F. Loisy (1857–1940); cf. Broderick 1987, 566 and Ghiberti 1992, 218.
49. Duhem 1954, 39 n. 9.

50. Duhem 1954, 39.
51. Duhem 1908, 136; cf. Duhem 1969, 113.
52. Duhem 1908, 140; cf. Duhem 1969, 117.
53. Duhem 1908, 135; cf. Duhem 1969, 112.
54. Duhem 1908, 128; cf. Duhem 1969, 106.
55. Duhem 1908, 135; cf. Duhem 1969, 112.
56. Duhem 1908, 127; cf. Duhem 1969, 106.
57. Duhem 1908, 128; cf. Duhem 1969, 106.
58. There are widely shared misgivings about whether such a tradition exists, as well as some reservations about whether such precepts can be attributed even to Bellarmine and Urban; see Popper 1956, 99 n. 6; Lloyd 1978; Morpurgo-Tagliabue 1981, 41–52; Jardine 1984, 225; Westfall 1989a, 16–17; Goddu 1990; McMullin 1990a; Finocchiaro 1992b, 291–94; Barker and Goldstein 1998. For more nuanced accounts of the issues, see Westman 1972, 1980; Biagioli 1993, 218–27.
59. Finocchiaro 1992b.
60. Müller 1909a,b; these were immediately translated into Italian and combined into one volume in Müller 1911.
61. Müller 1911, chapter 17.
62. Müller 1911, chapter 19.
63. Müller 1911, chapter 37.
64. Müller 1911, chapter 38.
65. Müller 1911, chapter 38.
66. Indeed, the myth seems to have disappeared in the twentieth century, if we except such stragglers as Sacchi (1913) and Carrara (1914), who may be deemed insignificant, as one may gather from Fabris 1986.
67. Cf. Grisar 1882.
68. Müller 1911, 139–40; cf. Reusch 1879, 55.
69. This part of such an account was later elaborated, more explicitly and with some new (if inconclusive) evidence, by Drake 1976, 1980, 1999 (1: 153–56); cf. also Finocchiaro 2002a.
70. Koestler 1959, 437; for a critique of the other aspects of Koestler's account of Galileo, see chapter 15.2.
71. Garzend 1912, 54 n. 91 bis; this specific reference is useful for the case of this particularly arcane detail.
72. Garzend 1912, 55 n. 92.
73. Garzend 1912, 55 n. 92.

CHAPTER 14

1. *L'Osservatore Romano,* 1–2 December 1941, 1.
2. *L'Osservatore Romano,* 1–2 December 1941, 4.
3. Gemelli 1941, 4; I have omitted about ten lines at the first ellipsis and about five at the second.
4. *Nel terzo centenario della morte di Galileo Galilei* 1942.
5. Gemelli gave no reference, but this statement can be found in Pastor 1938b, 62.
6. Gemelli 1942b, 1–2.

7. Pastor 1938b, 59–62.
8. Gemelli 1942b, 11. His notion of the "convergence of probabilities" is very important and revealing. It could be usefully compared and contrasted with Galileo's own analysis in his "Considerations on the Copernican Opinion" (Favaro 5: 351–70; Finocchiaro 1989, 70–86); with Foscarini's account (Foscarini 1615, 1635; Blackwell 1991, 217–63); with Riccioli's (1651) probabilistic approach (cf. Dinis 1989, 236–38); with Auzout's (1665) notion of reasonable demonstration (see chapter 5.2); and with Soccorsi's (1947) remarks on this point in section 3 of this chapter.
9. Here Gemelli (1942b, 11 n. 1) attributed this thesis to Maffi (1918b, 527).
10. There is also an insightful discussion of this question in Gemelli 1942a.
11. Gemelli 1942b, 14.
12. See Gemelli 1922a; a version of this thesis has recently been independently elaborated by Rowland (2003).
13. Gemelli 1942b, 16–17.
14. Gemelli 1942b, 18.
15. Gemelli 1942b, 22.
16. Gemelli 1942b, 26.
17. Biot 1858a–c; Chasles 1862; cf. chapter 11.4.
18. Gemelli 1942b, 27.
19. Gemelli 1942b, 27.
20. Simoncelli (1992, 68 n. 23) appears to deny this influence, but he is referring to Paschini's book (1964a,b, 1965), whereas I am referring to Paschini's article (1943); moreover, Simoncelli interprets Gemelli (1922a, 1941, 1942a,b) as more of a conservative and apologist than I do here.
21. Paschini 1943, 94.
22. Cf. Maccarrone 1979a, 76–77.
23. In Paschini 1943, 94.
24. That this was Paschini's own meaning is supported by a statement he made in a letter to his friend Giuseppe Vale, dated 4–5 July 1946: "One must have the courage to tell the truth even when it turns out to be bitter" (Paschini 1946c, 77 = Simoncelli 1992, 77); cf. chapter 16.1.
25. Paschini 1943, 96–97.
26. See, for example, Lazzari 1757 and chapter 7.2; Olivieri 1820c and chapter 10.3.
27. See, for example, Bellarmine 1615 and chapter 5.2.
28. See, for example, Madden 1863 and chapter 11.5.
29. Cf. Tiraboschi 1793 and chapter 8.4; Cooper 1838 and chapter 11.2; Marini 1850 and chapter 11.3.
30. Soccorsi 1946, 1947, 1963, 1964.
31. Simoncelli 1992, 81; cf. Soccorsi 1946.
32. Bertolla 1979, 177; Simoncelli 1992, 49.
33. Soccorsi 1947, 13; in this section, subsequent references to this work are given in parentheses in the text. The other editions of this book are essentially identical, the minor differences involving mostly editorial matters; for example, some of the very long and very substantive notes of the 1947 edition are incorporated into the text of the 1964 edition.

34. Quoted in Soccorsi 1947, 29. At the time of Soccorsi's writing, Bellarmine's Louvain lectures were unpublished; they were later published by Baldini and Coyne (1984); for this passage, see Baldini and Coyne 1984, 20.

35. Quoted here from Finocchiaro 1989, 68; cf. Soccorsi 1947, 31–32; Favaro 12: 172.

36. Here Soccorsi had a note elaborating the limitations of Galileo's arguments (1947, 103–5 n. 72).

37. Here Soccorsi had a note referring to his detailed, insightful, and original analysis of the Galilean arguments (1947, 70–81 n. 26).

38. For this sentence and the previous one, my translation is relatively free.

39. Here quoted from Finocchiaro 1989, 57–58; cf. Favaro 5: 295. The biblical allusion paraphrases Matthew 5:29 and 18:9.

40. Soccorsi 1947, 50–60.

41. For each of these two statements, Soccorsi had a lengthy discussion in the notes with detailed comments about the various Galilean arguments; see Soccorsi 1947, 103–5 n. 72, for the first statement and Soccorsi 1947, 103 n. 73, 70–81 n. 26, for the second.

42. Soccorsi 1947, 75–78; cf. Favaro 7: 372–83; G. Galilei 1953, 345–56.

43. Such criticism can be found in Strauss 1891, 555–56; Taylor 1938, 134–35; Koestler 1959, 477–78; and Langford 1971, 124–25.

44. Soccorsi (1947, 77–78) indicates that Müller (1897) also advanced a similar (although not identical) appreciative analysis of Galileo's argument. More recently, several scholars have followed up on such an appreciation with ever more nuanced and deeper analyses; cf. Drake 1970, 191–96; Finocchiaro 1980, 129–30, 246–53; A. Smith 1985; Hutchison 1990; Topper 1999, 2000, 2003.

45. Here quoted from Finocchiaro 1989, 292; cf. Favaro 19: 407.

46. Here Soccorsi referred to Franzelin 1882, 127ff.; Billot 1909, 434–39.

47. Here Soccorsi (1947, 102) again referred to Franzelin (1882) and Billot (1909).

48. Here Soccorsi (1947, 102) spoke of "falsity of the sentence" (*falsità della sentenza*), where *sentence* means the ecclesiastical "decree" declaring the "doctrine" in question false; so he is talking about the falsity of the decree that the doctrine is false, which is to say the truth of the doctrine. What makes Soccorsi's discussion confusing is that he uses the word *sentence (sentenza)* to refer to such things as doctrines (on various topics; for example, the Copernican doctrine); judgments about such doctrines issued by such institutions as the Inquisition; and theories about the nature and status of such judgments, such as the theories of Franzelin and Billot. This is why here I translate *falsità* as "truth"!

49. Soccorsi 1947, 100–103 n. 71.

CHAPTER 15

1. Ponsard 1867; cf. Redondi 1994, 102–8; chapter 11.4.

2. Llofriu y Sagrera 1875; cf. chapter 13.1.

3. Pieracci di Turricchi 1820 (cf. Redondi 1994, 104 n. 100); S. Brown 1850; and Glaser 1861, respectively.

4. Kelly 1948; Stavis 1966; Goodwin 1998; MacLachlan (forthcoming).

5. Raven 1860, 1869; Harsányi 1939; Sawyer 1992; G. Smith 1995.

6. Burtt 1932; Husserl 1936, 1962, 1970; cf. Agosti 1964; Gurwitsch 1967; Kvasz 2002.

7. Feyerabend 1985; cf. Finocchiaro 1986b and chapter 17.2.

8. In general my account relies on Brecht 1934, 1935, 1948, 1994a,c; Bentley 1966b; Hecht 1968; Hiley 1981; Willett and Manheim 1994, vi–xxii, 162–200.

9. Bentley 1966b, 14, 35; Hecht 1968, 180; Willett and Manheim 1994, ix, 162–64.

10. Willett and Manheim 1994, ix.

11. Bentley 1966b 154; Brecht 1953; Hecht 1968, 186; Willett and Manheim 1994, xiii–xvi, 160.

12. Willett and Manheim 1994, xvi.

13. Hecht 1968, 186; Willett and Manheim 1994, xviii.

14. Hecht 1968, 186.

15. Hecht 1968, 188.

16. Hecht 1968, 189.

17. Hecht 1968, 189; Simoncelli 1992, 121.

18. Hecht 1968, 186–92.

19. Hiley 1981, x, 1–4, 224.

20. Brecht 1994b, 7; in this section, subsequent references to this edition are given in parentheses in the text.

21. Favaro 16: 272–73.

22. Favaro 16: 450–52, 475–76.

23. Bentley (1966b, 9) and McMullin (1998, 271–77) have pointed this out, although in the context of different discussions.

24. Bentley (1966b, 9–10) has also pointed this out, although he stresses mostly Brecht's overlooking the importance of mathematics.

25. For some relevant arguments and references, see Finocchiaro 1980 (202–23), 1997 (344–48).

26. Bentley 1966b, 11.

27. Bentley 1966b, 11.

28. Bentley 1966b, 12.

29. Brecht 1994c, 117–18; cf. Hecht 1968, 13.

30. Brecht 1994c, 118; cf. Hecht 1968, 13.

31. For example, Genovesi 1966, 357–67.

32. Brecht 1994c, 118; cf. Hecht 1968, 14–15.

33. Brecht 1994c, 119; cf. Hecht 1968, 14–15.

34. Willett and Manheim 1994, xviii–xix.

35. Brecht 1994c, 126; cf. Hecht 1968, 12–13.

36. Brecht 1994c, 130; cf. Hecht 1968, 32–37.

37. Brecht 1994c, 127; cf. Hecht 1968, 13.

38. Brecht 1994c, 131 (italics added); cf. Hecht 1968, 37.

39. Cf. Galileo's *Dialogue,* in Favaro 7: 82–95, 289–90, 385–99; Finocchiaro 1980 (109–10, 125–26, 130–31), 1997 (257–64, 354–55).

40. Bentley 1966b, 18–19.

41. Bentley 1966b, 20.

42. Brecht 1994c, 148–49; cf. Hecht 1968, 40–77.

43. Bentley 1996b, 21.
44. See Koestler 1959; and cf. Koestler 1960a,b, 1964, 1981a,b; Santillana and Drake 1959, 1960; Santillana 1960; Dubarle 1961.
45. Snow 1964, 1–51.
46. Shusterman 1975, 25.
47. Shusterman 1975, 26.
48. Koestler 1959, 14; in this section, subsequent references to this work are given in parentheses in the text.
49. Cf. chapters 14.1 and 17.1.
50. In such a "modernist" interpretation of Galileo and antimodernist criticism of him, Koestler appears to have been part of a movement of critics of science that included such authors as Burtt (1932) and Husserl (1936, 1962, 1970), and he may have been influenced by them. The possibility of their historical connections and intellectual similarities would be worth further exploration.
51. Koestler's account is found mostly in Koestler 1959, 432–66, and relevant passages are quoted below. But one should not miss an equally explicit statement he gives in a lengthy note, where he criticizes the fact that "some of Galileo's biographers are anxious to give the impression that the decree of 5 March was not caused by Galileo's persistent provocations, but the result of a coldly planned inquisitorial campaign to stifle the voice of science" (Koestler 1959, 596 n. 41).
52. Koestler 1959, 449; note also that here Koestler comes close to endorsing Mallet du Pan's old myth.
53. Koestler 1959, 453; Koestler (1959, 595 n. 39) credits Butterfield (1949, 63) for having coined the phrase "secret weapon" to describe Galileo's argument from tides.
54. This thesis seems explicit from the claims that "Galileo had done everything in his power to provoke a showdown" (Koestler 1959, 455) and that "his defeat was really due to the fact that he had been unable to deliver the required proof" (Koestler 1959, 464).
55. Actually, Galileo's idea about a geokinetic explanation of the tides went back to the 1590s; see, for example, Drake 1978, 35–38.
56. Quoted by Koestler (1959, 451) from Drake 1957, 165–67; cf. Favaro 12: 183–85.
57. The document may be read in Favaro 19: 305; Pagano 1984, 68–69; Finocchiaro 1989, 135–36.
58. Quoted from Finocchiaro 1989, 122; cf. Favaro 5: 381; italics added.
59. See Finocchiaro 1992b and chapter 13.3; cf. Finocchiaro 1992a.
60. Quoted by Koestler (1959, 452) from Santillana 1955a, 112–13; cf. Favaro 12: 226–27.
61. It is surprising to also find the compromise-refusal thesis in a work by Drake (1957), but Drake's version does not survive critical scrutiny any better than Koestler's; in his later work, Drake (1978, 1980) himself did not seem to rely on that thesis and may be taken to have effectively abandoned it. Other scholars, such as Lerner (1998a), who appear to suggest the same thesis are not really doing so.
62. In Favaro 5: 351–70, especially 368–69; Finocchiaro 1989, 70–86, especially 85; cf. Finocchiaro 1986a.

63. See the Letters to Castelli and to Christina, in Favaro 5: 281–88, 309–48, and in Finocchiaro 1989, 49–54, 87–118. Koestler's analysis of them (1959, 432–39) is another example of the straw-man fallacy; for a detailed criticism of his interpretation, see Finocchiaro 1986a.

64. For example, in Galileo to Dini, May 1615, in Favaro 12: 183–85.

65. Koestler 1959, 478; Koestler's Galilean quotation is from G. Galilei 1953b, 407; cf. Favaro 7: 423–24.

66. Here quoted from Finocchiaro 1989, 261–62; cf. Favaro 19: 341; cf. chapter 12.1.

67. For Marini (1850), see chapter 11.3; for Müller (1909a,b, 1911), see chapter 13.4.

68. For an indirect criticism of Koestler's approach, interpretations, and evaluation, see Finocchiaro 1980, 1997.

69. Quoted by Koestler (1959, 478) from G. Galilei 1953b, 364.

70. Soccorsi 1947, 75–78; chapter 14.3. Cf. Santillana and Drake 1959, 1960; Koestler 1960a.

71. I speak of "psychic analysis" and not of "psychoanalysis" in order to avoid the unnecessary confusion that the technical Freudian connotation of the latter term would inject into the discussion. Moreover, it should be clear that my criticism of Koestler at this point does not presuppose a methodological (or principled) rejection of psychological analysis, which can be helpful in historiography; I am objecting rather to the innuendo in Koestler's account.

CHAPTER 16

1. There are three editions of the book, all with the same pagination: the first (Paschini 1964a) makes up volumes 1 and 2 of the three volumes of *Miscellanea Galileiana* and is the only one that includes Lamalle's introduction; the second (Paschini 1964b) is a freestanding edition in two volumes, without Lamalle's introduction but otherwise identical to the first; both were published by the Pontifical Academy of Sciences and have continuous pagination from the first to the second volume; the third (Paschini 1965) was edited by Michele Maccarrone with an introduction by him and is a facsimile reprint of the previous editions in one volume and with the typographical errors corrected.

2. My account relies on *Atti del convegno di studio su Pio Paschini nel centenario della nascita* 1979; Bertolla 1979; Blackwell 1998a, 361–66; Brandmüller 1992b, 20 n. 27; Fabris 1986, 8–10; Fantoli 1996, 503–5, 523–28; Lamalle 1964; Maccarrone 1979a,b; Nonis 1979; Simoncelli 1992; Tamburini 1990, 128–29. Other essential sources are Paschini's manuscripts held at the library of the Udine Seminary, especially the original manuscript of his book on Galileo and his correspondence with his friend Giuseppe Vale; I have not consulted these manuscripts but rely on the efforts of these scholars (especially Bertolla, Maccarrone, and Simoncelli), as found in their published works.

3. Maccarrone 1979a, 49.

4. Maccarrone 1979a, 52.

5. Broderick 1987, 566; Ghiberti 1992, 218.

6. Simoncelli 1992, 19.

7. Simoncelli 1992, 20.
8. For such details, see Maccarrone 1979a, 49–60.
9. Maccarrone 1979a, 64; for more details about the points mentioned in this paragraph, see pp. 60–71.
10. Simoncelli 1992, 23.
11. Simoncelli 1992, 34–38; for more details on the other points mentioned in this paragraph, see Maccarrone 1979a, 71–75.
12. Simoncelli, 1992, 39; cf. Paschini 1940.
13. Translated from the Italian text of Paschini 1941, as transcribed and published by Bertolla (1979, 175–76) from the Paschini manuscripts at the Udine Seminary library.
14. Cf. Mercati 1942; Paschini 1942; Blumenberg 1987c, 371.
15. Bertolla 1979, 176 n. 2.
16. Maccarrone 1979a, 78; Simoncelli 1992, 59–60.
17. Maccarrone 1979a, 78.
18. For further details about the facts summarized in this paragraph, see Maccarrone 1979a, 78–79; Simoncelli 1992, 59–79.
19. For more details, see Maccarrone 1979a, 79–84; Simoncelli 1992, 59–79.
20. Here and elsewhere in this letter, the ellipses are in the text as published by Maccarrone (1979b, 202–3).
21. Here Maccarrone states parenthetically that "in the manuscript this last clause is erased" (1979b, 203).
22. Translated from the Italian text of Paschini 1946a, as transcribed and published by Maccarrone (1979b, 202–3). Slightly more abridged versions of this letter were also quoted by Maccarrone (1979a, 82–83 n. 112) and Simoncelli (1992, 70–71), but the text I have quoted and translated here is itself abridged.
23. Cf. Galileo's "Considerations on the Copernican Opinion," especially part 3, sections 6–7, in Favaro 5: 368–69 and Finocchiaro 1989, 85.
24. Cf. Favaro 19: 348–60; Finocchiaro 1989, 262–76.
25. Translated from the Italian text of Paschini 1946b, as transcribed and published by Simoncelli (Milan: FrancoAngeli, 1992), 72–73, from the Paschini manuscripts at the Udine Seminary library; an abridged version of this letter is printed by Bertolla (1979, 180–81).
26. Cardinal Francesco Marchetti Selvaggini was the Vatican overseer of the Lateran University (or Roman Seminary) of which Paschini was president; cf. Simoncelli 1992, 76.
27. Cf. Müller 1909a,b, 1911, and chapter 13.4.
28. Cf. Paschini 1943 and chapter 14.2.
29. Translated from the Italian text of Paschini 1946c, as transcribed and published by Simoncelli (Milan: FrancoAngeli, 1992), 77–78, from the Paschini manuscripts at the Udine Seminary library; an abridged version of this letter is printed by Bertolla (1979, 181–82).
30. Maccarrone 1979b, 204; Simoncelli 1992, 79.
31. Paschini 1950; cf. Simoncelli 1992, 100–101.
32. Maccarrone 1979a, 87–88; Simoncelli 1992, 128–39; Fantoli 1996, 525–26 n. 41.
33. Fenu 1964; Gusdorf 1969, 1: 133–34; Masi 1964; Maccarrone 1979a, 88 n. 28.

34. For more information on the events reported in this paragraph, see Maccarrone 1979a, 88–90; Simoncelli 1992, 109–39; Fantoli 1996, 526–27; Beltrán Marí 1998, 95.

35. Lamalle 1964, viii.

36. Apparently the published introduction was a revised draft since Lamalle's first draft gave an even more negative and severe assessment of Paschini's work, and Maccarrone objected to it; cf. Maccarrone 1979b, 212–13 n. 130.

37. Lamalle 1964, xii.

38. Lamalle 1964, xii.

39. Lamalle 1964, xiii.

40. Fantoli 1996 (528 n. 42), 2001 (734 n.2); for more relevant information, see Koven 1980.

41. Fantoli 1996 (529 n. 44), 2001 (734 n. 2).

42. For the account that follows, I rely very heavily on Maccarrone 1979a, 90–93, which was also published in Maccarrone 1979b, 214–18; see also Beltrán Marí 1998, 99–100; Blackwell 1998a, 361–66; Fantoli 1996, 505–6, 528–31; Gusdorf 1969, 1: 133; *Le Monde* (Paris), 31 October 1964; Mayaud 1997, 297–98; Simoncelli 1992, 128–39.

43. Cf. Mayaud (1997, 297), who refers to *Acta Synodalia Sacrosancti Concilii Oecumenici Vaticani Secundi* (Rome, 1975), I, iii, 145.

44. Cf. Gusdorf (1969, 1: 133), who quotes *Le Monde* (Paris) of 31 October 1964.

45. Quoted by Maccarrone (1979a, 91) from *Acta Synodalia Sacrosancti Concilii Oecumenici Vaticani Secundi* (Rome, 1975), vol. 3, third period, part 6, general sessions 112–118, in session 114, 4 November 1964, p. 268.

46. Here quoted from Fantoli 1996, 528–29, where the passage is translated from the Latin original quoted by Maccarrone (1979a, 91), who quoted it from the archives of the Second Vatican Council.

47. Quoted by Maccarrone (1979a, 92) from the archives of the Second Vatican Council.

48. Maccarrone 1979a, 88; Simoncelli 1992, 35–36.

49. Here quoted from Blackwell 1998a, 365; cf. Maccarrone (1979a, 92–93), who quotes the Latin text from *Concilium Oecumenicum Vaticanum II: Constitutiones, Decreta, Declarationes* (Vatican City), p. 731, number 31; cf. Simoncelli (1992, 137–38), who translates the passage into Italian from "Constitutio Pastoralis de Ecclesia in Mundo Huius Temporis," in *Acta Apostolicae Sedis: Commentarium Officiale,* 1966, 59: 1025–1120, at p. 1054.

50. Maccarrone 1979a, 93; Simoncelli 1992, 138.

51. Maccarrone 1979b, 218.

52. Maccarrone 1979b, 217. In n. 141 on this page, Maccarrone wrote that in 1967 Paul VI, in connection with the question of the beatification of the Danish scientist Nicolaus Steno (1638–1686), was considering reopening Galileo's trial; the pope twice consulted Lamalle about this idea and received a negative recommendation; cf. also Maccarrone 1979a (88), 1979b (210 n. 127); Simoncelli 1992, 138–39.

53. Maccarrone 1979b, 217.

54. Maccarrone 1979b, 218.

55. *Atti del convegno di studio su Pio Paschini nel centenario della nascita* 1979.

56. Bertolla 1979; Maccarrone 1979a; Nonis 1979.

57. Bertolla 1979, 173.
58. Bertolla 1979, 185–208.
59. Quoted by Bertolla (1979, 193–94), who transcribed it from p. 385 of Paschini's manuscript held in the library of the Udine Seminary.
60. For Bellarmine's point, see Favaro 12: 172; Finocchiaro 1989, 68; and chapters 5.2 and 14.3. At this point, the printed book (Paschini 1965, 318 n. 70 [but really n. 71]) has a footnote containing some material unfavorable to Galileo; Bertolla (1979, 193–95) says nothing about this footnote; so although the unfavorable content suggests that this footnote was added by Lamalle, one would have to consult Paschini's original manuscript in order to be sure.
61. Paschini 1965, 317–18; cf. Bertolla 1979, 193–95. It is worth repeating here that the pagination is identical for the three editions of Paschini's book (1964a,b, 1965); their text is also identical, except for some typographical errors corrected by Maccarrone in Paschini 1965.
62. See, for example, Finocchiaro 1980, 1986a, 1997.
63. Quoted by Paschini (1965, 339) from Riccioli 1668, 33.
64. Quoted by Bertolla (1979, 196) from p. 412 of the manuscript of Paschini's book held at the Udine Seminary.
65. This work appears to be Lenoble 1957.
66. Paschini 1965, 339–40 n. 53. It is puzzling that in his comparison, Bertolla (1979, 196) mentioned the omission of Paschini's sentence, but did *not* indicate that the published book included this comment by Lamalle in a footnote.
67. Quoted by Bertolla (1979, 196) from p. 413 of the manuscript of Paschini's book held at the Udine Seminary.
68. Paschini 1965, 341; cf. Bertolla 1979, 196.
69. Quoted by Bertolla (1979, 203) from p. 666 of the manuscript of Paschini's book held at the Udine Seminary.
70. That is, Delannoy 1906, which was a review of Vacandard 1905b; cf. Simoncelli 1992, 125 n. 58.
71. We have seen that Voltaire's view (chapter 6.3) came close to such a caricature.
72. Paschini 1965, 548–49; cf. Bertolla 1979, 203–4.
73. Lenoble 1957, 475–76; cf. Paschini 1965, 549, and Bertolla 1979, 204.
74. Paschini 1965, 548 n. 41; cf. Bertolla 1979, 204.
75. Citing p. 338, which was a typographical error and should have been p. 358; cf. Delannoy 1906; Bertolla 1979, 203; Paschini 1965, 549 n. 41; and Maccarrone 1979b, 212–13 n. 130.
76. Paschini 1965, 549 n. 41; cf. Bertolla 1979, 204.
77. Beltrán Marí 1998, 95–99; Blackwell 1998a, 364–65; Fabris 1986, 8–10; Nonis 1979, 168–69 n. 1; Simoncelli 1992, 109–28.
78. Maccarrone 1979b, 212–13 n. 130; Tamburini 1990; Brandmüller 1992b, 20 n. 27.
79. Simoncelli 1992, 142–43, expresses this legitimate complaint, and mentions several examples, one of which is Zoffoli 1990.
80. Besides Maccarrone 1979a,b, see Maccarrone 1963.
81. Maccarrone 1979b, 213 n. 130.
82. Quoted in Maccarrone 1979b, 213 n. 130 from Paschini 1965, 549 n. 41.
83. Maccarrone 1979b, 213 n. 130. Here Maccarrone gives no hint of what he is referring to regarding Leo XIII, but he probably has in mind the statement in

Leo's encyclical *Saepenumero Considerantes* to the effect that "the first law of history is not to dare tell any falsehood, and then to dare tell the whole truth" (quoted in Jaugey 1888, 7).

84. Nonis 1979, 168 n. 1.
85. Maccarrone 1979b, 213 n. 130.

CHAPTER 17

1. Here I cite John Paul II 1979b, but the section numbers are identical in other editions, such as John Paul 1979a,c,d.
2. Cf. respectively chapters 14.1 and 16.2.
3. John Paul II 1979b, section 1.
4. John Paul II 1979b, section 2.
5. John Paul II 1979b, section 5.
6. John Paul II 1979b, section 5.
7. John Paul II 1979b, sections 6–8.
8. John Paul II 1979b, section 6.
9. To be sure, Pope Paul VI had mentioned Galileo by name in a favorable context on 10 June 1965 at a speech to the Eucharistic Congress in Pisa; however, there was no reference to the trial and condemnation, but rather Galileo was named as one of the great minds of Tuscany, together with Dante and Michelangelo; cf. Fantoli 1996, 530 n. 44; 2001, 734 n. 2; Brandmüller 1992b, 21.
10. John Paul II 1979b, section 6.
11. John Paul II 1979b, section 6.
12. John Paul II 1979b, section 6.
13. Gemelli 1942b; cf. chapter 14.1.
14. Cf. Pius XII 1943.
15. See, for example, De Santis 1979.
16. Sharratt 1994, 209; Rowland 2003, 227.
17. For good examples of the concrete approach, see Brooke 1991, 1996, 1998; Brooke and Cantor 1998.
18. "Vatican Reviewing Galileo's Conviction for Heresy," *New York Times,* 23 October 1980; "Vatican Opens Study on Clearing Galileo," *Los Angeles Times,* 24 October 1980.
19. Cf. Garrone 1984 and Beltrán Marí 1998, 101 n. 8.
20. Here Segre (1997, 501) translated Casaroli's clause "Non di revisione di un processo si tratta o di riabilitazioni" as "This is not the review of a trial or a rehabilitation," in which I have inserted "to be."
21. Quoted from Casaroli 1981, as translated by Segre (1997, 500–501).
22. Poupard 1983, 1984, 1987.
23. See, for example, Garrone 1984; Poupard 1987, ix–x.
24. See, for example, Golden 1984.
25. Segre 1997.
26. D'Addio 1985; cf. D'Addio 1983, 1984b, 1993.
27. D'Addio 1985, 51–52.
28. Poupard 1983, 1984, 1987; cf. Poupard 1992a–c, 1994a,b.
29. See, for example, the criticism in Finocchiaro 1986c.

30. Coyne, Heller, and Zycínski 1985; cf. Baldini and Coyne 1984; Coyne 1992.
31. Feyerabend 1985; cf. Feyerabend 1975, 1987, 1988. For a criticism of Feyerabend's view, see Finocchiaro 1986b, 2001b.
32. Pedersen 1983b, which was serialized as vol. 1, no. 1; cf. also Pedersen 1983a, 1985.
33. Baldini and Coyne 1984, which was serialized as vol. 1, no. 2.
34. Cf. Soccorsi 1947 and chapter 14.3; Duhem 1908 and chapter 13.3; Baldini 1984, 1990, 1992; and Westfall 1989a and the later discussion in this section.
35. Cf. Finocchiaro 1988a.
36. Coyne, Heller, and Zycínski 1985, which was serialized as vol. 1, no. 3; cf. Finocchiaro 1986b.
37. Zycínski 1988, which was serialized as vol. 1, no. 4; cf. Finocchiaro 1988b.
38. Westfall 1989b, which was serialized as vol. 1, no. 5; cf. the critical analysis in Finocchiaro 1990.
39. This last of Westfall's theses is, of course, his refutation of Redondi (1983, 1985a, 1987), which Westfall did with grace and goodwill.
40. Cf. Fantoli 1993, 1996, 1997.
41. For some details, see Finocchiaro 1995b.
42. This is even more obvious from Fantoli 2001, 2003.
43. Pagano 1984; for more details, see Finocchiaro 1985.
44. For another good illustration of these standards, see Pagano's (1994) critical review of Brandmüller and Greipl 1992.
45. See Fabris 1986.
46. See, for example, Fabris 1992a–d.
47. For a statement about this division of labor, see Brandmüller and Greipl 1992, 4.
48. The copyright page also bears an imprimatur that reads: "I am very glad to accept the work of Walter Brandmüller and Egon Johannes Greipl, published under the auspices of the Pontifical Academy of Sciences, as a contribution to the efforts of the Pontifical Commission on Galilean Studies. [Signed:] Paul Cardinal Poupard, Coordinator of the Efforts of the Pontifical Commission on Galilean Studies, Rome 2 October 1992." Several years earlier, Poupard had replaced Garrone as chair of the commission.
49. But the documentation is not beyond criticism, as Pagano (1994) has demonstrated.
50. Brandmüller 1982, 1987, 1992b.
51. Brandmüller 1992b, 161–84.
52. Brandmüller 1992b, 193–98.
53. Brandmüller 1992b, 193.
54. Brandmüller 1992b, 193.
55. Brandmüller 1992b, 195.
56. Cf. Soccorsi 1947; chapter 14.3; and Finocchiaro 1980, 1986a.
57. Paschini 1943; chapter 16.2; cf. Finocchiaro 1992a, 1997.
58. In Finocchiaro 1989, 119–33; cf. chapter 15.2.
59. Brandmüller 1992b, 196.
60. I cite Poupard 1992b, but the section numbers are identical in other editions, such as Poupard 1992a,c.
61. Poupard 1992b, section 1.

62. Poupard 1992b, section 2.
63. Brandmüller 1992b, 137–38.
64. Poupard 1992b, section 3; Brandmüller 1992b, 134–38.
65. Wallace 1987, 59–60; cf. Poupard 1984, 95–96; Wallace's conclusion is based on his own analysis of Galileo's theory of demonstration in MS 27 (cf. Wallace 1984a, 1992a,b) and on the analysis of the arguments in the *Dialogue* provided by Finocchiaro 1980.
66. Poupard 1992b, sections 3–4.
67. Brandmüller 1992b, 161–92.
68. For example, Poupard (1992b, section 3) says that "in 1741, in the face of the optical proof of the fact that the earth revolves round the sun, Benedict XIV had the Holy Office grant an imprimatur to the first edition of the *Complete Works of Galileo*"; we saw in chapter 7.1, however, that the rationale underlying the imprimatur for Galileo's *Dialogue* was the plan to change its geokinetic language from categorical to hypothetical; hence this imprimatur was not, as Poupard goes on to say in the next paragraph, an "implicit reform of the 1633 sentence" but rather a kind of reaffirmation of it, "correcting" the *Dialogue* in the way that the Index's decree of 1620 "corrected" Copernicus's book. Poupard (1992b, section 4) also says that "this implicit reform of the 1633 sentence became explicit in the decree of the Sacred Congregation of the Index that removed from the 1757 edition of the Catalogue of Forbidden Books works favoring the heliocentric theory"; but we have seen (in chapter 7.2) that the 1757 decision was still implicit and indirect, so much so that Galileo's *Dialogue* was still left on the *Index* and Settele's *Astronomy* in 1820 could run into difficulties (cf. chapter 10); moreover, the 1757 decision amounted to dropping the clause "all books teaching the earth's motion and sun's immobility" from the *Index,* and to describe this action as a "decree . . . that removed . . . works favoring the heliocentric theory" amounts to a sophistical use of equivocation; for what was being removed was not the listed heliocentric works (which would imply removing Galileo's *Dialogue,* Copernicus's R*evolutions,* etc.), but rather the clause "all heliocentric works" (which in fact left those specific works in the *Index*). Referring to the Settele affair, Poupard (1992b, section 4) asserts that "the unjustly censored author lodged an appeal with Pope Pius VII, from whom in 1822 he received a favorable opinion"; and here Poupard's chronology is careless at best, for we have seen (in chapter 10.1) that the favorable decision on Settele's personal case came in 1820, although it was indeed in 1822 that the general Inquisition ruling came; however, the 1822 decision was not implemented until the 1835 *Index* (and not in 1846, as Poupard misstates in the next paragraph).
69. Poupard 1992b, section 5.
70. Brandmüller 1992b, 138–41.
71. Here I cite John Paul II 1992a, but the section and paragraph numbers are identical in other editions, such as John Paul II 1992b,c.
72. John Paul II 1992a, p. 1, section 2.
73. John Paul II 1992a, p. 1, section 2.
74. From *Journal of the Franklin Institute,* vol. 221, no. 3, March 1936.
75. John Paul II 1992a, p. 2, section 14.
76. John Paul II 1992a, sections 4–12.

77. John Paul II 1992a, section 4.
78. John Paul II 1992a., section 4.
79. John Paul II 1992a, section 2, paragraph 2.
80. John Paul II 1992a, section 12.
81. John Paul II 1992a, p. 2, section 5, paragraph 4.
82. John Paul II 1992a, p. 2, section 6, paragraph 1.
83. See, for example, Westman 1980, 1986; Finocchiaro 1989, 7–8.
84. John Paul 1992a, sections 7–8.
85. Soccorsi 1947, 59–60; cf. chapter 14.3.
86. John Paul II 1992a, section 7, paragraph 2.
87. John Paul II 1992a, section 9, paragraph 1.
88. John Paul II 1992a, section 9, paragraph 3.
89. John Paul II 1992a, section 10.
90. See for example, chapter 6.3.
91. See chapter 4.3.
92. John Paul II 1992a, section 12, paragraph 1.
93. John Paul II 1992a, section 12.

EPILOGUE

1. For example, Di Canzio 1996, 321–30; Segre 1997, 1999; Fantoli 2001; Coyne (forthcoming).
2. Beltrán Marí 1998; Benítez 1999b.
3. Poupard 1992b; Brandmüller and Greipl 1992; chapter 17; Segre (1998) has stressed this unending character of the Galileo affair.
4. Motzkin 1989 may be viewed as a first step in this direction.
5. Cf. Morpurgo-Tagliabue 1984a,b; D'Addio 1984a; Pagano 1984, 43–48; Ferrone and Firpo 1985a,b, 1986; Redondi 1985b; Westfall 1989b; Artigas 2001; Martínez 2001; Mateo-Seco 2001.
6. Melissa Shogren, Redmond, WA, 13 October 2002, in Readers' Comments on Dava Sobel's *Galileo's Daughter,* at Amazon.com (consulted on 1 November 2002).
7. This is the impression I get from the comments titled "Penguin Targets Female Buyers (Was it Title VII?)," by M. Spencer, New York, dated 5 February 2001, in Readers' Comments on Dava Sobel's *Galileo's Daughter,* at Amazon.com (consulted on 1 November 2002).

Accarisius, J. 1637. *Terrae Quies, Solisque Motus Demonstratus Primum Theologicis, tum Plurimis Philosophicis Rationibus*. Rome.

Acloque, P. 1982. "L'Histoire des Expériences pour la Mise en Evidence du Mouvement de la Terre." *Cahiers d'Histoire et de Philosophie des Sciences*, new series, no. 4, 1–141.

Adams, J. Q. 1843. *An Oration, Delivered before the Cincinnati Astronomical Society*. Cincinnati.

Agassi, J. 1957. "Duhem versus Galileo." *British Journal for the Philosophy of Science* 8: 237–48.

———. 1971. "On Explaining the Trial of Galileo." *Organon* 8: 137–66.

Agnani, G. D. 1734. *Philosophia Neo-Palaea*. Rome.

Agosti, V. 1964. "Galilei visto da Husserl." *Giornale di metafisica* 19: 779–96.

Albèri, E., ed. 1842–1856. *Le opere di Galileo Galilei*. 15 tomes in 16 vols. Florence.

[Allan-Olney, Mary]. 1870. *The Private Life of Galileo*. London.

Amort, E. 1734. *Philosophia Pollingana*. Venice.

Andres, G. 1776. *Saggio della filosofia del Galileo*. Mantua.

———. 1782–1799. *Dell'origine, progresso e stato attuale d'ogni letteratura*. 7 vols. Parma.

Anfossi, Filippo. 1820a. *Le fisiche rivoluzioni della natura*. 2nd. ed. Rome.

———. 1820b. (May.) "Nota aggiunta dal R.mo P. Anfossi Maestro del S.P.A. alla ristampa del suo opuscolo *Le fisiche rivoluzioni*." In Anfossi 1820a, 93ff.; in Maffei 1987, 536–37; in Brandmüller and Greipl 1992, 182–84.

———. 1820c. (August.) "Motivi per cui il P. Maestro del S. Palazzo Apostolico ha creduto, e crede non doversi permettere al Signor Canonico Settele d'insegnare come tesi e non come semplice ipotesi a tenore del Decreto del 1620 la mobilità della Terra, e la stabilità del Sole nel centro del Mondo." In Maffei 1987, 548–54; in Brandmüller and Greipl 1992, 310–17.

———. 1820d. (September.) *Ragioni per cui il P. Maestro del S. Palazzo Apostolico ha creduto e crede che non si può permettere la Stampa del Manoscritto del Signor Canonico Set-*

tele, che incomincia "Movendosi la Terra intorno al Sole." Rome. In Maffei 1987, 452–63; in Brandmüller and Greipl 1992, 336–49.

———. 1820e. (November.) "Breve risposta al lunghissimo voto fatto stampare contro di lui dal P. Maurizio Benedetto Olivieri commissario e consultore." In Brandmüller and Greipl 1992, 380–84.

———. [1822]. *Se possa difendersi, ed insegnare, non come semplice ipotesi, ma come verissima e come tesi, la mobilità della terra, e la stabilità del sole da chi ha fatta la professione di fede di Pio IV.* Rome.

Angeli, S. degli. 1667. *Considerationi sopra la forza di alcune ragioni fisicomattematiche.* Venice.

———. 1668a. *Seconde considerationi sopra la forza dell'argomento fisicomattematico.* Padua.

———. 1668b. *Terze considerationi sopra una lettera del molto illustre et eccellentissimo Signor Gio. Alfonso Borelli.* Venice.

———. 1669. *Quarte considerationi sopra la confermatione d'una sentenza del Sig. Gio. Alfonso Borelli . . . e sopra l'Apologia del M.R.P. Gio. Battista Riccioli.* Padua.

L'apertura degli archivi del Sant'Uffizio romano. 1998. Rome: Accademia Nazionale dei Lincei.

Arduini, C. 1864. *La primogenita di Galileo Galilei.* Florence.

Argoli, A. 1644. *Pandosion Sphaericum.* Padua.

Ariew, R., and D. Garber, trans. and eds. 1989. *Philosophical Essays [of Leibniz].* Indianapolis: Hackett.

Arnauld, A. 1691. "Difficultés Proposées à M. Steyaert: IX Partie: XCIV Difficulté: Quinzième Exemple [La condamnation des livres de Galilée]." In Arnauld 1775–1783, 9: 307–14.

———. 1775–1783. *Oeuvres de Messire Antoine Arnauld.* 49 vols. Paris.

Artigas, M. 2001. "Un nuovo documento sul caso Galileo: EE 291." *Acta Philosophica* 10: 199–214.

Atti del convegno di studio su Pio Paschini nel centenario della nascita, 1878–1978. [1979]. Udine: Pubblicazioni della Deputazione di Storia Patria per il Friuli.

Auzout, Adrien. 1665. *Lettre à Monsieur l'Abbé Charles sur le "Ragguaglio di due nuove osservazioni" da* [sic] *Giuseppe Campani.* Paris. In *Memoires de l'Academie Royale des Sciences, depuis 1666 jusqu'à 1699* (Paris, 1729–1733), vol. 7, part 1, 1–68.

Bacon, F. 1624. *De Dignitate, ed Augmentis Scientiarum.* Paris.

Bailly, J. S. 1785. *Histoire de l'Astronomie Moderne.* New ed. 3 vols. Paris.

Baldigiani, Antonio. 1678. Baldigiani to Viviani, 18 July. In Favaro 1887b, 143–44.

———. 1693. Baldigiani to Viviani, 25 January. In Favaro 1887b, 155–56.

Baldini, Ugo. 1984. "L'astronomia del cardinale Bellarmino." In Galluzzi 1984, 292–305.

———. 1990. "Bellarmino tra vecchia e nuova scienza." In *Roberto Bellarmino,* ed. G. Galeota, 2: 629–96. Capua: Istituto Superiore di Scienze Religiose. Rpt. in Baldini 1992, 305–44.

———. 1992. *Legem Impone Subactis.* Rome: Bulzoni.

———. 1994. "Fabroni, Angelo." *Dizionario biografico degli italiani* 44: 2–12.

———. 1996a. "La formazione scientifica di G. B. Riccioli." In Pepe 1996, 123–82.

———. 1996b. "Sul contesto storico e scientifico del caso Settele." *Pontificia Academia Scientiarum, Commentarii* 3(34): 21–58. Vatican City: Ex Aedibus Academicis in Civitate Vaticana.

————. 2000a. "Filosofia naturale e scienza nell'Accademia di religione Cattolica (1799–1846)." In *Rosmini e Roma,* ed. L. Malusa and P. De Lucia, 173–225. Stresa: Centro Internazionale di Studi Rosminiani.

————. 2000b. *Saggi sulla cultura della Compagnia di Gesù (secoli XVI–XVIII).* Padua: CLEUP Editrice.

————. 2000c. "Teoria boscovichiana, newtonismo, eliocentrismo: dibattiti nel Collegio Romano e nella Congregazione dell'Indice a metà Settecento." In Baldini 2000b, 281–347.

Baldini, U., and G. V. Coyne, eds. 1984. *The Louvain Lectures (Lectiones Lovanienses) of Bellarmine and the Autograph Copy of His 1616 Declaration to Galileo.* Vatican City: Specola Vaticana.

Baldini, U., and L. Spruit. 2001. "Nuovi documenti galileiani degli archivi del Sant'Ufficio e dell'Indice." *Rivista di storia della filosofia* 56: 661–99.

Baliani, G. B. 1638. *De Motu Naturali Gravium Solidorum.* Genoa.

————. 1646. *De Motu Naturali Gravium Solidorum et Liquidorum.* Genoa.

Barberini, Antonio. 1633a. A. Barberini to Modena's Inquisitor, 2 July. In Favaro 15: 169.

————. 1633b. A. Barberini to Siena's Inquisitor, 2 July. In Pagano 1984, 244–45.

————. 1633c. A. Barberini to Venice's Inquisitor, 2 July. In Riccioli 1651, 2: 497.

Barberini, Maffeo. 1620a. M. Barberini to Galileo, 28 August. In Favaro 13: 48–49.

————. 1620b. "Adulatio Perniciosa." In M. Barberini 1634, 278–82; in G. Galilei 1656, vol. 1; in M. Barberini 1726; in Venturi 1818–1821; in Pieralisi 1875, 22–25.

————. 1634. *Poemata.* Antwerp.

————. 1726. *Poemata.* Oxford.

Barbier, Antoine A. 1811. Barbier to Napoleon, 12 March. In Favaro 1887a, 198.

Barenghi, G. 1638. *Considerazioni sopra il Dialogo.* Pisa.

Baretti, Giuseppe. 1757. *The Italian Library.* London.

Barker, P., and B. R. Goldstein. 1998. "Realism and Instrumentalism in Sixteenth Century Astronomy." *Perspectives on Science* 6: 232–58.

Barni, J. 1862a. "Huitième Leçon: Jordano Bruno—Campanella—Vanini—Galilée." In Barni 1862b, 206–37.

————. 1862b. *Les Martyrs de la Libre Pensée.* Geneva.

Baruzi, J. 1907. *Leibniz et l'Organisation Religieuse de la Terre d'après des Documents Inedits.* Paris: Alcan.

Bellarmine, Robert. 1615. Bellarmine to Foscarini, 12 April. In Berti 1876a, 121–25; in Favaro 12: 171–72; in Finocchiaro 1989, 67–69.

Belli, T. P. 1633. "Aviso per il Santo Uffitio," Rimini, 27 September. In Dibner 1967, 169.

Beltrán Marí, A. 1998. "'Una Reflexión Serena y Objectiva'." *Arbor* 160(629): 69–108.

————. 2001a. *Galileo, Ciencia y Religión.* Barcelona: Paidós.

————. 2001b. "Tratos Extrajudiciales, Determinismo Procesal y Poder." In Montesinos and Solís 2001, 463–90.

Bemis, S. F. 1956. *John Quincy Adams and the Union.* New York: Knopf.

Benedict XIV. 1748. *Bref au Grand Inquisiteur d'Espagne.* N.p. 31 July.

Benítez, H. H. 1999a. *Ensayos sobre Ciencia y Religión.* Santiago, Chile: Bravo y Allende.

―――. 1999b. "El Mito de la Rehabilitación de Galileo." In Benítez 1999a, 85–110.

Bentley, E., ed. 1966a. *Galileo/Bertolt Brecht.* New York: Grove Press.

―――. 1966b. "Introduction: The Science Fiction of Bertolt Brecht." In Bentley 1966a, 7–42.

Benzenberg, J. F. 1804. *Versüche über das Gesetz des Falls, über den Widerstand der Luft und über die Umdrehung der Erde.* Dortmund.

―――. 1845. *Versüche über die Umdrehung der Erde aufs neue Berechnet.* Düsseldorf.

Bérault-Bercastel, A. H. de. 1778–1790. *Histoire de l'Église.* 24 vols. Paris.

―――. 1790. "L'Affaire de Galilée avec l'Inquisition." In Bérault-Bercastel 1778–1790, 21: 140–46; rpt. in Tiraboschi 1782–1797, 10: 362–64 n. 1; English trans. in Madden 1863, 144–47.

Beretta, F. 1999a. "Le Procès de Galilée et les Archives du Saint-Office." *Revue des Sciences Philosophiques et Théologiques* 83: 441–90.

―――. 1999b. "La Siège Apostolique et l'Affaire Galilée." *Roma moderna e contemporanea* 7: 421–61.

―――. 2001. "Urbain VIII Barberini Protagoniste de la Condamnation de Galilée." In Montesinos and Solís 2001, 549–74.

Berggren, L., and L. Sjöstedt. 1996. *L'ombra dei grandi.* Rome: Artemide Edizioni.

Bergier, N. S. 1788–1790a. "Galilée." Rpt. in Bergier 1823, vol. 3, column 464.

―――. 1788–1790b. "Sciences Humaines." Rpt. in Bergier 1823, vol. 7, 365–70.

―――. 1823. *Dictionnaire de Théologie: Extrait de l'Encyclopédie Méthodique.* 8 vols. Toulouse.

Bernini [or Bernino], Domenico. 1709. *Historia di tutte l'heresie.* Vol. 4. Rome.

Berti, Domenico. 1876a. *Copernico e le vicende del sistema copernicano in Italia.* Rome.

―――. 1876b. *Il processo originale di Galileo Galilei pubblicato per la prima volta.* Rome.

―――. 1877. "La critica moderna e il processo di Galileo Galilei." *Nuova antologia,* year 12, 2nd series, vol. 4, 5–34.

―――. 1878. *Il processo originale di Galileo Galilei: Nuova edizione accresciuta, corretta e preceduta da un'avvertenza.* Rome.

―――. 1882. "Antecedenti al processo galileiano e alla condanna della dottrina copernicana." *Atti della R. Accademia dei Lincei,* 1881–1882, 3rd series, *Memorie della classe di scienze morali, storiche e filosofiche* 10: 49–96.

Bertolla, Pietro. 1979. "Le vicende del 'Galileo' di Paschini (dall'Epistolario Paschini-Vale)." In *Atti del convegno di studio su Pio Paschini* 1979, 173–208.

Bertoloni Meli, D. 1988. "Leibniz on the Censorship of the Copernican System." *Studia Leibnitiana* 20: 19–52.

―――. 1992. "St. Peter and the Rotation of the Earth." In *The Investigation of Difficult Things,* ed. P. T. Harman and A. E. Shapiro, 421–48. Cambridge: Cambridge University Press.

Bertoni, G. 1937. "Tiraboschi, Girolamo." *Enciclopedia italiana* 33: 908.

Besomi, O., and M. Helbing, eds. 1998. *Dialogo sopra i due massimi sistemi del mondo, tolemaico e copernicano.* 2 vols. Padua: Antenore.

Biagioli, M. 1993. *Galileo Courtier.* Chicago: University of Chicago Press.

―――. 1996. "Playing with the Evidence." *Early Science and Medicine* 1: 70–105.

―――. 2000. "Replication or Monopoly?" *Science in Context* 13: 547–90.

―――. 2003. "Stress in the Book of Nature." *MLN: Modern Language Notes* 118: 557–85.

Biancani, G. 1620. *Sphaera Mundi.* Bologna.

Billot, L. 1898. *Tractatus de Ecclesia Christi.* Rome.

Biot, Jean B. 1816. "Galilée." In *Biographie Universelle* (52 vols. Paris, 1811–1828), 16: 318–37. Rpt. in Biot 1858b, 2: 427–50.

———. 1858a. "Une Conversation au Vatican." *Journal des Savants,* 137–42. Rpt. in Biot 1858b, 2: 451–59.

———. 1858b. *Mélanges Scientifiques et Littéraires.* 3 vols. Paris.

———. 1858c. "La Verité sur le Procès de Galilée." *Journal des Savants,* 397–406, 461–71, 543–51, and 607–20. Rpt. in Biot 1858b, 3: 1–49.

Blackburne, F. 1770. *The Confessional.* 3rd ed. London.

Blackwell, R. J. 1991. *Galileo, Bellarmine, and the Bible.* Notre Dame: University of Notre Dame Press.

———, ed. and trans. 1994. *A Defense of Galileo, the Mathematician from Florence, . . . by Thomas Campanella.* Notre Dame: University of Notre Dame Press.

———. 1998a. "Could There Be Another Galileo Case?" In Machamer 1998, 348–66.

———. 1998b. *Science, Religion, and Authority.* Milwaukee: Marquette University Press.

Blind, K. 1889. "Giordano Bruno and the New Italy." *The Nineteenth Century* 26: 106–19.

Blumenberg, H. 1987a. "Experiences with the Truth: Galileo." In Blumenberg 1987b, 386–430.

———. 1987b. *The Genesis of the Copernican World.* Trans. R. M. Wallace. Cambridge, MA: MIT Press.

———. 1987c. "Not a Martyr for Copernicanism: Giordano Bruno." In Blumenberg 1987b, 353–85.

Bocchini Camaiani, B., and A. Scattigno, eds. 1998. *Anima e paura.* Macerata: Quodlibet.

Boffito, G. 1943. *Bibliografia galileiana, 1896–1940.* Rome: Libreria dello Stato.

Bonora, T., ed. 1872. *Di Copernico e di Galileo: Scritto postumo del P. Maurizio-Benedetto Olivieri.* Bologna.

Borgato, M. T. 1996. "La prova fisica della rotazione della Terra e l'esperimento di Guglielmini." In Pepe 1996, 201–61.

Borgato, M. T., and A. Fiocca. 1994. "Introduzione." In Guglielmini 1994, 1–14.

Bouix, D. 1869. *Tractatus de Papa, Ubi et de Concilio Oecumenico.* 3 vols. Paris.

Boulliau, I. 1639. *Philolai, sive Dissertationis de Vero Systemate Mundi, Libri IV.* Amsterdam.

———. 1645. *Astronomia Philolaica.* Paris.

Boyer, C. B. 1970. "Boulliau, Ismaël." *Dictionary of Scientific Biography* 2: 348–49.

Bradley, J. 1729. "A Letter Giving an Account of a New-Discovered Motion of the Fix'd Stars." *Philosophical Transactions of the Royal Society of London* 35(406): 637–61.

———. 1748. "A Letter Concerning an Apparent Motion Observed in Some of the Fixed Stars." *Philosophical Transactions of the Royal Society of London* 45(485): 1–43.

———. 1832–1833. *Miscellaneous Works and Correspondence.* 2 parts. Ed. S. P. Rigaud. Oxford.

Brandmüller, Walter. 1982. *Galilei und die Kirche, oder das Recht auf Irrtum.* Regensburg: F. Pustet.

———. 1987. *Galileo y la Iglesia.* Madrid: RIALP.

————. 1992a. "Commento." In Brandmüller and Greipl 1992, 15–130.

————. 1992b. *Galilei e la Chiesa, ossia il diritto di errare.* Vatican City: Libreria Editrice Vaticana.

————, and E. J. Greipl, eds. 1992. *Copernico, Galilei e la Chiesa.* Florence: Olschki.

Brecht, Bertolt. 1934. "Dichter sollen die Wahrheit schreiben." *Pariser Tageblatt,* 12 December.

————. 1935. "Dichter sollen die Wahrheit schreiben." *Unsere Zeit* (Paris), vol. 8, nos. 2–3, April.

————. 1948. "Writing the Truth." Trans. by Richard Winston. *Twice a Year,* tenth anniversary issue, 1948. Rpt. in Bentley 1966a, 133–50.

————. 1953. *Galileo.* In *From the Modern Repertoire, Series Two,* ed. E. Bentley. Bloomington: Indiana University Press.

————. 1994a. "Building up a Part." In Brecht 1994b, 131–58.

————. 1994b. *Life of Galileo.* Trans. J. Willett. Ed. J. Willett and R. Manheim. New York: Arcade.

————. 1994c. "Texts by Brecht." In Brecht 1994b, 115–61.

Brenna, L. 1778. "De Vita et Scriptis Galilaei Galilaei." In Fabroni 1778–1805, 1: 1–230.

Brenner, A. A. 1990. *Duhem: Science, Réalité et Appearance.* Paris: Vrin.

Brewster, David. 1835. "Galileo." In *Eminent and Literary Scientific Men of Italy, Spain, and Portugal,* 3 vols., 2: 1–62. In *The Cabinet Cyclopedia,* ed. Dionysius Lardner. London.

————. 1841. *The Martyrs of Science, or the Lives of Galileo, Tycho Brahe, and Kepler.* London.

Brice, C., and A. Romano, eds. 1999. *Sciences et Religions de Copernic à Galilée (1540–1610).* Rome: École Française de Rome.

Broderick, R. C., ed. 1987. *The Catholic Encyclopedia.* New York: Thomas Nelson.

Brooke, J. H. 1991. *Science and Religion.* Cambridge: Cambridge University Press.

————. 1996. "Religious Belief and Natural Science." In Van der Meer 1996, 1: 1–26.

————. 1998. "The Historiography of Religion and Science Interaction." Paper presented at the Conference "Science in Theistic Contexts," Pascal Centre for Advanced Studies in Faith and Science, Redeemer College, Ancaster, Ontario, Canada, 21–25 July.

Brooke, J., and G. Cantor. 1998. *Reconstructing Nature.* Edinburgh: T. & T. Clark.

Brown, H. 1974. "Peiresc, Nicolas Claude Fabri de." *Dictionary of Scientific Biography* 10: 488–92.

Brown, S. 1850. *The Tragedy of Galileo Galilei.* Edinburgh.

Brucker, J. J. 1742–1744. *Historia Critica Philosophiae.* 4 vols. Leipzig.

————. 1766–1767. *Historia Critica Philosophiae.* 2nd ed. 6 vols. Leipzig.

Bucciantini, M. 1994a. "Dopo il *Sidereus Nuncius.*" *Nuncius* 9: 15–35.

————. 1994b. "Galileo e la Chiesa." *Memorie domenicane,* new series, 25: 471–76.

————. 1995. *Contro Galileo.* Florence: Olschki.

————. 1997. "Scienza e filologia." *Giornale critico della filosofia italiana* 76: 424–45.

————. 1998. "Celebration and Conservation." In Hunter 1998, 21–34.

————. 1999. "Teologia e nuova filosofia." In Brice and Romano 1999, 411–52.

————. 2001. "Novità celesti e teologia." In Montesinos and Solís 2001, 795–808.

Buonamici, Giovanfrancesco. 1633. "Relazione." In Favaro 19: 407–11.

Bursill-Hall, P., ed. 1993. *R. J. Boscovich.* Rome: Istituto della Enciclopedia Italiana.

Burtt, E. A. 1932. *The Metaphysical Foundations of Modern Physical Science.* 2nd ed. London: Routledge.

Butterfield, H. 1949. *The Origins of Modern Science, 1300–1800.* London: Bell.

Cabeo, N. 1646. *In Quatuor Libros Meteorologicorum Aristotelis Commentaria . . . Tomus Primus-Secundus.* Rome.

Calandrelli, Giuseppe. 1806a. *Osservazioni e riflessioni sulla parallasse annua dell'Alfa della Lira.* In Calandrelli and Conti 1803–1824.

———. 1806b. *Risultato di varie osservazioni sopra la parallasse annua di Wega, o Alfa della Lira.* Rome.

———, and A. Conti. 1803–1824. *Opuscoli astronomici.* 8 vols. Rome.

Calmet, A. 1720. "Dissertation sur le Système du Monde des Anciens Hébreux." In idem, *Dissertations Qui Peuvent Servir de Prolégomènes de l'Écriture Sainte* 1: 438–59. Paris.

———. 1734. *Prolegomena et Dissertationes in Omnes et Singulos Scripturae Libros.* 2 vols. Venice.

———. 1744. "Dissertazione sovra il sistema del mondo degli antichi ebrei." In G. Galilei 1744a, 4: 1–20.

Campanella, T. 1622. *Apologia pro Galilaeo.* Frankfurt.

———. 1637. *Disputationum in Quatuor Partes Suae Philosophiae Realis Libri Quatuor.* Paris.

Canone, E. 1997. *Brunus Redivivus.* Pisa: Istituti Editoriali e Poligrafici Internazinali.

Cantor, M. 1864. "Galileo Galilei." *Zeitschrift für Mathematik und Physik* 9(3): 172–97.

———. 1877. "Die Actenfälschung im Processe gegen Galileo Galilei." *Die Gegenwart,* nos. 44–45.

Carafa, Petrus A. 1633. [Notification of the Condemnation of Galileo]. In Monchamp 1893, 14–17; in Favaro 19: 412–13.

Caramuel Lobkowitz, J. 1644. *Sublimium Ingeniorum Crux.* Louvain.

Carlen, C., ed. 1981. *The Papal Encyclicals, 1740–1981.* 5 vols. Wilmington: McGrath.

Carli, A., and A. Favaro, eds. 1896. *Bibliografia galileiana (1568–1895).* Rome.

Carniciero, G. C. 1816. *La Inquisicion Justamente Restebleida.* 3 vols. Madrid.

Carrara, B. 1914. *La S. Scrittura, i SS. Padri e Galileo sopra il moto della terra.* Milan.

Carroll, W. E. 1995. "The Legend of Galileo." *Catholic Dossier* 1(2): 14–19.

———. 1997. "Galileo, Science, and the Bible." *Acta Philosophica* 6: 5–37.

———. 1999. "Galileo and the Interpretation of the Bible." *Science and Education* 8: 151–87.

———. 2001. "Galileo and Biblical Exegesis." In Montesinos and Solís 2001, 677–92.

Casanovas, J. 1989. "G. Settele and the Final Annulment of the Decree of 1616 against Copernicanism." *Memorie della Società Astronomica Italiana* 60(4): 791–805.

Casaroli, A. 1981. Casaroli to Poupard, 3 July. In Segre 1997, 499–501.

Casini, P., ed. 1985. *Paolo Frisi, Elogi: Galilei, Newton, D'Alembert.* Rome: Edizioni Theoria.

———. 1987. "Frisi tra illuminismo e rivoluzione scientifica." In *Ideologia e scienza nell'opera di Frisi,* ed. G. Barbarisi, 1:15–33. 2 vols. Milan: FrancoAngeli.

———. 1993. "Boscovich and the *Hypothesis Terrae Motae.*" In Bursill-Hall 1993, 229–35.

Castagnetti, G., ed. 2000. "Appendix: A Forgotten Controversy." *Science in Context* 13: 591–691.

Castagnetti, G., and M. Camerota. 2000a. "Antonio Favaro and the *Edizione Nazionale* of Galileo's Works." *Science in Context* 13: 627–31.

———. 2000b. "Raffaello Caverni and His *History of the Experimental Method in Italy*." *Science in Context* 13: 597–609.

Cavalieri, B. 1633. Cavalieri to Galileo, 18 May. In Favaro 15: 354–55.

———. 1642. "Sphaera, seu Doctrinae Sphaericae Tractatus." Bologna University Library, ms. lat. 1858, no. 9.

Caverni, R. 1891–1900. *Storia del metodo sperimentale in Italia.* 6 vols. Florence.

Chambers, E. 1749. *Dizionario universale delle arti e delle scienze . . . Traduzione esatta ed intiera dall'Inglese.* Vol. 8. Venice.

Chaney, E. 1991. "The Visit to Vallombrosa." In Di Cesare 1991, 113–46.

Charpentier, J. 1560. *Descriptionis Universae Naturae.* Paris.

Chasles, Philarète. 1862. *Galileo Galilei: Sa Vie, Son Procès et Ses Contemporaines.* Paris.

Châtelet, Madame du. 1743. *Istituzioni di fisica.* Venice.

Chiaramonti, S. 1633. *Difesa al suo Antiticone, e libro delle tre nuove stelle.* Florence.

Christianson, J. R., et al., eds. 2002. *Tycho Brahe and Prague.* Frankfurt: Harri Deutsch.

Cifres, A. 1998. "L'archivio storico della Congregazione per la Dottrina della Fede." In *L'apertura degli archivi del Sant'Uffizio romano* 1998, 73–84.

Cinti, D. 1957. *Biblioteca galileiana.* Florence: Sansoni Antiquariato.

Cioni, M. 1908. *I documenti galileiani del S. Uffizio di Firenze.* Florence: Libreria Editrice Fiorentina.

———. 1996. *I documenti galileiani del S. Uffizio di Firenze.* 2nd. ed. Florence.

Collins, J., ed. 1961. *Philosophical Readings on Cardinal Newman.* Chicago: Henry Regnery.

Comte, A. 1830–1842. *Cours de Philosophie Positive.* 6 vols. Paris.

———. 1835. "Considérations Générales sur le Mouvement de la Terre," vol. 2, lesson 22 of *Cours de Philosophie Positive.* In Comte 1975a, 348–62.

———. 1842. Selection from "Appréciation Générale du Développement Fondamental Propre aux Divers Eléments Essentiels de l'État Positif de l'Humanité," vol. 6, lesson 56 of *Cours de Philosophie Positive.* In Comte 1975b, 557–62.

———. 1844. *Traité Philosophique d'Astronomie Populaire.* Paris.

———. 1975a. *Philosophie Premiere* (vols. 1–3 of *Cours de Philosophie Positive*). Ed. M. Serres, F. Dragognet, and A. Sinaceur. Paris: Hermann.

———. 1975b. *Physique Sociale* (vols. 4–6 of *Cours de Philosophie Positive*). Ed. J.-P. Enthoven. Paris: Hermann.

Cooper, Peter. 1838. "Galileo—The Roman Inquisition." *Dublin Review* 5(9): 72–116.

———. 1844. *Galileo—The Roman Inquisition.* Cincinnati.

Copernicus, N. 1976. *On the Revolutions of the Heavenly Spheres.* Trans. A. M. Duncan. Newton Abbot: Davis & Charles.

———. 1992a. *On the Revolutions.* Trans. and ed. E. Rosen. Baltimore: Johns Hopkins University Press.

———. 1992b. *Minor Works.* Trans. and ed. E. Rosen. Baltimore: Johns Hopkins University Press.

Couturat, L., ed. 1903. *Opuscules et Fragments Inédits de Leibniz.* Paris.

Coyne, G. V. 1992. "Address to the Pontifical Aademy of Sciences." *L'Osservatore Romano*, 1 November, p. 9.

————. Forthcoming. "The Church's Most Recent Attempt to Dispel the Galileo Myth." In McMullin (forthcoming).

————, and U. Baldini. 1985. "The Young Bellarmine's Thoughts on the World Systems." In Coyne, Heller, and Zycínski 1985, 103–11.

————, M. Heller, and J. Zycínski, eds. 1985. *The Galileo Affair.* Vatican City: Specola Vaticana.

Crombie, A. C. 1956a. *Galilée devant les Critiques de la Posterité.* Les Conferences du Palais de la Decouverte; Serie D: Histoire des Sciences, no. 45. Paris.

————. 1956b. "Galileo Galilei: A Philosophical Symbol." *Actes du VIIIe Congrès International d'Histoire des Sciences* 11: 1089–95.

————. 1961. *Augustine to Galileo.* 2nd ed. Cambridge, MA: Harvard University Press.

D'Addio, M. 1983. "Considerazioni sui processi a Galileo (pt. I)." *Rivista di storia della chiesa in Italia* 37: 1–52.

————. 1984a. "Alcune fasi dell'istruttoria del processo a Galileo." *L'Osservatore Romano*, 2 March, p. 2.

————. 1984b. "Considerazioni sui processi a Galileo (pt. II)." *Rivista di storia della chiesa in Italia* 38: 47–114.

————. 1985. *Considerazioni sui processi a Galileo.* Rome: Herder.

————. 1993. *Il caso Galilei.* Rome: Edizioni Studium.

D'Alembert, Jean. 1751a. "Antipodes." In Diderot and D'Alembert 1751–1780, 1: 512–14.

————. 1751b. "Astronomie." In Diderot and D'Alembert 1751–1780, 1: 790.

————. 1751c. "Discours Preliminaire." In Diderot and D'Alembert 1751–1780, vol. 1, i–xlv.

————. 1754. "Copernic." In Diderot and D'Alembert 1751–1780, 4: 173–174.

————. 1764(?). "Samos, Aristarque." In Diderot and D'Alembert 1751–1780, 14: 600.

————. 1963. *Preliminary Discourse to the Encyclopedia of Diderot.* Trans. and ed. R. N. Schwab. Indianapolis: Bobbs-Merrill.

Daville, L. 1909. *Leibniz Historien.* Paris: Alcan.

Delambre, Jean B. 1820. Delambre to Venturi, 30 June. In Favaro 1887a, 224–26.

————. 1821. *Histoire de l'Astronomie Moderne.* 2 vols. Paris.

Delannoy, P. 1906. [Review of Vacandard's *Études*]. *Revue d'Histoire Ecclésiastique* 7: 354–61.

Del Prete, A. 2001. "Tra Galileo e Descartes." In Montesinos and Solís 2001, 719–32.

De Santis, F. 1979. "Papa Wojtyla riabilita Galileo." *Corriere della Sera*, 11 November, p. 6.

Descartes, René. 1633. Descartes to Mersenne, November. In Favaro 15: 340–41; in Descartes 1897–1913, 1: 270–73.

————. 1634a. Descartes to Mersenne, February. In Favaro 16: 56; in Descartes 1897–1913, 1: 280–84.

————. 1634b. Descartes to Mersenne, April. In Favaro 16: 88–89; in Descartes 1897–1913, 1: 284–91; in Kenny 1970, 25–27.

————. 1637. *Discours de la Méthode.* Leiden. In Descartes 1897–1913, 6: 1–78.

————. 1644. *Principia Philosophiae*. Amsterdam.

————. 1647. *Les Principes de la Philosophie*. Trans. Abbé C. Picot. Paris.

————. 1664. *Le Monde, ou le Traité de la Lumière*. Paris.

————. 1897–1913. *Oeuvres*. 13 vols. Ed. C. Adam and P. Tannery. Paris: Cerf.

————. 1955. *Philosophical Works*. 2 vols. Trans. and ed. E. S. Haldane and G. R. T. Ross. New York: Dover.

————. 1991. *Principles of Philosophy*. Trans. V. R. Miller and R. P. Miller. Dordrecht: Kluwer.

Deusingius, A. 1643. *De Vero Systemate Mundi Dissertatio Mathematica*. Amsterdam.

De Waard, C., and A. Beaulieu, eds. 1932–1988. *Correspondance du P. Marin Mersenne*. 17 vols. Paris: Editions du Centre National de la Recherche Scientifique.

Dibner, B. 1967. "Of Martyrs, Books, and Science." In Lehmann-Haupt 1967, 163–82.

Di Canzio, A. 1996. *Galileo: His Science and His Significance for the Future of Man*. Dover: Adasi.

Di Cesare, M. A., ed. 1991. *Milton in Italy*. Binghamton, N.Y.: Medieval & Renaissance Texts & Studies.

Diderot, D., and J. D'Alembert, eds. 1751–1780. *Encyclopédie, ou Dictionnaire Raisonné des Sciences, des Arts et des Métiers*. 35 vols. Paris. Rpt., Stuttgart–Bad Cannstatt: Frommann, 1966.

Digby, K. 1644. *Two Treatises*. Paris.

Dinis, A. de Oliveira. 1989. "The Cosmology of Giovanni Battista Riccioli (1598–1671)." Ph.D. diss., University of Cambridge.

Divini, Eustachio. 1660. *Brevis Annotatio in Systema Saturnium Christiani Eugenii*. Rome.

————. 1661. *Pro Sua Annotatione in Systema Saturnium Christiani Eugenii Adversus Eiusdem Assertiones*. Rome.

Doncel, M. G. 2001. "Juan Pablo II y los 'Studi Galileiani.'" In Montesinos and Solís 2001, 753–64.

Drake, S. 1957. *Discoveries and Opinions of Galileo*. Garden City: Doubleday.

————. 1965. "The Galileo-Bellarmine Meeting: A Historical Speculation." In Geymonat 1965, 205–20.

————. 1970. *Galileo Studies*. Ann Arbor: University of Michigan Press.

————. 1976. *Galileo against the Philosophers*. Los Angeles: Zeitlin & Ver Brugge.

————. 1978. *Galileo at Work*. Chicago: University of Chicago Press.

————. 1980. *Galileo*. New York: Hill and Wang.

————. 1999. *Essays on Galileo and the History and Philosophy of Science*. 3 vols. Ed. N. M. Swerdlow and T. H. Levere. Toronto: University of Toronto Press.

Draper, J. W. 1875. *History of the Conflict between Religion and Science*. New York.

Dreyer, J. J. E. 1953. *A History of Astronomy from Thales to Kepler*. 2nd ed. New York: Dover.

Drinkwater Bethune, J. E. 1832. *Life of Galileo Galilei*. Boston.

Dubarle, D. 1961. "Autour de Galilée." *Critique* 17: 445–57.

————. 1964. "Le Dossier Galilée." *Signes du Temps* 14: 21–26.

Dubois, I. 1655. *Veritas et Authoritas Sacra in Naturalibus et Astronomicis Asserta et Vindicata*. Utrecht.

Duhem, Pierre. 1908. *Sozein ta Phainomena: Essai sur la Notion de Theorie Physique de Platon à Galilée*. Paris: Hermann.

————. 1909. *Le Mouvement Absolu et le Mouvement Relatif.* Montligeon: Imprimerie-Librairie de Montligeon.

————. 1954. *The Aim and Structure of Physical Theory.* Trans. P. Wiener. Princeton: Princeton University Press.

————. 1969. *To Save the Phenomena.* Trans. E. Doland and C. Maschler. Chicago: University of Chicago Press.

Dutens, L., ed. 1768. *Leibnitii Opera Omnia.* Geneva. Rpt., Hildesheim: Georg Olms Verlag, 1989.

Einstein, A. 1953. "Foreword." In G. Galilei 1953, vi–xx.

"Epigrafi ed offese."1887. *L'Osservatore Romano,* 23 April

Erythraeus, Janus N. [Giovanni Vittorio de' Rossi]. 1643. *Pinacotheca Imaginum Illustrium, Doctrinae vel Ingenii Laude, Virorum.* Coloniae Agrippinae.

Estève, P. 1755. *Histoire Générale et Particulière de l'Astronomie.* 3 vols. Paris.

Fabri, Honoré. 1660. *Brevis Annotatio in Systema Saturnium Christiani Eugenii.* Rome.

————. 1661. *Pro Sua Annotatione in Systema Saturnium Christiani Eugenii adversus Eiusdem Assertionem.* Rome.

Fabris, R. 1986. *Galileo Galilei e gli orientamenti esegetici del suo tempo.* Vatican City: Pontifical Academy of Sciences.

————, ed. 1992a. *La Bibbia nell'epoca moderna e contemporanea.* Bologna: Edizioni Dehoniane.

————. 1992b. "Introduzione Generale." In Fabris 1992a, 5–19.

————. 1992c. "Strumenti e sussidi per lo studio della Bibbia nei secoli XV–XVII." In Fabris 1992a, 43–73.

————. 1992d. "Lo sviluppo e l'applicazione del metodo storico-critico nell'esegesi biblica (secoli XVII–XIX)." In Fabris 1992a, 103–45.

Fabroni, Angelo, ed. 1773–1775. *Lettere inedite di uomini illustri.* 2 vols. Florence.

————, ed. 1778–1805. *Vitae Italorum Doctrina Excellentium.* 20 vols. Pisa.

————. 1802. "Elogio di Giuseppe Toaldo." In Toaldo 1802, 1: vii–xxxvi.

Fahie, J. J. 1903. *Galileo: His Life and Work.* London: John Murray.

————. 1929. *Memorials of Galileo Galilei, 1564–1642.* London: Courier Press.

Falco, G. 1964. "Introduzione." In Muratori 1964a, xv–xxxiii.

Fantoli, A. 1993. *Galileo: Per il copernicanesimo e per la chiesa.* Vatican City: Specola Vaticana.

————. 1996. *Galileo: For Copernicanism and for the Church.* 2nd ed. Trans. G. V. Coyne. Vatican City: Vatican Observatory Publications.

————. 1997. *Galileo: Per il copernicanesimo e per la chiesa.* 2nd ed. Vatican City: Vatican Observatory Publications.

————. 2001. "Galileo e la Chiesa cattolica." In Montesinos and Solís 2001, 733–52.

————. 2003. *Il caso Galileo.* Milan: Rizzoli.

Favaro, Antonio. 1885a. "Ragguaglio dei manoscritti galileiani nella Collezione Libri-Ashburnham presso la Biblioteca Mediceo-Laurenziana di Firenze." In Favaro 1885b, 21–30.

————. 1885b. "Documenti inediti per la storia dei manoscritti galileiani nella Biblioteca Nazionale di Firenze." *Bullettino di bibliografia e di storia delle scienze matematiche e fisiche* 18: 1–112, 151–230.

————. 1887a. "Documenti per la storia del processo originale di Galileo." In Favaro 1887b, 178–228.

————. 1887b. *Miscellanea galileiana inedita.* Venice.

————. 1887c. "Sulla pubblicazione della sentenza contro Galileo, e sopra alcuni tentativi del Viviani per far revocare la condanna dei Dialoghi galileiani." In Favaro 1887b, 97–156.

————. 1887–1888. "Il testo originale della condanna di Galileo." *Atti e memorie dell'Accademia di scienze, lettere ed arti in Padova*, 3rd series, 4: 115–16.

————. 1888. "Galileo e Diodati." *Memorie del Reale Istituto Veneto di scienze, lettere ed arti* 22: 851–71.

————, ed. 1890–1909. *Le opere di Galileo Galilei*. 20 vols. Florence: Barbèra. Rpt. 1929–1939, 1968.

————. 1891a. *Galileo Galilei e Suor Maria Celeste*. Florence.

————. 1891b. *Nuovi studi galileiani*. Venice.

————. 1891c. "Tre consulti in favore di Galileo." In Favaro 1891b, 373–88.

————. 1891d. "L'ultima fase della lotta contro il sistema copernicano." In Favaro 1891b, 419–30.

————. 1902a. "Amici e corrispondenti di Galileo Galilei: IV. Alessandra Bocchineri—V. Francesco Rasi—VI. Giovanfrancesco Buonamici." *Atti del Reale Istituto Veneto di scienze, lettere ed arti*, academic year 1901–1902, vol. 61, part 2, 665–701.

————. 1902b. "I documenti del processo di Galileo." *Atti del Reale Istituto Veneto di scienze, lettere ed arti*, series 8, tome 4, vol. 61, part 2, no. 10, 757–806.

————, ed. 1902c. *Il processo di Galileo*. Florence: Barbèra.

————. 1905. "Amici e corrispondenti di Galileo Galilei: XII. Vincenzo Renieri." *Atti del Reale Istituto Veneto di scienze, lettere ed arti*, academic year 1904–1905, vol. 64, part 2, 111–95.

————. 1906. "L'episodio di Gustavo Adolfo di Svezia nei racconti della vita di Galileo." *Atti del Reale Istituto Veneto di scienze, lettere ed arti*, academic year 1905–1906, vol. 65, part 2, 16–39.

————. 1907a. "Amici e corrispondenti di Galileo Galilei: XX. Fulgenzio Micanzio." *Nuovo archivio veneto*, new series, year 7, vol. 13, part 1, 34–67.

————. 1907b. "Antichi e moderni detrattori di Galileo." *La Rassegna Nazionale*, 16 February, 29(153): 577–600.

————, ed. 1907c. *Galileo e l'inquisizione: Documenti del processo galileiano*. Florence: Barbèra.

————. 1908. "Galileo Galilei e la determinazione del peso dell'aria." *Rivista di fisica, matematica e scienze naturali* 9(108): 577–88.

————. 1911a. "Alla ricerca delle origini del motto 'E pur si muove.'" *Atti del Reale Istituto Veneto di scienze, lettere ed arti*, vol. 70, part 2, 1219–32.

————. 1911b. "E pur si muove." *Il Giornale d'Italia*, 12 July, 4.

————. 1912. "Amici e corrispondenti di Galileo Galilei: XXIX. Vincenzio Viviani." *Atti del Reale Istituto Veneto di science, lettere ed arti*, academic year 1912–1913, vol. 72, part 2, 1–155.

————. 1915a. "Amici e corrispondenti di Galileo Galilei: XXXII. Francesco di Noailles." *Atti e memorie della Reale Accademia di scienze, lettere ed arti in Padova*, new series, 31: 99–125.

————. 1915b. "Sulla veridicità del *Racconto istorico della vita di Galileo* dettato da Vincenzio Viviani." *Archivio storico italiano* 73: 323–80.

————. 1916a. "Adversaria galilaeiana: Serie prima." *Atti e memorie della Reale Accademia di scienze, lettere ed arti in Padova*, new series, 32: 123–49.

————. 1916b. "Amici e corrispondenti di Galileo Galilei: XXXIII. Mattia Berneg-ger." *Atti del Reale Istituto Veneto di science, lettere ed arti,* academic year 1915–1916, vol. 75, part 2, 29–53.

————. 1916c. "Amici e corrispondenti di Galileo Galilei: XXXIV–XXXVI. Bonaventura, Abramo e Lodovico Elzevier." *Atti del Reale Istituto Veneto di science, lettere ed arti,* academic year 1915–1916, vol. 75, part 2, 481–514.

————. 1916d. "La condanna di Galileo e le sue conseguenze per il progresso degli studi." *Scientia,* year 10, vol. 20, no. 51, 1–11.

————. 1917a. "Amici e corrispondenti di Galileo Galilei: XXXVIII. Marino Mer-senne." *Atti del Reale Istituto Veneto di scienze, lettere ed arti,* academic year 1916–1917, vol. 76, part 2, 35–92.

————. 1917b. "Amici e corrispondenti di Galileo Galilei: XXXIX. Niccolò Fabri di Peiresc." *Atti del Reale Istituto Veneto di scienze, lettere ed arti,* academic year 1916–1917, vol. 76, part 2, 591–636.

————. 1920. "Galileo Galilei, Benedetto Castelli e la scoperta delle fasi di Venere." *Archeion* 1: 283–96.

————, G. Lorenzoni, and A. Minich. 1890. "Relazione della giunta del R. Istituto Veneto." In Caverni 1891–1900, 1: 5–20.

Feldhay, R. 1995. *Galileo and the Church.* Cambridge: Cambridge University Press.

————. 2000. "Recent Narratives on Galileo and the Church." *Science in Context* 13: 489–507.

Feller, François X. de. 1782. [Article on Galileo]. In idem, *Dictionnaire Historique,* 3: 174. Augsburg.

————. 1797. [Article on Galileo]. In idem, *Dictionnaire Historique,* 2nd ed., 4: 251–53. Liège.

————. 1832. "Galilée-Galilei." In idem, *Dictionnaire Historique,* 8th ed., 6: 26–28. Lille.

Fenu, E. 1964. "Polemica galileiana." *L'Osservatore Romano,* 15 February.

Ferngren, G. B., ed, 2002. *Science and Religion.* Baltimore: Johns Hopkins University Press.

Ferri, G. 1785. "Apologie de Galilée." *Mercure de France,* 8 January, 54–63.

Ferrone, V. 1982. *Scienza, natura, religione.* Naples: Jovene.

Ferrone, V., and M. Firpo. 1985a. "Galileo tra inquisitori e micro-storici." *Rivista storica italiana* 97: 177–238.

————. 1985b. "Replica." *Rivista storica italiana* 97: 957–68.

————. 1986. "From Inquisitors to Microhistorians." *Journal of Modern History* 58: 485–524.

Feyerabend, P. K. 1975. *Against Method.* London: NLB.

————. 1985. "Galileo and the Tyranny of Truth." In Coyne, Heller, and Zycínski 1985, 155–66.

————. 1987. *Farewell to Reason.* London: Verso.

————. 1988. *Against Method.* Revised ed. London: Verso.

Finocchiaro, M. A. 1980. *Galileo and the Art of Reasoning.* Dordrecht: Reidel.

————. 1985. Review of Pagano's *I documenti del processo di Galileo Galilei. Isis* 76: 380–81.

————. 1986a. "The Methodological Background to Galileo's Trial." In Wallace 1986, 241–72.

————. 1986b. Review of Coyne, Heller, and Zycínski's *The Galileo Affair. Isis* 77: 192.

————. 1986c. "Toward a Philosophical Interpretation of the Galileo Affair." *Nuncius* 1: 189–202.

————. 1988a. Review of Baldini and Coyne's *Louvain Lectures (Lectiones Lovanienses) of Bellarmine* and Coyne, Heller, and Zycínski's *The Galileo Affair. Journal of the History of Philosophy* 26: 149–51.

————. 1988b. Review of Zycínski's *The Idea of Unification in Galileo's Epistemology. Isis* 79: 734–35.

————, trans. and ed. 1989. *The Galileo Affair.* Berkeley: University of California Press.

————. 1990. Review of Westfall's *Essays on the Trial of Galileo. Nuncius* 5: 301–7.

————. 1992a. "Galileo's Copernicanism and the Acceptability of Guiding Assumptions." In *Scrutinizing Science,* ed. A. Donovan, L. Laudan, and R. Laudan, 49–67. Baltimore: Johns Hopkins University Press.

————. 1992b. "To Save the Phenomena." *Revue Internationale de Philosophie* 46: 291–310.

————. 1995a. "Methodological Judgment and Critical Reasoning in Galileo's *Dialogue.*" In Hull, Forbes, and Burian 1995, 2: 248–57.

————. 1995b. Review of Fantoli's *Galileo* and Reston's *Galileo. Isis* 86: 486–88.

————, trans. and ed. 1997. *Galileo on the World Systems.* Berkeley: University of California Press.

————. 1999. "The Galileo Affair from John Milton to John Paul II." *Science and Education* 8: 189–209.

————. 2001a. "Aspects of the Controversy about Galileo's Trial." In Montesinos and Solís 2001, 491–512.

————. 2001b. "Science, Religion, and the Historiography of the Galileo Affair." *Oriris,* 2nd series, 16: 114–32.

————. 2002a. "Drake on Galileo." *Annals of Science* 59: 83–88.

————. 2002b. "Galileo as a 'Bad Theologian.'" *Studies in History and Philosophy of Science* 33: 753–91.

————. 2002c. "Philosophy versus Religion and Science versus Religion." In Gatti 2002, 51–96.

Fiorani, L. 1969. *Onorato Caetani.* [Rome]: Istituto di Studi Romani.

————. 1973a. "Caetani, Francesco." *Dizionario biografico degli italiani* 16: 168–70.

————. 1973b. "Caetani, Onorato." *Dizionario biografico degli italiani* 16: 209–12.

Firpo, L., ed. 1993. *Il processo di Giordano Bruno.* Ed. D. Quaglioni. Rome: Salerno Editrice.

Flammarion, C. 1872. *Vie de Copernic et Histoire de la Découverte du Système du Monde.* Paris.

Foscarini, Paolo A. 1615. *Lettera sopra l'opinione de' Pittagorici e del Copernico.* Naples.

————. 1635. *Epistola circa Pythagoricum et Copernici Opinionem.* Trans. E. Diodati. In G. Galilei 1635b.

Foucault, Léon. 1851. [Report]. In *Procès Verbaux de la Société Philomatique,* 68. Paris.

————. 1878. *Recueil des Travaux Scientifiques.* Ed. C. M. Gariel. Paris.

Franzelin, J. B. 1882. *Tractatus de Divina Traditione et Scriptura* (1870). Rome.

Frisi, Paolo. 1756. *De Motu Diurno Terrae Dissertatio.* Pisa.

————. 1766. "Saggio sul Galileo." *Il Caffè*, 2(3): 17–27.

————. 1774–1784. Letters to Angelo Fabroni. In Fabroni 1778–1805, 20: 214–89.

————. 1775. *Elogio del Galileo*. Milan and Leghorn. Rpt. in Casini 1985, 31–92.

————. 1777. "Galilée, Philosophie de." In *Supplément à l'Encyclopédie* 3: 172–76. Amsterdam.

Froidmont, L. 1631. *Ant-Aristarchus*. Antwerp.

————. 1634. *Vesta, sive Ant-Aristarchi Vindex*. Antwerp.

"A Further Account by Monsieur Auzout of Signor Campani's Book, and Performances about Optick-Glasses." 1665. *Philosophical Transactions of the Royal Society of London* 1(4): 70–75, 5 June.

[Gaetani, Onorato]. 1785. [Apocryphal letter by Galileo to Renieri, December 1633]. In Tiraboschi 1782–1797, 8 (1785): 147–49 n.

Galilée: Aspects de Sa Vie et de Son Oeuvre. 1968. Paris: Presses Universitaires de France.

Galilei, Galileo. 1632a. *Dialogo sopra i due massimi sistemi del mondo, tolemaico e copernicano*. Florence.

————. 1632b. Galileo to F. Barberini, 13 October. In Favaro 14: 406–10.

————. 1634a. Galileo to Diodati, 7 March. In Favaro 16: 58–60.

————. 1634b. Galileo to Diodati, 25 July. In Favaro 16: 115–19.

————. 1634c. *Les Mécaniques de Galilée*. Trans. M. Mersenne. Paris.

————. 1635a. Galileo to Peiresc, 21 February. In Favaro 16: 215–16.

————. 1635b. *Systema Cosmicum*. Strasbourg.

————. 1636. *Nov-antiqua Sanctissimorum Patrum, & Probatorum Theologorum Doctrina de Sacrae Scripturae Testimoniis*. Ed. M. Bernegger. Trans. E. Diodati. Strasbourg.

————. 1638. *Discorsi e dimostrazioni matematiche intorno a due nuove scienze attenenti alla mecanica et i movimenti locali*. Leiden.

————. 1639. *Les Nouvelles Pensées de Galilei*. Trans. M. Mersenne. Paris.

————. 1641. *Systema Cosmicum*. Lyons.

————. 1649a. Galileo to Castelli, 21 December 1613. In Gassendi 1649, appendix, 65–78.

————. 1649b. Galileo to Christina, 1615. In Gassendi 1649, appendix, 1–60.

————. 1649c. Galileo to Dini, 23 March 1615. In Gassendi 1649, appendix, 79–95.

————. 1661a. "Epistle to the Grand Duchesse Mother." In Salusbury 1661–1665, vol. 1, part 1, 425–60.

————. 1661b. *System of the World*. In Salusbury 1661–1665, vol. 1, part 1, 1–424.

————. 1710. *Dialogo sopra i due massimi sistemi del Mondo*. Florence [Naples].

————. 1718. *Opere di Galileo Galilei*. 3 vols. Ed. T. Bonaventuri and G. Grandi. Florence.

————. 1744a. *Opere*. 4 vols. Ed. G. Toaldo. Padua.

————. 1744b. *Dialogo sopra i due massimi sistemi del mondo*. Ed. G. Toaldo. Padua.

————. 1813–1814. [Reply to Ingoli]. *Giornale enciclopedico* (Florence), vol. 6, no. 62: 122–30, no. 63: 172–89, no. 65: 3–60.

————. 1842–1856. *Opere*. 15 vols. with 1-vol. supplement. Ed. E. Albèri. Florence.

————. 1890–1909. *Le Opere di Galileo Galilei*. 20 vols. National Edition by A. Favaro et al. Florence: Barbèra. Rpt. 1929–1939, 1968.

————. 1953a. *Dialogue Concerning the Two Chief World Systems*. Trans. and ed. S. Drake. Berkeley: University of California Press. (2nd ed., 1967).

————. 1953b. *Dialogue on the Great World Systems*. Trans. and ed. G. de Santillana. Chicago: University of Chicago Press.

————. 1997. *Galileo on the World Systems*. Trans. and ed. M. A. Finocchiaro. Berkeley: University of California Press.

————. 1998. *Dialogo sopra i due massimi sistemi del mondo*. 2 vols. Critical edition by O. Besomi and M. Helbing. Padua: Antenore.

————. 2000. *Lettera a Cristina di Lorena*. Ed. F. Motta. Genoa: Marietti.

————. 2001. *Dialogue Concerning the Two Chief World Systems*. Trans. S. Drake. Ed. J. L. Heilbron. New York: Modern Library.

Galilei, Maria Celeste. 1633. M. C. Galilei to Galileo, 3 October. In Favaro 15: 292–93.

————. 1883. *Lettere al padre*. Ed. G. Morandini. Turin.

————. 1992. *Lettere al padre*. Ed. G. Ansaldo. Genoa: Blengino.

————. 2001. *Letters to Father*. Trans. and ed. D. Sobel. New York: Walker & Company.

Galileo a Padova, 1592–1610: Celebrazioni del IV centennario. 1995. 5 vols. Trieste: Lint.

Galluzzi, P. 1977. "Galileo contro Copernico." *Annali dell'Istituto e Museo di Storia della Scienza* 2: 87–148.

————, ed. 1984. *Novità celesti e crisi del sapere*. Florence: Giunti Barbèra.

————. 1993a. "Gassendi e *l'affaire Galilée* delle leggi del moto." *Giornale critico della filosofia italiana* 72/74: 86–119.

————. 1993b. "I sepolcri di Galileo." In *Il pantheon di Santa Croce a Firenze*, ed. L. Berti, 145–82. Florence: Cassa di Risparmio di Firenze.

————. 1998. "The Sepulchers of Galileo." In Machamer 1998, 417–47.

————. 2000. "Gassendi and l'*Affaire Galilée* of the Laws of Motion." *Science in Context* 13: 509–45.

Gamba, B, ed. 1819. *Lettere descrittive di celebri italiani alla studiosa gioventù proposte*. 2nd ed. Venice.

Garcia, S. 2000. "L'Edition Strasbourgeoise du *Systema Cosmicum* (1635–1636)." *Bulletin de la Société de l'Histoire du Protestantisme Français* 146: 307–34.

————. 2001. "Elie-Diodati-Galilée." In Montesinos and Solís 2001, 883–94.

Gardair, J. M. 1984. "Elia Diodati e la diffusione europea del *Dialogo*." In Galluzzi 1984, 391–98.

Garin, E. 1984. "Il caso galileiano nella cultura moderna." In Galluzzi 1984, 5–14.

Garrone, G.-M. 1984. "Prefazione." In Poupard 1984, 5–6.

Garzend, Léon. 1911–1912. "Si Galilée Pouvait, Juridiquement, Etre Torturé." *Revue des Questions Historiques*, 90(1911): 353–89, and 91(1912): 36–67.

————. [1912]. *L'Inquisition et l'Hérésie*. Paris: Desclée de Brouwer.

Gassendi, Pierre. 1641. *Viri Illustris Claudii Fabricii de Peiresc Vita*. Paris.

————. 1642. *De Motu Impressu a Motore Translato*. Paris.

————. 1646. *De Proportione Qua Gravia Decidentia Accelerantur Epistolae Tres*. Paris.

————. 1649. *Apologia in Io. Bap. Morini Librum, cui Titulus, Alae Telluris Fractae*. Lyons.

————. 1654. *Tychonis Brahei, Equitis Dani, Astronomorum Coryphaei Vita*. Paris.

————. 1657. *The Mirrour of True Nobility & Gentility*. Trans. W. Rand. London.

Gatti, H. 1997. "Bruno nella cultura inglese dell'ottocento." In Canone 1997, 19–66.

————, ed. 2002. *Giordano Bruno: Philosopher of the Renaissance*. Aldershot: Ashgate.

Gaukroger, S. 1995. *Descartes: An Intellectual Biography*. Oxford: Clarendon.

Gauss, C. F. 1803. Gauss to Benzenberg, 2 February, and 8 March. In idem, *Werke* 5: 495–503. 12 vols. Hildesheim: Georg Olms, 1981.

Gebler, Karl von. 1876. *Galileo Galilei und die römische Curie.* Stuttgart. (= Gebler 1876–1877, vol. 1).

———. 1876–1877. *Galileo Galilei und die römische Curie.* 2 vols. Stuttgart.

———. 1877. *Die Acten des Galilei'schen Processes, nach der Vaticanischen Handschrift.* Stuttgart. (= Gebler 1876–1877, vol. 2).

———. 1878. "Ist Galilei gefoltert worden?" *Die Gegenwart,* vol. 13, nos. 18, 19, 24, and 25.

———. 1879a. *Galileo Galilei and the Roman Curia.* Trans. Mrs. G. Sturge. London.

———. 1879b. *Galileo Galilei e la curia romana.* 2 vols. Trans. G. Prato. Florence.

Gemelli, Agostino. 1922a. "Il processo e la condanna di Galilei." In Gemelli 1922b, 305–70.

———. 1922b. *Religione e scienza.* 2nd ed. Milan: Società Editrice "Vita e Pensiero."

———. 1941. "La relazione di Agostino Gemelli al Santo Padre." *L'Osservatore Romano,* 1–2 December, 3–4.

———. 1942a. "Galileo ha data la dimostrazione che la terra gira intorno al sole?" *Vita e pensiero* 28: 289–94.

———. 1942b. "Scienza e fede nell'uomo Galilei." In *Nel terzo centenario della morte di Galileo Galilei,* 1–27.

Genovesi, E. 1966. *Processi contro Galileo.* Milan: Ceschina.

Gerhardt, C. I., ed. 1849–1863. *Leibnizens Mathematische Schriften.* 7 vols. Halle.

Gerdil, G. S. 1806. *Storia delle sette dei filosofi.* In Gerdil 1806–1821, vol. 1.

———. 1806–1821. *Opere edite ed inedite del Cardinale G. S. Gerdil.* 20 vols. Rome.

Geymonat, L. 1962. *Galileo Galilei* (1957). 2nd ed. Turin: Einaudi.

———. 1965. *Galileo Galilei.* Trans. S. Drake. New York: McGraw-Hill.

Gherardi, Silvestro. 1870. *Il processo di Galileo riveduto sopra documenti di nuova fonte.* Florence.

———. 1872. *Sulla dissertazione del dott. Emilio Wohlwill "Il processo di Galileo Galilei."* Florence.

Ghiberti, G. 1992. "Lettura e interpretazione della Bibbia dal Vaticano I al Vaticano II." In Fabris 1992a, 187–245.

Giacchi, O. 1942. "Considerazioni giuridiche sui due processi contro Galileo." In *Nel terzo centenario della morte di Galileo Galilei,* 383–406.

Gillespie, C. C. 1974. "Voltaire, François Marie Arouet de." *Dictionary of Scientific Biography* 14: 82–85.

Gingerich, O. 1981. "The Censorship of Copernicus's *De Revolutionibus.*" *Annali dell'Istituto e Museo di Storia della Scienza* 6(2): 45–61.

———. 1982. "The Galileo Affair." *Scientific American,* August, 132–143.

Giovasco, L. M. 1742. [Consultant's Report to the Inquisition]. In Mayaud 1997, 146–49.

Glaser, A. 1861. *Galileo Galilei: Trauerspiel in fünf Acten.* Berlin.

Goddu, A. 1990. "The Realism that Duhem Rejected in Copernicus." *Synthese* 83: 301–16.

Golden, F. 1984. "Rehabilitating Galileo's Image." *Time,* 12 March.

Goodwin, R. N. 1998. *The Hinge of the World.* New York: Farrar, Straus, and Giroux.

Gorman, M. J. 1996. "A Matter of Faith?" *Perspectives on Science* 4: 283–320.

Govi, G. 1872. "Il Sant'Uffizio, Copernico e Galileo." *Atti della reale accademia delle scienze di Torino* 7:565–90, 808–38.

Grandi, A. M. 1820a. (9 August.) "Voto." In Maffei 1987, 539–42; in Brandmüller and Greipl 1992, 294–98.

———. 1820b. (November.) "Voto." In Brandmüller and Greipl 1992, 386–93.

Grisar, H. 1878a. "Der Galileische Process." *Zeitschrift für katholische Theologie* 2: 65–128.

———. 1878b. "Die römische Congregationsdekrete." *Zeitschrift für katholische Theologie* 2: 673–736.

———. 1882. *Galileistudien*. Regensburg.

Grua, G, ed. 1948. *Textes Inédits*. 2 vols. Paris: Presses Universitaires de France.

Guasco, M., E. Guerriero, and F. Traniello, eds. *Storia della Chiesa, Vol. XXIII*. Cinisello Balsamo: Edizioni Paoline.

Guasti, C. 1873. "Le relazioni di Galileo con alcuni pratesi." *Archivio storico italiano*, 3rd series, 17: 32–75.

Guglielmini, Giambattista. 1789. *Riflessioni sopra un nuovo esperimento in prova del diurno moto della terra*. Rome. Rpt. in *Opuscoli scelti sulle scienze e sulle arti* 12 (1789): 422–28; 22 vols.; Milan, 1778–1803.

———. 1792. *De Diurno Terrae Motu Experimentis Physico-Mathematicis Confirmato Opusculum*. Bologna.

———. 1994. *Carteggio: De Diurno Terrae Motu*. Ed. M. T. Borgato and A. Fiocca. Florence: Olschki.

Guicciardini, Piero. 1616. Guicciardini to Cosimo II, 4 March. In Favaro 12: 241–43; in Fabroni 1773–1775, 1: 53–57.

Guiducci, Mario. 1633a. Guiducci to Galileo, 20 August. In Favaro 15: 230–31.

———. 1633b. Guiducci to Galileo, 27 August. In Favaro 15: 240–42.

Gurwitsch, A. 1967. "Galilean Physics in the Light of Husserl's Phenomenology." In McMullin 1967, 388–401.

Gusdorf, G. 1969. *La Révolution Galiléenne*. 2 vols. Paris: Payot.

Haeckel, E. H. 1878–1879. *Gesammelte populäre Vorträge aus dem Gebiete der Entwicklungslehre*. Bonn.

Hagen, J. H. 1911. *La Rotation de la Terre*. Rome: Specola Astronomica Vaticana.

Haldane, E. S., and G. R. T. Ross, trans. and eds. 1955. *The Philosophical Works of Descartes*. 2 vols. New York: Dover.

Hall, A. R. 1979. "Galileo nel XVIII secolo." *Rivista di filosofia* 15: 367–90.

———. 1980. "Galileo in the Eighteenth Century." In *Transactions of the Fifth International Congress on the Enlightenment*, ed. H. Mason, 1: 81–99. Oxford: The Voltaire Foundation at the Taylor Institution.

Hall, E. H. 1903. "Do Falling Bodies Move South?" *Physical Review* 17:179–90.

———. 1904. "Experiments on the Deviations of Falling Bodies." *Proceedings of the American Academy of Arts and Sciences* 39: 341–49.

———. 1910. "Air Resistance to Falling Inch Spheres." *Proceedings of the American Academy of Arts and Sciences* 45: 379–84.

Hallam, H. 1837–1839. *Introduction to the Literature of Europe*. London.

Harris, N. 1985. "Galileo as Symbol." *Annali dell'Istituto e Museo di Storia della Scienza di Firenze* 10(2): 3–29.

Harris, S. n.d. "Bibliographical Essay." Unpublished manuscript.

Harsányi, Z. de. 1939. *The Star Gazer*. Trans. P. Tabor. New York: Putnam.

Hatfield, G. 1990. "Metaphysics and the New Science." In Lindberg and Westman 1990, 93–166.

Haynes, R. 1970. *Philosopher King*. London: Weidenfeld & Nicolson.

Hecht, W, ed. 1968. *Materialien zu Brechts 'Leben des Galilei.'* Frankfurt: Suhrkamp.

Heilbron, J. L. 1999. *The Sun in the Church*. Cambridge, MA: Harvard University Press.

———. 2001. "Introduction." In G. Galilei 2001, xiii–xxi.

Hevelius, J. 1647. *Selenographia sive Lunae Descriptio*. Danzig.

Hiley, J. 1981. *Theatre at Work*. London: Routledge.

Hobbes, T. 1642. ["De Mundo"]. Unpublished ms. First published in Hobbes (1973).

———. 1973. *Critique du "De Mundo" de Thomas White*. Ed. J. Jacquot and H. W. Jones. Paris: Vrin/CNRS.

Hooykaas, R. 1976. "The Reception of Copernicanism in England and the Netherlands." In *The Anglo-Dutch Contribution to the Civilization of Early Modern Society*, ed. C. Wilson et al., 33–55. Oxford.

Horace. 1863. *The Works of Horace*. Trans. C. Smart. Notes by Theodore A. Buckley. New York: Harper.

Howell, K. J. 1966a. "Copernicanism and the Bible in Early Modern Science." In Van der Meer 1996, 4: 261–84.

———. 1996b. "Galileo and the History of Hermeneutics." In Van der Meer 1996, 4: 245–60.

———. 2002. *God's Two Books*. Notre Dame: University of Notre Dame Press.

Hull, D., M. Forbes, and R. M. Burian, eds. 1995. *PSA 1994*. 2 vols. East Lansing: Philosophy of Science Association.

Hunter, M., ed. 1998. *Archives of the Scientific Revolution*. Woodbridge: Boydell.

Husserl, E. 1936. "Die Krisis der europäischen Wissenschaften und die transzendentale Phänomenologie." *Philosophia* (Belgrade) 1: 77–176.

———. 1962. *Die Krisis der europäischen Wissenschaften und die transzendentale Phänomenologie*. Ed. W. Biemel. The Hague: Martinus Nijoff.

———. 1970. "Galileo's Mathematization of Nature." In idem, *The Crisis of European Sciences and Transcendental Phenomenology*, trans. D. Carr, 23–59. Evanston: Northwestern University Press.

Hutchison, K. 1990. "Sunspots, Galileo, and the Orbit of the Earth." *Isis* 81: 68–74.

Huygens, C. 1646. "De Motu Naturaliter Accelerato." In Huygens 1888–1950, 11: 68–75.

———. 1660. *Brevis Assertio Systematis Saturnii Sui*. Florence.

———. 1698. *Kosmotheoros, sive de Terris Coelestibus, Earumque Ornatu, Conjecturae*. Hagae Comitum.

———. 1888–1950. *Oeuvres Complètes*. 22 vols. The Hague: Nijhoff.

Inchofer, M. 1633. *Tractatus Syllepticus*. Rome.

———. 1635. "Vidiciarum S. Sedi Apostolicae, Sacrorum Tribunalium et Authoritatum adversos Neopythagoraeos Terrae Motores et Solis Statores [Libri Duo]." Unpublished ms. Rome: Biblioteca Casanatense, MS 182.

Jamin, N. 1782. *Pensieri teologici relativi agli errori de' nostri tempi*. 2nd ed. Milan.

Jardine, N. 1984. *The Birth of History and Philosophy of Science*. Cambridge: Cambridge University Press.

Jaugey, J.-B. 1888. *Le Procès de Galilée et la Théologie*. Paris.

Jemolo, A. C. 1923. "Il pensiero religioso di Ludovico Antonio Muratori." *Rivista trimestrale di studi filosofici e religiosi* 4: 23–78.

John Paul II. 1979a. [Sull'armonia profonda tra verità della fede e verità della scienza]. *L'Osservatore Romano*, 11 November, 1–2.

———. 1979b. "Deep Harmony Which Unites the Truths of Science with the Truths of Faith." *L'Osservatore Romano*, weekly edition in English, 26 November, 9–10.

———. 1979c. "La grandezza di Galileo è a tutti nota." In Poupard 1984, 271–77.

———. 1979d. "The Greatness of Galileo is Known to All." In Poupard 1987, 195–200.

———. 1992a. "Faith Can Never Conflict with Reason." *L'Osservatore Romano*, weekly edition in English, 4 November, 1–2.

———. 1992b. "Discours à l'Académie Pontificale des Sciences, 31 octobre 1992." In Poupard 1994a, 99–107.

———. 1992c. "Lessons of the Galileo Case." *Origins: CNS Documentary Service*, 12 November, vol. 22, no. 22, 369, 371–74.

Keill, J. 1742. *Introductiones ad Veram Physicam, et Veram Astronomiam*. Milan.

Kellison, M. 1633. Kellison to Lagonissa, 7 September. In Favaro 19: 392–93; Pagano 1984, 206; French trans. in Monchamp 1892, 121; Italian trans. in Genovesi 1966, 295–96.

Kelly, M. 1948. *Error in the Universe*. Radio play aired on BBC Third Programme, 16 October.

Kenny, A., ed. 1970. *Philosophical Letters of R. Descartes*. Minneapolis: University of Minnesota Press.

Kepler, J. 1619a. "Admonitio ad Bibliopolas Exteros." French trans. in Segonds 1984, 267–68.

———. 1619b. *Harmonices Mundi*. Lincii Austriae.

———. 1635. "Perioche ex Introductione in Martem." In G. Galilei 1635b, 459–64.

———. 1661. "Reconcilings of Texts of Sacred Scripture That Seem to Oppose the Doctrine of the Earth's Mobility." In Salusbury 1661–1665, vol. 1, part 1, 461–67.

Koestler, Arthur. 1959. *The Sleepwalkers*. New York: Macmillan.

———. 1960a. "Sleepwalkers and Vigilantes." *Isis* 51: 73–77.

———. 1960b. *Les Somnambules*. Paris.

———. 1964. "The Greatest Scandal in Christendom." *Observer*, 2 February, 21, 29.

———. 1981a. *I sonnambuli*. Trans. G. Giacometti. Milan.

———. 1981b. *Los Sonámbulos*. Mexico City: CONACYT.

Koven, R. 1980. "World Takes Turn in Favor of Galileo." *Washington Post*, 24 October.

Kuhn, T. S. 1977. *The Essential Tension*. Chicago: University of Chicago Press.

Kvasz, L. 2002. "Galilean Physics in Light of Husserlian Phenomenology." *Philosophia Naturalis* 39:209–33.

Lämmel, R. 1927. *Galilei im Lichte des 20. Jahrhunderts (Menschen, Völker, Zeiten)*. Berlin.

———. 1928. "Untersuchung der Dokumente des Galileischen Inquisitionsprozesses." *Archiv für Geschichte der Mathematik, der Naturwissenschaften und der Technik* 10: 405–19.

Lagonissa, Fabio de. 1633. Lagonissa to Jansenius, 1 September. In Favaro 15: 245.

Lalande, Joseph J. 1771. *Astronomie* (1764). 2nd ed. 3 vols. Paris.

Lamalle, E. 1964. "Nota introduttiva all'opera." In Paschini 1964a, 1: vii–xv.

Langford, J. J. 1971. *Galileo, Science and the Church*. Rev. ed. Ann Arbor: University of Michigan Press.

Lansbergen, Jacob van. 1633. *Apologia pro Commentationibus Philippi Lansbergii in Motum Terrae Diurnum et Annum*. Middleburg.

Lansbergen, Philip van. 1630. *Commentationes in Motum Terrae Diurnum et Annum*. Middleburg.

Laplace, P. S. 1803. "Mémoire sur le Mouvement d'un Corps Qui Tombe d'une Grande Hauteur." *Bulletin de la Société Philomatique* 3: 109–15. Rpt. in Laplace 1878–1912, 14: 266–77.

———. 1805. "De la Chute des Corps Qui Tombent d'une Grande Hauteur." In idem, *Traité de Mécanique Céleste*, 4: 294–305. 5 vols. Paris. Rpt. in Laplace 1878–1912, 4: 303–6.

———. 1878–1912. *Oeuvres Complètes*. 14 vols. Paris: Gauthier-Villars.

Lazzari, Pietro. 1757. "Riflessioni sopra l'articolo 'Libri omnes docentes mobilitatem terrae et immobilitatem solis.'" In Baldini 2000c, 307–28.

Le Cazre, P. 1645a. *Physica Demonstratio Qua Ratio, Mensura, Modus ac Potentia Accelerationis Motus in Naturali Descensu Gravium Determinantur*. Paris.

———. 1645b. *Vindiciae Demonstrationis Physicae de Proportione Qua Gravia Decidentia Accelerantur*. Paris.

Lehmann-Haupt, H., ed. 1967. *Homage to a Bookman*. Berlin: Gebr. Mann.

Leibniz, Gottfried W. 1679–1686. "Apologia Fidei Catholicae ex Recta Ratione." In Grua 1948, 1: 30–34.

———. 1684. Leibniz to Landgrave Ernst, 10/20 October, no. xxvii. In Rommel 1847, 2: 44–48.

———. 1688. Leibniz to Landgrave Ernst, July or August, no. lii, postscript. In Rommel 1847, 2: 200–202.

———. 1689a. "Praeclarum Ciceronis dictum est, . . ." In Robinet 1988, 107–110.

———. 1689b. "Cum geometricis demonstrationibus. . . ." In Robinet 1988, 111–114.

———. 1689c. "On Copernicanism and the Relativity of Motion." In Leibniz 1989, 90–94.

———. 1704a. "De l'erreur." In *Nouveaux Essais sur l'Entendment Humain*, book 4, chapter 20. In Leibniz 1923ff, vol. 6, part 6, 509–21.

———. 1704b. "De l'erreur." In *Nouveaux Essais sur l'Entendment Humain*, ed. J. Brunschvig, 452–63. Paris: Garnier-Flammarion, 1966.

———. 1704c. "Of Error." In Remnant and Bennett 1997, 509–21.

———. 1768. *Opera Omnia*. Ed. L. Dutens. Geneva. Rpt., Hildesheim: Georg Olms, 1989.

———. 1923ff. *Sämtliche Schriften und Briefe*. Darmstadt, Leipzig, and Berlin.

———. 1948. *Textes Inédits*. Ed. G. Grua. 2 vols. Paris: PUF.

———. 1989. *Philosophical Essays*. Ed. and trans. R. Ariew and D. Garber. Indianapolis: Hackett.

———. 1997. *New Essays on Human Understanding*. Trans. and ed. P. Remnant and J. Bennet. Cambridge: Cambridge University Press.

Leman, A. 1920. *Recueil des Instructions Générales aus Nonces Ordinaires de France de 1624 à 1634*. Paris: Giard-Champion.

Lenoble, R. 1943. *Mersenne; ou, la Naissance du Mécanisme.* Paris: Vrin.

———. 1957. "Origines de la Pensée Scientifique Moderne." In *Histoire de la Science,* ed. M. Daumas, 367–534. Paris: Gallimard.

Leo XIII. 1893. *Providentissimus Deus.* In Carlen 1981, 2: 325–39.

L'Epinois, Henri de. 1867. "Galilée: Son Procès, Sa Condamnation d'après des Documents Inédits." *Revue des Questions Historiques* 3:68–171.

———. 1877. *Les Pièces du Procès de Galilée Précédées d'un Avant-propos.* Paris.

———. 1878. *La Question de Galilée.* Paris.

Lerner, M.-P. 1998a. "Essay Review: 'Copernicus Is Not Susceptible to Compromise.'" *Studies in the History and Philosophy of Science* 29: 663–72.

———. 1998b. "Pour une Edition Critique de la Sentence et de l'Abjuration de Galilée." *Revue des Sciences Philosophiques et Théologiques* 82: 607–29.

———. 1999. "L' 'Hérésie' Héliocentrique." In Brice and Romano 1999, 69–91.

———. 2001a. "La Réception de la Condamnation de Galilée en France au XVIIe Siècle." In Montesinos and Solís 2001, 513–48.

———, ed. 2001b. *Tommaso Campanella, Apologia pro Galileo.* Paris: Belles Lettres.

———. 2001c. "Introduction." In Lerner 2001b, ix–clxv.

———. 2001d. "Le Moine, le Cardinal et le Savant." *Les Cahiers de l'Humanisme* 2: 71–94.

———. 2002a. "Aux Origines de la Polémique Anticopernicienne (I)." *Revue des Sciences Philosophiques et Théologiques* 86: 681–721.

———. 2002b. "Tycho Brahe Censured." In Christianson et al. 2002, 95–101.

Lessl, T. S. 1999. "The Galileo Legend as Scientific Folklore." *Quarterly Journal of Speech* 85: 146–68.

Le Tenneur, L. A. 1646. "Disputatio Physico-Mathematica." Paris: National Library, Ms. Fonds Lat. 6740.

———. 1649. *De Motu Naturaliter Accelerato.* Paris.

Libri, Guglielmo. 1838–1841. *Histoire des Sciences Mathématiques en Italie.* 4 vols. Paris.

———. 1840–1841. Review of Brewster's *Life of Galileo. Journal des Savants,* 1840, 556–69, 589–602; 1841, 157–71, 203–23.

———. 1841a. *Essai sur la Vie et les Travaux de Galilée.* Paris.

———. 1841b. *Galileo, sua vita e sue opere.* Milan.

———. 1841c. *Histoire des Sciences Mathématiques en Italie.* Vol. 4. Paris.

———. 1842. *Galileo Galilei zu seinem Gedächtniss im zweiten Säcularjahr seines Todes, sein Leben und seine Werke.* Trans. and ed. F. W. Carove. Siegen and Wiesbaden.

Limborch, P. van. 1692. *Historia Inquisitionis.* Amsterdam.

———. 1731. *The History of the Inquisition.* 2 vols. Trans. S. Chandler. London.

Lindberg, D. C. 2003. "Galileo, the Church, and the Cosmos." In Lindberg and Numbers 2003, 33–60.

Lindberg, D. C., and R. L. Numbers, eds. 1986. *God and Nature.* Berkeley: University of California Press.

———. 1987. "Beyond War and Peace." *Perspectives on Science and Christian Faith* 39: 140–49.

———, eds. 2003. *When Science and Christianity Meet.* Chicago: University of Chicago Press.

Lindberg, D. C., and R. S. Westman, eds. 1990. *Reappraisals of the Scientific Revolution.* Cambridge: Cambridge University Press.

Linemannus, A. 1635. *Disputatio Theorematica.* Königsberg.

Lipsius, I. 1604. *Physiologiae Stoicorum Libri Tres.* Paris.

Llofriu y Sagrera, E. 1875. *Galileo: Episodio Dramático en un Acto y en Verso.* Madrid.

Lloyd, G. E. R. 1978. "Saving Appearances." *Classical Quarterly* 28: 202–22.

Lorenzoni, G. 1913. "Ricordi intorno a Giuseppe Toaldo." In *Atti e memorie,* ed. Accademia delle Scienze (Padua), 29(2): 271–316.

Maccagni, C., ed. 1972. *Saggi su Galileo Galilei.* Vol. 3, tome 2. Florence: Barbèra.

Maccarrone, M. 1963. "Mons. Pio Paschini (1870–1962)." *Rivista di storia della Chiesa in Italia* 17: 181–304.

———. 1979a. "Mons. Paschini e la Roma ecclesiastica." In *Atti del convegno di studio su Pio Paschini* 1979, 49–93.

———. 1979b. "Mons. Paschini e la Roma ecclesiastica." *Lateranum* 45: 154–218.

Machamer, P., ed. 1998. *The Cambridge Companion to Galileo.* Cambridge: Cambridge University Press.

MacLachlan, J. Forthcoming. *Catch a Falling Star.*

MacLaurin, C. 1749. *Exposition des Découvertes de M. le Chevalier Newton.* Paris.

Maculano, Vincenzo. 1633a. Maculano to F. Barberini, 22 April. Vatican City, Archivio della Congregazione per la Dottrina della Fede, Fondo Sant'Uffizio, St. st. N 3 f, first fascicle, f. 185. In Beretta 2001, 571.

———. 1633b. Maculano to F. Barberini, 28 April. In Pieralisi 1875, 197–98; in Favaro 15: 106–7.

Madden, Richard R. 1853. *The Life and Martyrdom of Savonarola.* 2 vols. London.

———. 1863. *Galileo and the Inquisition.* London.

Maffei, Paolo. 1975. "Il sistema copernicano dopo Galileo e l'ultimo conflitto per la sua affermazione." *Giornale di astronomia* 1: 5–12.

———. 1987. *Giuseppe Settele, il suo Diario e la questione galileiana.* Foligno: Edizioni dell'Arquata.

Maffi, P. 1918a. *Lettere pastorali, omelie e discorsi.* Turin: Libreria Editrice Internazionale.

———. 1918b. "La questione galileiana." In Maffi 1918a, 525–30.

Magalotti, Filippo. 1632. Magalotti to Guiducci, 7 August. In Favaro 14: 368–71.

Maiocchi, R. 1985. *Chimica e filosofia.* Florence: Nuova Italia.

———. 1990. "Pierre Duhem's *The Aim and Structure of Physical Theory.*" *Synthese* 83: 385–400.

Maistre, J. de. 1837. *Lettre à un Gentilhomme Russe sur l'Inquisition Espagnole.* Lyons.

Mallet du Pan, Jacques. 1784. "Mensognes Imprimées au Sujet de la Persécution de Galilée." *Mercure de France,* 17 July, 121–30.

Mamachi, T. M. 1785. "Annotazione per la pag. 165 e segg." In Tiraboschi 1782–1797, tome 8, viii.

Manfredi, E. 1729. *De Annuis Inerrantium Stellarum Aberrationibus.* Bologna.

Margolis, H. 1987. *Patterns, Thinking, and Cognition.* Chicago: University of Chicago Press.

———. 1991. "Tycho's System & Galileo's *Dialogue.*" *Studies in History and Philosophy of Science* 22: 259–75.

———. 1993. *Paradigms and Barriers.* Chicago: University of Chicago Press.

———. 2002. *It Started with Copernicus.* New York: McGraw-Hill.

Marini, Marino. 1817a. Marini to Richelieu, 23 July. In Favaro 1887a, 212–13.

———. 1817b. Marini to Pasquier, 11 September. In Favaro 1887a, 216–17.

———. 1850. *Galileo e l'Inquisizione.* Rome.

Martin, R. N. D. 1987. "Saving Duhem and Galileo." *History of Science* 25: 301–19.

———. 1991. *Pierre Duhem*. La Salle, Ill.: Open Court.

Martin, T. H. 1868. *Galilée, les Droits de la Science et la Méthode des Sciences Physiques*. Paris.

Martínez, R. 2001. "Il manoscritto ACDF, *Index, Protocolli,* vol. EE, f. 291r–v." *Acta Philosophica* 10: 215–42.

Martini, C. M. 1972. "Galileo e la teologia." In Maccagni 1972, 441–51.

———, G. Ghiberti, and M. Pesce. 1995. *Cento anni di cammino biblico*. Milan: Vita e Pensiero.

Masi, R. 1964. "La più grande scoperta: il metodo sperimentale." *L'Osservatore Romano,* 15 February.

Masini, Eliseo. 1621. *Sacro arsenale overo Prattica dell'officio della Santa Inquisizione.* Genoa. Rpt., Bologna, 1665; Rome, 1716.

Mateo-Seco, L. F. 2001. "Galileo e l'Eucaristia." *Acta Philosophica* 10: 243–56.

Mauri, A. 1833. Editorial Preface. In Tiraboschi 1833, 1: v–viii.

Mayaud, Pierre-Noël. 1997. *La Condamnation des Livres Coperniciens et sa Révocation à la Lumière de Documents Inédits des Congrégations de l'Index et de l'Inquisition.* Rome: Editrice Pontificia Università Gregoriana.

McColley, G. 1938. "The Ross-Wilkins Controversy." *Annals of Science* 3: 153–89.

McMullin, E., ed. 1967. *Galileo: Man of Science.* New York: Basic Books.

———. 1980. Letter to the Editor. *New York Times,* 10 November.

———. 1990a. "Comment: Duhem's Middle Way." *Synthese* 83: 421–30.

———. 1990b. "Conceptions of Science in the Scientific Revolution." In Lindberg and Westman 1990, 27–92.

———. 1998. "Galileo on Science and Scripture." In Machamer 1998, 271–347.

———, ed. 2005. *The Church and Galileo.* Notre Dame: University of Notre Dame Press.

Mendoza, R. G. 1995. *The Acentric Labyrinth.* Shaftesbury: Element Books.

Mercati, A. 1926–1927. "Come e quando ritornò a Roma il codice del processo di Galileo." *Atti della Pontificia Accademia delle scienze dei Nuovi Lincei* 80: 58–63.

———, ed. 1942. *Il sommario del processo di Giordano Bruno.* Vatican City: Biblioteca Apostolica Vaticana.

Mersenne, Marin. 1623. *Quaestiones Celeberrimae in Genesim.* Paris.

———. 1634. *Les Questions Théologiques, Physiques, Morales, et Mathématiques.* Paris.

———. 1636. *Harmonie Universelle.* Paris.

———. 1647. *Novarum Observationum Physico-Mathematicarum Tomus Tertius.* Paris.

———. 1932–1988. *Correspondance.* 17 vols. Ed. C. de Waard and R. Pintard. Paris: Éditions du Centre National de la Recherche Scientifique.

———. 1985. *Questions Inouyes.* Ed. A. Pessel. Paris: Fayard.

Micanzio, F. 1639. Micanzio to Galileo, 17 September. In Favaro 18: 104–5.

Miller, P. N. 2000. *Peiresc's Europe.* New Haven: Yale University Press.

Miller, V. R., and R. P. Miller, eds. and trans. 1991. *Principles of Philosophy by René Descartes.* Dordrecht: Kluwer.

Milliet de Chales, C. F. [alias C. F. Milliet de Challes or C. F. M. Dechales]. 1674. *Cursus seu Mundus Mathematicus.* 3 vols. Lyons.

Milton, John. 1644. *Areopagitica.* London.

———. 1959. *Areopagitica.* In Milton 1953–1982, 2: 485–570.

————. 1953–1982. *Complete Prose Works.* 8 vols. New Haven: Yale University Press.

Miscellanea galileiana, 3 vols. Pontificiae Academiae Scientiarum Scripta Varia, no. 27. Vatican City: Ex Aedibus Adademicis in Civitate Vaticana, 1964.

Mivart, St. George Jackson. 1885. "Modern Catholics and Scientific Freedom." *Nineteenth Century* 18: 30–47.

Monchamp, Georges. 1892. *Galilée et la Belgique.* Saint-Trond, Brussels, and Paris.

————. 1893. *Notification de la Condamnation de Galilée Datée de Liège, 20 Septembre 1633, Publiée par le Nonce de Cologne.* Cologne and Saint-Trond.

Monsagrati, G. 2000. "Giordani, Pietro." *Dizionario biografico degli italiani* 55: 219–26.

Montesinos, J., and C. Solís, eds. 2001. *Largo Campo di Filosofare.* La Orotava: Fundación Canaria Orotava de Historia de la Ciencia.

Montfaucon, B. de. 1706. *Collectio Nova Patrum et Scriptorum Graecorum.* Vol. 2. Paris.

Montucla, J. E. 1758a. "Examen des Objections de Divers Genres qu'on a Proposée contre le Mouvement de la Terre." In Montucla 1758b, 1: 529–42.

————. 1758b. *Histoire des Mathématiques.* 2 vols. Paris.

————. 1799–1802. *Histoire des Mathématiques.* New augmented ed. 4 vols. Paris.

Morin, J. B. 1631. *Famosi et Antiqui Problematis de Telluris Motu vel Quiete, Hactenus Optata Solutio.* Paris.

————. 1634. *Responsio pro Telluris Quiete ad Jacobi Lansbergii Doct. Med. Apologiam pro Telluris Motu.* Paris.

————. 1640. *Astronomia iam a Fundamentis Integre et Exacte Restituta.* Paris.

————. 1642. *Tycho Brahaeus in Philolaum pro Telluris Quiete.* Paris.

————. 1643. *Alae Telluris Fractae.* Paris.

————. 1650. *Résponse à une Longue Lettre de Monsieur Gassend.* Paris.

Morpurgo-Tagliabue, G. 1981. *I processi di Galileo e l'epistemologia.* Rome: Armando.

————. 1984a. "Galileo eretico?" *Tempo Presente,* nos. 43–44, 30–36.

————. 1984b. "Galileo: Quale Eresia?" *Rivista di storia della filosofia* 39: 741–50.

Moss, J. D. 1986. "The Rhetoric of Proof in Galileo's Writings on the Copernican System." In Wallace 1986, 179–204.

————. 1993. *Novelties in the Heavens.* Chicago: University of Chicago Press.

Mothu, A., ed. 2000. *Révolution Scientifique et Libertinage.* Turnhout: Brepols.

Motta, Franco. 1993. "La Ricezione della Condanna di G. Galilei nel XVII Secolo." Thesis in the History of Christianity, Faculty of Political Science, University of Bologna.

————. 1996. Review of Two 1992 Books by W. Brandmüller. *Rivista di storia e letteratura religiosa* 32: 673–83.

————. 1997a. "Bellarminiana." *Rivista di storia e letteratura religiosa* 33: 131–60.

————. 1997b. "Copernico, i gesuiti, le sorgenti del Nilo." In Venturi Barbolini 1977, 109–70.

————. 1997c. "'Geographia Sacra.'" *Annali di storia dell'esegesi* 14: 477–506.

————, ed. 2000. *Galileo Galilei, Lettera a Cristina di Lorena.* Genoa: Marietti.

————. 2001. "I criptocopernicani." In Montesinos and Solís 2001, 693–718.

Motzkin, G. 1989. "The Catholic Response to Secularization and the Rise of the History of Science as a Discipline." *Science in Context* 3: 203–26.

Mousnier, P. 1646. *Tractatus Physicus de Motu Locali.* Lyons.

Mousnier, P., and H. Fabri. 1648. *Metaphysica Demonstrativa*. Lyons.

Müller, Adolf. 1897. "Die Sonnenflecke im Zusammenhang mit dem Copernicanischen Weltsystem." *Stimmen aus Maria Laach,* vol. 52, 4 April.

———. 1909a. *Der Galilei-Prozess (1632–1633) nach Ursprung, Verlauf und Folgen.* Freiburg im Breisgau: Herdersche Verlagshandlung.

———. 1909b. *Galileo Galilei und das Kopernikanische Weltsystem.* Freiburg im Breisgau: Herder.

———. 1911. *Galileo Galilei.* Trans. P. Perciballi. Rome: Max Bretschneider.

Muratori, L. A. 1714. *De Ingeniorum Moderatione in Religionis Negotio*. Paris.

———. 1779. *De Ingeniorum Moderatione in Religionis Negotio.* New revised ed. Augsburg.

———. 1964a. *Opere.* 2 vols. Ed. G. Falco and F. Forti. Milan: Ricciardi.

———. 1964b. "Le prime polemiche sul 'De Ingeniorum Moderatione.' " In Muratori 1964a, 326–35.

———. 1964c. [Selections from *De Ingeniorum Moderatione,* book 1, chapters 21–24]. In Muratori 1964a, 296–325.

Navarro Brotons, V. 1995. "The Reception of Copernicus in 16th Century Spain." *Isis* 86: 52–78.

———. 2001. "Galileo y España." In Montesinos and Solís 2001, 809–30.

Nellen, H. J. M. 1994. *Ismaël Boulliau (1605–1694).* Amsterdam: APA–Holland University Press.

Nelli, G. B. C. 1793. *Vita e commercio letterario di Galileo Galilei.* 2 vols. Lausanne [Florence].

Nel terzo centenario della morte di Galileo Galilei. 1942. Ed. Università Cattolica del Sacro Cuore. Milan: Società Editrice "Vita e Pensiero."

Newman, John H. 1858. Newman to E. B. Pusey. In Newman 1961ff, 18: 321–22.

———. 1861. "An Essay on the Inspiration of Holy Scripture." In Seynaeve 1953, appendix 4, 60–70.

———. 1877. "Preface to the Third Edition of *Via Media.*" In Newman 1918, 1: xv–xciv; in Newman 1990, 10–57.

———. 1884. Newman to St. George Jackson Mivart, 8 May. In Newman 1961ff, 30 [1976]: 358–59.

———. 1918. *The Via Media.* 2 vols. New impression. London: Longmans, Green.

———. [1945]. *Christianity and Science.* Dublin: [Browne and Nolan].

———. 1961. "Galileo, Revelation, and the Educated Man." In Collins 1961, 284–91.

———. 1961ff. *Letters and Diaries.* Ed. I. Ker and T. Gotnall. Oxford: Clarendon.

———. 1990. *The Via Media of the Anglican Church.* Ed. H. D. Weidner. Oxford: Clarendon.

Newton, I. 1739–1742. *Philosophiae Naturalis Principia Mathematica.* 3 vols. Ed. T. Le Seur and F. Jacquier. Rev. ed., Cologne, 1760.

Nieuwentijd, B. 1727. *L'Existence de Dieu, Demontrée par le Merveilles de la Nature.* Amsterdam.

Nonis, P. 1979. "L'ultima opera di Paschini: Galilei." In *Atti del convegno di studio su Pio Paschini* 1979, 158–72.

Nonnoi, G. 2000. *Saggi galileiani.* Cagliari: AM&D Edizioni.

Numbers, R. L. 1985. "Science and Religion." *Osiris,* 2nd series, 1: 59–80.

Observations sur le Bref de N.S.P. le Pape Benoit XIV au Grand-Inquisiteur d'Espagne. N.p.: 4 March 1749.

Olivieri, Maurizio B. 1820a. (June 10.) "Riflessioni sopra i Motivi pe' quali il R.mo P. Filippo Anfossi, Maestro del Sacro Palazzo Apostolico, dice di aver negato l'imprimatur ad uno scritto in cui s'insegna come positiva la mobilità della Terra e immobilità del Sole." In Maffei 1987, 471–533; in Brandmüller and Greipl 1992, 225–87.

———. 1820b. (September.) "Commenti sui 'Motivi' di Filippo Anfossi." In Brandmüller and Greipl 1992, 317–25.

———. 1820c. (November.) "Ristretto di Ragione, e di Fatto." In Maffei 1987, 427–50; in Brandmüller and Greipl 1992, 351–80.

———, ed. 1820d. (November.) "Sommario di Documenti, e di Allegazioni." In Maffei 1987, 451–580.

———. 1821a. Olivieri to Anfossi, 24 May. In Brandmüller and Greipl 1992, 397–405.

———. 1821b. Olivieri to Anfossi, undated. In Brandmüller and Greipl 1992, 406–11.

———. 1822. (August?) "Voto riguardante la pubblicazione di un'cstratto dell'*Astronomia*." In Brandmüller and Greipl 1992, 412–26.

———. 1823a. (September.) "Riflessioni sopra la lettera di Filippo Anfossi a Pietro Odescalchi." In Brandmüller and Greipl 1992, 432–38.

———. 1823b. (October–November.) "Voto." In Brandmüller and Greipl 1992, 440–80.

———. 1840. *Di Copernico e di Galileo: Scritto postumo.* In Bonora 1872, 1–133.

[Olivieri, Maurizio B.]. 1841a. "Galilée et l'inquisition romaine." *L'Université Catholique,* 1st series, 11: 219–27.

[Olivieri, Maurizio B.]. 1841b. "Der heilige Stuhl gegen Galilei und das astronomische System der Copernicus." *Historische-politische Blätter für das katholische Deutschland,* vol. 7, 385ff., 449ff., 513ff., and 577ff.

———. 1841c. "Meriti dei Romani Pontefici verso l'astronomia." *Annali delle scienze religiose* 14: 95–96.

O'Meara, D. J., ed. 1981. *Studies in Aristotle.* Washington: Catholic University of America Press.

Pagano, S. M., ed. 1984. *I documenti del processo di Galileo Galilei.* Vatican City: Pontificia Academia Scientiarum.

———. 1994. Review of Brandmüller and Greipl's *Copernico, Galilei e la Chiesa. Barnabiti studi* 11: 270–81.

Paladini, L. A., ed. 1890. *Lettere di ottimi autori sopra cose familiari.* Florence.

Palmerino, C. R. 1999. "Infinite Degrees of Speed." *Early Science and Medicine* 4: 269–328.

Pantin, I. 1999. "New Philosophy and Old Prejudices." *Studies in History and Philosophy of Science* 30: 237–62.

———. 2000. "'Dissiper les Ténèbres Qui Restent Encore à Percer.'" In Mothu 2000, 11–34.

———. 2001. "Libert Froidmont et Galilée." In Montesinos and Solís 2001, 615–36.

[Parasin, Matthias M.]. 1648. *Systema Mundi.* Stockholm.

Parchappe, M. 1866. *Galilée: Sa Vie, Ses Decouvertes et Ses Travaux.* Paris.

Pardo Tomás, J. 1991. *Ciencia y Censura.* Madrid: Consejo Superior de Investiga-
ciones Cientificas.

Pascal, Blaise. 1657. "Letter No. XVIII," *The Provincial Letters.* In Pascal 1744, 2:
278–318; 1962, 2: 34–55; 1967, 279–98.

———. 1744. *The Life of Mr. Paschal, with His Letters Relating to the Jesuits.* Trans. W. A.
2 vols. London.

———. 1962. *Les Provinciales.* 2 vols. Ed. J. Steinmann. Paris: Librairie Armand
Colin.

———. 1967. *The Provincial Letters.* Trans. and ed. A. J. Krailsheimer. Harmonds-
worth: Penguin.

Paschini, Pio. 1940. *Roma nel Rinascimento.* Bologna: Cappelli.

———. 1941. Paschini to Vale, 4 December. In Bertolla 1979, 175–76.

———. 1942. Review of Mercati's *Sommario del processo di Giordano Bruno. Studium*
38:226–29.

———. 1943. "L'insegnamento di Galileo: non temere la verità." *Studium,* April, 39:
94–97.

———. 1946a. Paschini to Montini, 12 May. In Maccarrone 1979a, 82–83 n. 112;
1979b, 202–3.

———. 1946b. Paschini to Vale, 15 May 1946. In Simoncelli 1992, 72–73.

———. 1946c. Paschini to Vale, 4–5 July 1946. In Simoncelli 1992, 77–78.

———. 1950. "Galileo Galilei." *Enciclopedia cattolica,* vol. 5, columns 1871–1880.
Vatican City: Ente per l'Enciclopedia cattolica e per il libro cattolico.

———. 1964a. *Vita e opere di Galileo Galilei.* Ed. E. Lamalle. 2 vols. In *Miscellanea
Galileiana,* vols. 1 and 2.

———. 1964b. *Vita e opere di Galileo Galilei.* 2 vols. Vatican City: Pontificia Accademia
delle Scienze.

———. 1965. *Vita e opere di Galileo Galilei.* Ed. M. Maccarrone. Rome: Herder.

Pastor, L. von. 1891–1953. *History of the Popes from the Close of the Middle Ages.* 40 vols.
Vols. 1–2, London: Hodges, 1891. Vols. 3–40, London: Kegan Paul, 1894–1953.
Also vols. 1–40, St. Louis: Herder, 1898–1953.

———. 1937a. "Galileo and the Roman Inquisition." In Pastor 1937b, vol. 25, chapter 7, 285–309.

———. 1937b. *History of the Popes from the Close of the Middle Ages,* vols. 25–26, *Leo XI
and Paul V (1605–1621).* Trans. Dom Ernest Graf. St. Louis: Herder.

———. 1938a. *History of the Popes from the Close of the Middle Ages,* vols. 27–29, *Gregory
XV and Urban VIII (1621–1644).* Trans. Dom Ernest Graf. St. Louis: Herder.

———. 1938b. "The Roman Inquisition and the Trial of Galileo." In Pastor 1938a,
vol. 29, chapter 1, 34–62.

Pedersen, O. 1983a. "Galileo and the Council of Trent." *Journal for the History of
Astronomy* 14: 1–29.

———. 1983b. *Galileo and the Council of Trent.* Vatican City: Specola Vaticana.

———. 1985. "Galileo's Religion." In Coyne, Heller, and Zycínski 1985, 75–
102.

Peiresc, Nicolas C. F. de. 1634. Peiresc to F. Barberini, 5 December. In Favaro 16:
169–71.

———. 1635. Peiresc to F. Barberini, 31 January. In Favaro 16: 202.

Pepe, L., ed. 1996a. *Copernico e la questione Copernicana in Italia dal XVI and XIX secolo.* Florence: Olschki.

———. 1996b. "Ferrara e le celebrazioni copernicane, 1871–1973." In Pepe 1996a, 281–91.

Pera, M. 1998. "The God of the Theologians and the God of the Astronomers." In Machamer 1998, 367–88.

Pesce, Mauro. 1987. "L'interpretazione della Bibbia nella Lettera di Galileo a Cristina di Lorena e la sua ricezione." *Annali di storia dell'esegesi* 4: 239–84.

———. 1989. "Esegesi storica ed esegesi spirituale nell'ermeneutica biblica cattolica dal pontificato di Leone XIII a quello di Pio XII." *Annali di storia dell'esegesi* 6: 261–91.

———. 1991a. "Momenti della ricezione dell'ermeneutica biblica galileiana e della *Lettera a Cristina* nel XVII secolo." *Annali di storia dell'esegesi* 8: 55–104.

———. 1991b. "Una nuova versione della lettera di G. Galilei a B. Castelli." *Nouvelles de la République des Lettres,* no. 2, 89–122.

———. 1991c. "Il rinnovamento biblico." In Guasco, Guerriero, and Traniello 1991, 575–610.

———. 1992a. "Il *Consensus veritatis* di Christoph Wittich e la distinzione tra verità scientifica e verità biblica." *Annali di storia dell'esegesi* 9: 53–76.

———. 1992b. "Le redazioni originali della Lettera 'copernicana' di G. Galilei a B. Castelli." *Filologia e critica* 17: 394–417.

———. 1995a. "L'indisciplinabilità del metodo e la necessità politica della simulazione e della dissimulazione in Galilei dal 1609 al 1642." In Prodi 1995, 161–84.

———. 1995b. "Dalla enciclica biblica di Leone XIII 'Providentissimus Deus' 1893 a quella di Pio XII 'Divino Afflante Spiritu' 1943." In Martini, Ghiberti, and Pesce 1995, 39–100.

———. 1996. "Una rinnovata difesa dell'esegesi storica ed esigenza di un'interpretazione teologica," *Studia patavina* 43: 25–42.

———. 1998. "Il primo Galileo e l'ermeneutica biblica." In Bocchini Camaiani and Scattigno 1998, 331–45.

———. 2000. "Introduzione." In Motta 2000, 7–66.

———. 2001. "Gli ingegni senza limiti e il pericolo per la fede." In Montesinos and Solís 2001, 637–60.

Pessel, A., ed. 1985. *Questions Inouyes, Questions Harmoniques, Questions Theologiques, les Mécaniques de Galilée, les Préludes de l'Armonie Universelle.* Paris: Fayard.

Pieracci di Turricchi, V. 1820. *Galileo Galilei: Commedia in versi in cinque atti.* In idem, *Commedie,* 143–86. Florence.

Pieralisi, Sante, ed. 1858. *Breve discorso della istituzione di un principe.* Rome.

———. 1875. *Urbano VIII e Galileo Galilei.* Rome.

———. 1876. *Correzioni al libro "Urbano VIII e Galileo Galilei," con osservazioni sopra "Il processo originale di Galileo Galilei" pubblicato da Domenico Berti.* Rome.

———. 1879. *Sopra la nuova edizione del processo originale di Galileo Galilei fatta da Domenico Berti.* Rome.

Pieroni, G. 1637. Pieroni to Galileo, 9 July. In Favaro 17: 130–32.

Pino, D. 1802. *Esame del newtoniano sistema intorno al moto della terra.* 3 vols. Como.

———. 1806. *L'incredibilità del moto della terra brevemente esposta.* Milan.

Pius XII. 1943. *Divino Afflante Spiritu.* In Carlen 1981, 4: 65–79.

Poisson, N.-J. 1670a. *Commentaire ou Remarques sur la Méthode de René Descartes.* Vendôme.

———. 1670b. "Réponse à la Lettre d'un Amy, Touchant l'Ame des Bestes." In Poisson 1670a, 235–37; in Lerner 2001a, 546–47.

Polacco, G. 1644. *Anticopernicus Catholicus.* Venice.

Ponsard, F. 1867. *Galilée: Drame en Trois Actes en Vers.* Paris. Rpt. in Ponsard 1875, 3: 116–83.

———. 1875. *Oeuvres Complètes.* 3 vols. Paris.

Ponzio, P., ed. 1997. *Tommaso Campanella, Apologia pro Galileo.* Milan: Rusconi.

Popper, K. R. 1956. "Three Views of Human Knowledge." Rpt. in idem, *Conjectures and Refutations,* 97–119. New York: Harper, 1963.

Portolano, M. 2000. "John Quincy Adams's Rhetorical Crusade for Astronomy." *Isis* 91: 480–503.

Poupard, Paul, ed. 1983. *Galileo Galilei: 350 Ans d'Histoire.* Tournai: Desclée International.

———, ed. 1984. *Galileo Galilei, 350 anni di storia.* Rome: Edizioni Piemme di Pietro Marietti.

———, ed. 1987. *Galileo Galilei: Toward a Resolution of 350 Years of Debate.* Trans. I. Campbell. Pittsburgh: Duquesne University Press.

———. 1992a. "Compte Rendu des Travaux de la Commission Pontificale d'Études de la Controverse Ptoléméo-Copernicienne aux XVIe–XVIIe Siècles." In Poupard 1994a, 93–97.

———. 1992b. "'Galileo Case' Is Resolved." *L'Osservatore Romano,* weekly wdition in English, 4 November, 8.

———. 1992c. "Galileo: Report on Papal Commission Findings." *Origins: CNS Documentary Service,* 12 November, 22(22): 374–75.

———, ed. 1994a. *Après Galilée.* Paris: Desclée de Brouwer.

———. 1994b. "Avant-propos." In Poupard 1994a, 9–15.

Prodi, P., ed. 1995. *Disciplina dell'anima, disciplina del corpo e disciplina della società.* Bologna: Il Mulino.

[Purcell, John B., ed.]. 1844a. *Galileo—The Roman Inquisition.* Cincinnati. (= Cooper 1844.)

———. 1844b. "Introduction." In Purcell 1844, 3–28.

Purgotti, S. 1823. *Riflessioni sopra un opuscolo.* Pergola.

Ratzinger, J. 1998. "Le ragioni di un'apertura." In *L'apertura degli archivi del Sant'Uffizio romano* 1998, 181–89.

Raven, M. 1860. *Galileo Galilei: Ein geschichtlicher Roman.* 2 vols. Leipzig.

———. 1869. *Galileo Galilei: Romanzo storico.* 2 vols. Trans. and ed. G. Strafforello. Turin.

Redondi, P. 1983. *Galileo eretico.* Turin: Einaudi.

———. 1985a. *Galilee Hérétique.* Trans. M. Aymard. Paris: Gallimard.

———. 1985b. "*Galileo eretico:* Anatema." *Rivista storica italiana* 97: 934–56.

———. 1987. *Galileo Heretic.* Trans. R. Rosenthal. Princeton: Princeton University Press.

———. 1994. "Dietro l'immagine." *Nuncius* 9: 65–116.

Remnant, P., and J. Bennet, trans. and eds. 1997. *New Essays on Human Understanding.* Cambridge: Cambridge University Press.

Renaudot, Théophraste, ed. 1633. [Abridgment of Inquisition's Sentence against Galileo]. *Gazette* (Paris), no. 122, 531–32.

———, ed. 1634. *Recueil des Gazettes Nouvelles et Relations de Toute l'Année 1633*. Paris.

———, ed. 1636. [Abridgment of Inquisition's Sentence against Galileo]. *Mercure François*, 19: 696–700.

Renieri, V. 1639. *Tabulae Mediceae Secundorum Mobilium Universales*. Florence.

———. 1647. *Tabulae Motuum Caelestium Universales*. Florence.

Renn, J. 2000a. "Editor's Introduction." *Science in Context* 13: 271–80.

———. 2000b. "A Forgotten Controversy." *Science in Context* 13: 593–95.

———, ed. 2001. *Galileo in Context*. Cambridge: Cambridge University Press.

Restiglian, M. 1982. "Nota su Giuseppe Toaldo e l'edizione toaldina del *Dialogo* di Galileo." In *Galileo Galilei e Padova*, 235–39. Padua: Studia Patavina, Rivista di scienze religiose.

Reumont, A. von. 1849. "Galilei und Rom." *Berliner Calender für 1849*, 139–240.

———. 1853. "Galilei und Rom." In idem, *Beiträge zur italianischen Geschichte* 1: 303–424. 6 vols. Berlin, 1853–1857.

Reusch, F. H. 1875. "Der Galilei'sche Process." *Historische Zeitschrift* 17(3): 121–43.

———. 1879. *Der Process Galilei's und die Jesuiten*. Bonn.

Riccioli, Giovanni Battista. 1651. *Almagestum Novum*. 2 vols. Bologna.

———. 1668. *Argomento fisicomattematico*. Bologna.

Rizza, C. 1961. "Galileo nella corrispondenza di Peiresc." *Studi francesi* 15: 431–51.

———. 1965. *Peiresc e l'Italia*. Turin: Giappichelli.

Roberts, W. W. 1870. *The Pontifical Decrees against the Motion of the Earth, Considered in their Bearing on Advanced Ultramontanism*. London.

———. 1885. *The Pontifical Decrees against the Doctrine of the Earth's Movement, and the Ultramontane Defence of Them*. London.

Robinet, A. 1988. *G. W. Leibniz: Iter Italicum*. Florence: Olschki.

Rocco, A. 1633. *Esercitationi filosofiche*. Venice.

Rommel, C. von, ed. 1847. *Leibniz und Landgraf Ernst von Hessen-Rheinfels*. 2 vols. Frankfurt.

Roselli, S. M. 1777–1783. *Summa Philosophica ad Mentem Angelici Praeceptoris S. Thomae Aquinatis*. 4 vols. Rome.

———. 1785. *Summa Philosophica ad Mentem Angelici Praeceptoris S. Thomae Aquinatis*. 2nd ed. 4 vols. Rome.

Rosini, G. 1841. *Descrizione della tribuna innalzata da Leopoldo II alla memoria di Galileo*. Florence.

Ross, A. 1634. *Commentum De Terrae Motu Circulari*. London.

———. 1636. *Novus Planeta Non Planeta*. London.

———. 1646. *The New Planet No Planet*. London.

Rossi, P. 1978. "Galileo Galilei e il libro dei Salmi." *Rivista di filosofia* 69: 45–71.

Rowland, W. 2003. *Galileo's Mistake*. Revised ed. New York: Arcade.

Russell, B. 1935. *Religion and Science*. New York: Oxford University Press, 1997.

Russell, J. L. 1989. "Catholic Astronomers and the Copernican System after the Condemnation of Galileo." *Annals of Science* 46: 365–86.

———. 1995. "What Was the Crime of Galileo?" *Annals of Science* 52: 403–10.

Russell, R. J., W. R. Stoeger, and G. V. Coyne, eds. 1990. *John Paul II on Science and Religion*. Vatican City: Vatican Observatory Publications.

Sacchi, G. 1913. *Sulla condanna del sistema copernicano: Appunti di esegesi*. Turin.

Sackenreiter-Zeyssolff, M. T. 1984. *Foi et Loi à Strasbourg*. Strasbourg.

Salusbury, Thomas. 1661a. Foreword to Galileo's *System of the World*. In Salusbury 1661–1665, vol. 1, part 1.

———, ed. and trans. 1661b. *System of the World*. In Salusbury 1661–1665, vol. 1, part 1, 1–424.

———, ed. and trans. 1661–1665. *Mathematical Collections and Translations*. 2 vols. London.

Salvini, S. 1717. *Fasti consolari dell'accademia fiorentina*. Florence.

Sandonnini, T. 1886. "Ancora di due controversie sul processo galileiano." *Rivista storica italiana* 3: 673–726.

Santillana, G. de. 1955a. *The Crime of Galileo*. Chicago: University of Chicago Press.

———. 1955b. "Il dramma di Galileo." *Ponte* 11: 1076–86.

———. 1960. "Galileo e i moderni." *Tempo presente* 5: 322–28.

Santillana, G. de, and S. Drake. 1959. "Arthur Koestler and His *Sleepwalkers*." *Isis* 50: 255–60.

———. 1960. [Reply to Koestler]. *Isis* 51: 77–79.

Sarpi, P. 1616. "Consulto intorno alla proibizione del libro di Copernico, 7 May." In Berti 1876b, 151–53.

Saverio, F., and M. Rossi. 1984. *Galileo Galilei nelle lettere della figlia Suor Maria Celeste*. Lanciano: Rocco Carabba.

Sawyer, R. J. 1992. *Far-Seer*. Ace Books.

Scartazzini, J. A. 1877–1878. "Il processo di Galileo Galilei e la moderna critica tedesca." *Rivista europea*, year 8, 4: 829–61; year 9, 5: 1–15 and 221–49, 6: 401–23, and 10: 417–53.

Schlereth, T. J. 1978. "John Zahm." Unpaginated preface to Zahm 1978.

Schoppe, K. 1600. Schoppe to Rittershausen, 17 February. In Firpo 1993, 348–55.

Schütt, H.-W. 2000. "Emil Wohlwill, *Galileo and His Battle for the Copernican System*." *Science in Context* 13: 641–63.

Schwab, R. N., trans. and ed. 1963. *Jean Le Rond D'Alembert, Preliminary Discourse to the Encyclopedia of Diderot*. Indianapolis: Bobbs-Merrill.

Segonds, A., trans. and ed. 1984. *J. Kepler, Secret du Monde*. Paris: Belles Lettres.

Segre, M. 1989. "Viviani's Life of Galileo." *Isis* 80: 207–32.

———. 1991a. *In the Wake of Galileo*. New Brunswick: Rutgers University Press.

———. 1991b. "Science at the Tuscan Court, 1642–1667." In *Physics, Cosmology and Astronomy, 1300–1700*, ed. S. Unguru, 295–308. Dordrecht: Kluwer.

———. 1997. "Light on the Galileo Case?" *Isis* 88: 484–504.

———. 1998. "The Never-Ending Galileo Story." In Machamer 1998, 388–416.

———. 1999. "Galileo: A 'Rehabilitation' That Has Never Taken Place." *Endeavour* 23(1): 20–23.

Serry, J. H. 1742. *Praelectiones Theologicae Dogmaticae Polemicae Scholasticae Habitae in Celeberrima Patavina Academia*. Venice.

Settele, Giuseppe. 1818 [actual publication date 1819]. *Elementi di ottica e di astronomia*, vol. 1, *Ottica*. Rome.

———. 1819 [actual publication date 1821]. *Elementi di ottica e di astronomia*, vol. 2, *Astronomia*. Rome.

———. 1820a. (March.) "Supplica presentata in Marzo p.p. a S. Santità." In Maffei 1987, 463–70; in Brandmüller and Greipl 1992, 167–77.

————. 1820b. (1 August.) "Seconda supplica a Sua Santità." In Maffei 1987, 535–38; in Brandmüller and Greipl 1992, 288–89.

————. 1820c. (23 August.) "Paragrafo d'inserzione." In Maffei 1987, 543–45; in Brandmüller and Greipl 1992, 303–5.

————. 1820–1833. *Diario.* In Maffei 1987, 283–421.

————. 1821. [Nota sulla condanna di Galileo]. In Settele 1819, 130–33, n. 1.

Sfondrati, C. 1695. *Cursus Philosophici Monasterii S. Calli Tomus III Physica Pars Posterior, cum Metaphysica.* S. Gallen.

Sharratt, M. 1974. "Copernicanism at Douai." *Durham University Journal,* December, 67: 41–48.

————. 1994. *Galileo, Decisive Innovator.* Cambridge, MA: Blackwell.

Shea, W. R. 1991. *The Magic of Numbers and Motion.* Canton: Science History Publications.

Shea, W. R., and M. Artigas. 2003. *Galileo in Rome.* Oxford: Oxford University Press.

Shusterman, D. 1975. *C. P. Snow.* Boston: Twayne.

Simoncelli, Paolo. 1988. "Inquisizione romana e riforma in Italia." *Rivista storica italiana* 100: 5–125.

————. 1992. *Storia di una censura.* Milan: FrancoAngeli.

————. 1993. "Galileo e la curia." *Belfagor* 48: 23–40.

Sirluck, E. 1959. "Introduction." In Milton 1953–1982, 2: 1–216.

Smith, A. M. 1985. "Galileo's Proof for the Earth's Motion from the Movement of Sunspots." *Isis* 76: 543–51.

Smith, G. 1995. *Galileo: A Dramatised Life.* London: Janus.

Snow, C. P. 1964. *The Two Cultures and a Second Look.* Cambridge: Cambridge University Press.

Sobel, D. 1999. *Galileo's Daughter.* New York: Walker & Company.

————, trans. and ed. 2001. *Letters to Father.* New York: Walker & Company.

Soccorsi, Filippo. 1946. "Il processo di Galileo." *La Civiltà Cattolica* 97: 175–84, 429–38.

————. 1947. *Il processo di Galileo.* Rome: Edizioni La Civiltà Cattolica.

————. 1963. *Il processo di Galileo.* Rome: Edizioni La Civiltà Cattolica.

————. 1964. *Il processo di Galileo.* In *Miscellanea Galileiana* 3: 849–929.

Spreafico, S., ed. 2003. *Scienza, coscienza e storia nel "caso Galilei."* Milan: FrancoAngeli

Stabile, G. 1994. "Linguaggio della natura e linguaggio della scrittura in Galilei." *Nuncius* 9: 37–64.

Stavis, B. 1966. *Lamp at Midnight.* New York: Bantam.

Stevart, A. 1871. *Procès de Martin Etienne Van Helden Professeur à l'Université de Louvain.* Brussels.

————. 1890. *Copernic & Galilée devant l'Université de Louvain.* Paris.

Stimson, D. 1917. *The Gradual Acceptance of the Copernican Theory.* New York: Baker and Taylor.

Stoffel, J.-F. 2001. "Pierre Duhem Interprète de l' 'Affaire Galilée.'" In Montesinos and Solís 2001, 765–82.

Strauss, E., trans. and ed. 1891. *Dialog über die beiden hauptsächlichsten Weltsysteme.* Stuttgart.

Symposium internazionale di storia, metodologia, logica e filosofia della scienza. 1967. Florence: Gruppo Italiano di Storia delle Scienze.

Tabarroni, G. 1983. "Giovanni Battista Guglielmini e la prima verifica sperimentale della rotazione terrestre (1790)." *Angelicum* 60: 462–86.

Tadini, G. A. 1796a. "Opuscolo intorno alla deviazione de' corpi cadenti dall'alto." *Avanzamenti della medicina e della fisica* [Giornale Brugnatelli], no. 1, 171–93.

———. 1796b. "Volgarizzamento del calcolo della deviazione orientale." *Avanzamenti della medicina e della fisica* [Giornale Brugnatelli], no. 3, 123–42.

———. 1796c. "Della deviazione australe e d'altri minuti articoli di calcolo." *Avanzamenti della medicina e della fisica* [Giornale Brugnatelli], no. 4, 146–72.

———. 1815. *Quotidiana Terrae Conversio, Devio Corporum Casu Demonstrata.* Milan.

Tamburini, F. 1990. "La riforma della Penitenzieria nella prima metà del secolo XVI." *Rivista di storia della chiesa in Italia* 44: 110–40.

Targioni Tozzetti, G. 1780. *Notizie degli aggrandimenti delle scienze fisiche accaduti in Toscana.* 3 tomes in 4 vols. Florence.

Tassoni, A. 1620. *Pensieri diversi.* Carpi.

Taylor, F. S. 1938. *Galileo and the Freedom of Thought.* London: Watts.

Thorndike, L. 1924. "*L'Encyclopédie* and the History of Science." *Isis* 6: 361–86.

Tiraboschi, Girolamo. 1772–1782. *Storia della letteratura italiana.* 9 tomes in 13 vols. Modena.

———. 1778. Review of Paolo Frisi's *Elogio di Bonaventura Cavalieri. Continuazione del Nuovo giornale de' letterati d'Italia* 14: 191–229.

———. 1779. Review of Paolo Frisi's *Elogio del cavaliere Isacco Newton. Continuazione del Nuovo giornale de' letterati d'Italia* 18: 95ff.

———. 1780. [Account of Galileo], *Storia della letteratura italiana,* tome 8, book 2, chapter 2, sections vi–xviii. In Tiraboschi 1772–1782, 8: 123–44; 1782–1797, 8:143–72; 1822–1826, 14: 248–99.

———. 1782. "Manifesto." *Gazzetta universale* (Florence).

———. 1782–1797. *Storia della letteratura italiana.* 10 tomes in 13 vols. Rome.

———. 1785. "Lettera al Reverendissimo Padre N. N. autore delle annotazioni aggiunte alla edizione romana della Storia della letteratura italiana." In Tiraboschi 1782–1797, 10:388–440.

———. 1787–1794. *Storia della letteratura italiana.* 2nd. ed. 9 tomes in 16 vols. Modena.

———. 1792. "Sui primi promotori del sistema copernicano." In Tiraboschi 1782–1797, 10: 362–73.

———. 1793. "Sulla condanna del Galileo e del sistema copernicano." In Tiraboschi 1782–1797, 10: 373–83.

———. 1822–1826. *Storia della letteratura italiana.* 2nd ed. 9 tomes in 16 vols. Milan.

———. 1900. *On the Condemnation of Galileo Galilei and the Copernican System.* London.

Toaldo, G. 1744a. "A chi legge." In G. Galilei 1744c, vol. 1.

———. 1744b. "A chi legge." In G. Galilei 1744c, vol. 4.

———, ed. 1744c. *Opere di Galileo Galilei.* 4 vols. Padua.

———. 1802. *Completa raccolta di opuscoli, osservazioni e notizie diverse.* 4 vols. Venice.

———. 1996. [Preface to Galilei 1744a, vol. 4]. In Fantoli 1996, 495–96.

Topper, D. 1999. "Galileo, Sunspots, and the Motions of the Earth." *Isis* 90: 757–67.

———. 2000. "On Clarifying a Passage in Galileo's *Dialogue.*" *Centaurus* 42: 288–96.

———. 2003. "Colluding with Galileo." *Journal for the History of Astronomy* 34: 75–78.

Torricelli, E. 1644. *De Motu Gravium Naturaliter Descendentium, et Proiectorum.* In idem, *Opera geometrica,* 95–243 (irregular pagination). Florence.

Vacandard, E. 1905a. "La Condamnation de Galilée." In Vacandard 1905b, 293–387.

———. 1905b. *Études de Critique et d'Histoire Religieuse*. Paris: Le Couffre.

Valles, F. 1587. *De Iis, Quae Scripta Sunt Physice in Libris Sacris*. Turin.

Van der Meer, J. M., ed. 1996. *Facets of Faith and Science*. 4 vols. Lanham: University Press of America.

Varenius, B. 1650. *Geographia Generalis in qua Affectiones Generales Telluris Explicantur.* Amsterdam.

Venturi, Giambattista, ed. 1818–1821. *Memorie e lettere inedite finora o disperse di Galileo Galilei*. 2 vols. Modena.

———. 1820a. Venturi to Delambre, 25 April. In Favaro 1887a, 222–23.

———. 1820b. [History of Galileo's Trial]. In Venturi 1818–1821, 2: 192–97.

———. 1820c. Venturi to Delambre, 4 August. In Favaro 1887a, 226–27.

Venturi Barbolini, A. R., ed. 1997. *Girolamo Tiraboschi*. Modena: Biblioteca Estense Universitaria.

Viganò, M. 1969. *Il mancato dialogo fra Galileo e i teologi*. Rome: Edizioni La Civiltà Cattolica.

Viviani, Vincenzio. 1654. *Racconto istorico della vita di Galileo*. In Favaro 19: 599–632.

———. 1690. Viviani to Baldigiani [after 22 August]. In Favaro 1887b, 153–55.

———. 1701. *De Locis Solidis*. Florence

———. 1717. *Racconto istorico della vita di Galileo*. In Salvini 1717, 397–431.

———. 1992. *Vita di Galileo, con appendice di testi e di documenti*. Ed. L. Borsetto. Bergamo.

Voltaire. 1734a. "Descartes and Newton," in *Essays on Literature, Philosophy, Art, History*. In Voltaire 1901, 37: 164–71.

———. 1734b. "Sur Descartes et Newton," letter 14 of *Lettres Philosophiques*. In Voltaire 1877–1883, 22 (1879): 127–32.

———. 1751a. "Science," chapter 31 of *The Age of Louis XIV*. In Voltaire 1951, 352–56.

———. 1751b. "Progress of the Sciences," chapter 29 of *The Age of Louis XIV*. In Voltaire 1901, 23: 277–86.

———. 1751c. "Des Sciences," chapter 31 of *Le Siècle de Louis XIV*. In Voltaire 1877–1883, 14 (1878): 534–39.

———. 1753a. "Customs of the Fifteenth and Sixteenth Centuries." chapter 30 of part 3 of *The General History and State of Europe*. In Voltaire 1754–1757, part 3, chapter 30, 207–15.

———. 1753b. "Customs of the Fifteenth and Sixteenth Centuries, and the State of the Liberal Arts," chapter 100 of *Ancient and Modern History*. In Voltaire 1901, 26: 310–20.

———. 1753c. "State of the Polite Arts in the Sixteenth Century," chapter 1 of part 4 of *The General History and State of Europe*. In Voltaire 1754–1757, part 4, chapter 1, 1–6.

———. 1753d. "Usages des XVe. et XVIe. Siècles, et de l'État des Beaux Arts," chapter 121 of *Essai sur les Moeurs et l'Esprit des Nations et sur les Principaux Faits de l'Histoire depuis Charlemagne jusqu'à Louis XIII*. In Voltaire 1877–1883, 12: 241–50.

———. 1754–1757. *The General History and State of Europe*. 6 parts in 3 vols. London.

———. 1756a. "Newton," in *Essays on Literature, Philosophy, Art, History*. In Voltaire 1901, 37: 172–76.

————. 1756b. "Newton and Descartes," in *A Philosophical Dictionary*. In Voltaire 1824, 5: 107–19.

————. 1756c. "Newton et Descartes, Section II," in *Dictionnaire Philosophique*. In Voltaire 1877–1883, 20: 120–22.

————. 1770a. "Authority," in *A Philosophical Dictionary*. In Voltaire 1824, 1: 365–66.

————. 1770b. "Autorité," in *Dictionnaire Philosophique*. In Voltaire 1877–1883, 19: 501–2.

————. 1824. *A Philosophical Dictionary*. 2nd ed. 6 vols. London.

————. 1877–1883. *Oeuvres Complètes*. 52 vols. Ed. L. Moland. Paris.

————. 1901. *The Works of Voltaire*. 42 vols. Trans. W. F. Fleming. Ed. T. Smollett et al. Paris: Du Mont.

————. 1951. *The Age of Louis XIV*. Trans. M. P. Pollack. London: J. M. Dent.

————. 1968ff. *Oeuvres Complètes*. 100+ vols. Ed. T. Besterman. Geneva: Institut et Musée Voltaire.

Wallace, W. A. 1981a. "Does Galileo's Trial Beg for Reopening?" *Los Angeles Times*, 11 April, Part I-B, 4.

————. 1981b. *Prelude to Galileo*. Dordrecht: Reidel.

————. 1981c. "Aristotle and Galileo." In O'Meara 1981, 47–77.

————. 1982. "Science and Religion." Moreau Lecture, King's College, Wilkes-Barre, PA, 25 March. Unpublished ms.

————. 1983a. "Galileo and Aristotle in the *Dialogo*." *Angelicum* 60: 311–32.

————. 1983b. "Galileo's Science and the Trial of 1633." *The Wilson Quarterly* 7: 154–64.

————. 1984a. *Galileo and His Sources*. Princeton: Princeton University Press.

————. 1984b. "Galileo's Early Arguments for Geocentrism and His Later Rejection of Them." In Galluzzi 1984, 31–40.

————, ed. 1986. *Reinterpreting Galileo*. Washington: Catholic University of America Press.

————. 1987. "Galileo and the Professors of the Collegio Romano." In Poupard 1987, 44–60.

————, trans. and ed. 1992a. *Galileo's Logical Treatises*. Dordrecht: Kluwer.

————. 1992b. *Galileo's Logic of Discovery and Proof*. Dordrecht: Kluwer.

————. 1995. "Galileo's Trial and the Proof of the Earth's Motion." *Catholic Dossier* 1(2): 7–13.

————. 1996. *The Modeling of Nature*. Washington: Catholic University of America Press.

————. 1999. "Galilei, Galileo." In *Encyclopedia of the Renaissance*, ed. P. F. Grendler, 3: 2–9. 6 vols. New York: Scribner's.

Ward, S. 1635. *In Ismaelis Bullialdi Astronomiae Philolaicae Fundamenta, Inquisitio Brevis*. Oxford.

Ward, W. G. 1865. "Doctrinal Decrees of a Pontifical Congregation." *Dublin Review*, new series, 5(10): 376–424.

————. 1866a. "Appendix to the October Article on Galileo." *Dublin Review*, new series, 6(11): 260–67.

————. 1866b. *The Authority of Doctrinal Decisions Which Are not Definitions of Faith*. London.

————. 1871a. "Copernicanism and Pope Paul V." *Dublin Review*, new series, 16(32): 351–68.

————. 1871b. "Galileo and the Pontifical Congregations." *Dublin Review*, new series, 17(33): 140–69.

Wegg-Prosser, F. R. 1889. *Galileo and His Judges*. London.

Wendelen, G. 1644. *Eclipses Lunares ab Anno 1573 ad 1643 Observatae*. Antwerp.

————. 1647. *De Causis Naturalibus Pluviae Purpureae Bruxellensis*. 2nd ed. Brussels.

Westfall, R. S. 1989a. "Bellarmine, Galileo, and the Clash of Two World Views." In Westfall 1989b, 1–30.

————. 1989b. *Essays on the Trial of Galileo*. Vatican City: Vatican Observatory Publications.

Westman, R. S. 1972. "Kepler's Theory of Hypothesis and the 'Realist Dilemma.'" *Studies in History and Philosophy of Science* 3: 233–64.

————, ed. 1975a. *The Copernican Achievement*. Berkeley: University of California Press.

————. 1975b. "The Melanchthon Circle, Rheticus, and the Wittenberg Interpretation of the Copernican Theory." *Isis* 66: 165–93.

————. 1975c. "Three Responses to the Copernican Theory." In Westman 1975a, 285–345.

————. 1975d. "The Wittenberg Interpretation of the Copernican Theory." In *The Nature of Scientific Discovery*, ed. O. Gingerich. Washington: Smithsonian Institution.

————. 1980. "The Astronomer's Role in the Sixteenth Century." *History of Science* 18: 105–47.

————. 1984. "The Reception of Galileo's 'Dialogue.'" In Galluzzi 1984, 329–71.

————. 1986. "The Copernicans and the Churches." In Lindberg and Numbers 1986, 76–113.

————. 1987. "La Préface de Copernic au Pape." *History and Technology* 4: 359–78.

————. 1990. "Proof, Poetics, and Patronage." In Lindberg and Westman 1990, 167–205.

————. Forthcoming. *The Copernican Question, 1470–1610*. Chicago: University of Chicago Press.

Whewell, W. 1837. "The Copernican System Opposed on Theological Grounds." In idem, *History of the Inductive Sciences*, 1st ed., 1: 397–404. 3 vols. London.

————. 1847. "Case of Galileo." In idem, *Philosophy of the Inductive Sciences Founded upon their History*, 2nd ed., 1: 696–700. 2 vols. London.

————. 1857a. "The Copernican System Opposed on Theological Grounds." In idem, *History of the Inductive Sciences*, 3rd ed., 1: 303–12. 3 vols. London.

————. 1857b. "Were the Papal Edicts against the Copernican System Repealed?" In idem, *History of the Inductive Sciences*, 3rd ed., "Additions to the Third Edition," 1: 393–94. 3 vols. London.

White, A. D. 1896. *A History of the Warfare of Science with Theology in Christendom*. 2 vols. New York.

White, T. 1642. *De Mundo Dialogi Tres*. Paris.

Wilkins, J. 1638. *The Discovery of a World in the Moone*. London.

————. 1640a. *A Discourse concerning a New Planet*. In Wilkins (1640b).

————. 1640b. *A Discourse concerning a New World & Another Planet in 2 Books*. London.

Willett, J., and R. Manheim, eds. 1994. *Bertolt Brecht, Life of Galileo*. New York: Arcade.

Wilson, D. B. 1996. "On the Importance of Eliminating *Science* and *Religion* from the History of Science and Religion." In Van der Meer 1996, 1: 27–48.

———. 1999. "Galileo's Religion *Versus* the Church's Science?" *Physics in Perspective* 1: 65–84.

———. 2002. "The Historiography of Science and Religion." In Ferngren 2002, 13–30.

Wittich, C. 1659. *Consensus Veritatis.* Leiden.

Wohlwill, Emil. 1870. *Der Inquisitionsprocess des Galileo Galilei.* Berlin.

———. 1877b. *Ist Galilei gefoltert worden?* Leipzig.

———. 1909. *Galilei und sein Kampf für die Kopernickanischen Lehre.* vol. 1. Hamburg and Leipzig: Voss.

———. 1926. *Galilei und sein Kampf für die Kopernickanischen Lehre.* vol. 2. Leipzig: Voss.

———. 2000. "The Discovery of the Parabolic Shape of the Projectile Trajectory." *Science in Context* 13: 645–80.

Wolff, C. 1735. *Elementa Astronomiae.* 2nd ed. Halae Magdeburgicae.

Wolynski, A. 1872–1873. "Relazioni di Galileo Galilei colla Polonia." *Archivio storico italiano,* 3rd series, 16: 63–94, 231–71; 17: 3–31, 262–80, 434–41.

———. 1877. "Francesco de Noailles e Galileo Galilei." *Rivista europea,* year 8, 3: 688–94.

———. 1878. *Nuovi documenti inediti del processo di Galileo Galilei.* Florence.

Zahm, J. A. 1896. *Evolution and Dogma.* Chicago.

———. 1978. *Evolution and Dogma.* Rpt. of 1896 edition. New York: Arno Press.

Zoffoli, E. 1990. *Galileo: Fede nella ragione: Ragioni della fede.* Bologna: Edizioni Studio Domenicano.

Zúñiga, Diego de. 1584. *In Job Commentaria.* Toledo.

———. 1591. *In Job Commentaria.* Rome.

———. 1636. [Commentary on Job 9:6]. In G. Galilei 1636.

———. 1661. "An Abstract of Some Passages in the Commetaries of Didacus à Stunica of Salamanca upon Job." In Salusbury 1661–1665, vol. 1, part 1, 468–70.

———. 1991. "Commentary on Job 9:6." In Blackwell 1991, 185–86.

Zycínski, J. M. 1988. *The Idea of Unification in Galileo's Epistemology.* Vatican City: Specola Vaticana.

Indexer: Andrew Joron
Compositor: Sheridan Books, Inc.
Text: 10/12 Baskerville
Display: Baskerville